Simulation and Optimization Strategies: Theory and Experiments

Simulation and Optimization Strategies: Theory and Experiments

Edited by
Tommy Haynes

Willford Press
www.willfordpress.com

Published by Willford Press,
118-35 Queens Blvd., Suite 400,
Forest Hills, NY 11375, USA

ISBN: 978-1-68285-332-0

Cataloging-in-publication Data

Simulation and optimization strategies : theory and experiments / edited by Tommy Haynes.
p. cm.
Includes bibliographical references and index.
ISBN 978-1-68285-332-0
1. System engineering. 2. Structural optimization. 3. Engineering systems--Computer simulation. 4. Computer simulation.
I. Haynes, Tommy.
TA168 .S56 2017
620.001 171--dc23

For information on all Willford Press publications
visit our website at www.willfordpress.com

Willford Press

Contents

Preface

This book presents the complex subject of multidisciplinary design optimization in the most comprehensible and easy to understand language. This discipline is a field of engineering that uses optimization methods to solve design problems occurring in varied fields. This field of study is also known as multidisciplinary system design optimization. The aim of this book is to present researches that have transformed this discipline and aided its advancement. The text elaborately discusses the various applications of design optimization in different areas like, electronics and computers amongst others. The extensive content of this book provides the readers with a thorough understanding of the subject. It is beneficial text for engineers, researchers and students alike.

This book aims to highlight the current researches and provides a platform to further the scope of innovations in this area. This book is a product of the combined efforts of many researchers and scientists from different parts of the world. The objective of this book is to provide the readers with the latest information in the field.

I would like to express my sincere thanks to the authors for their dedicated efforts in the completion of this book. I acknowledge the efforts of the publisher for providing constant support. Lastly, I would like to thank my family for their support in all academic endeavors.

Editor

1

Interpretability and variability of metamodel validation statistics in engineering system design optimization: a practical study

Husam Hamad[1,a,*], Awad Al-Zaben[2] and Rami Owies[3]

[1] Electronics and Communications Engineering Department, College of Engineering at Al-Lith, Umm Al-Qura University, Makkah Al-Mukarramah, KSA
[2] Biomedical Engineering Department, Hijjawi College of Engineering Technology, Yarmouk University, Irbid, Jordan
[3] Biomedical Engineering Department, College of Engineering, Jordan University of Science and Technology, Irbid, Jordan

Abstract – Prediction accuracy of a metamodel of an engineering system in comparison to the simulation model it approximates is one fundamental criterion that is used in metamodel validation. Many statistics are used to quantify prediction accuracy of metamodels in deterministic simulations. The most frequently used ones include the root-mean-square error (RMSE) and the *R*-square metric derived from it, and to a lesser degree the average absolute error (AAE) and its derivates such as the relative average absolute error (RAAE). In this paper, we compare two aspects of these statistics: interpretability of results returned by these statistics and their sample-to-sample variations, putting more emphasis on the latter. We use the difference-mode to common-mode ratio (DMCMR) as a measure of sample-to-sample variations for these statistics. Preliminary results are obtained and discussed via a number of analytic and electronic engineering examples.

Key words: Simulation, Modeling, Metamodel validation.

1. Introduction

Simulators are used in the design of many engineering systems. For example, the design of electronic integrated circuits usually involves the use of transistor-level simulators such as PSPICE. Because such simulators are often expensive in terms of simulation times, different approximation models-that are often termed metamodels- may be used to replace simulation models. Metamodels are built and validated using simulation results for samples of data points in the input space. Two fundamental criteria are used as the basis for accepting or rejecting a metamodel: efficiency and accuracy. Efficiency is indicative of how expeditiously predictions can be obtained; accuracy is indicative of how good these predictions are.

Efficiency of a metamodel can be determined prior to metamodel construction, and without any computational cost in terms of the simulation runs needed, e.g., the time taken to evaluate a second-order polynomial metamodel in a given number of dimensions is the same regardless of the underlying simulation model. On the other hand, determining the accuracy of a metamodel is closely linked to the number of data points used in error calculations.

The accuracy of a metamodel is determined using quantitative methods which are mostly based on average statistics, or subjective methods using data displays such as box plots as in Sargent [1] and Kleijnen and Deflandre [2]. Hamad and Al-Hamdan use circle, ordinal, and marksman plots in [3] and [4].

Two of the most popular quantitative measures used in deterministic simulations to validate metamodels in terms of their accuracy of prediction are the root-mean-square error (RMSE) and the average absolute error (AAE), or some of their derivates. Calculation of these statistics is obtained usually using all of the available points in validation test samples, but in some techniques cross-validation methods are employed using subsets of the available test data; see Martin and Simpson [5] and Meckesheimer et al. [6].

Derivates of the RMSE and AAE statistics which are in essence relative error averages are sometimes used in the literature to give more interpretability to the results returned by RMSE and AAE, or otherwise to enable comparisons of metamodels especially when responses from different disciplines are approximated. Two of the mostly used ones include *R*-square and relative average absolute error (RAAE) or the like; see for example Jin et al. [7]. The *R*-square metric is in essence the square of the RMSE relative to the variance of the response data in the test sample, while RAAE may be obtained by

---

[a] Husam Hamad is on a sabbatical leave from Yarmouk University, Jordan
*e-mail: husam@yu.edu.jo

relating AAE to the standard deviation (defining equations for these statistics are given later). Note that some applications use AAE and RMSE relative to the average response instead, e.g., Qu et al. [8]. Other work relates RMSE to the range of response values in the test data samples, e.g., Eeckelaert et al. [9].

An important assumption for the validity of the results obtained by average-based statistics is related to the number of data points used in test samples. Of course, results of these statistics are meaningful only if the data used is sufficient in number. It is often the case that obtaining a sufficient number of observations is impractically expensive for complex simulation models. For such cases, average-based metrics such as RMSE and AAE may be "sensitive" to the number of observations used.

This paper provides a comparative study of four of the average-based statistics used for metamodel assessment in terms of prediction accuracy. The statistics used in this study are the RMSE and R-square, and AAE and RAAE. Two aspects of these metrics are considered: interpretability of results and sample-to-sample variation in the results. We put more emphasis on the issue of sample-to-sample variation, introducing a measure to quantify this variation. The term given to this measure is the common-mode to difference-mode ratio (DMCMR), as defined in the next section.

The remainder of this paper is organized as follows. In Section 2, the four statistics mentioned above are defined and contrasted in terms of their results interpretability and variability, after defining what we mean by these terms. Preliminary results are presented via examples in Section 3 with a discussion in Section 4. The paper is then concluded by Section 5.

2 Statistics for metamodel prediction accuracy

In this section we define and compare the four statistics of RMSE, R-square, AAE, and RAAE used for expressing prediction accuracy of metamodels in relation to their respective simulation models. We compare two aspects of these statistics: interpretability of the results they return and the sample-to-sample variation of these results. We start by defining these four statistics. We then clarify the term interpretability used in this context, followed by the definition of DMCMR – the measure we use to quantify sample-to-sample variations.

2.1 RMSE and *R*-square

Two of the more important measures used for model accuracy assessment including deterministic simulation models are RMSE and R-square. They are defined by

$$\mathrm{RMSE} = \sqrt{\frac{\sum_{i-1}^{n} (\hat{y}_i - y_i)^2}{n}}, \tag{1}$$

$$R^2 = 1 - \frac{\mathrm{MSE}}{\sigma^2}, \tag{2}$$

where MSE is the mean square error and σ^2 is the variance. In these equations and the discussion to follow, y_i is used to denote the response modeled by $\hat{y}_i$ for the ith data point in a validation test sample having n observations. Note that R-square is essentially derived from RMSE by squaring it then relating the result to the variance of y_i data. Theoretical thresholds for best accuracies are zero and unity for RMSE and R-square, respectively.

2.2 AAE and RAAE

The other two statistics that are compared in this paper are the AAE and its derivate RAAE. AAE is defined by

$$\mathrm{AAE} = \frac{1}{n} \sum_{i=1}^{n} |\hat{y}_i - y_i|. \tag{3}$$

Rather than relating AAE to the standard deviation of the response data y_i as in Jin et al. [7], we obtain RAAE by relating AAE to the average of absolute values of the response data y_i in the validation sample. Defining RAAE this way gives more interpretability to the results as explained shortly and provides a simpler means for comparing results for different responses. RAAE is then defined as

$$\mathrm{RAAE} = \frac{\mathrm{AAE}}{\frac{1}{n}\sum_{i=1}^{n} |y_i|} \times 100\,\%. \tag{4}$$

As can be seen from these defining equations, RAAE can be thought of as representing a measure for the percentage error in the metamodel. The lower thresholds indicative of best accuracies for both AAE and RAAE are zero.

2.3 Interpretability of results

Assume that in a test problem RMSE is found to be 0.05 and 50 in another, then which of the two metamodels has better prediction accuracy for its respective simulation model? Assume on the other hand that the statistics for R-square are given instead as 0.78 and 0.98 for the two metamodels, respectively, then which one is better?

Note that results returned by RMSE cannot be interpreted without referring to the context of the problem. For the example given, if the response values are within the range 0.04–0.06 for the first metamodel with RMSE = 0.05, and 10,000–20,000 for the second metamodel having RMSE = 50, then obviously the second metamodel is superior. On the other hand, the closer R-square result for a metamodel is to unity the better its fit quality is regardless of response values. Looking at interpretability of results of these statistics from another perspective, what thresholds can we set on the values of the four statistics defined above for adequate metamodels? Note that thresholds can be readily set for R-square and RAAE, but not for RMSE and AAE without reference to the context of the problem. For example, the lower and upper thresholds in a given situation may be set at 0.95 and 5% for R-square and RAAE, respectively. However, thresholds for RMSE and AAE cannot be easily set without reference to the context of the problem.

Based on this discussion, we use two classifiers in this paper for interpretability of statistical results in terms of prediction accuracy. The results are either: (1) interpretable, or (2) not interpretable without context. Hence, R-square and RAAE as defined in equations (3) and (4) above are classified as having "interpretable results", while results returned by RMSE and AAE are classified as "not interpretable without context".

2.4 Sample-to-sample variation using DMCMR

Another probably more important issue related to the results returned by these statistics concerns the sample-to-sample variations in the results. Ideally, variations should approach zero for deterministic simulation models. However, achieving such ideal results comes at the cost of increased test samples sizes. In this work, we quantify sample-to-sample changes in results using DMCMR, the difference-mode to common-mode ratio, where DMCMR for the two quantities ζ_1 and ζ_2 is defined by

$$\text{DMCMR} = \frac{\zeta_1 - \zeta_2}{\frac{1}{2}(\zeta_1 + \zeta_2)} \times 100\,\%, \tag{5}$$

where absolute values are taken to calculate the common-mode component in the dominator of equation (5).

Variations with sample size n in the statistics defined by equations (1)–(4) above are expected to be noticeable for test sample sizes which are not adequate for taking averages, like all other average-based quantities. However, since the sample size n appears in different forms in these equations (RMSE $\propto n^{-1/2}$, AAE $\propto n^{-1}$, while n cancels out in the numerator and dominator for R-square and for RAAE), then variations with n are expected to be different for these four statistics, as will be demonstrated by the examples of the next section.

3 Examples

We compare in this section variation with sample size for RMSE and R-square on one hand, and AAE and RAAE and on the other hand, via three examples. For these examples:

- Polynomials metamodels are used. The number of coefficients q for a polynomial in k dimensions with a degree d is $q = (k+d)!/k!d!$. These q coefficients are determined by the method of least squares in the examples.
- Latin hypercube validation test samples are used to determine the four statistics above. Sample sizes of ωq are used, where the number of coefficients multiplier ω is varied in steps of 1 starting at $\omega = 1$. Latin hypercube sampling is used to provide flexibility with sample sizes and good uniformity over the input space.

The examples are taken from Hamad [10]. Example 1 uses a one-dimensional analytic function for the response. For this example, two metamodels having different number of coefficients q are studied. In Example 2, a two-dimensional function that is frequently used in the literature is modeled, also via two metamodels with different complexities. The third example involves simulation results for an electronic circuit with three design variables (inputs).

3.1 Example 1

The following response is defined for the space $x \in [-1,1]$:

$$y = 1 \times 10^{-5}\left(e^{20x} - 1\right) + 1000. \tag{6}$$

Two metamodels are derived for this response: the first one is a second-order polynomial built using a minimum bias design having four points, and the second metamodel is a fifth-order polynomial derived using another minimum bias design with 10 points.

Accuracy tests are carried out using 50 samples for each metamodel. The number of observations for these samples are ωq; $\omega = 1, 2, \ldots, 50$. The number of coefficients q for the second-order and fifth-order polynomials is three and six, respectively. Calculations of RMSE, R-square, AAE, and RAAE are carried out for the second-order polynomial metamodel using the 50 test samples in turn. Results are shown in Figure 1 for one Latin hypercube sampling trial for each of the 50 samples.

Note from the figure that each metric settles at a "final" value shown by solid lines crossing the plots in the middle. These final values for RMSE in part (a) of the figure and AAE in part (c) are at approximately 750 and 250, respectively. Based on these two results, is the second-order polynomial an acceptable metamodel? Moving on to parts (b) and (d) of the figure for R-square and RAAE final values of approximately 0 and 19%, respectively, now is the metamodel acceptable?

Note that results from RMSE and AAE cannot be interpreted without reference to the context of the problem to examine the range of response values. On the other hand, merely taking into consideration that the value of R-square is much lower than the upper limit of one, then it is concluded that there is a problem with the metamodel prediction capability. Similarly, RAAE results can be interpreted by the mere consideration of the values returned. A value of 19% for RAAE also indicates that there is a problem. Note that some kind of quantification to the size of the problem may also be inferred from RAAE results; it can be further concluded that the size of the problem is such that the prediction performance of the metamodel is erroneous by 19% on average. In light of the discussion given so far, we may therefore say that results returned by R-square and RAAE are classified as "interpretable" by comparison to RMSE and AAE results which are "not interpretable without context".

Sample-to-sample variations measured by DMCMR defined in equation (5) above are depicted in Figures 2 and 3. Figure 2a shows variation results for RMSE superimposed on those for R-square. Similarly, Figure 2b shows superimposition of results for AAE and RAAE. Figure 3 shows sample-to-sample variations for all four metrics for sample sizes of $10q$ or less.

Note from Figure 2a that the minimum sample size is at approximately $25q$ for RMSE to settle to within ±10% of the final value (the two dotted lines running across the figure),

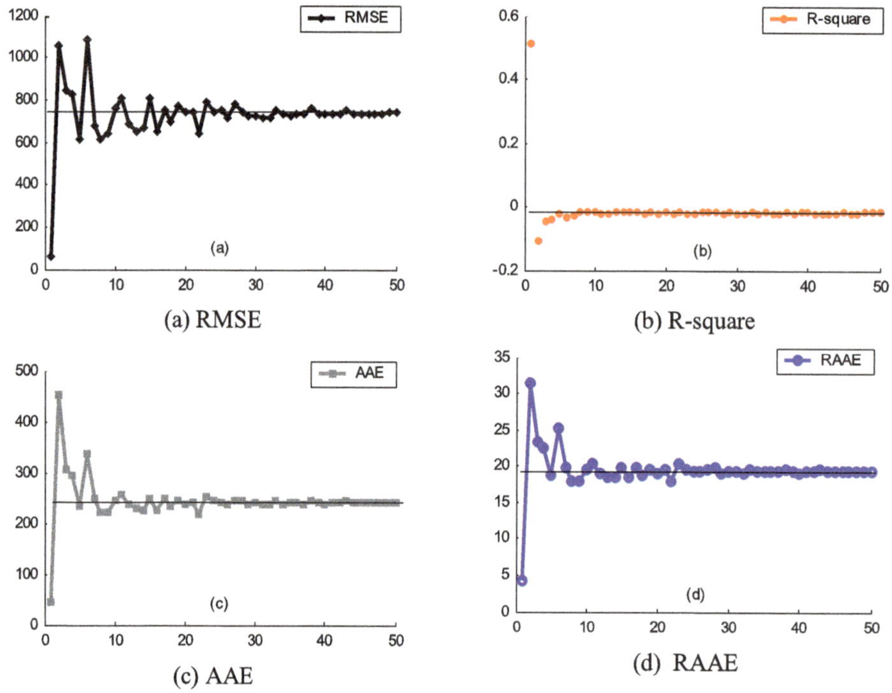

(a) RMSE (b) R-square

(c) AAE (d) RAAE

Figure 1. Validation results for the second-order polynomial metamodel vs. the number of coefficients multiplier ω.

and approximately $30q$ for R-square. For smaller sample sizes of $10q$ or less, Figure 3 reveals that as a whole R-square performs worse than RMSE vis-à-vis sample-to-sample variations. This unexpected result may be explained by noting that the response y in equation (6) above is almost constant for most of the input space and increases sharply for $x > 0.8$, a condition which is not favorable for R-square calculation. To explain, refer to equation (2) above. It can be seen that the dominator of the second term in equation (2) is small for nearly constant response values, and even if the metamodel returns nearly zero MSE, the value of the second term in equation (2) resulting from dividing two small numbers is not without numerical problems. A similar situation leading to questioning the validity of R-square results that is dealt with in the literature rises for the case when n is close to the number of coefficients q in the metamodel. For such cases, R-square is "adjusted" to accommodate the relative size of n to q; see Kleijnen and Deflandre [2].

Sample-to-sample variations for AAE and RAAE can be compared by reference to Figures 2b and 3 for the smaller sample sizes. It can be seen from both figures that RAAE performs slightly better that AAE. Note from Figure 2b that both AAE and RAAE settle down to ±10% variations at a minimum sample size of around $25q$, which is the same result for RMSE as discussed above. Referring to Figure 3, it can be seen that RMSE behaves only marginally worse that AAE or RAAE.

In order to investigate the sample-to-sample variations of the four metrics in more detail, 10 trials are carried out for each of the 50 Latin hypercube samples and DMCMRs are calculated for each metric. Then, minimum sample sizes after which the corresponding metric is confined within DMCMR levels of ±10% are noted and plotted in Figure 4. The average of such minimum sample sizes for the 10 trials for each metric are also calculated and given in Table 1. Figure 5 explains how the minimum size is determined for trial 1 for DMCMR calculations for RMSE.

As can be seen from Figure 4, R-square performs consistently worse for all 10 trials, while RAAE performs consistently the best. The results in Table 1 also show that RAAE has the best performance on average in terms of sample-to-sample variations in its results, with minimum average size of $11.7q$. This means that a test sample of size $11.7q$ on average is considered "adequate" in the sense that the results returned will not be more than ±10% of the "true" value that would be obtained if the test sample size were infinite. By the same token, adequate sample sizes for RMSE, R-square, and AAE are $21q$, $29.2q$, and $16.2q$ respectively, as given in Table 1. Note that for this example R-square unexpectedly performs worse than RMSE. The reason for the unexpected performance of R-square was mentioned above, where it was explained that R-square results cannot be used in two situations: (1) if the number of observations n is close to the number of coefficients q, and (2) if the response is nearly constant for a sizable portion of the input space. Note in addition that R-square is close to zero for most test samples; see Figure 1b above.

In summary, RAAE outperforms the other three statistics for Metamodel 1 of Example 1 because: (1) it has the best interpretable results, and (2) it has the smallest adequate sample size of $11.7q$, i.e., the minimum sample size for ±10% deviation from the "true" result is $11.7q$ or nearly 35 observations for this case.

3.1.1 Metamodel 2

The order of the metamodel polynomial is changed to five. The final values are approximately 450, 0.65, 170, and 13% for RMSE, R-square, AAE, and RAAE, respectively.

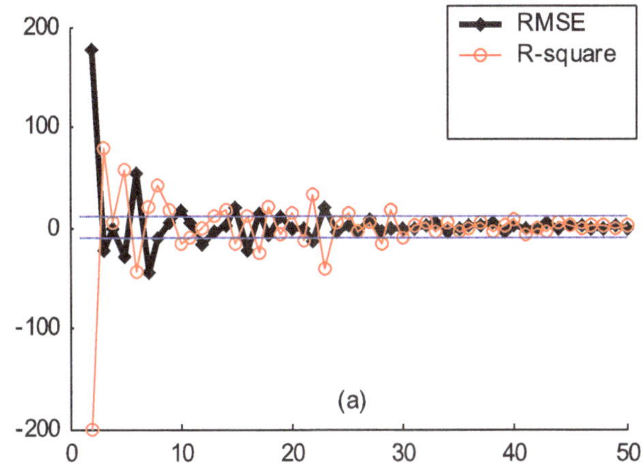

(a) RMSE and R-square

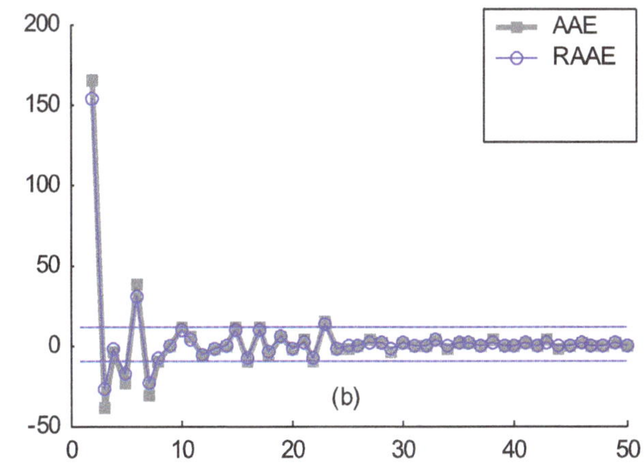

(b) AAE and RAAE

Figure 2. Sample-to-sample variations in DMCMR vs. ω.

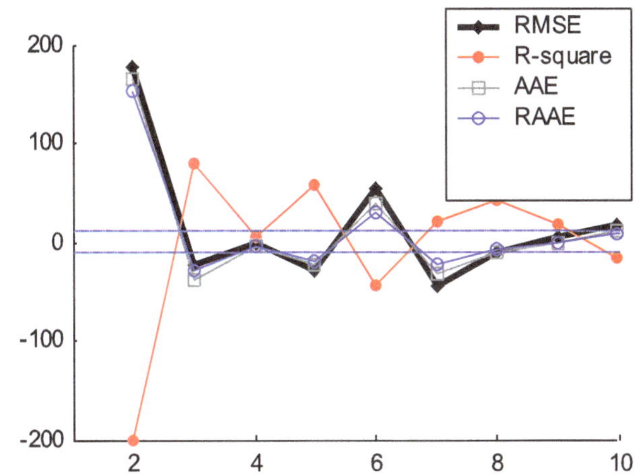

Figure 3. Sample-to-sample variations in DMSMR vs. ω for "smaller" samples.

Sample-to-sample variations for the four metrics are calculated, again using 50 Latin hypercube samples. Results for DMCMRs are shown for one sampling trial in Figure 6 superimposed for

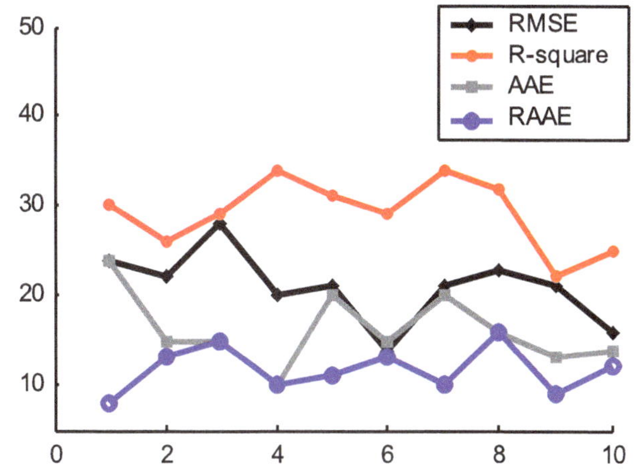

Figure 4. Minimum sample sizes for confinement within ±10% DMCMR using 10 trials for each of the 50 test samples.

Table 1. Average of minimum sample sizes for ±10% DMCMR levels confinement using 10 trials for each test sample.

Statistic	Metamodel 1	Metamodel 2
RMSE	$21q$	$24.9q$
R-square	$29.2q$	$14.6q$
AAE	$16.2q$	$17.5q$
RAAE	$11.7q$	$16.7q$

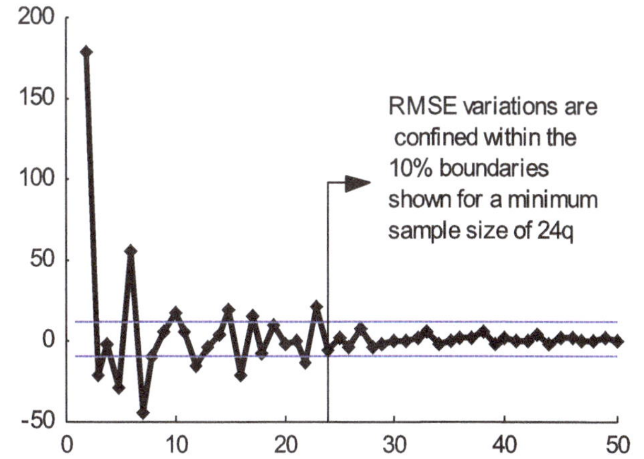

Figure 5. RMSE sample-to-sample variation for Latin hypercube sampling trial 1 showing the minimum sample size of $24q$ for confinement within ±10% DMCMR.

RMSE and *R*-square in part (a), and AAE and RAAE in part (b) of the figure.

As seen in Figure 6a, performance of *R*-square in terms of sample-to-sample variation is overall better than RMSE performance, as would be expected. Note for example that the minimum sample size for ±10% confinement of DMCMR level variations for the sampling trial shown is $28q$ for RMSE and $17q$ for *R*-square. The situation for AAE and RAAE is similar to that for Metamodel 1 above, with RAAE performing marginally better than AAE as seen in Figure 6b. Sample-to-sample variations for the four metrics are shown in Figure 7 for sample

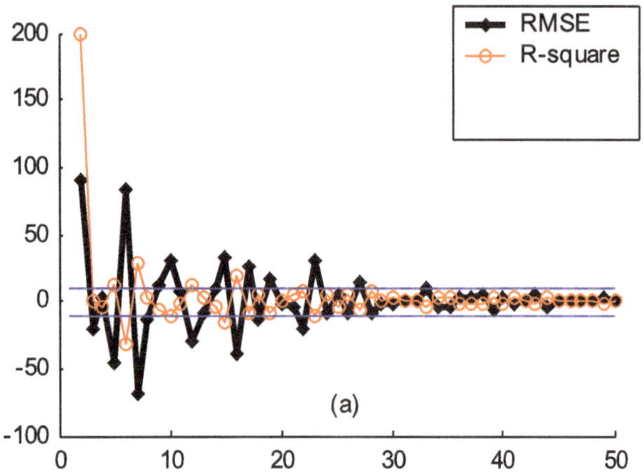

(a) RMSE and R-square

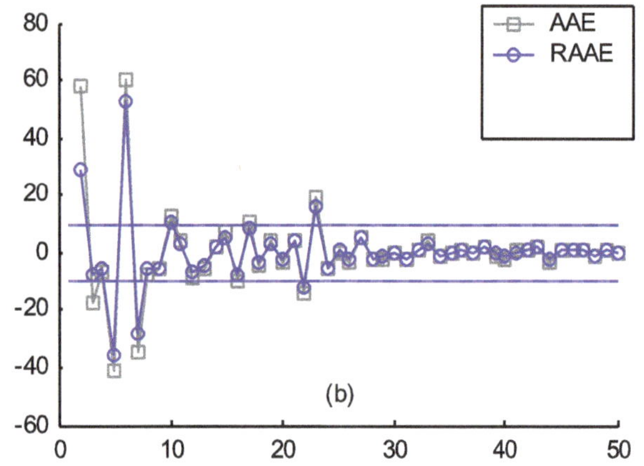

(b) AAE and RAAE

Figure 6. Sample-to-sample variations in DMCMR vs. ω.

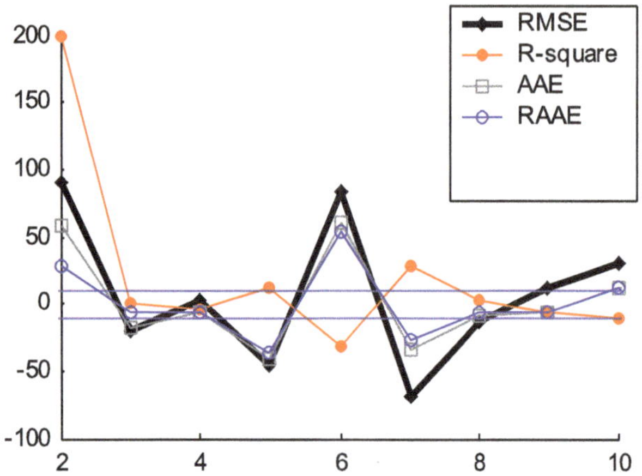

Figure 7. Sample-to-sample variations vs. ω for "smaller" samples.

sizes of $10q$ or smaller. It can be seen from the figure that for the smaller samples with sizes between $2q$ and $3q$ the variability is largest for R-square, while variability becomes largest for RMSE for sample sizes of $4q$ up to $10q$.

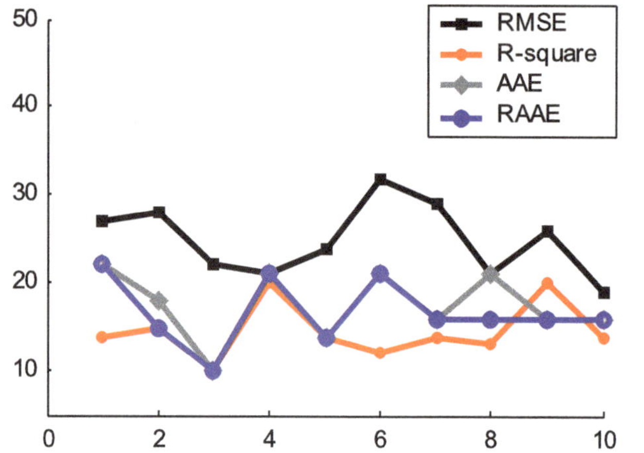

Figure 8. Minimum sample sizes for confinement within ±10% DMCMR using 10 trials for each of the 50 test samples.

In order to investigate the sample-to-sample variations of the four metrics in more detail, the scenario used for Metamodel 1 above is followed here where 10 sampling trials are carried out for each of the 50 Latin hypercube samples and DMCMRs are calculated for each metric. Then, minimum sample sizes after which the corresponding metric is confined within DMCMR levels of ±10% are noted and plotted in Figure 8. The average of such minimum sample sizes for the 10 trials for each metric are also calculated and given in Table 1 above.

Note from Figure 8 that RMSE performs consistently worse for all 10 trials. It can also be seen from the figure that R-square performs marginally better than AAE and RAAE for all 10 trials except trial 9. RAAE performs slightly better than AAE for all 10 trials; see Figure 8. These are the same conclusions in relation to the average minimum sample sizes in Table 1, where RMSE has the worst average of $24.9q$, and the other three metrics having nearly equal averages, with R-square slightly the best.

Therefore, we can conclude for Metamodel 2 of Example 1 that RMSE has the worst performance in relation to sample-to-sample variations. Also, interpretability of its results does not put it up in front of the other three metrics either. While the sample-to-sample variation for RAAE is slightly higher than R-square ($16.7q$ for RAAE and $14.6q$ for R-square as given in Table 1), however, its interpretability of results may make it the best choice in this situation as well. Thus, while the grounds for rejecting a metamodel because of an R-square of 0.65 are shaky, but on the other hand a metamodel with a RAAE of 13% is understood to have an error of 13% on average, may be giving clearer indications of whether to accept or to reject the metamodel.

3.2 Example 2

The second example involves the two dimensional response studied in other work including Martin and Simpson [5], Jin et al. [7], and Hamad et al. [11]; given by:

$$y = \cos(6(x_1 - 0.5)) + 3.1 \times |x_1 - 0.7| + 2(x_1 - 0.5) + \sin\left(\frac{1}{|x_1 - 0.7| + 0.31}\right) + 0.5x_2 \quad x_1, x_2 \in [0, 1]. \tag{7}$$

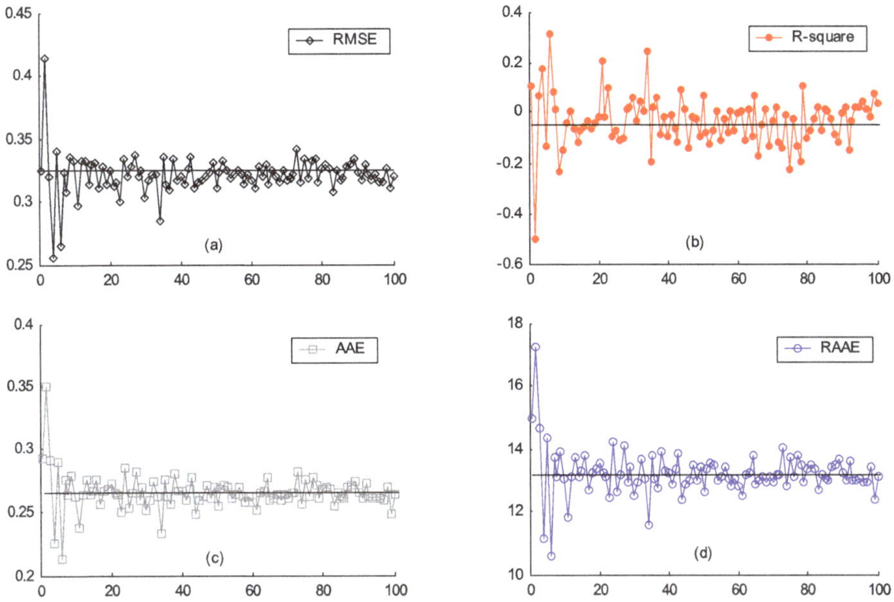

Figure 9. Validation results for Metamodel 1 vs. the number of coefficients multiplier ω.

Two metamodels are derived and tested for this response. Metamodel 1 is a second-order polynomial having $q = 6$, while Metamodel 2 is a piecewise metamodel consisting of two second order polynomials: one for each part of the input space partitioned along x_1 in two halves, and with $q = 6$ for each polynomial; see Hamad et al. [11].

3.2.1 Metamodel 1

Figure 9 shows validation results for Metamodel 1 using one-hundred Latin hypercube samples with sizes $\omega q = 1q$, $2q$, …, $100q$ observations. Part (a) of the figure shows that most of the one-hundred test samples have RMSEs around 0.32, while AAEs for most samples are around 0.27 as depicted in part (c); what can be inferred about the prediction validity of Metamodel 1 from these results? Again, prediction accuracy cannot be judged by merely referring to the results for RMSE or AAE.

Referring to *R*-square results of Figure 9b, however, it can be immediately concluded that the metamodel is not accepted in terms of its prediction merits because *R*-square is far from the upper threshold of unity for all test samples; what is wrong with Metamodel 1? This question can be answered by interpreting RAAE results of Figure 8d. RAAE results show that for the one-hundred samples tested, their RAAE levels are between 10% and 17%, with most points having 12–14% RAAE as depicted in the figure, i.e., the error for observations in these samples is between 12% and 14% on average.

Sample-to-sample variations for the four metrics are calculated, again using one-hundred Latin hypercube samples. Results for DMCMRs are shown in Figure 10 for one sampling trial for the first 50 samples superimposed for RMSE and *R*-square in part (a), and AAE and RAAE in part (b) of the figure. Superimposition of the four metrics is shown in Figure 11 for the first 10 test samples. As can be seen from these figures, *R*-square performance is poor; however, the results for the other three metrics are comparable.

In order to investigate the sample-to-sample variations of the four metrics in more detail, we use 10 sampling trials as before for each of the first 50 Latin hypercube samples and DMCMRs are calculated for each metric. Then, minimum sample sizes after which the corresponding metric is confined within DMCMR levels of ±10% are noted and plotted in Figure 12. The average of such minimum sample sizes for the 10 trials for each metric are also calculated and given in Table 2.

It can be seen from Figure 12 and Table 2 that sample-to-sample variation is worst for *R*-square, with RMSE slightly better that AAE and RAAE. Note from Figure 9b that *R*-square is close to zero; this is a similar situation to that obtained for Metamodel 1 of Example 1 and shown in Figure 1b.

The response considered in this example is waving in the direction of x_1, and the global second-order Metamodel 1 cannot follow this waving response consistently leading to increased sample sizes as shown in Table 2. Results are improved by using the piecewise Metamodel 2 as demonstrated below.

3.2.2 Metamodel 2

The piecewise Metamodel 2 is validated to test its prediction accuracy using the same scheme outlined above for Metamodel 1. Results are shown in Figure 13 for *R*-square and RAAE only in order to save space. Part (a) of the figure shows that *R*-square is around 0.82 for most of the test samples indicating that the piecewise second-order metamodel is a good improvement relative to the global Metamodel 1. With this improved *R*-square value of 0.82, is Metamodel 2 acceptable? RAAE provides more information that helps in answering this question. Refer to Figure 12b, it is seen that RAAE is 4–5% for most samples, meaning that Metamodel 2 has 4–5% error on average. This allows for a more informed basis for judging the acceptability of the metamodel.

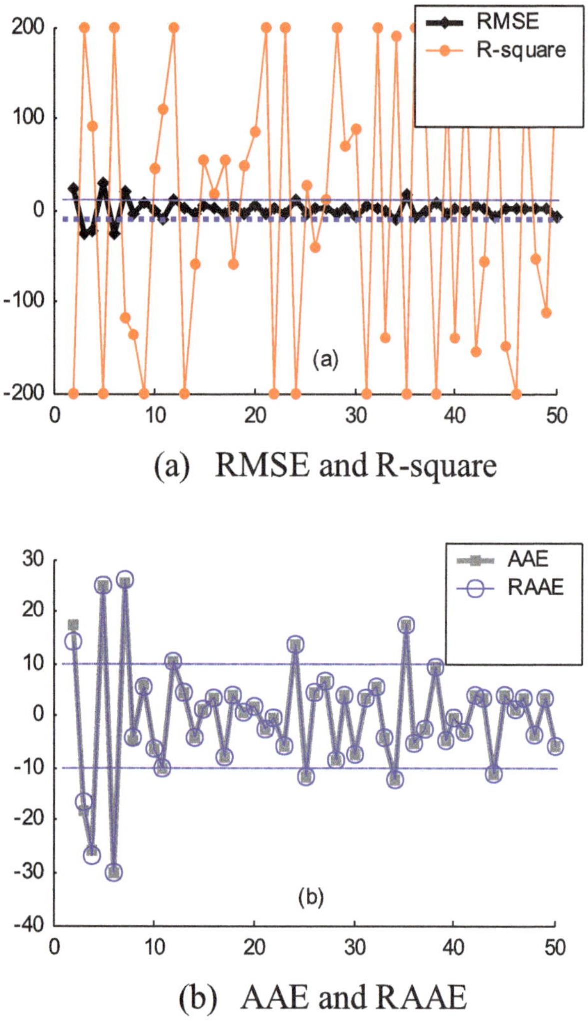

Figure 10. Sample-to-sample variations in DMCMR vs. ω.

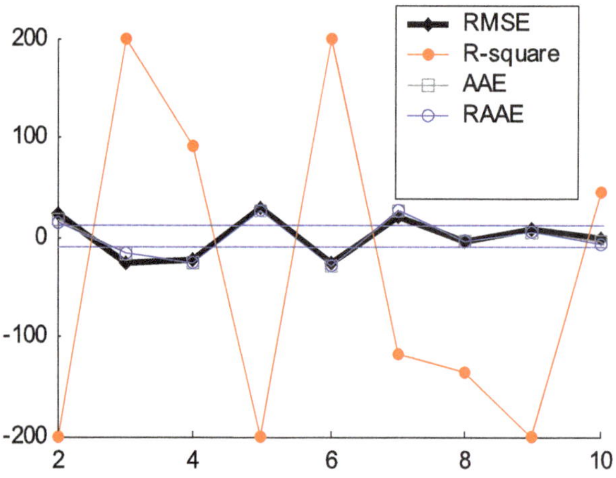

Figure 11. Sample-to-sample variations vs. ω for "smaller" samples.

Sample-to-sample variations are shown in Figure 14a for RMSE and R-square, and for AAE and RAAE in Figure 14b, and Figure 15 shows these variations for the smaller samples with sizes of $10q$ or less. To investigate these variations in more

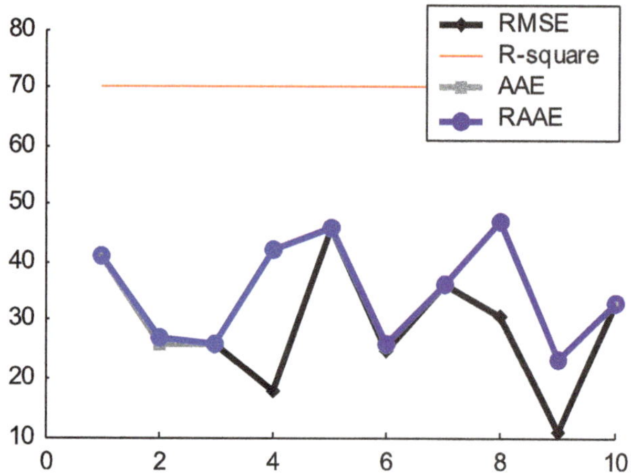

Figure 12. Min. sample sizes for confinement within ±10% DMCMR using 10 trials for each of the 50 test samples.

Table 2. Average of minimum sample sizes for ±10% DMCMR levels confinement using 10 trials for each test sample.

Statistic	Metamodel 1	Metamodel 2
RMSE	$29.4q$	$6.4q$
R-square	$>100q$	$4.6q$
AAE	$34.6q$	$6.4q$
RAAE	$34.7q$	$6.4q$

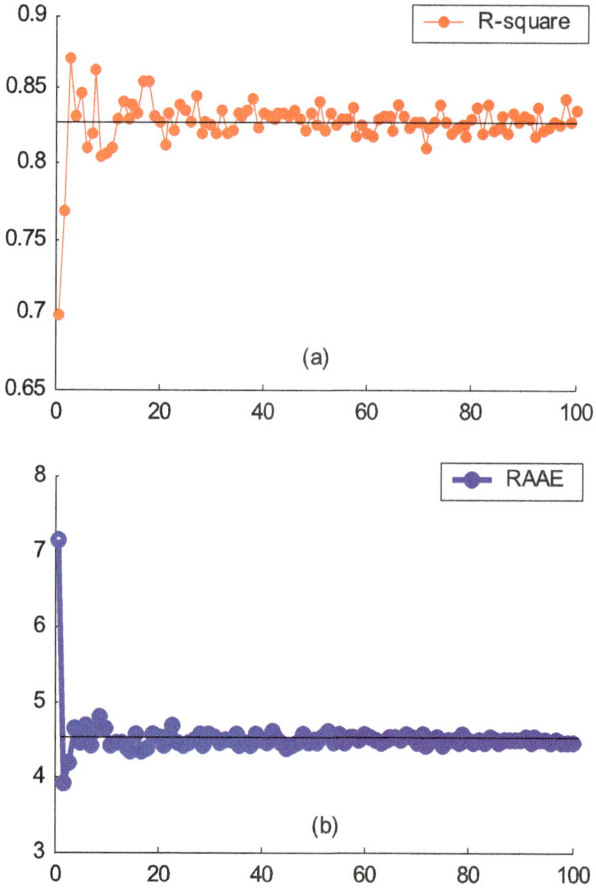

Figure 13. Validation statistics for Metamodel 2 vs. the number of coefficients multiplier ω.

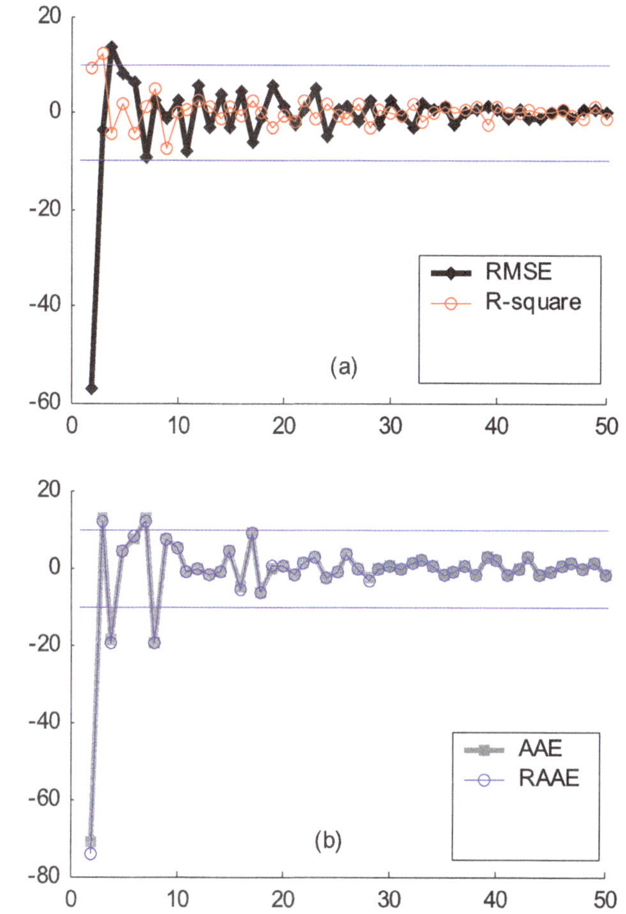

Figure 14. Superimposition of DMCMR results.

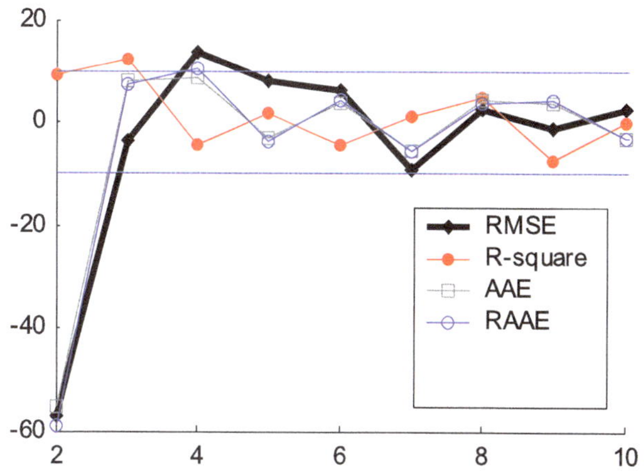

Figure 15. Variations vs. ω for "smaller" samples.

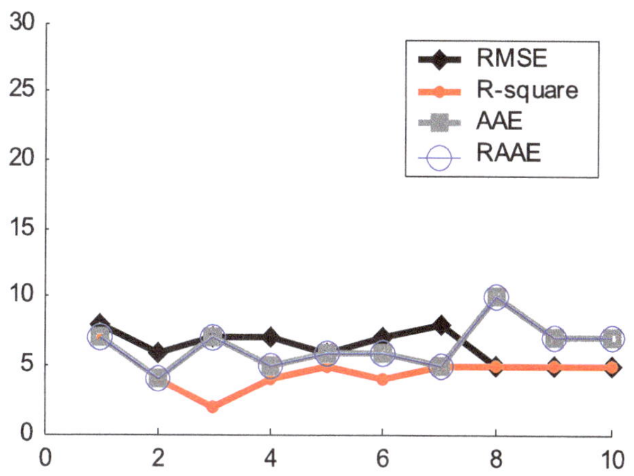

Figure 16. Minimum sample sizes for confinement within ±10% DMCMR.

details, the method used above for Metamodel 1 is repeated here for Metamodel 2, where 10 sampling trials for each of the 50 Latin hypercube test samples are examined for minimum sizes after which the respective metrics are confined to ±10% variations. The results are summarized by Figure 16, and Table 2.

It can be seen from Figure 16 and Table 2 that *R*-square performs marginally better than the other three metrics. RAAE performs the same as AAE for all 10 trials, and RMSE is slightly worse for the first seven trials; see Figure 16.

Similar conclusions can be made for Metamodel 2 in this example as those given above for Metamodel 2 in Example 1, where it was mentioned that while the sample-to-sample variation for RAAE is slightly worse than *R*-square, however, its interpretability of results may make it the best choice. Thus, while the grounds for rejecting a metamodel because of an *R*-square of 0.82 are shaky (see Figure 13a), but on the other hand a metamodel with a RAAE of 4–5% is understood to have an error of 4–5% on average, may be giving clearer indications of whether to accept or to reject the metamodel.

3.3 Example 3

The three-dimensional problem in this subsection is an electronic engineering problem that relates the portion H of the input signal that appears as an output in the circuit of Figure 17.

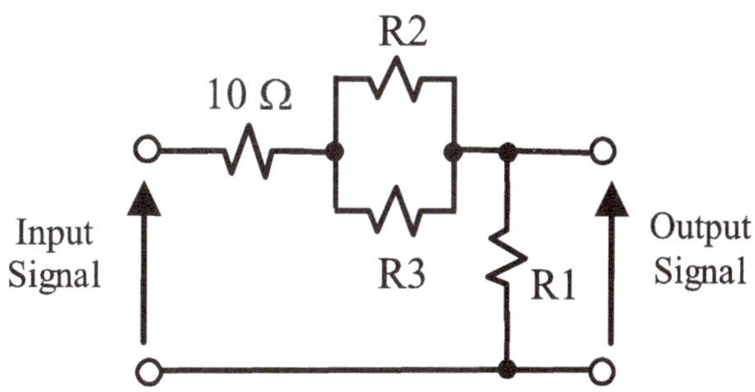

Figure 17. Electric circuit for Example 3.

The portion H is dependent upon the three design variables R_1, R_2, and R_3 connected as shown in the figure. Using a circuit simulator gives results which are identical to those given by the following equation obtained from elementary circuit analysis techniques

$$H = \frac{R_1R_2 + R_1R_3}{R_1R_2 + R_1R_3 + R_2R_3 + 10R_2 + 10R_3}. \tag{8}$$

A second-order polynomial is constructed from a minimum bias experimental design having seventeen points in the space

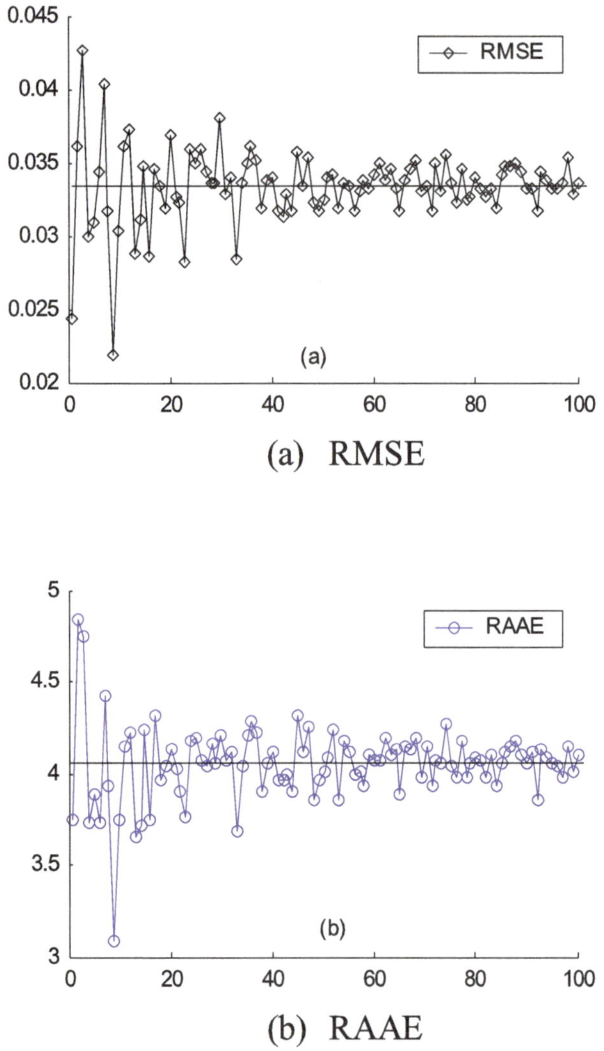

Figure 18. Validation statistics for Example vs. the number of coefficients multiplier ω.

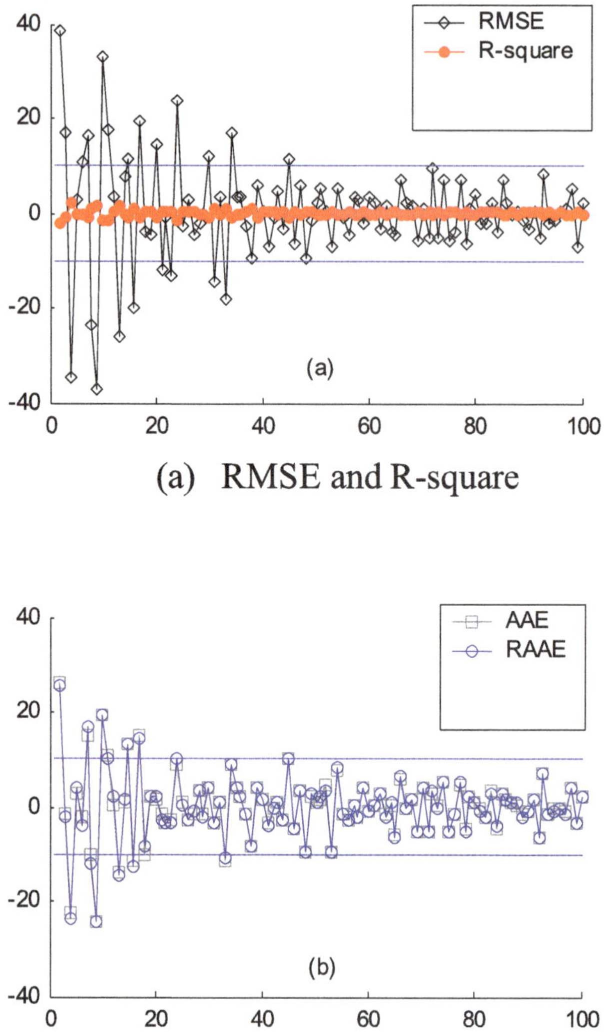

Figure 19. Superimposition of DMCMR results.

$[1,100]^3$. Accuracy tests are then carried out using one-hundred Latin hypercube samples. The number of observations for these samples are ωq; $\omega = 1, 2, \ldots, 100$. The number of coefficients q for this case is 10. Calculations of RMSE, R-square, AAE, and RAAE are carried out for the metamodel using the one-hundred test samples in turn. Results are shown in Figure 18 for RMSE and RAAE using one sampling trial for each of the one-hundred Latin hypercube test samples used. Results for AAE and R-square are omitted to save space.

Figure 18a shows that RMSE varies from a little above 0.02 to nearly 0.043, being approximately 0.033 ± 0.002 for most samples; is this metamodel accurate based on these RMSE results? Note that in equation (8) above $0 < H < 1$ for the entire input space; now the quality of the metamodel can be determined after this information about the response is known. Interpreting the results from R-square can be carried out without reference to the context of the problem; R-square values (not shown) are between 0.959 and 0.991 for all of the one-hundred test samples, a performance which is close to the upper theoretical threshold of unity.

Note that interpreting the results for RAAE shown in Figure 18b gives more information about the prediction accuracy of the metamodel. The figure reveals that for all of the one-hundred test samples RAAE is 3.1–4.8%, indicating that the error in metamodel predictions is on average between 3.1% and 4.8%.

Sample-to-sample variations are depicted by Figure 19a for RMSE and R-square, and by Figure 19b for AAE and RAAE, while Figure 20 depicts these variations superimposed for small sizes having $10q$ observations or less. The same scales are used in both parts of Figure 19 for easy comparison. Note from the figures that sample-to-sample variation is worst for RMSE, with the R-square performance being up in front of the other three statistics.

These sample-to-sample variations are investigated further by carrying out 10 sampling trials for each the first 50 of the 100 Latin hypercube samples and DMCMRs are calculated for each of the four metrics considered. Minimum sample sizes

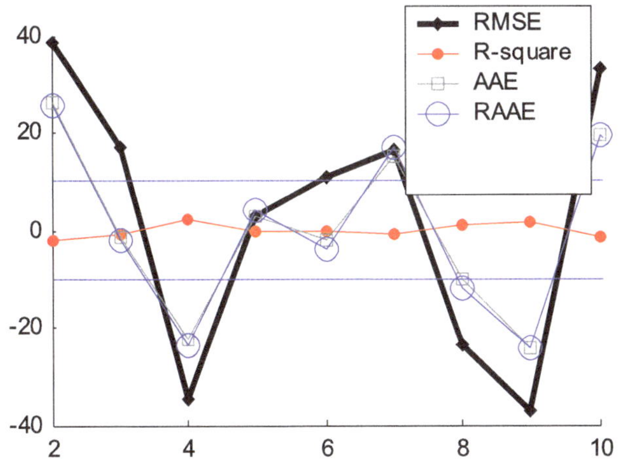

Figure 20. Sample-to-sample variations vs. ω for "smaller" samples.

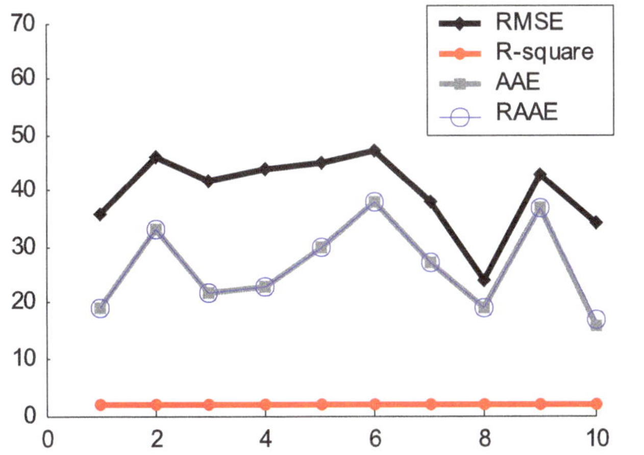

Figure 21. Minimum sample sizes for confinement within ±10% DMCMR.

Table 3. Average of minimum sample sizes for ±10% DMCMR levels confinement.

Statistic	Average size
RMSE	$39.9q$
R-square	$2.0q$
AAE	$26.4q$
RAAE	$26.5q$

after which the corresponding metric is confined within DMCMR levels of ±10% are plotted in Figure 21. The average of such minimum sample sizes for the 10 sampling trials for each metric are also calculated and given in Table 3.

It can be seen from Figure 21 and Table 3 that *R*-square far outperforms the other three metrics with an average minimum size of $2q$, and RMSE showing worst results in terms of sample-to-sample variations with an average minimum size of nearly $40q$. RAAE performs the same as AAE for nearly all 10 trial at averages of about $26q$. Thus, *R*-square is up in the front being robust against sample-to-sample variations. Also, the results returned by it can be interpreted without context, as compared to those results returned by RMSE or AAE.

4. Discussion

Five metamodels are considered in this comparative study: two metamodels for each problem in Examples 1–2, and one metamodel in Example 3. These metamodels are validated in terms of their prediction accuracy using four statistics: RMSE and its derivate *R*-square, and AAE and its derivate RAAE. Results obtained using these four statistics are compared based on two criteria: interpretability and sample-to-sample variations. With regard to interpretability of results:

- RMSE and AAE results are not interpretable without referring to the problems context; a metamodel with RMSE of 0.05 and AAE of 0.02, say, may be an accurate model or otherwise depending on the response values used.
- On the other hand, there is no need to refer to the problem's context for *R*-square and RAAE results; an *R*-square of 0.999 is always welcome and a RAAE of 0.5% is also always taken positively. However, there is more "substance" in the information returned by RAAE since it points to the size of average error in prediction and not merely giving a yes/no answer to the metamodel acceptability as in the case for *R*-square.

Therefore, RAAE has some preference over *R*-square in terms of interpretability of results, with both RMSE and AAE way behind in this respect. This leading role for RAAE vis-à-vis interpretability does not necessarily guarantee the first position for it among the other three statistics; the other more important aspect related to sample-to-sample variability is still to be considered.

The performance of the four statistics in terms of sample-to-sample variation is thoroughly studied in this work: at least 50 test samples are used for any of the five metamodels and the number of observations used in these samples being $1q$, $2q$, up to $50q$, with each of the 50 test samples undergoing 10 Latin hypercube sampling trials. Variability is quantified by the newly introduced measure referred to as difference-mode to common-mode ratio DMCMR. The minimum sample size for a given sampling trial for which DMCMR is confined within ±10% (see Figure 5) is noted for the 10 sampling trials and plotted for the five metamodels in the five Figures 4, 8, 12, 16, and 21. Averages of results in these figures are summarized in three tables: Tables 1–3. These five figures and three tables, summarized in Figure 22 for convenience, are the basis upon which preliminary results for sample-to-sample variability are given. The following can be concluded by reference to the summary in Figure 22:

- Sample-to-sample variability for RAAE shows marginal improvement over AAE variability in worst cases. Another more important merit for RAAE over AAE is related to interpretability of results. Therefore, if the choice has to be made between AAE and RAAE, then RAAE would be a better choice since it conveys more information about the metamodel prediction validity with reduced sample-to-sample variability.

<table>
<tr><th></th><th>Trial-by-trial results</th><th>Averages</th></tr>
<tr><td>Example 1:
Metamodel 1
(see Fig. 4)</td><td>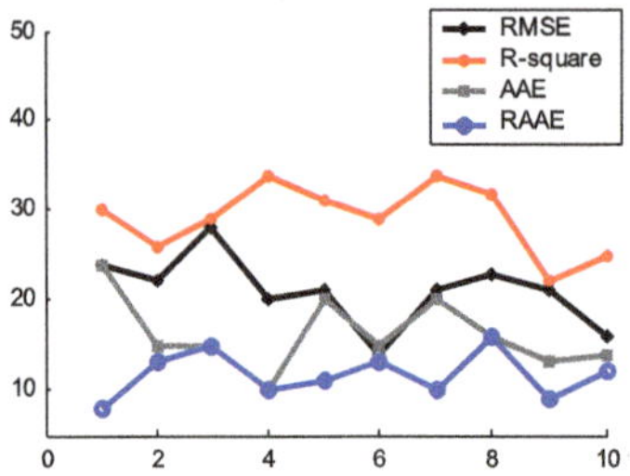
</td><td>21q
29.2q
16.2q
11.7q</td></tr>
<tr><td>Example 1:
Metamodel 2
(see Fig. 8)</td><td>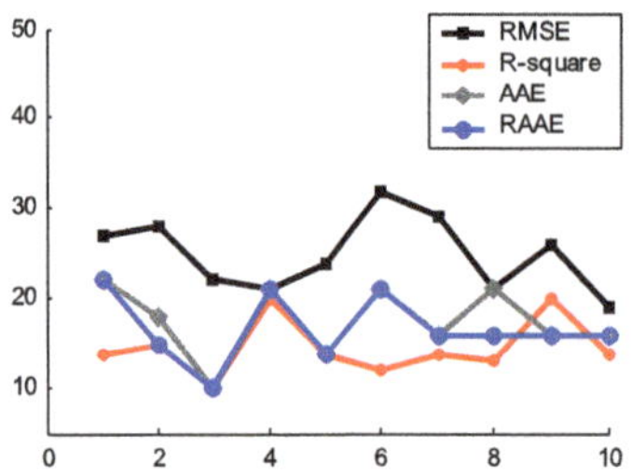
</td><td>24.9q
14.6q
17.5q
16.7q</td></tr>
<tr><td>Example 2:
Metamodel 1
(see Fig. 12)</td><td>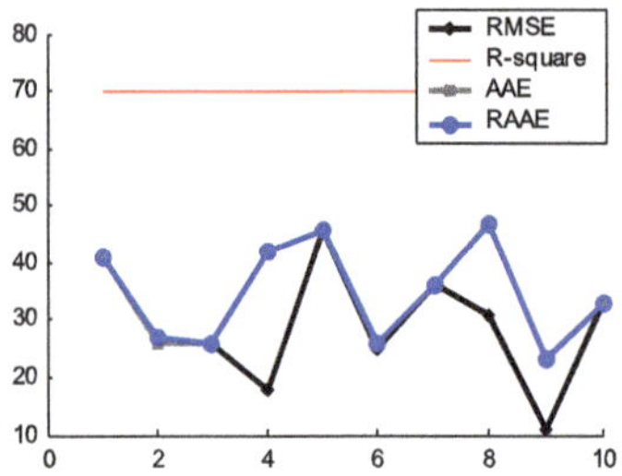
</td><td>29.4q
>100q
34.6q
34.7q</td></tr>
<tr><td>Example 2:
Metamodel 2
(see Fig. 16)</td><td>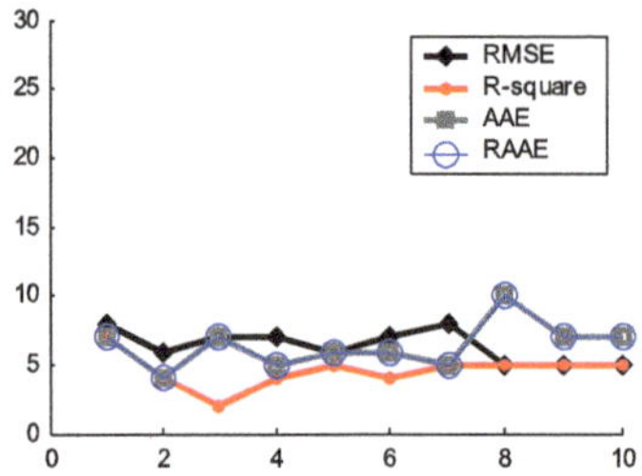
</td><td>6.4q
4.6q
6.4q
6.4q</td></tr>
<tr><td>Example 3
(see Fig. 21)</td><td>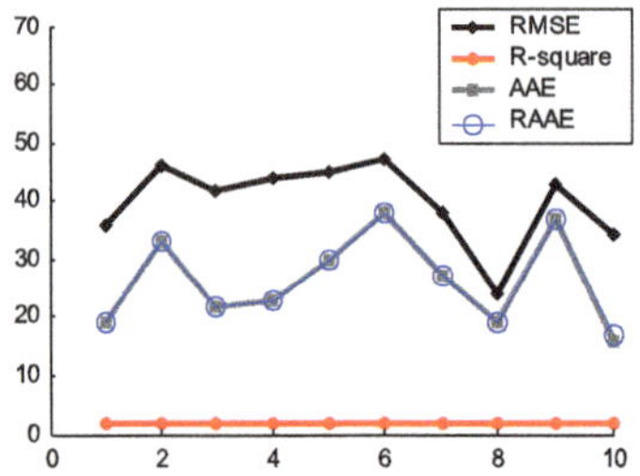
</td><td>39.9q
2.0q
26.4q
26.5q</td></tr>
</table>

Figure 22. Summary of sample-to-sample variations results (averages in the last column are for, from top: RMSE, *R*-square, AAE, and RAAE).

- R-square has the best performance in relation to sample-to-sample variability provided that the following conditions are satisfied by the response data used in its calculation: (1) The number of such data points is larger than the number of coefficients in the metamodel, and more importantly (2) The data variance is not close to zero.

Based on this preliminary study we are inclined to conclude that using both R-square and RAAE in metamodel predictability validation serves two purposes at the same time: (1) it minimizes the number of observations in test samples for robustness against sample-to-sample variations, and (2) it allows for a more solid ground for judging the acceptability of the metamodel. To clarify this latter point, recall the results obtained for Metamodel 2 in Example 2 above, with R-square values in Figure 13a close to 0.82 and RAAE results are 4–5% as shown in Figure 13b. Using R-square results alone only permits a choice between either yes or no for acceptability of the metamodel. However, if in addition it is known that the prediction error is 4–5% on average, then this may constitute a better grounds for accepting (or rejecting) the metamodel.

Finally, it can be seen by referring to the summary in Figure 22 that minimum sample sizes for confinement within ±10% variability are impractically high for most cases. For instance, the average of minimum sample sizes for the 10 sampling trials is $4.6q$ for R-square results related to Metamodel 2 in Example 2, and as can be seen from the figure, this is the second best case among the 20 averages given. This result means that the required sample size is nearly twenty-eight observations for the two-dimensional problem considered in the example with six coefficients used for the corresponding metamodel. This then raises the question about the need to develop a "substitute" statistic which can be exclusively used in validating metamodels for deterministic simulations. This statistic should be chosen to reduce the minimum size of adequate validation samples. The Metamodel Acceptance Score (MAS) discussed in [10] is one such statistic. MAS sample-to-sample variability is reduced since MAS value depends on error "count" in a sample rather than error average.

5. Conclusions

This paper presented a comparative study of four statistics used in validating metamodels in deterministic simulations: RMSE along with its derivate R-square, and AAE and its derivate RAAE. RMSE and AAE are "purely" average-based, while their derivates are "less" dependant on averages through the cancellation of the sample size terms in the numerators and dominators of their defining equations. It is shown that the derivates are less prone to variations in sample sizes by comparison to their respective statistic, at least marginally as in the case of AAE and RAAE, provided that the derivate is used correctly in the first place, e.g., the variance of response data used for R-square calculations not approaching zero. In addition to being more robust against sample size variations, the derivate statistics provide better interpretability of results without reference to the context of the problem. Based on the results of this preliminary study, we are not hesitant to advocate the use of R-square and RAAE together in reporting metamodel validation results for deterministic simulations instead of RMSE or AAE. However, due to the excessive sample sizes required for ±10% sample-to-sample variations, we are even less hesitant to report the need to develop a substitute statistic which can be used exclusively for deterministic simulations.

References

1. Sargent R. 2004. Validation and verification of simulation models. In Proceedings of IEEE Winter Simulation Conference. IEEE, p. 13–24.
2. Kleijnen J, Deflandre D. 2006. Validation of regression metamodels in simulation: bootstrap approach. European Journal of Operational Research, 170, 120–131.
3. Hamad H, Al-Hamdan S. 2005. Two new subjective validation methods using data displays. In Proceedings of IEEE Winter Simulation Conference. IEEE, p. 2542–2545.
4. Hamad H, Al-Hamdan S. 2007. Discovering metamodels' quality-of-fit via graphical techniques. European Journal of Operational Research, 178, 543–559.
5. Martin J, Simpson T. 2005. Use of Kriging models to approximate deterministic computer models. American Institute of Aeronautics and Astronautics Journal, 43, 853–863.
6. Meckesheimer M, Booker A, Barton R, Simpson T. 2002. Computationally inexpensive metamodel assessment strategies. American Institute of Aeronautics and Astronautics Journal, 40, 2053–2056.
7. Jin R, Chen W, Simpson T. 2002. Comparative studies of metamodeling techniques under multiple modeling criteria. Journal of Structural Optimization, 23, 1–5.
8. Qu X, Venter G, Haftka R. 2004. New formulation of a minimum-bias central composite experimental design and Gauss quadrature. Structural and Multidisciplinary Optimization, 28, 231–242.
9. Eeckelaert T, Daems W, Gielen G, Sansen W. 2004. Generalized simulation-based posynomial model generation for analog integrated circuits. Analog Integrated Circuits and Signal Processing, 40, 193–203.
10. Hamad H. 2011. Validation of metamodels in simulation: a new metric. Engineering with Computer, 27, 309–317.
11. Hamad H, Al-Hamdan S, Al-Zaben A. 2010. Space partitioning in engineering design: a graphical approach. Structural and Multidisciplinary Optimization, 41, 441–452.

2

Numerical simulation of vertical Bridgman solidification of CdZnTe

Hanen Jamai[1,*], Bernard Pateyron[2], Habib Sammouda[1], and M. El Ganaoui[3]

[1] LabEM (LR11ES34), University of Sousse, ESSTH Sousse, Rue Lamine Abbassi, 4011 Hammam Sousse, Tunisia
[2] CNRS-SPCTS/CEC, 12 rue Atlantis, 87068 Limoges, France
[3] Université de Lorraine, Lab. Energétique de Longwy – LERMAB/FJV, Longwy, France

Abstract – The solidification of the singular CdZnTe in axisymetric Vertical Bridgman (VB) cavity is investigated in the current study. The numerical approach is developed by using COMSOL Multiphysics software, which is based on finite elements method (FEM) with Arbitrary Lagrangian-Euleurian (ALE) formulation allowing treating the moving interface. Numerical aspects of different parameters that affect the solidification interface and present an important factor during the crystal growth's process are presented and analyzed. Especial attention will be paid to investigate the thermophysical properties such as thermal diffusivity, heat capacity. Thermal conductivity and the effect of thermal condition specially the effect of cold temperature T_c which affects significantly the concavity and the convexity of the interface for specific ranges in the solidification of CdZnTe phenomena, but it is of great importance the crystal growth experts are not it given importance. It has been shown that this properties, the effect of T_c, and the variation of interface velocity affect the solid/liquid interface shape and the crystal growth process.

Key words: CdZnTe solidification, Axisymetric Bridgman Vertical (BV), ALE formulation.

1 Introduction

The vertical Bridgman method for crystal growth from the melt was firstly developed in 1925. Nowadays, it is the most useful technique for the growth of crystals. This technique is explained in Figure 1. Its perfections to control temperature gradients and to provide provides pure crystals of large sizes by allowing testing the boundary conditions of mass and heat transfer. Hence, such technique is widely utilized in order to produce II–VI compound semiconductor crystals. Their industrial applications are mostly concentrated in detectors and solar cell production fields. Among these semiconductors, the Cadmium zinc telluride (CdZnTe) is considered as one of the most important and promising applied materials for room-temperature γ and X-ray spectroscopy. Despite the fact that great efforts have been made to produce high quality of CdZnTe dominated by the Bridgman method, its production still suffers from small device size and low quality.

The solid-liquid interface shape is specially determined by heat transfer in melt crystal growth process. The search focus in heat transfer during Bridgman growth are presented in references [1–5], these others studied the factors affecting the shape of the liquid-solid interface, such as the such as the thermal conductivities of the crystal, melt, and ampoule, simple furnace characteristics and latent heat generation. In particular, Adornato and Brown [6] examined the effect of furnace profiles and ampoule materials on interface shape and satisfactorily compared their results to experiments of Wang et al. [7]. Jasinski and Witt [8, 9] clarified notable insights on how interface shape may be controlled in Bridgman processes and considered the growth of CdTe as an example system.

In the aim of better understanding the solidification phenomenon and solving the physical problem during the crystal growth, researchers have chosen to test them by implementing a number of different numerical methods that can predict both the shape and position of the phase change front that are compatible with the governing conservation equations in each phase [10]. For instance, one of the most used methods is the front tracking approach which solves weak formulations of these equations by interface tracking by remeshing both states at every temporal step [11]. However, it is known that such method is not appropriate to study solidification process of alloys that undergo a frequent "mushy region" instead of sharp interface. Furthermore, another used method is the enthalpy method which is an approach that fixes the energy equation by using coordinates transformation. On another hand, the

*e-mail: hanene.jamai@etu.unilim.fr

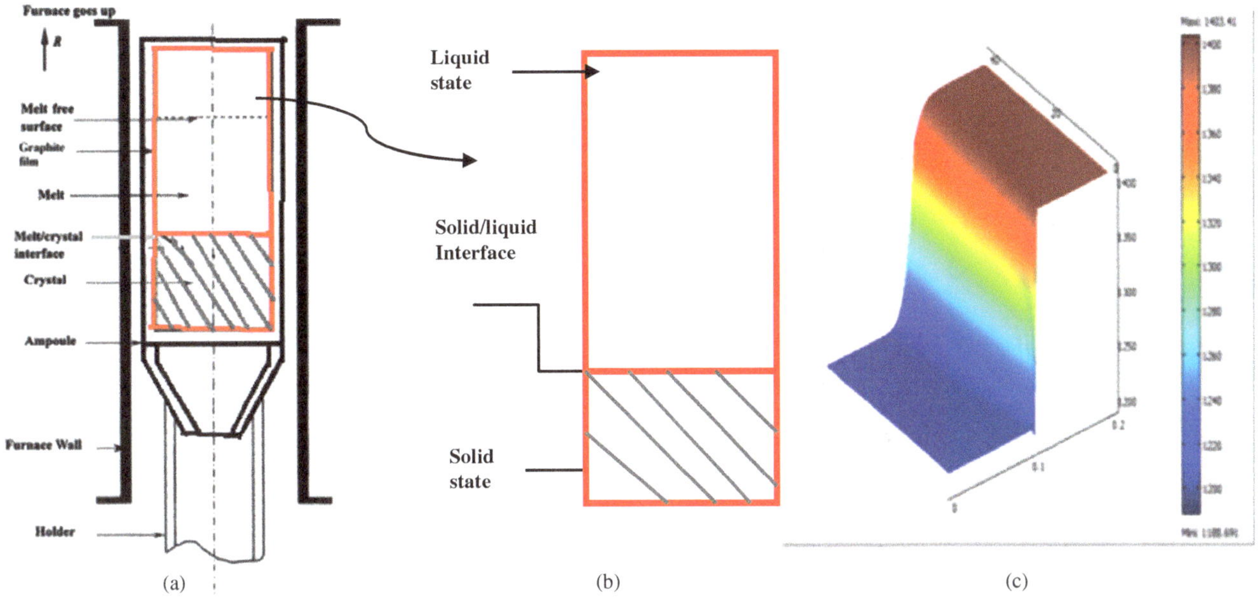

Figure 1. System. (a) Vertical Bridgman, (b) design domain, and (c) isotherms profile.

Table 1. Physical properties of CdZnTe.

Property	Temperature	Polynomial
Conductivity, W/(K m)	500 K < T < 1200 K	$5.337 - 2.263 \times 10^{-3}$ T
	1200 K < T < 1364 K	$-38.31 + 6.78 \times 10^{-2}$ T $- 2.811 \times 10^{-5}$ T2
	T > 1364 K	$9.987 - 5.676 \times 10^{-3}$ T
Specific heat Cp, J/(Kg K)	298 K < T < 1600 K	$177.8 + 1.663 \times 10^{-1}$ T -6.607×10^{-5} T2 $+ 1.703 \times 10^{-8}$ T3
Heat of fusion ΔH, J/Kg	209 000	209 000
Expansion coefficient	T > 525 K	6.3629×10^{-6}
Melting point T_m, K	1364	–
Density Kg/m^3	–	5680

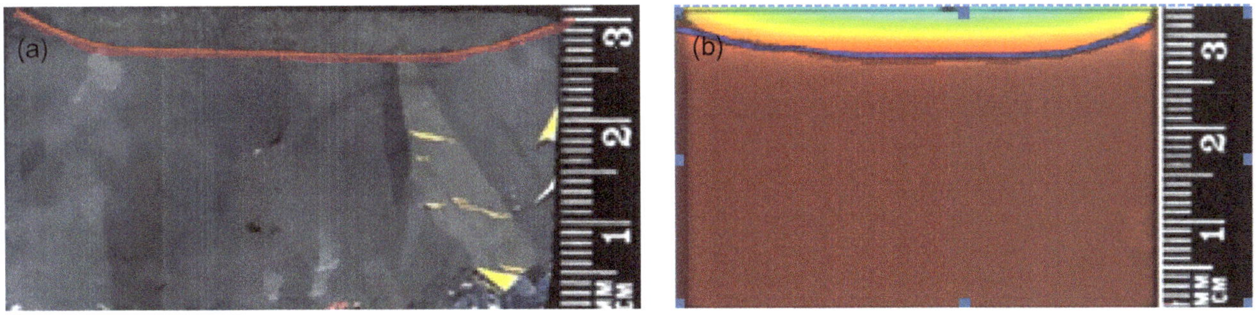

Figure 2. Comparison between experimental data obtained by Sen et al. [18] (a) and our numerical results (b) in terms of the interface growth of CdZnTe.

interface position is recovered a posteriori [11] and an extension of such a front fixing approach has been proposed in literature in order to introduce and predict the convection phenomenon within the two mushy region phases [12–14]. El Ganaoui et al. [15–17] improved this method by implementing a time-dependent enthalpy-porosity formulation for directional solidification, in which the energy equation is introduced by using the enthalpy variable computing implicitly the thermal Stefan condition at the interface with an implicit procedure.

The current work uses COMSOL Multi-physics software, to simulate solidification in Bridgman Vertical (BV) configuration during CdZnTe, which is an II–VI compound crystal growth. To the best knowledge of the authors, it seems that thermophysical properties such as thermal diffusivity, heat capacity,

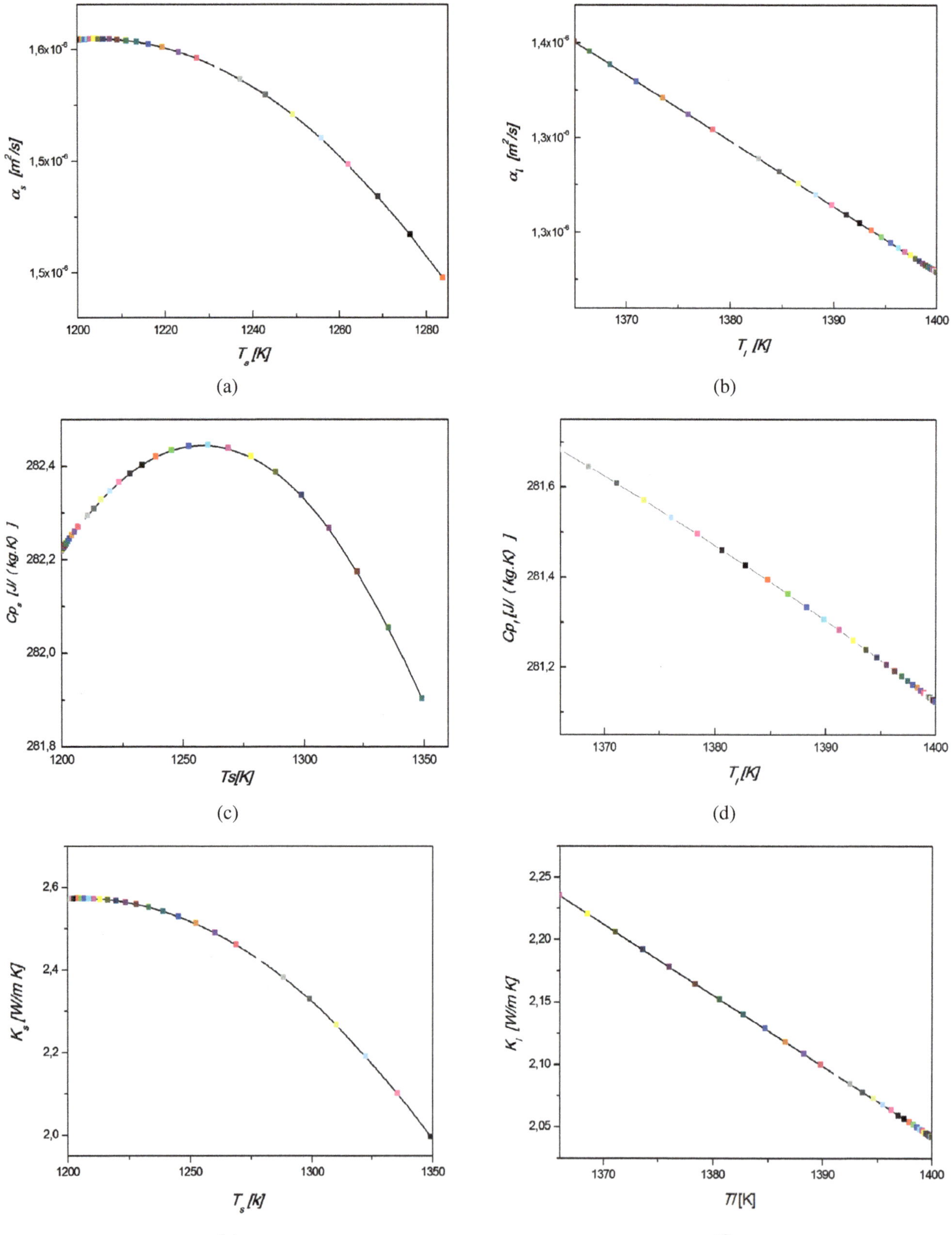

Figure 3. Thermal diffusivity, Specific heat and Thermal conductivity of CdZnTe relatively to the solid domain (a) and the liquid one (b), respectively for fixed parameters T_c = 1200 K, T_m = 1364 K, and T_h = 1400 K.

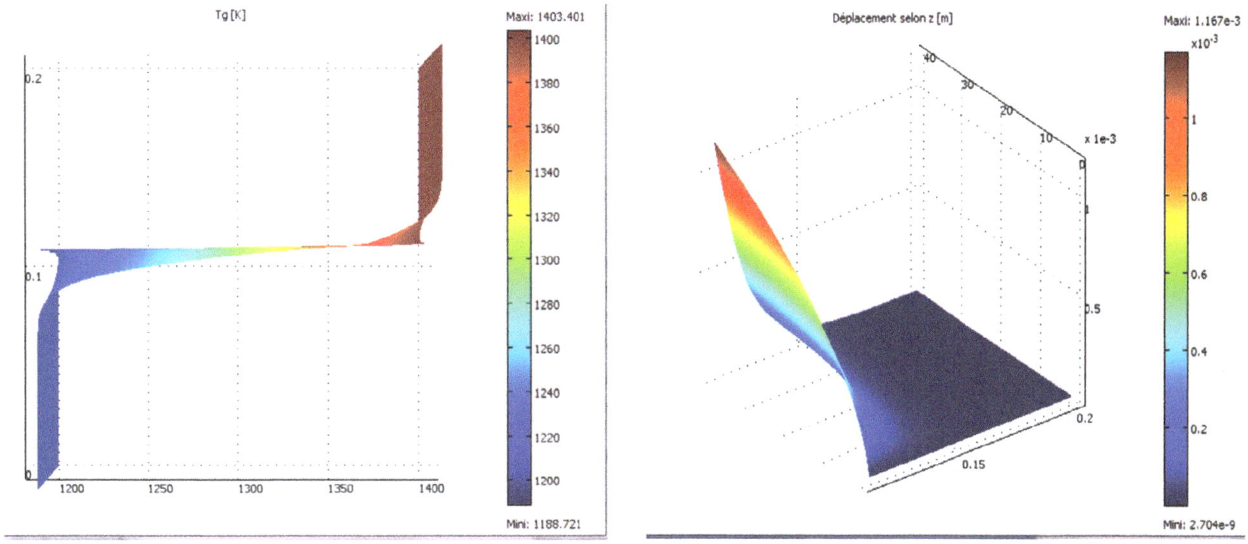

(a) Thermal profile (b) Interface profile

Figure 4. Vizualisation of temperature field, and the spatial and temporal variations of the interface evolution in both solid and liquid domains.

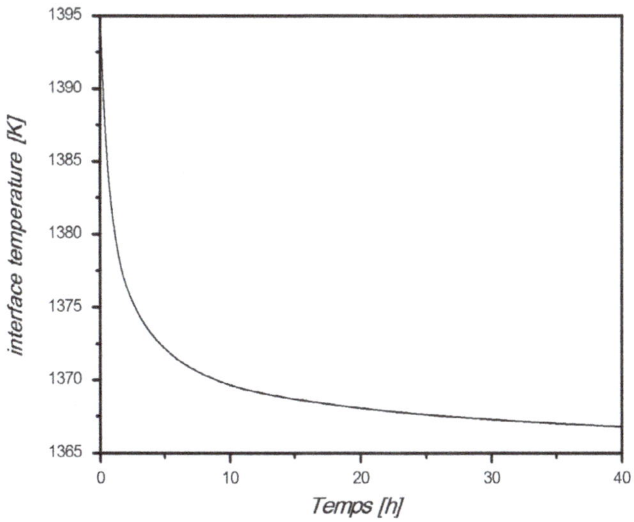

Figure 5. Temporal variation relative to the temperature at the interface for T_h = 1400 K, T_c = 1200 K, and T_m = 1364 K.

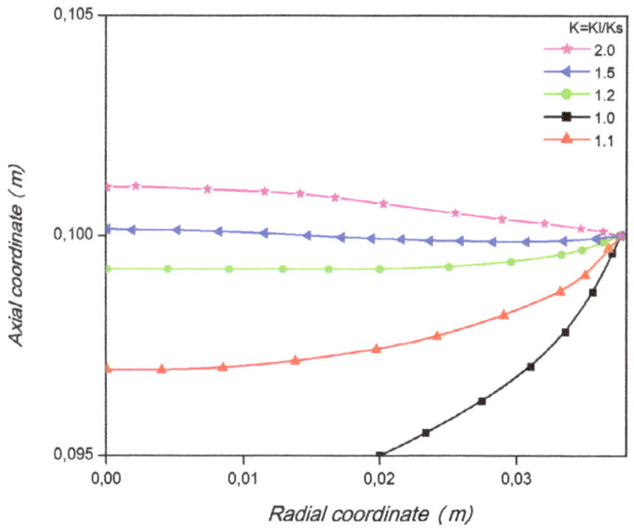

Figure 6. Influence of thermal conductivity ratio K = K_l/K_s on the interface shape for T_c = 1200 K, T_h = 1400 K, and T_m = 1364 K.

and thermal conductivity have not been well investigated. In fact, these thermophysical properties have important factors in the Stefan equation and therefore have an effect on the solidification, interface of CdZnTe. It is worth mentioning that both governing parameters the temperature of the solid domain is not well studied in literature and the growth rate phenomena still need better understanding and we propose to address such issue in the present work.

From this point of view, the main idea of the current investigation is to control the shape interface, in solidification process. In particular, we will focus on the various parameters that affect the concavity and convexity of the interface that are also an important factor in crystal growth, namely, the thermal conductivity, temperature of the solid domain, and growth rate.

2 Materials and methods

2.1 Physical properties of the materials

The physical properties of CdZnTe are listed in Table 1 [18, 19]. It should be noted that these properties depend on the given temperature, and they are in accordance with the thermal boundary conditions, for each domains.

2.2 Physics model

The Bridgman crystal system is schematically displayed in Figure 1a. The furnace uses a tapered heating element to establish a linear temperature profile Figure 1b. The ampoule is

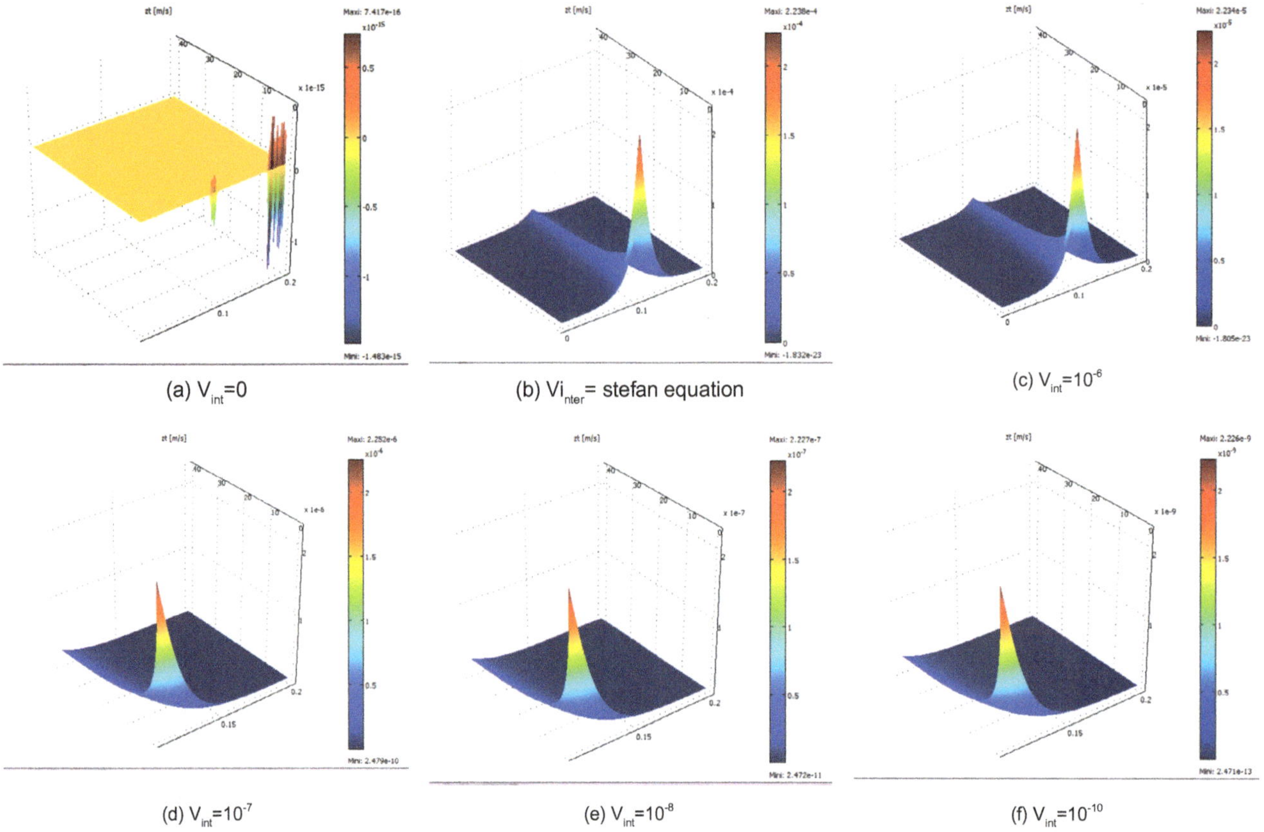

Figure 7. Temporal and spatial variations of isovalues relative to the interface velocity.

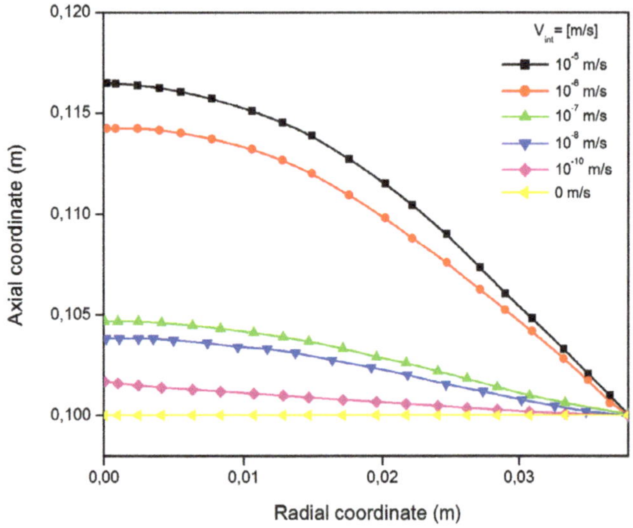

Figure 8. Effect of the interface velocity on the interface shape for T_c = 1200 K, T_h = 1400 K, and T_m = 1364 K.

axially positioned at the center of the growth furnace. During the process, the cylindrical ampoule containing the material in the solid and liquid phase is fixed at the constant rate $V_t = 0$. Because of the symmetry of the system, the right half of Figure 1c is taken as the calculation domain only supposed.

There is only heat conduction between the melt and crystal.

We can assume that this problem can be considered as a two dimensional axis-symmetry one, the temperature field, the Stefan equations are expressed in the cylindrical coordinates system.

The energy equations for the melt and the crystal given by:

$$\nabla(K_l \nabla T_l) - \rho Cp_l \frac{\partial T_l}{\partial t} = 0, \quad (1)$$

$$\nabla(K_s \nabla T_s) - \rho Cp_s \frac{\partial T_s}{\partial t} = 0. \quad (2)$$

The shape of the solid-liquid interface is a compulsory controlling factor that allows producing a single crystal in vertical Bridgman enclosure. This interface shape is predicted by heat transfer and primarily by the direction of the heat flux near the interface.

This interface is governed by the Stefan's equation expressed as:

$$(K_s \nabla T_s) \times n - (K_l \nabla T_l) \times n = \rho \Delta H V_{int}. \quad (3)$$

The aforementioned parameters in the above equations K_s, K_l, ρ, Cp_l, and Cp_s are the thermal conductivities, the density, specifics heats, in solid (denoted by s) and liquid (denoted by l), respectively. Here, ΔH expresses the latent heat and V_{int} denoted the velocity at the interface.

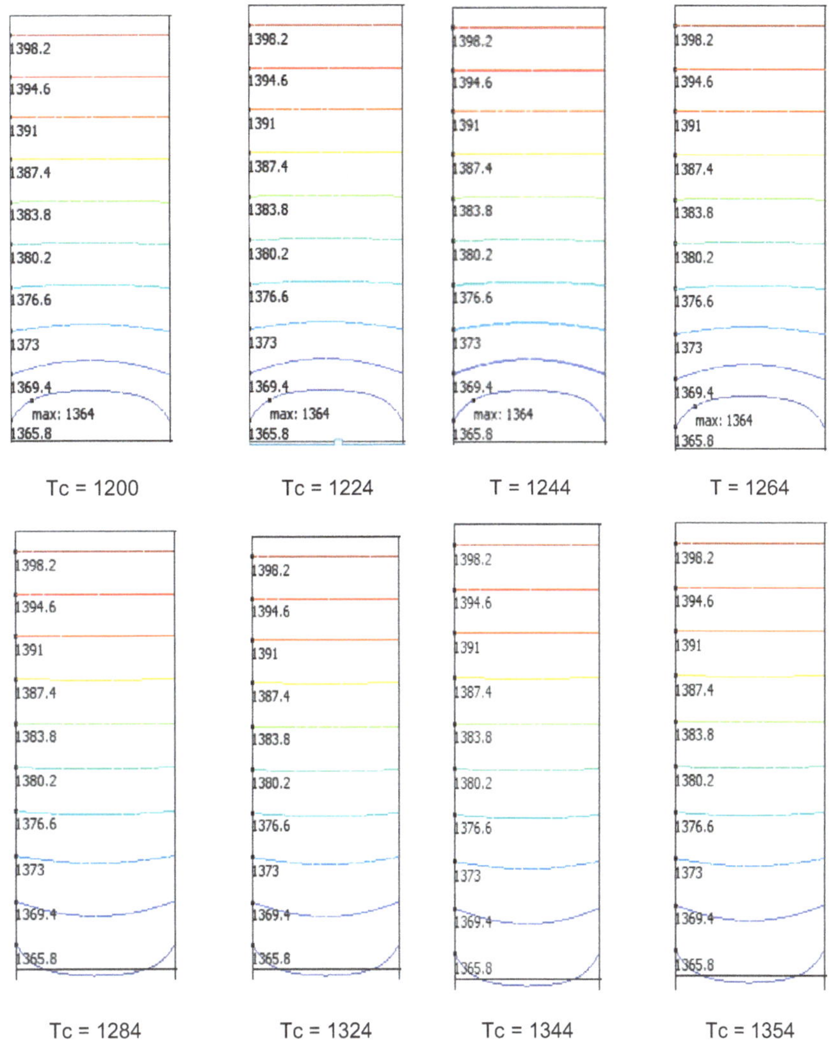

Figure 9. Isotherm patterns with respect to different values of the cold temperature T_c for T_h = 1400 K and T_m = 1364 K.

The boundaries conditions corresponding to the temperature field are as follows: the left and right boundaries are kept adiabatic while the upper, the lower, and the interface position are maintained at T_h = 1400 K, T_c = 1200, and T_m = 1364 K, respectively.

3 Numerical tools

For solving the mathematical model industrial software has been used, namely COMSOL which essentially integrated MATLAB which includes mathematics and computation; data acquisition; modeling, algorithm development; simulation, and prototyping; data analysis, exploration, and visualization. On another hand, the MATLAB language is particularly well suited to designing predictive mathematical models and developing specific algorithms [20, 21]. Where, these algorithms contain operations for linear algebra, partial differential, matrix manipulation, or differential equation solving, data analysis and statistics.

First results concerned the validation of the computational approach versus the mathematical model. For that a test validation has been defined and conducted. It refers to available experimental results with considering shaped solid/liquid interface.

3.1 Test code validation

The validation herein concerns the physical problem of CdZnTe solidification in Vertical Bridgman (BV). In fact, the choice of this configuration is based on the experimental data [22–24] for CdZnTe crystal growth: R = 0.0375 m, L = 0.20 m, the temperature of the Bridgman furnace the upper of them being at a higher temperature than the melting point while the lower one is below the melting point. Thus, the system is cooled from below, and the solidification process begins from the bottom of the ampoule.

We noticed the same shape interface between our numerical computations (Figure 2b) and experimental ones (Figure 2a) in which the authors [18] presented the half of longitudinal section of grown ingot of CdZnTe. In fact, they mentioned, that the interface shape at the end of growth is concave toward the crystal, with interface bending of 3 mm Figure 2a. The same results

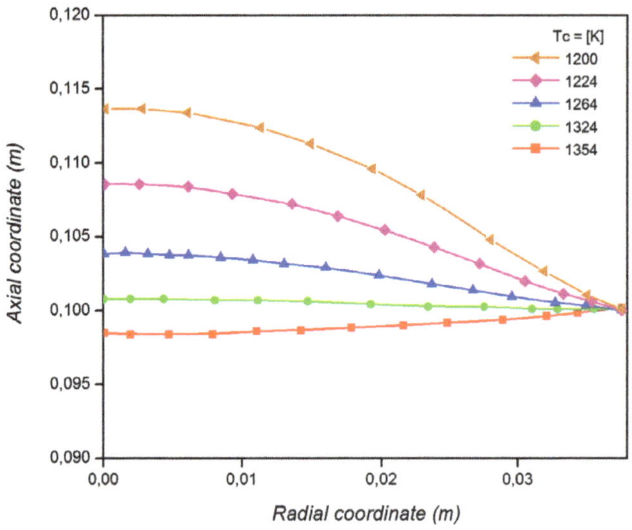

Figure 10. Effect of the cold temperature on the interface shape at fixed parameter values $T_c = 1200$ K, $T_h = 1400$ K, and $T_m = 1364$ K.

are shown numerically Figure 2b and a good agreement is observed with those obtained by Sen et al. [18]. In fact, the same results are revealed by the current simulation in which the shape interface is concaved and the progress rate is found to be around 3 mm in 9 h.

4 Results and discussion

Our results concerned the solidification process of CdZnTe in the Vertical Bridgman (VB) configuration. In fact, we will present the different parameters which affect the solidification of CdZnTe, with the aid of numerical simulations.

As aforementioned, one of the main ideas of our investigation is to present a general study of different parameters that affect the quality of the single CdZnTe crystal. Especially, we will give a great importance to study the thermo-physical properties of CdZnTe in each domain (solid or liquid) such as the thermal diffusivity α, the specific heat Cp (at constant pressure), and the thermal conductivity, that all depend on the temperature and which are measured for both solid and liquid CdZnTe. In addition, thermal diffusivity is an important factor in the current investigation. In fact, not only it offers a convenient and accurate method of determining the thermal conductivity (k), but also it is a crucial parameter in the conduction equation which governs heat transfer during crystal growth [18]. In addition, these parameters have a great impact on the Stefan equation that interferes in the solidification interface process.

In fact, as shown in Figures 3a and 3b illustrating the profile of thermal diffusivity in both solid and liquid domains, we note that this diffusivity increases for temperature in the solid domain while it decreases with temperature in the liquid state. Indeed, one can note that the diffusivity goes through a sharp minimum in the region of melting point; the value in the liquid is higher than that in the solid just before the onset of melting. The same results are showed by Sen et al. [18], who treated the influence of thermal diffusivity for CdTe and Cd0.96Zn0.04Te. However, their results are significantly different from those reported by Marina et al. [22] who claimed that thermal diffusivity increased sharply above the liquid temperature for $Hg_{1-x}Cd_xTe$.

In the same context, we study the shape of the heat capacity shown in Figures 3c and 3d in both the solid and liquid domain. Besides, as shown in Figure 3a, the heat capacity is observed to be progressively augmented in the solid domain until reaching a maximum of temperature, precisely at the value $T_c = 1260$ K. Consequently, by approaching the melting temperature, the profile of thermal capacity undergoes a decreasing trend until reaching a constant value in the liquid domain Figure 3b.

Similarly, we propose now to study the thermal conductivity and its influence on the interface shape and position as determined, in this case. In fact, we observed that the profile of thermal conductivity Figures 3e and 3f is similar to that relative to thermal diffusivity in both solid and liquid domain as revealed in Figures 3a and 3b. This makes good sense because thermal conductivity is proportional to thermal diffusivity which expression is defined by $\alpha = k/\rho Cp$. In fact, it was found that the profile of solid thermal conductivity K_s is more enhanced compared to K_l in liquid state.

Figure 4a shows the temperature according to the solidification of CdZnTe along the both liquid-solid state plotted as function of axial coordinate, where $z = -0.01$ m denotes the bottom tip of solid domain, and $z = 0.20$ m corresponds to both liquid and liquid states. This thermal profile is based on two linear segments designed to yield a gradient in the melt and a steeper cooling rate in the crystal [25]. And, the visualization of interface evolution denoted by the Figure 4b, proved that is increased with increasing time.

Figure 5 shows the interface temperature parameter as function of time. In fact, as note that after the initial transient, the temperature presented a peak which decreases steadily until a mild increase near the end of the run. The decreasing peak temperature coincides with decreasing melt height and volume as growth occurs. It is worth mentioning that similar results have been revealed by Cerny et al. [26], which mentioned that the time development of interface temperatures reveals well the establishment of the pseudo-steady regime of crystal growth after 15 h, and by Zang et al. [27] in the problem of solidification of CdZnTe crystal growth from the melt.

Thereafter, we will present the effect of the thermal conductivities ratio defined by K in the shape interface. We found that it is of great importance, because it changes not only the sign of curvature and consequently the concavity and the convexity of the interface shape, but also the curvature magnitude. Therefore, it was argued that for $T_h = 1400$ K, $T_c = 1200$ K and $T_m = 1364$ K, the parameter K ensures the achievement of a convex interface when it is less than 1.1 and over which the interface becomes concave Figure 6. Therefore, it is to conclude that the conductivity ratio is of great importance in the CdZnTe solidification because it helps controlling the process of crystal growth.

Among the problems encountered in numerical simulation is the interface computation. These problems are generally called Stefan problems (Eq. (3)) that connects the axial temperature gradient in the liquid and the solid domain by the velocity growth. In fact, we devote this section to study the evolution of velocity interface for different values taken in the solidification of CdZnTe. The main issue in this section is to give a better visualization of the velocity growth and surface evolution which will allow to better understand the phenomenon of growth, especially the evolution and of the interface shape Figure 7 of great interest in this portion is to vary the velocity interface while keeping the speed of the bulb at zero, in other words the bulb is held fixed. The velocity profile depicted in Figure 7 demonstrates that the velocity value determined by the Stefan equation is strengthened compared to that computed during the growth of CdZnTe especially in the solidification of the material. This value is seen to be nearly 10^{-5} m/s given by COMSOL after resolution and compilation of the Stefan equation while respecting the requirement for thermal boundary conditions s in both liquid and solid areas.

It was also noted that the decrease of the interface velocity from 10^{-5} to 10^{-10} m/s has a great effect on the evolution and the interface shape which becomes progressively weakened in Figure 8. Therefore one can conclude that the velocity variation is an important factor in the industrialization of crystal growth.

In this part we attempt to focus on predicting the effect of the variation of cold temperature T_c on the isotherms in the liquid state which they are affected by the advancement of the shape interface and consequently the concavity and the convexity magnitude. In fact, as we mentioned before in the introduction, although this factor is also important in the understanding of crystal growth industrialization, it has not been well investigated until now. As revealed from Figure 9, we can mention that the variation of T_c affects the concavity and the convexity of the interface. In fact, for $T_c \in [1200\ \text{K} - 1264\ \text{K}]$, the interface remains concave while it becomes convex for $T_c \in [1265 - 1354\ \text{K}]$ especially, when the cold temperature reaches the value $T_c = 1265$ k according to the fixed values $T_h = 1400$ K, $T_m = 1364$ K, and $R = 0.037$. However, by increasing T_c over the value 1354 K, the problem is seen to diverge because the temperature approaches the specific value of T_m.

We can finally conclude that the variation of T_c is of great importance such as the thermal conductivities ratios and the velocity interface variation, and this parameter can help to solve some problems encountered during the growth of CdZnTe, especially the problem of the concavity and convexity of the interface shape which affects the quality of crystal growth.

Do not forget also that the purpose of any study is to solve the problems not only in the semiconductor factory, but also to obtain crystals of good qualities. Investigations must help to choose the value of the cold temperature which allow operators to obtain a concave or convex interface

In the same as mentioned in Figure 10 the variation of axial coordinate with radial coordinate for different values of cold temperature which reveals the same interpretation proved in Figure 9.

Our numerical findings may contribute in controlling the parameters for the growth process, and it represents a strong support to resolve difficulties for experimentalists.

5 Conclusion

In the current investigation, solidification of the singular CdZnTe in axisymmetric Vertical Bridgman (VB) cavity is studied. All numerical computations are obtained using COMSOL Multiphysics software, which is based on finite elements method (FEM) with ALE (Arbitrary Lagrangian-Euleurian) formulation allowing treating the interface displacement. The visualization of temperature fields in terms of isotherms and spatial and temporal profiles and according to the solidification of CdZnTe is presented and discussed with the help of COMSOL to provide a better understanding of phase change. Especial attention has been attributed to study the thermo-physical properties of CdZnTe in each domain (solid or liquid) such as the thermal diffusivity α, the specific heat Cp (at constant pressure), and the thermal conductivity, which are measured for both solid and liquid CdZnTe. In this context, it is found that iso-surfaces of thermal diffusivity in both solid and liquid domains show that diffusivity increases for temperature in the solid domain while it is lowered with temperature in the liquid state. Furthermore, the diffusivity goes through a sharp minimum in the region of melting point and the value in the liquid is higher than that in the solid just before the onset of melting. In addition, the heat capacity is observed to progressively augment in the solid domain until reaching a maximum of temperature, at exactly the value $T_c = 1260$ K. On another hand, it is found that the variation of T_c affects the concavity and the convexity of the interface. In fact, it is seen that for specific interval of T_c the interface remains concave while it becomes convex for another one. Concerning the effect of the interface velocity variation, it is found that the decrease of the interface velocity from 10^{-5} to 10^{-10} m/s has a great influence on the evolution and the interface shape which becomes progressively weakened. In this vein, analysis of the results shows that the velocity variation is seen to be an important factor in the industrialization of crystal growth.

Acknowledgements. The fist author thanks Profs. R. Bennacer (ENS Cachan) and A. Aydi (university of sfax) for fruitful discussion.

References

1. Chang CE, Wilcox WR. 1974. Control of interface shape in the vertical Bridgman-Stockbarger technique. Journal of Crystal Growth, 21, 135–140.
2. Sen S, Wilcox WR. 1975. Influence of crucible on interface shape, position and sensitivity in the vertical Bridgman-Stockbarger technique. Journal of Crystal Growth, 28, 26–40.
3. Fu T-W, Wilcox WR. 1980. Influence of insulation on stability of interface shape and position in the vertical Bridgman-Stockbarger technique. Journal of Crystal Growth, 48, 416–424.
4. Naumann RJ. 1982. An analytical approach to the thermal modeling of Bridgman-type crystal growth. II. Two-dimensional analysis. Journal of Crystal Growth, 58, 569–584.
5. Naumann RJ, Lehoczky SL. 1983. Effect of variable thermal conductivity on isotherms in Bridgman growth. Journal of Crystal Growth, 61, 707–710.

6. Adornato P, Brown RA. 1987. Convection and segregation in directional solidification of dilute and non-dilute binary alloys: effects of ampoule and furnace design. Journal of Crystal Growth, 80, 155–190.
7. Wang CA, Witt AF, Carruthers JR. 1984. Analysis of crystal growth characteristics in a conventional vertical Bridgman configuration. Journal of Crystal Growth, 66, 299–308.
8. Jasinski T, Witt AF. 1985. On control of the crystal melt interface shape during growth in a vertical Bridgman configuration. Journal of Crystal Growth, 71, 295–304.
9. Jasinski T, Witt AF, U.S. Patent: Apparatus for Growing Crystals, 1986.
10. El Ganaoui M, Lamazouade A, Bontoux P. 2002. Computational solution for fluid flow under solid/liquid phase change conditions. Computers & Fluids, 31, 539–556.
11. Crank IJ. 1984. Free and moving boundary problems. Clarendon Press: Oxford, UK.
12. Voller R, Prakash C. 1987. A fixed grid numerical modelling methodology for convection diffusion mushy region phase change problems. International Journal of Heat and Mass Transfer, 30(8), 1709–1719.
13. Bennon WD, Incropera FP. 1987. A continuum model for momentum, heat and species transport in binary solid-liquid phase change systems: 1. Model formulation. International Journal of Heat and Mass Transfer, 30(10), 2161–2170.
14. Voller R, Cross M. 1980. Accurate solutions of moving boundary problems using the enthalpy method. International Journal of Heat and Mass Transfer, 24, 545–556.
15. El Ganaoui M. Modélisation numérique de la convection thermique instationnaire en présence d'un front de solidification déformable. Thèse de l'Université d'Aix-Marseille, Octobre, 1997.
16. El Ganaoui M, Mazhorova OS, Bontoux P. 2000. Computer simulation of pure and alloys melt growth. Microgravity Quart Rev, 7(4), 171–178.
17. Elganaoui M, Bontoux P, Morvan D. 1999, Localisation d'un front de solidification en interaction avec un bain fondu instationnaire, Paris: CR Acad Sciences, Série II b, t. 327, p. 41–48.
18. Sen S, Konkel WH, Tighe SJ, Bland LG, Sharma SR, Taylor RE. 1988. Crystal growth of large-area single-crystal CdTe and CdZnTe by the computer-controlled vertical modified-Bridgman process. Journal of Crystal Growth, 86, 111.
19. Bruder M, Figgemeier H, Schmitt R, Maier H. 1993. Mat. Res. Eng. B, 16, 40.
20. Kuppurao S, Brandon S, Derby JJ. 1995. Journal of Crystal Growth, 155, 93.
21. MATLAB. The Language of Technical Computing – Programming Version 7, MA, USA: The Mathworks, Inc, 2005
22. Marchenko MP, Golyshev VD, Bykova SV. 2007. Investigation of $Cd_{1-x}Zn_xTe$ composition inhomogeneity at crystal growth by AHP-method. Journal of Crystal Growth, 303, 193–198.
23. Huang ZH, Conway PP, Thomson RC, Dinsdale AT, Robinson JAJ. 2008. A computational interface for thermodynamic calculations software MTDATA. Computer Coupling of Phase Diagrams and Thermochemistry, 32, 129–134.
24. Bruder M, Figgemeier H, Schmitt R, Maier H. 1993. Mat. Res. Eng. B, 16, 40.
25. Fang HS, Wang S, Zhou L, Zhou NG, Lin MH. 2012. Influence of furnace design on thermal stress during directional solidification of multicristalline silicon. Journal of Crystal Growth, 346, 5–11.
26. Cerny C, Kalbac A, Prikryl P. 2000. Computational modeling of CdZnTe crystal growth from the melt. Computational Materials Science, 1734–1760.
27. Zhang N, Yeckel A, Derby JJ. 2012. Maintaining convex interface shapes during electrodynamic gradient freeze growth of cadmium zinc telluride using a dynamic, bell-curve furnace profile. Journal of Crystal Growth, 355, 113–121.

3

Efficient harmony search optimization for preventive-maintenance-planning for nuclear power systems

Abdelkader Rami, Habib Hamdaoui, Houari Sayah, and Abdelkader Zeblah*

University of Djillali Liabes, Faculty of Technology, Sidi Bel Abbes, Algeria

Abstract – This paper combines the universal generating function UGF with harmony search (HSO) meta-heuristic optimization method to solve a preventive maintenance (PM) problem for series-parallel system. In this work, we consider the situation where system and its components have several ranges of performance levels. Such systems are called multi-state systems (MSS). To enhance system availability or (reliability), possible schedule preventive maintenance actions are performed to equipments and affect strongly the effective age. The MSS measure is related to the ability of the system to satisfy the demand. The objective is to develop an algorithm to generate an optimal sequence of maintenance actions providing system working with the desired level of availability or (reliability) during its lifetime with minimal maintenance cost rate. To evaluate the MSS system availability, a fast method based on UGF is suggested. The harmony search approach can be applied as an optimization technique and adapted to this PM optimization problem.

Key words: Universal generating function harmony search, Optimization, Preventive maintenance.

Nomenclature

C_{Mj}	Minimal repair cost
A_{iv}	Availability of *j*th MSS devices
h_{j}	Hazard function
Q_{j}	Probability of failure of *j*th devices
Ξ_{iv}	Performance of *j*th devices of version *v*
W	Demand levels
$\Im$	Distributive operator
δ	Operator for parallel devices

Abbreviations

HSO	Harmony search optimization
PM	Preventive maintenance
MSS	Multi states system
UMGF	Universal moment generating function

1 Introduction

A necessary precondition for high production is availability of the technical equipment. In addition, reliability engineers have to build a reliable and efficient production system. The system reliability affects essentially the reliability of its equipments. This characteristic is a function of equipment age on system's operation life. In this work, we consider series-parallel systems. To keep the desired levels of availability, strongly performs a preventive maintenance actions to components are best than breakdown maintenance. This suggestion is supported by a number of case studies demonstrating the benefits of PM in [1]. In this case, the task is to specify how PM activity should be scheduled. One of the commonly used PM policies is called periodic PM, which specifies that systems are maintained at integer multiple of some fixed period. Another PM is called sequential PM, in which the system is maintained at a sequence of interval that have unequal lengths. The first kind of PM is more convenient to schedule. Contrary the second is more realistic when the system require more frequent maintenance at it age. A common assumption used in both these PM is that minimal repair is conducted on system if it fails between successive PM activities. In other words, minimal repairs do not change the hazard function or the effective age of the system.

Traditionally PM models assume that the system after PM is either as good as new state in this case is called perfect PM or simply replacement, as bad as old state the same as minimal repair, where he only restores the function of the system, this concept is well understood in the literature [2]. The more realistic assumption is that the system after PM not return at zero age and remains between as good as new and as bed as old. This kind of PM is called imperfect PM. The case when

*e-mail: zeblaha@yahoo.fr

equipment fails (damage), a corrective maintenance (CM) is performed which returns equipment to operating condition, in fact specially, the task of preventive maintenance actions served to adjust the virtual age of equipment. Our particular interest is under investigation to present an harmony search algorithm which determines the optimal intervals of PM actions to minimize maintenance-cost rate or maximize mission reliability.

1.1 Summary of previous work

Years ago, much work was reported on policy optimization of preliminary planned PM actions with minimal repair as in [3, 4]. Most of these researches are based on two popular approaches to determine the optimal intervals for a PM sequence. The first is reliability-based method and the second is optimization method.

In the first one the PM is performed whenever the system availability or the hazard function of the system reaches a predetermined level and the optimal PM intervals will be selected. The second is finding the optimal intervals as a decision variable in the optimization problem. [5] presents an algorithm to determine the optimal intervals based on the reliability-based method and in there models the effective age reduction and hazard function are combined. [6] presents a genetic algorithm which determine a minimal cost plan of the selecting PM actions which provides the required levels of power system reliability. A list of possible PM actions available for each MSS, are used. Each PM action is associated with cost and reduction age coefficient of its implementation.

1.2 Approach and outlines

The proposed approach is based on the optimization method using harmony search algorithm, which determines the intervals sequence of PM actions to minimize the maintenance-cost subject to availability or (reliability) constraints. The goal of the proposed approach is to know when, where, to which component and what kind of available PM actions among the set of available PM actions should be implemented. To evaluate the reliability and the effect of PM actions of series-parallel MSS, UGF method is applied. It's proved to be effective at solving problem of MSS redundancy and maintenance in [7–9].

The rest of this paper is outlined as follows. We start in Section 2 with the general description of the preventive maintenance model. Next, we describe the optimization problem formulation in Section 3. A description of availability estimation based on UGF method is presented in Section 4. In Section 5, we present the harmony search algorithm. Conclusion is drawn in Section 6.

2 Preventive maintenance

It has been shown that the incorporation of the preventive maintenance has a benefit and success. Also it was observed that the impact of the decrease of component failure rate and improvement of component reliability is vital to maintain efficiency of production. The major subject of maintenance is

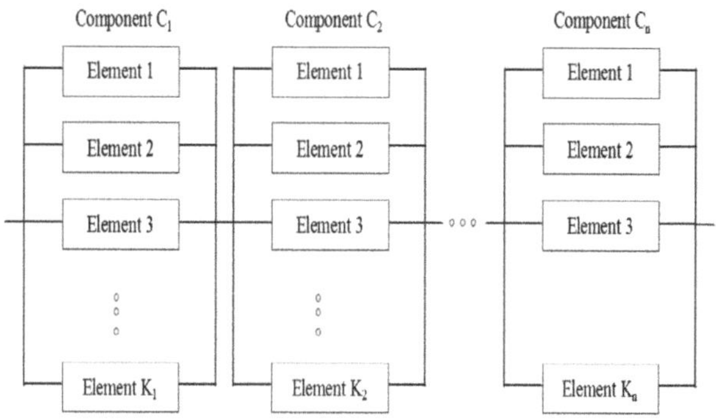

Figure 1. Series-parallel power system.

focused on the planning maintenance service of the power system. Such as cleaning, adjustment and inspection performed on operation's lifetime are classed as a preventive maintenance policy. However, all actions of PM not capable to reduce age component to zero age is imperfect. There are two main alternatives for modeling an imperfect PM activity. The first one assumes that PM is equivalent to minimal repair with probability p and $1 - p$ is the equivalent to replacement in [10]. The second model where the imperfect PM directly analyzes how the hazard function or the effective age change after PM as in [5]. The proposed model is based on reduction age concept. Let consider the series-parallel MSS system shown in Figure 1.

If the component j undergoes on PM actions calendar at chronological times as follows:

$$t_{j1}, \ldots, t_{jn} \tag{1}$$

Based on the second model description, the effective age after i-th PM actions may be written as:

$$\tau_j(t) = \tau_j^+(t) + (t - t_{ji}) \text{ for } t_{ji} < t < t_{ji+1}, \ (1 \le i \le n), \tag{2}$$

$$\text{and } \tau_j^+(t_{ji}) = \varepsilon_i \tau_j(t_{ji}) = \varepsilon_i\left(\tau_j^+(t_{ji-1}) + (t_{ji} - t_{ji-1})\right),$$

where $\tau_j^+(t_{ji})$ is the age of component immediately after the ith PM action which ranges in the interval [0, 1]. By definition, we assume that $\tau_j(0) = 0$, $t_{j0} = 0$ and ε_i is the age reduction coefficient. Two limits for PM actions is, where $\varepsilon_i = 1$ and $\varepsilon_i = 0$. In the first case the component at least be restored to "as bed as old" state which assumes that PM does not affect the effective age. In the second case the model reduce the component age to "as good as new", which means that the component age reaches zero age (replacement). In fact, all PM actions which improve the component age are imperfect. As it be mentioned and demonstrated in [5], the hazard function of component j, as function of its actual age, can be calculated as:

$$h_j^* = h_j(\tau_j(t)) + h_{j0}, \tag{3}$$

where $h_j(t)$ is the hazard function is defined when equipment does not undergo PM actions and h_{j0} correspond to the initial age of equipment. The reliability of the equipment j in the interval between PM actions i and $i + 1$ can be written as:

$$\begin{aligned} r_j(t) &= \exp\left(-\int_{\tau_j^+(t_{ji})}^{\tau_j(t)} h_j^*(x)\mathrm{d}x\right) \\ &= \exp\left(H_j(\tau_j^+(t_{ji})) - H_j(\tau_j(t))\right). \end{aligned} \tag{4}$$

$H_j(\tau)$ represents the accumulative hazard function. Clearly if $t = t_{ji}$ in equation (4) the reliability reaches the maximum and is equal to 1.

The Minimal repairs are performed if MSS equipment fails between PM actions, and there cost expected in interval [0, t] can be given as:

$$C_{Mj} = c_j \int_0^t h_j(x)\mathrm{d}x. \quad (5)$$

Possible equipment j, undergoes PM actions at each chronological time $t_{j1}, \ \ldots, \ t_{jn_j}$, in this case, the total minimal repair cost is the sum of all cost can be written as :

$$C_{Mj} = c_j \sum_{i=0}^{n_j} \int_{\tau_j^+(t_{ji})}^{\tau_j(t_{ji+1})} h_j(x)\mathrm{d}x =$$

$$c_j \sum_{i=0}^{n_j} \left(H(\tau_j(t_{ji+1})) - H_j(\tau_j^+(t_{ji})) \right), \quad (6)$$

where $t_{j0} = 0$ and $t_{jn_j+1} = T$ where T represents the lifetime.

3 Optimization problem

Let consider a power system organized with components connected in series arrangement. Each component contains different component put in parallel. Components are characterized by their nominal performance rate Ξj, hazard function $hj(t)$ and associated minimal repair cost Cj. The system is composed of a number of failure prone components, such that the failure of some components leads only to a degradation of the system performance. This system is considered to have a range of performance levels from perfect working to complete failure. In fact, the system failure can lead to decreased capability to accomplish a given task, but not to complete failure. An important MSS measure is related to the ability of the system to satisfy a given demand.

When applied to electric power systems, reliability is considered as a measure of the ability of the system to meet the load demand (W), i.e., to provide an adequate supply of electrical energy (Ξ). This definition of the reliability index is widely used for power systems: see e.g., [11, 12, 13, 14, 15]. The Loss of Load Probability index (LOLP) is usually used to estimate the reliability index [16]. This index is the overall probability that the load demand will not be met. Thus, we can write R = Probab(ΞMSS $\geq W$) or $R = 1$ – LOLP with LOLP = Probab(ΞMSS $< W$). This reliability index depends on consumer demand W.

For reparable MSS, a multi-state steady-state instantaneous availability A is used as Probab(ΞMSS $\geq W$). While the multi-state instantaneous availability is formulated by equation (7):

$$A_{\mathrm{MSS}}\{t, W\} = \sum_{\Sigma_j \geq D} P_j(t), \quad (7)$$

where ΞMSS (t) is the output performance of MSS at time t. To keep system reliability at desired level, preventive and curative maintenance can be realized on each MSS. PM actions modify components reliability and CM actions does not affect it. The effectiveness of each PM actions is defined by the age reduction coefficient ε ranging from 0 to 1. As in [6], the structure of the system as defined by an available list of possible PM actions (v) for a given MSS. In this list each PM actions (v) is associated with the cost of its implementation $C_p(v)$, and $M(v)$ is the number of equipment affected corresponding to their age reduction $\varepsilon(v)$. Commonly the system lifetime T is divided into y unequal lengths, and each interval have duration θ_y $1 \leq y \leq Y$, at each end of this latter an PM action is performed. This action will be performed if the MSS reliability $R(t, w)$ becomes lower than the desirable level R_0.

Let us remark that the increase in the number of intervals increases solution precision. On the other hand, the number of intervals can be limited for technical reasons.

All the PM actions performed to maintain the MSS reliability are arranged and presented by a vector V as they appear on the PM list. Each time the PM is necessary to improve the system reliability; the performed following action is defined by the next number from this vector. When the scheduled PM action v_i was insufficient to improve reliability, automatically the v_{i+1} action should be performed at the same time and so on.

For a given vector V, the total number n_j and chronological times of PM action in equation (1) are determined for each component $j1 \leq j \leq J$. For all scheduled PM actions $v_i \in V$. The total cost of PM actions can be expressed as:

$$C_p(V) = \sum_{i=1}^{N} C_p(v_i) \quad (8)$$

and the cost of minimal repair can be calculated as:

$$C_M(V) = \sum_{j=1}^{J} c_j \sum_{i=0}^{n_j} \left(H(\tau_j(t_{ji+1})) - H(\tau_j^+(t_{ji})) \right). \quad (9)$$

The optimization problem can be formulated as follows: find the optimal sequence of the PM actions chosen from the list of available actions which minimizes the total maintenance cost while providing the desired MSS availability. That is,

Minimize:

$$f(V) \rightarrow C = C_p(V) + C_M(V). \quad (10)$$

Subject to:

$$A\theta(V, D, t) \geq R_0. \quad (11)$$

To solve this combinatorial optimization problem, it is important to have an effective and fast procedure to evaluate the availability index. Thus, a method is developed in the following section to estimate the system availability.

4 Reliability estimation based on Ushakov's method

The last few years have seen the appearance of a number of works presenting various methods of quantitative estimation of systems consisting of devices that have a range of working levels in [17, 18]. Usually one considers reducible systems. In general forms the series connection, the level of working is determined by the worst state observed for any one of the devices, while for parallel connection is determined by the best state. However, such the approach is not applicable for the majority of real systems.

In this paper the procedure used is based on the universal z-transform, which is a modern mathematical technique introduced in [19]. This method, convenient for numerical implementation, is proved to be very effective for high dimension combinatorial problems. In the literature, the universal z-transform is also called UMGF or simply u-transform. The UMGF extends the widely known ordinary moment generating function [11]. The UMGF of a discrete random variable Ξ is defined as a polynomial:

$$u(z) = \sum_{j=1}^{J} P_j z^{\Xi_j}. \tag{12}$$

The probabilistic characteristics of the random variable Ξ can be found using the function $u(z)$. In particular, if the discrete random variable Ξ is the MSS stationary output performance, the availability A is given by the probability Proba($\Xi \geq W$) which can be defined as follows:

$$\text{Proba}(\Xi \geq W) = \Phi\left(u(z)z^{-W}\right), \tag{13}$$

where Φ is a distributive operator defined by expressions (14) and (15):

$$\Phi(Pz^{\sigma-W}) = \begin{cases} P, & if\ \sigma \geq W \\ 0, & if\ \sigma < W \end{cases} \tag{14}$$

$$\Phi\left(\sum_{j=1}^{J} P_j z^{\Xi_j - W}\right) = \sum_{j=1}^{J} \Phi\left(P_j z^{\Xi_j - W}\right). \tag{15}$$

It can be easily shown that equations (14)–(15) meet condition Proba($\Xi \geq W$) = $\sum_{\Xi_j \geq W} P_j$. By using the operator Φ, the coefficients of polynomial $u(z)$ are summed for every term with $\Xi j \geq W$, and the probability that Ξ is not less than some arbitrary value W is systematically obtained.

Consider single devices with total failures and each device i has nominal performance Ξ_i and reliability A_i. The UMGF of such an device has only two terms can be defined as:

$$u_i(z) = (1 - A_i)z^0 + A_i z^{\Xi_i} = (1 - A_i) + A_i z^{\Xi_i}. \tag{16}$$

To evaluate the MSS availability of a series-parallel system, two basic composition operators are introduced. These operators determine the polynomial $u(z)$ for a group of devices.

Parallel devices: let consider a system device m containing Jm devices connected in parallel. The total performance of the parallel system is the sum of performances of all its devices. In power systems, the term capacity is usually used to indicate the quantitative performance measure of an device in [20]. Examples: generating capacity for a generator, carrying capacity for an electric transmission line, etc. Therefore, the total performance of the parallel unit is the sum of capacity (performances) in [19]. The u-function of MSS device m containing Jm parallel devices can be calculated by using the $\Im$ operator:

$u_p(z) = \Im_n(u_1(z), u_2(z), \ldots, u_n(z))$, where $\Im(\Xi_1, \Xi_2, \ldots, \Xi_n) = \sum \Xi_i$.

Therefore for a pair of devices connected in parallel:

$$\Im(u_1(z), u_2(z) = \Im\left(\sum_{i=1}^{n} P_i z^{a_i}, \sum_{j=1}^{m} Q_j z^{b_j}\right) = \sum_{i=1}^{n}\sum_{j=1}^{m} P_i Q_j z^{a_i + b_j}.$$

The parameters a_i and b_j are physically interpreted as the performances of the two devices. n and m are numbers of possible performance levels for these devices. P_i and Q_j are steady-state probabilities of possible performance levels for devices. One can see that the $\Im$ operator is simply a product of the individual u-functions. Thus, the device UMGF is: $u_p(z) = \prod_{j=1}^{J_m} u_j(z)$. Given the individual UMGF of devices defined in equation (11), we have: $u_p(z) = \prod_{j=1}^{J_m}(1 - A_j + A_j z^{\Xi_i})$.

Series devices: when the devices are connected in series, the device with the least performance becomes the bottleneck of the system. This device therefore defines the total system productivity. To calculate the u-function for system containing n devices connected in series, the operator δ should be used: $u_s(z) = \delta(u_1(z), u_2(z), \ldots, u_m(z))$, where $\delta(\Xi_1, \Xi_2, \ldots, \Xi_m) = \min\{\Xi_1, \Xi_2, \ldots, \Xi_m\}$ so that

$$\delta(u_1(z), u_2(z)) = \delta\left(\sum_{i=1}^{n} P_i z^{a_i}, \sum_{j=1}^{m} Q_j z^{b_j}\right) = \sum_{i=1}^{n}\sum_{j=1}^{m} P_i Q_j z^{\min\{a_i, b_j\}}.$$

Applying composition operators $\Im$ and δ consecutively, one can obtain the UMGF of the entire series-parallel system. To do this we must first determine the individual UMGF of each device.

Devices with total failures: let consider the usual case where only total failures are considered and each subsystem of type i and version vi has nominal performance Ξ_{iv} and availability A_{iv}. In this case, we have: Proba($\Xi = \Xi_{iv}$) = A_{iv} and Proba($\Xi = W$) = 1 − A_{iv}. The UMGF of such an device has only two terms can be defined as in equation (11) by $u^*_i(z) = (1 - A_{iv})z^0 + A_{iv}z^{\Xi_{iv}} = 1 - A_{iv} + A_{iv}z^{\Xi_{iv}}$. Using the $\Im$ operator, we can obtain the UMGF of the i-th system device containing k_i parallel devices $u_i(z) = (u^*_i(z))^{k_i} = (A_{iv}z^{\Xi_{iv}} + (1 - A_{iv}))^{k_i}$.

The UMGF of the entire system containing n devices connected in series is:

$$u_s(z) = \delta\begin{pmatrix} (A_{1v}z^{\Xi_{1v}} + (1 - A_{1v}))^{k_1}, \\ (A_{2v}z^{\Xi_{2v}} + (1 - A_{2v}))^{k_2}, \ldots, \\ (A_{nv}z^{\Xi_{nv}} + (1 - A_{nv}))^{k_n} \end{pmatrix}. \tag{17}$$

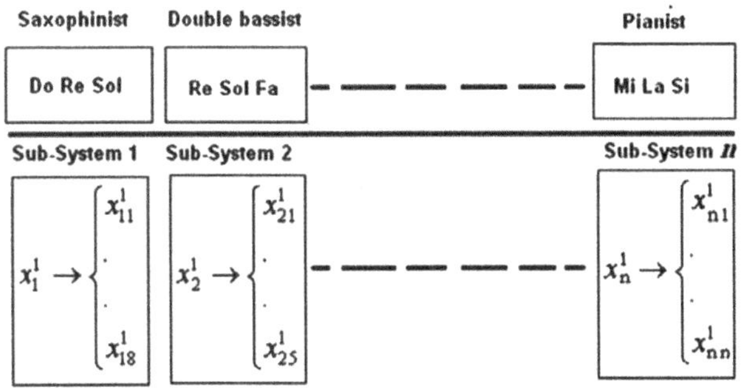

Figure 2. Synoptic modeling HS for optimization.

To evaluate the probability Proba($\Xi \geq W$) for the entire system, the operator Φ is applied to equation (18):

$$\text{Proba}(\Xi \geq W) = \Phi\left(u_s(z)z^{-W}\right). \tag{18}$$

5 The harmony search approach

The problem formulated is a complicated NP-hard complex problem. The total number of different solutions to be examined is very large. An exhaustive examination of the enormous number of possible solutions is not feasible given reasonable time limitations. Thus, because of the search- space size of the problem. Adopting the idea that existing evolutionary or meta-heuristic algorithms are found in the paradigm of natural processes, a new algorithm can be conceptualized from a musical performance process (say, a jazz trio) in [21, 22] involving searching for a better harmony. Musical performance seeks a best state (fantastic harmony) determined by aesthetic estimation, as the optimization process seeks a best state (global optimum: minimum cost; minimum error; maximum benefit; or maximum efficiency) determined by objective function evaluation. Aesthetic estimation is done by the set of the pitches sounded by joined instruments, as objective function evaluation is done by the set of the values produced by composed variables; the aesthetic sounds can be improved practice after practice, as the objective function values can be improved iteration by iteration in [23].

Figure 2 shows the structure of the Harmony Memory (HM) that is the core part of the HS algorithm. Consider a jazz trio composed of saxophone, double bass, and guitar. There exist certain amount of preferable pitches in each musician's memory: saxophonist, {Do, Fa, Mi, Sol, Re}; double bassist, {Si, Do, Si, Re, Sol}; and guitarist, {La, Sol, Fa, Mi, Do}. If saxophonist randomly plays {Sol} out of its memory {Do, Fa, Mi, Sol, Re}, double bassist {Si} out of {Si, Do, Si, Re, Sol}, and guitarist {Do} out of {La, Sol, Fa, Mi, Do}, the new harmony (Sol, Si, Do) becomes another harmony (musically chord). And if this new harmony is better than existing worst harmony in the HM, the new harmony is included in the HM and the worst harmony is excluded from the HM [24]. This procedure is repeated until fantastic harmony is found.

6 Harmony search algorithm

Step 1. Initialize:
Set N_istrument:=N_sybsystem {N is the integer number},
Set MHCR:=0.7{harmony memory considering rate},
Set PAR:=0.35{Pitch adjustment rate},
Set PAD {Pitch adjustment decision},
Set NI:=75{improvisation number},
Set t:=0 {t is the time counter},
Set Interval_t:=0 {θ is Interval time},
Set List PM_actions [L_{n1}]:= $\left[P^{\upsilon}_{n1}\right]$ {Available PM_actions},
Set PM_Action:=0 {m is the time counter},
For every Components (i,j) set an initial value $x^1_{ij} \to \left\{x^1_{11}, \ldots, x^1_{n1}\right\}x^1_{ij}$,
Set For All components ($1 \leq j \leq J$):
Set Effectve_age:= τj
Set $H_j(\tau_j^+) = 0$
For i:=1 to n **do**
Mat$_k$ (HM):= i {starting component is the first element of the **Mat** list of the kth instrument},
The HM matrix is filled with randomly generated as the HMS.
Step 2. Improvise a New list **Mat**$_k$ until is full {this step will be repeated ($n - 1$) times}, $x^{n\upsilon} \to \left\{x^{n\upsilon}_{11}, x^{n\upsilon}_{12}, \ldots, x^{n\upsilon}_{nk}\right\}$
2.0. y:=Interval_time+1
2.1. T:= Interval_time +t'
2.2. Effectve_age:=τj+t
2.3. New PM_action:= $x^{n\upsilon} \in \left[P^{\upsilon}_{n1}\right]$
Step 3. Compute For j:=1 To J
$H_j(\tau_j)$ According equation (3)
End
Compute For j:=1 To J
$r_j(\tau_j)$ According equation (4)
Step 4. Compute **For** j:=1 To n do {for every kth instrument on Subsystem i}
Choose the PM_Action with probability

$$PAD\ for\ x^{n\upsilon}_{ij} \begin{cases} Yes\ If\ ran_1 \leq PAR \\ No\ If\ ran_1 \succ (1 - PAR) \end{cases}$$

Then $x^{n\upsilon}_{ij} \quad x^{n\upsilon}_{gi} \pm ran_i() * bw$
With Cost j, corresponding to Equation (8) and (9)
Increment to the kth instrument on the i subsystems
Insert PM_Action and Cost j in **MAT**$_k$ (s).
Step 5. **If** R (t,Ξ) < R_0 increment and define the new PM_Action to perform, add the new cost to total Cost.
Recalculate the reliability r
Else Goto Step 2.
If R (t,Ξ) >= evaluate the cost of minimal repair for all components ($1 \leq j \leq J$): and add these costs to the total cost.
Print minimal total cost to the corresponding reliability and **Stop**.

7 Illustrative example

Let consider a series-parallel MSS (Nuclear power systems) consisting of five subsystems connected in series arrangement

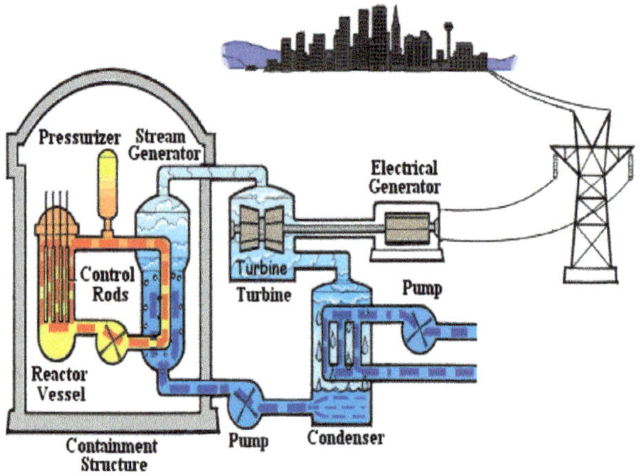

Figure 3. Detailed series-parallel nuclear power system.

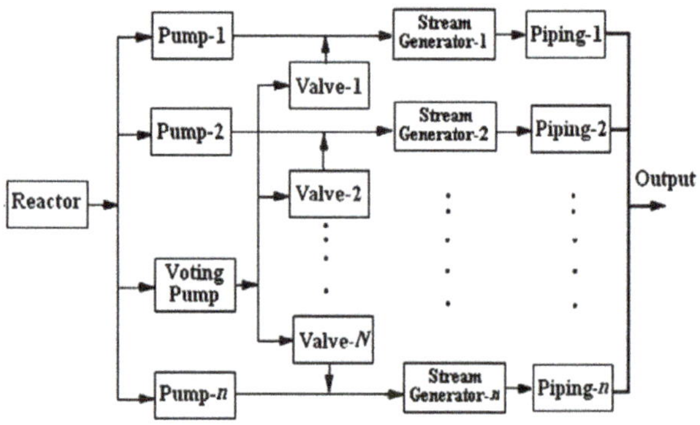

Figure 4. Synoptic of series-parallel nuclear power system.

Table 1. Parameter of components.

	λ	δ	h_0	Min_cost	Ξ%
1	0.050	1.8	0.001	1.02	97
2	0.050	1.8	0.003	0.90	80
3	0.050	1.8	0.003	0.90	80
4	0.050	1.8	0.003	0.90	80
5	0.020	1.3	0.002	0.70	70
6	0.020	1.3	0.002	0.60	70
7	0.020	1.3	0.002	0.70	70
8	0.080	1.6	0.006	0.80	85
9	0.070	1.8	0.005	0.60	89
10	0.050	1.8	0.003	0.80	90
11	0.080	1.7	0.002	0.80	90
12	0.030	1.6	0.001	0.90	95
13	0.004	1.5	0.001	0.80	87
14	0.010	1.6	0.003	0.85	80
15	0.200	1.6	0.005	0.75	92

as depickted in Figures 3 and 4. The system contains 15 equipments with different performance and reliability (parameters) as given in Tables 1 and 2, the process is done with basic components to transmit the stream energy to the electrical generator. The reliability of each component is defined by veibull function: $h(t) = \lambda^{\delta}\delta(\tau(t))^{(\delta - 1)} + h_0$.

Table 2. Parameter of PM_actions.

PM_actions	Components	PM_cost $103
01	Reactor	10.2
02	Pump-1	02.9
03	Pump-2	02.8
04	Pump-3	02.8
05	Voting-pump	03.6
06	Servo-valve-1	02.7
07	Servo-valve-2	02.7
08	Servo-valve-3	02.7
09	Stream-generator-1	08.1
10	Stream-generator-2	08.1
11	Bi-stream-generator	07.5
12	Pipe-1-2-3-4	02.7

Table 3. The best PM plan for $R(t, 0.82) > 0.9$.

t	PM_Actions	Affected components	$R(t, 0.82)$
12.250	05,00	(4)	0.923
16.652	08,00	(6,7,8)	0.922
20.000	09,10	(9,10)	0.954
22.110	04,00	(10;11)	0.911
25.640	12,00	(13,14,15)	0.937

MSS lifetime is 10 years. The time for possible PM_actions are spaced intervals of $\theta = 1.5$ months as given in Table 3. The problem is to guarantee a PM plan which provide the system work during its lifetime with a performance, reliability not less than Ξ0, R_0 and the age reduction is the same of all components $\varepsilon = 0.56$.

8 Conclusion

In this paper we formulated the problem of imperfect maintenance optimization for series-parallel nuclear power system structure. This work focused on selecting the optimal sequence of intervals to perform PM actions to improve the availability. The model analyzes cost and reliability, to construct a strategy to select the optimal maintenance intervals, formulating a complex problem. An exhaustive examination of all possible solution is not realistic, considering reasonable time limitations. Because of this, an efficient meta-heuristic can be applied (Harmony Search Algorithm) to solve the formulated problem. More specifically, the harmony search approach is a good solution for such a combinatorial problem.

References

1. Gude D, Schmidt KH. 1993. Preventive maintenance of advanced manufacturing systems: a laboratory experiment and its implications for human centered approach. International Journal of Human Factors in Manufacturing, 3, 335–350.
2. Brown M, Proschan F. 1983. Imperfect repair. Journal of Applied probability, 20, 851–859.
3. Zhao M. 2003. On preventive maintenance policy of critical reliability level for system subject to degradation. Reliability Engineering & System Safety, 79(3), 301–308.

4. Borgonovo E, Marseguerra M, Zio E. 2000. A Monte Carlo methodological approach to plant availability modeling with maintenance, aging and obsolescence. Reliability Engineering & System Safety, 67(1), 61–73.
5. Lin D, Zuo MJ, Yam RCM. 2000. General sequence imperfect preventive maintenance models. International Journal of Reliability, Quality and Safety Engineering, 7(3), 253–266.
6. Levitin G, Lisniaski A. 2000. Optimization of imperfect preventive maintenance for multi-state systems. Reliability Engineering & System Safety, 67, 193–203.
7. Monga A, Toogood R, Zuo MJ. 1997. Reliability-based design of systems considering preventive maintenance and minimal repair. International Journal of Reliability, Quality and Safety Engineering, 4, 55–71.
8. Levitin G, Lisniaski A. 1999. Optimal multistage modernization of power system subject to reliability and capacity requirements. Electric Power System Research, 50, 183–190.
9. Ushakov IA, Levitin G, Lisnianski A. 2002. Multi-state system reliability: from theory to practice. Proc. of 3 Int. Conf. on Mathematical Methods in Reliability, MMR 2002, Trondheim, Norway, p. 635–638.
10. Nakagawa T. 1999. Sequential imperfect preventive maintenance policies. IEEE Transactions on Reliability, 37(3), 295–298.
11. Ross SM. 1993. Introduction to probability models. Academic press.
12. Murchland J. 1975. Fundamental concepts and relations for reliability analysis of multi-state systems, in Reliability and Fault Tree Analysis, Barlow R, Fussell J, Singpurwalla N, Editors. SIAM: Philadelphia.
13. Levitin G, Lisnianski A, Ben-Haim H, Elmakis D. 1998. Redundancy optimization for series-parallel multi-state systems. IEEE Transactions on Reliability, 47(2), 165–172.
14. Levitin G, Lisniaski A, Ben-haim H, Elmakis D. 1996. Power system structure optimization subject to reliability constraints. Electric Power System Research, 40, 145–152.
15. Levitin G, Lisnianski A, Ben-Haim H, Elmakis D. 1997. Structure optimization of power system with different redundant elements. Electric Power Systems Research, 43(1), 19–27.
16. Billinton R, Allan R. 1990. Reliability evaluation of power systems. Pitman. IEEE Transactions PWRS-S, 1-11.
17. Reinschke B. 1985. System of elements with many states. Radio i svyaz: Moscow.
18. El-Neweihi E, Proschan F. 1984. Degradable systems: a survey of multistates system theory. Communications in Statistics – Theory and Methods, 13(4), 405–432.
19. Ushakov IA. 1986. Universal generating function. Soviet Journal of Computer and System Sciences, 24(5), 118–129.
20. Chern MS. 1992. On the computational complexity of reliability redundancy allocation in a series system. Operations Research Letters, 11, 309–315.
21. Geem ZW, Tseng C-L, Park Y. 2005. Harmony search for generalized orienteering problem: best touring in China. Book Advanced in Natural Computation, vol. 361, Springer: Berlin/Heidelberg, p. 741–750.
22. Mahdavi M, Fesanghary M, Damangir E. 2007. An improved harmony search algorithm for solving optimization problems. Applied Mathematics and Computation, 188, 1567–1579.
23. Fesanghary M, Mahdavi M, Minary-Jolandan M, Alizadeh Y. 2008. Hybridizing harmony search algorithm with sequential quadratic programming for engineering optimization problems. Computer Methods in Applied Mechanics and Engineering, 197, 3080–3091.
24. Omran MGH, Mahdavi M. 2008. Global-best harmony search. Applied Mathematics and Computation, 198, 643–656.

4

Comparison of 1D and 2D models predicting a packed bed drying

Souad Messai[1,*], Mohamed El Ganaoui[2], Jalila Sghaier[3], Laurent Chrusciel[2], and Slimane Gabsi[1]

[1] Unité de Recherche Environnement, Catalyse et Analyse des Procédés URECAP, Ecole Nationale d'Ingénieurs de Gabès, Rue Omar Ibn-Elkhattab, 6029 Gabès, Tunisie

[2] Laboratoire d'Etudes et de Recherche sur le Matériau Bois LERMAB, Université de Lorraine-Faculté des Sciences et Technologies, Boulevard des Aiguillettes, 54506 Vandœuvre-lès-Nancy Cedex, France

[3] Laboratoire d'Energétique et des Transferts Thermiques et Massiques, Faculté des Sciences de Tunis, Campus Universitaire, 2092 El Manar, Tunisie

Abstract – A numerical study for the modelling of porous particles packed bed drying in one and two dimensions (1D and 2D) is presented in this work. This model is based on the averaging volume approach using two scale changes. Taking into account the non validity of the thermal equilibrium between the gas phase and the porous particle, a two temperatures model is then developed. The numerical resolution of macroscopic equations describing heat and mass transfer during the drying of a packed bed is carried out by the method of finite volume. Experimental data for spherical porous alumina particles reported in the literature were used for the validation of the model. A good agreement was found. The influence of particles thermophysical properties and operating variables is tested. A set of time-space evolution describes the progression of these variables with time, in the height and radius of the packed bed. The model results allow a better understanding of the heat and mass transfer mechanisms involved in a packed bed drying process.

Key words: Superheated steam, Drying, Averaging volume, 1D and 2D model, Porous media.

Nomenclature

A	Specific surface, m^2/m^3
Cp_g	Gas specific heat, J kg^{-1} K^{-1}
Cp_S	Solid specific heat, J kg^{-1} K^{-1}
d	Particle diameter, m
F_m	Mass flux, kg $m^{-2}s^{-1}$
H	Height of the packed bed, m
h_{sg}	Convective heat transfer coefficient at solid-gas interface, W m^{-2} K^{-1}
h_0	Heat transfer coefficient at the inlet of the bed, W m^{-2} K^{-1}
h_1	Heat transfer coefficient at the outlet of the bed, W m^{-2} K^{-1}
$\dot{m}$	Evaporation rate (mass flow rate of water vapor per m^3 of drying gas) kg m^{-3} s^{-1}
m_g	Gas mass flux, kg m^{-2} s^{-1}
$\boldsymbol{n}$	Unit normal vector
$\boldsymbol{n}_{gs}$	Unit normal vector at the solid-gas interface
P	Pressure, Pa
t	Time, s
T	Temperature, K
R	Radius of the bed, m
X	Absolute moisture content kg of water/kg of dry mass
V_g	Gas velocity, m s^{-1}
Z	Axial coordinate, m

Greek symbols

L_v	Latent heat of vaporization, J kg^{-1}
ρ	Density, kg m^{-3}
μ	Dynamic viscosity, Pa s
λ	Thermal conductivity, W m^{-1} K^{-1}
ε	Porosity

Subscripts

eff	Effective
g	Gas phase
s	Solid phase
sg	Solid gas interface
l	Liquid phase
v	Vapor phase

Superscripts

g	Gas
s	Solid
l	Liquid

*Corresponding author: souad.messai@gmail.com

Introduction

Solid-fluid contact technique is widely applied in agriculture and chemical engineering in the form of packed bed drying. Indeed, modeling simultaneous heat and mass transfer occurring between the particles and the drying medium during the circulation through the packed bed, is complex. Because of its great technological applications, packed bed drying has been studied for many years. In fact, one of the first models for packed bed dryers, the one-phase model proposed by Thompson [1] was significant at that time for dryer design. Currently, a large variety of models has been developed for describing the heat and mass transfer inside packed bed dryers. Riede and Schlünder [2] described the mass transfer in a single particle with a modified shrinking core model for drying of fixed beds. Many assumptions are used to simplify the equations describing the transport mechanisms. Saastamoinen and Impola [3] developed a fixed bed drying model for the description of heat and mass transfer in a single particle considering the capillary pressure effects. Khan et al. [4] incorporated a front particle model into the modeling of a packed bed of alumina porous particles during superheated steam drying. This model has been validated with their experimental results. They assumed that capillary action had no effect on moisture transfer in the material and the drying process was controlled by internal heat transfer. In the similar way, Hager et al. [5] studied numerically and experimentally the steam drying process of a bed of porous particles (ceramic and alumina). The process is analyzed in terms of the hybrid model, that apply the volume averaged transport equation for the fluid flowing in the fixed bed and the local equations for heat and mass transfer taking place in the particle. Chen et al. [6] integrated a single particle model with a two-phase hydrodynamic to simulate the drying of coal in a fluidized bed with superheated steam.

Other studies investigated fixed bed steam drying of foodstuff by dividing the bed in three zones (dry zone, drying zone and wet zone) limited by two interfaces [7]. A dry front separates the dry zones from the ones and a wet front partition the drying and wet zones. The moisture content and temperature of the dry zone are equal to their equilibrium moisture content and steam temperature, respectively.

On other way, many researchers have developed mathematical models for humid air drying especially for low and average temperature. Arnaud and Fohr [8] proposed a partial differential equations model for the slow drying of a deep bed at low temperature. They proved the existence of a drying front which moves in the granular media without distortions. Sözen and Vafai [9] studied the transient forced convective condensing gas flux through a packed bed. The non validity of local thermal equilibrium assumption between the solid and fluid phases is assumed. They examined two different types of boundary conditions. The condensation in vapor phase has been also studied. Mhimid et al. [10] presented a two temperatures 2D-model for the drying of a cylindrical and granular bed with imposed hot air flux and conductive heat flux at the wall. The time-space evolution of temperature has been also studied. Ratti and Mujumdar [11] developed a model to simulate batch drying. They studied numerically the effect of airflow reversal for the batch drying of foodstuff in a deep bed. Wang and Chen [12] proposed a model of fixed bed drying of moist porous particles. They considered the distribution effects of moisture content, temperature and gas pressure of the particle and the conjugate effect between gas and particles. They divided the bed into series of thin layers. Kato et al. [13] studied a packed bed dryer using air as drying agent and developed a model of packed bed drying. With combination of heat transfer data, they predicted the drying process of activated alumina.

Otherwise modeling of heat and mass transfer in 1D and 2D configuration is also available in the literature. Izadifar et al. [14] presented a 1D-model for packed bed drying using the local volume averaging (LVA) approach, with local thermal equilibrium in each elementary volume in order to derive transient heat and mass transfer equations, solved by means of an implicit numerical method. Stakić et al. [15] developed a 1D-model for the case of convective drying of hygroscopic material in a packed bed. The assumption of local thermal equilibrium is assumed. The drying rate of the specific product is calculated by applying the concept of a drying coefficient, called a global mass transfer and often used to characterize mass transfer in chemical processes. Zili and Ben Nasrallah [16] developed a 2D-model to simulate forced convection air drying of granular products at low and average temperatures. The authors presented a general model taking into account conduction, convection and diffusion using variable coefficients. The assumption of local thermal equilibrium is not assumed. They demonstrated the effect of variable porosity and intermittent flux on the drying process. Recent models by Sitompul et al. [17] and Istadi and Sitompul [18] developed a 2D-model for spherical grains deep-bed drying taking into account transport phenomena within the bed. Moreover, the effect of air velocity distribution (pressure field) on the drying process was studied. Mainly the same model [19] was used for a different dryer geometry (large diameter column). Boukadida et al. [20] developed a 2D model to investigate heat and mass transfer during convective drying of clay brick. The effect of gaseous pressure is taken into account. The authors studied the effect of drying agent properties on drying kinetics and space-time evolution of drying parameters.

In this work, we developed a 2D-model describing heat and mass transfer during drying of a packed bed in superheated steam. The important advantages of superheated steam drying that are of interest to industries include the absence of oxidative reactions due to the absence of oxygen and high drying rates in both constant and falling rate periods [21–24].

The mathematical model is developed using two change of scale. Convective heat transfer is assumed between the granular bed and steam. To take into account the thermal non-equilibrium between the porous and the fluid phases a two-temperature macroscopic model is used to describe heat transfer. The mass transfer is introduced in the form of drying kinetics deduced from a single particle model. The predicted results from the packed bed model from the 1D and 2D models were compared to the experimental data in the literature. The comparison of time-space evolution of drying parameters shows the existence of drying front for 2D model.

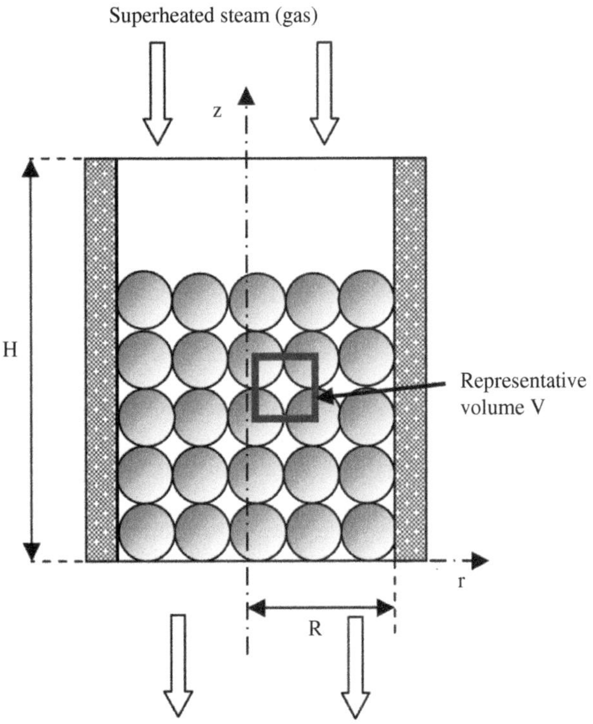

Figure 1. Geometrical configuration of the fixed bed and averaging volume of the bed.

Mathematical model

Knowledge of the heat transfer characteristics and spatial temperature distribution in packed beds is of great importance to the design and analysis of the packed beds dryers. Hence, an attempt is made to develop a model based on the averaging approach tacking into account the non validity of the thermal equilibrium. The packed bed is made of an inert and rigid solid matrix phase (spherical porous particles) and a gaseous phase (drying agent). Thus, the bed is a discontinuous medium. The theoretical formulation of heat and mass transfer in the packed bed is obtained by a change of scale [25–28]. At microscopic scale, the size of the representative volume V is small with respect to the pore sizes (Figure 1). At macroscopic scale the size of the representative volume is large with respect to the pore sizes. The change of scale allows converting the real discontinuous medium to an equivalent fictional continuous one.

Governing equations of heat and mass transfer

The macroscopic differential equations are obtained by taking the average of the microscopic equations over the averaging volume and using closing assumptions. Several simplifying assumptions are made in order to obtain a closed set of macroscopic governing equations:

- the porous medium is from a macroscopic point of view homogeneous and isotropic,
- non validity of the thermal equilibrium,
- the bed porosity is uniform in all the directions,
- the compression work and viscous dissipation are negligible,
- the dispersion and tortuosity terms are modeled as diffusion fluxes,

For an arbitrary α-phase scalar or tensor variable ψ_α, we note:

Superficial average:

$$\langle\psi_\alpha\rangle = \frac{1}{v}\int_{v_\alpha(t)} \psi_\alpha \mathrm{d}v. \tag{1}$$

Intrinsic average:

$$\langle\psi_\alpha\rangle^\alpha = \frac{1}{v_\alpha(t)}\int_{v_\alpha(t)} \psi_\alpha \mathrm{d}v. \tag{2}$$

The superficial and intrinsic average are related by:

$$\langle\psi_\alpha\rangle = \varepsilon_\alpha\langle\psi_\alpha\rangle^\alpha, \tag{3}$$

where $\varepsilon_\alpha = V_\alpha/V$ is the α-phase volume fraction or porosity.

The macroscopic equations governing heat and mass transfers in the packed bed are:

Energy conservation equation in the gas phase:

$$\begin{aligned} C_{pg}\varepsilon\langle\rho_g\rangle^g \frac{\partial\langle T_g\rangle^g}{\partial t} + \langle\rho_g\rangle^g\langle V_g\rangle C_{pg}\nabla\langle T_g\rangle^g \\ = \nabla\cdot\left[\lambda_{\mathrm{geff}}\nabla\langle T_g\rangle^g\right] - \dot{m}\, C_{pg}\left(\langle T_g\rangle^g - \langle T_s\rangle^s\right) \\ - h_{gs}A\left(\langle T_g\rangle^g - \langle T_s\rangle^s\right). \end{aligned} \tag{4}$$

Energy conservation equation in the solid phase s:

$$\begin{aligned} (1-\varepsilon)C_{ps}\langle\rho_s\rangle^s \frac{\partial\langle T_s\rangle^s}{\partial t} = \nabla.(\lambda_{\mathrm{seff}}\nabla\langle T_s\rangle^s) - \dot{m}\, L_v \\ + h_{gs}A\left(\langle T_g\rangle^g - \langle T_s\rangle^s\right), \end{aligned} \tag{5}$$

$\dot{m}$ is the evaporation rate defined as: $\dot{m} = A F_m$ A is the specific surface (defined as the ratio apparent surface of the particle divided by the apparent volume of particle), λ_{seff} and λ_{geff} are respectively the effective thermal conductivity of solid and gas phases and h_{gs} is the gas-particle convective heat transfer coefficient.

$$h_{\mathrm{gs}} = \frac{Nu\lambda_{\mathrm{geff}}}{d}, \tag{6}$$

where $Nu = 2 + 0.66Re^{0.52}\, Pr^{0.33}$ [28]

Initial Conditions

Initially, temperatures and moisture contents are constant:

$$\langle T_g\rangle^g(0,r,z) = \langle T_s\rangle^s(0,r,z) = T_i, \tag{7}$$

$$X(0,r,z) = X_i. \tag{8}$$

Thermal boundary conditions

Several simplifying assumptions are made in order to obtain a closed set of macroscopic governing equations:

- The lateral surface of the bed is adiabatic.
- At the inlet of the cylinder defining the volume of the fixed bed, the temperature of the gas is assumed constant, the solid and gas temperatures are related by:

$$-\lambda_{\text{seff}} \frac{\partial \langle T_s \rangle^s}{\partial z}(t,r,0) = h_0\left(\langle T_s \rangle^s - \langle T_g \rangle^g\right). \tag{9}$$

- At the outlet of the cylinder, a heat transfer coefficient is introduced to describe the transfer:

solid phase

$$-\lambda_{\text{seff}} \frac{\partial \langle T_s \rangle^s}{\partial z}(t,r,H) = h_1\left(\langle T_s \rangle^s - \langle T_g \rangle^g\right), \tag{10}$$

h_0 and h_1 are the heat transfer coefficients at the entry of the cylinder and that of the wall to bed, respectively.

The main difference between 1D and 2D models is the boundary conditions. In fact, for the 1D model, the heat transfer coefficients at the entry of the cylinder and that of the wall to bed are null. Thus, radial transfer is not taken into account.

Dying kinetics

In a previous paper [29], mathematical model for the superheated steam drying of a single porous particle was developed and compared with experiment. The single particle model is integrated to simulate the continuous drying of a porous particles packed bed with superheated steam. Hence, the mass flux calculated from the single porous particle model is incorporated in the bed model. In fact, the phenomenon of water evaporation and mass transfer in a solid particle and the surrounding gas phase is expressed by drying kinetic. The evaporated rate $\dot{m}$ is given by:

$$-(1-\varepsilon)\langle \rho_s \rangle^s \frac{\partial X}{\partial t} = \dot{m}, \tag{11}$$

$$\dot{m} = AF_{\text{m}}. \tag{12}$$

A is the specific surface, as mentioned above F_{m} is the mass flux obtained by the numerical resolution of a single particle model [29, 30].

The correlations of mass flux are obtained from a set of numerical tests carried out for various values of temperatures and steam velocity of a single particle model. These correlations are incorporated in the bed model.

The parameters that influence the mass flux in the constant rate period are the gas temperature (T_g) and the gas velocity (V_g) [29]:

$$F_{\max} = F_{\max}\left(T_g, V_g\right). \tag{13}$$

$F_{\max}$ is the mass flux value in the constant rate period (CRP).

First, correlation of maximum mass flux and moisture content are deduced as function of steam temperature for a given steam velocity. Then, the steam temperature is constant, we vary the steam velocity. Different correlations are also obtained according to the steam velocity for a given steam temperature. Different correlation parameters are illustrated in Figure 2:

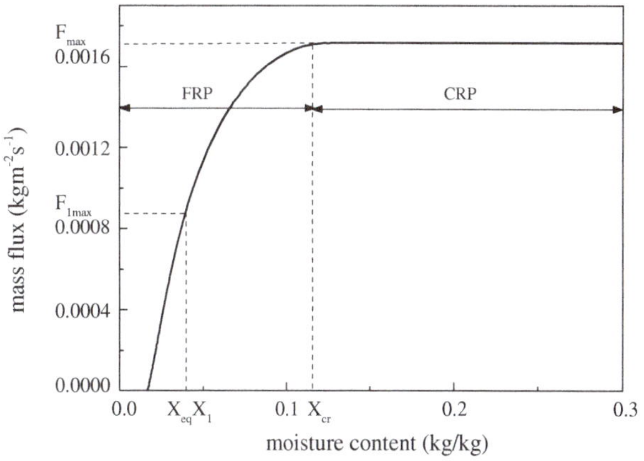

Figure 2. Different parameters of mass flux correlation.

$$F_{\max} = \frac{\left(a_1 T_g - b_1\right)\ \left(c_1 V_g{}^{d_1}\right)}{e_1}. \tag{14}$$

$F_{1\max}$, is the mass flux corresponding to the point of intersection of two under-periods of the falling-rate period (FRP) given by:

$$F_{1\max} = \frac{\left(a_2 T_g - b_2\right)\ \left(c_2 V_g{}^{d2}\right)}{e_2}. \tag{15}$$

The equilibrium moisture content X_{eq} is independent of steam velocity. Its expression is given according to the steam temperature:

$$X_{\text{eq}} = a_3\left(T_g\right)^{b3}. \tag{16}$$

X_{cr} the critical moisture content indicating the start of the falling-rate period is given by:

$$X_{\text{cr}} = \frac{\left(a_4 T_g - b_4\right)\ \left(c_3 V_g + d_3\right)}{e_3}, \tag{17}$$

X_1 is the moisture content which corresponds to the point of intersection of two under-periods of the falling-rate period:

$$X_1 = \frac{\left(a_5 T_g{}^{b_5}\right)\ \left(c_4 V_g + d_4\right)}{e_4}, \tag{18}$$

a_i, b_i, c_i, d_i and e_i are coefficients which depend of the medium used.

The values of different constants are listed in Table 1.

Final correlations of different parameters are:

$$X > X_{\text{cr}} \longrightarrow F_{\text{m}} = F_{\max}, \tag{19}$$

$$X_1 < X < X_{\text{cr}} \longrightarrow F_m = \left(F_{\max} - F_{1\max}\right)\left(\frac{X - X_1}{X_{\text{cr}} - X_1}\right) + F_{1\max}, \tag{20}$$

$$X_{\text{eq}} < X < X_1 \longrightarrow F_m = F_{1\max}\left(\frac{X - X_{\text{eq}}}{X_1 - X_{\text{eq}}}\right). \tag{21}$$

Table 1. Values of constants for drying kinetic with superheated steam.

Constants	Values
a_1	2.257×10^{-5}
a_2	8.76×10^{-6}
a_3	27.83×10^{-3}
a_4	1.92×10^{-4}
a_5	0.449×10^{2}
b_1	8.483×10^{-3}
b_2	0.3294×10^{-2}
b_3	-48.83×10^{-2}
b_4	5.381×10^{-2}
b_5	-140.9×10^{-2}
c_1	13.448×10^{-4}
c_2	0.050872×10^{-4}
c_3	6.592×10^{-3}
c_4	-1.418×10^{-3}
d_1	27.28×10^{-2}
d_2	14.14×10^{-2}
d_3	2.56×10^{-2}
d_4	1.0886×10^{-2}
e_1	10.04×10^{-4}
e_2	5×10^{-4}
e_3	30×10^{-3}
e_4	9×10^{-3}

These correlations of mass flux are integrated on the scale of the granular bed.

Numerical resolution

The system of equations is solved by a finite volume method based on the notion of a control domain [31] with a no regular mesh. To insure the convergence, an upwind scheme is used to evaluate convection terms on faces of the control domain. The discretization of the conservation equations leads to a system of algebraic equations. This system is strongly coupled and is solved numerically by the iterative method of Gauss Seidel.

Validation

The validity of the model to predict the evolution of temperature and drying rate throughout the granular bed during convective superheated steam drying is confirmed by using sets of data available for the drying of porous particles [28]. The authors investigated experimentally superheated steam drying of porous alumina and ceramic particles packed bed with a diameter of 9.5 and 10 mm, respectively. A high-resolution load cell continuously registered the weight of the entire test section, and the steam inlet and outlet temperatures were monitored as well as the centre temperature of a varying number of spheres. The internal bed vessel is a cylinder with a diameter of 0.3 m. The bed height was varied from 100 to 200 mm. All the experiments were performed at atmospheric pressure. Figures 3 and 4 illustrate the comparison of the model calculations with

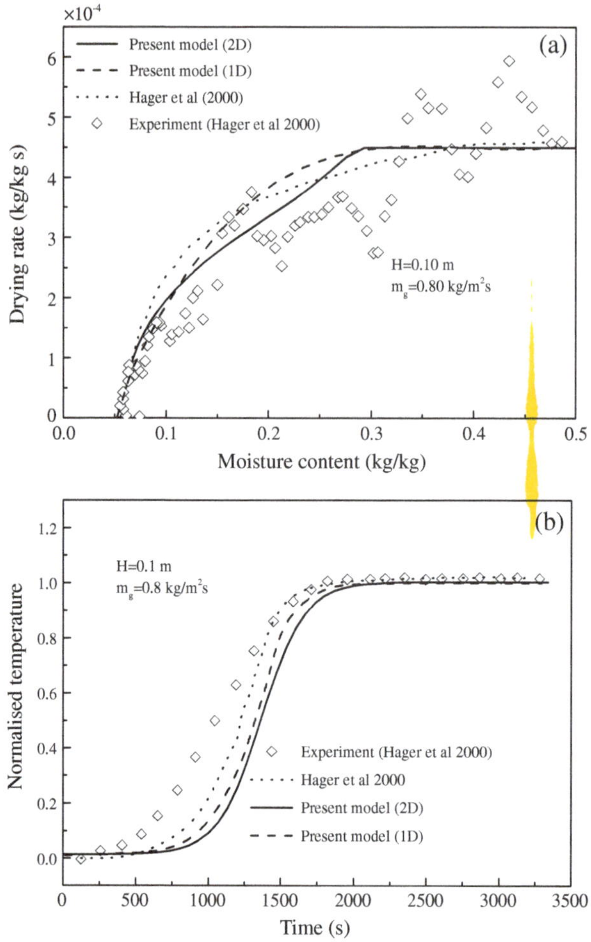

Figure 3. Comparison of experimental and numerical data of drying rate (a) and outlet steam temperature (b) of alumina particles packed bed for various operating parameters.

experimental data of the drying rate and the steam outlet temperature for varying height and steam mass flux. The curves indicate that the bed depth affects the amount of drying time. The bed having the thinnest depth (0.1 m) reaches the gas temperature and the equilibrium drying rate more quickly. Whereas, when we increase the steam mass flux for the same bed depth Reynolds number increases and consequently the convective heat transfer coefficient also increases. Thus we accelerate transfers and then the drying time decreases. As can be seen the agreement is good between predictions and experimental data, even though there is slight discrepancies that can be attributed to the position of thermocouples and errors in the measurement. Figures 3 and 4 show also the comparison of the present model in 1D and 2D configurations with the model of Hager et al. [28]. We note that there is a difference between the two models especially for the value of evaporation rate during the constant rate period. A possible explanation for discrepancies between the two models this is the approach used for communication between the particle and the packed bed. Indeed, Hager et al. [28] developed the so called hybrid model. The particle tortuosity is the only adjustable parameter. They showed that

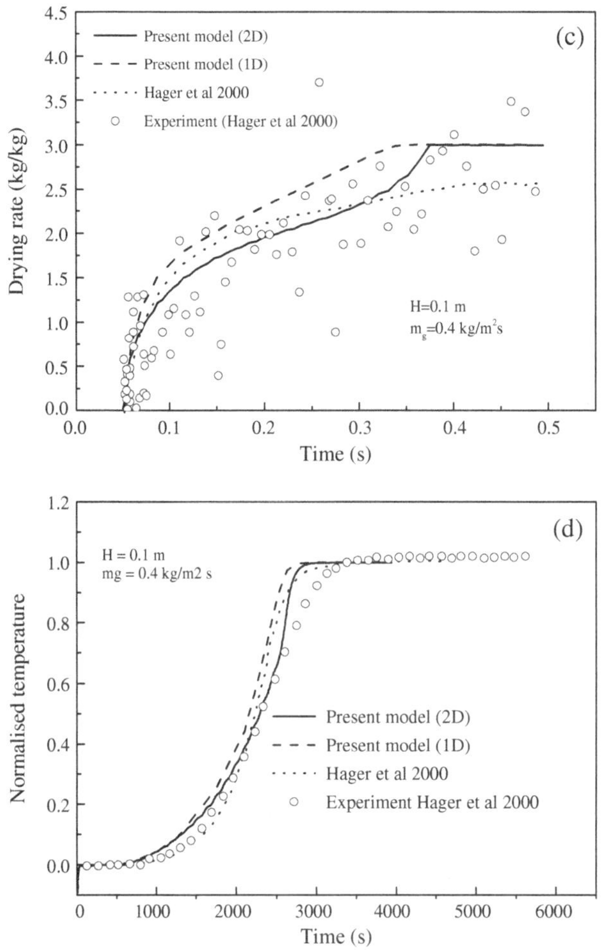

Figure 4. Comparison of experimental and numerical data of drying rate (a) and outlet steam temperature (b) of alumina particles packed bed for various operating parameters.

the variation of this parameter does not affect the drying process. The particle model is incorporated in the bed model by the temperature and the evaporation rate at the porous particle surface. The porous particle model of Hager et al. is based on the thermodynamic approach [32]. This model takes into account the effect of gravity and pressure gradient. The phenomenological coefficients involved in the transfer equations were determined experimentally or numerically. This is the disadvantage of this model since the determination of these coefficients is very complicated. Our porous particle model is based on the volume averaging approach. We used an iterative method (dichotomy) to calculate the mass flow while satisfying the boundary conditions [29, 30]. Thus, in our model, we need only physical properties of the porous medium. Experimentally, these parameters are easily accessible. The main difference between our model and that of Hager et al. is the incorporation of new correlations of mass flux and moisture content in the packed bed. These correlations are calculated from the drying of a porous particle, parameters are calculated during the constant rate period and the falling rate period.

The comparison of the 1D and 2D models shows that the 2D model fits experimental data better than the 1D model. In subsequent section, more explanation of differences between the two models is detailed.

Sensitivity analysis

Knowledge of the parameters that have significant impact on drying behavior is useful in designing drying operations. Figure 5 illustrates the effect of particle diameter on the drying process. We noted that particle diameter exerts a significant influence on the drying behavior. Larger particles require a much longer time to reach the equilibrium moisture content due to the fact that more energy is needed to heat them up. We also note that for the same drying temperatures and steam velocities, the predicted moisture content decreases much more rapidly when the diameter decreases. This is due to the fact that the Reynolds number and thus the heat transfer coefficient increases with smaller particles. As we would expect, when we increase the gas velocity, for the same drying temperature, the Reynolds number increases and consequently the convective heat transfer coefficient increases (Figure 6) leading to a reduction of the drying time. To analyze the effect of drying temperature on drying curves, we represented the evolution of the drying kinetics for four values of the gas temperature (Figure 7). As can be expected, at lower temperatures the moisture content decreases more slowly. While the temperature increases, the medium reaches the drying temperature more quickly. This is because the higher temperature gradient in the medium resulted in a higher heat flux [33].

Time-space evolution of particles temperature and moisture content

The temperature field evolution within the medium is the key to the drying phenomenon. The time-space evolution of drying parameters allows a better understanding of the heat and mass transfer mechanisms. Figure 8 shows the time-space evolution of the moisture content of the granular medium during superheated steam drying. We note the existence of a drying front which moves from the outlet to the inlet of the cylinder in the direction of fluid flow drying. This front divides the granular medium into two regions: wet and dry regions. As the drying process progresses, the dry region widens and the moisture content decreases to the equilibrium value. Indeed, an increase in the temperature of the solid during drying results in an increase of the evaporation rate which leads to a decrease in the moisture content. Figures 9 illustrates the time-space evolution of the temperature of the particle (solid phase of the packed bed) during superheated steam drying. Initially, the drying agent temperature is greater than that of the granular medium. Therefore, the temperature increases from the inlet of the cylinder ($z = H$) to the outlet ($z = 0$) and from the wall to the center, following the direction of the gas flow. Moreover the time-space evolution of temperature indicates the existence of the two classic drying periods: the constant rate period (with a level of 100 °C) and the falling rate period (when the temperature exceeds 100 °C). Drying process is achieved when moisture content and particle temperature reach the equilibrium moisture content and the drying agent temperature, respectively.

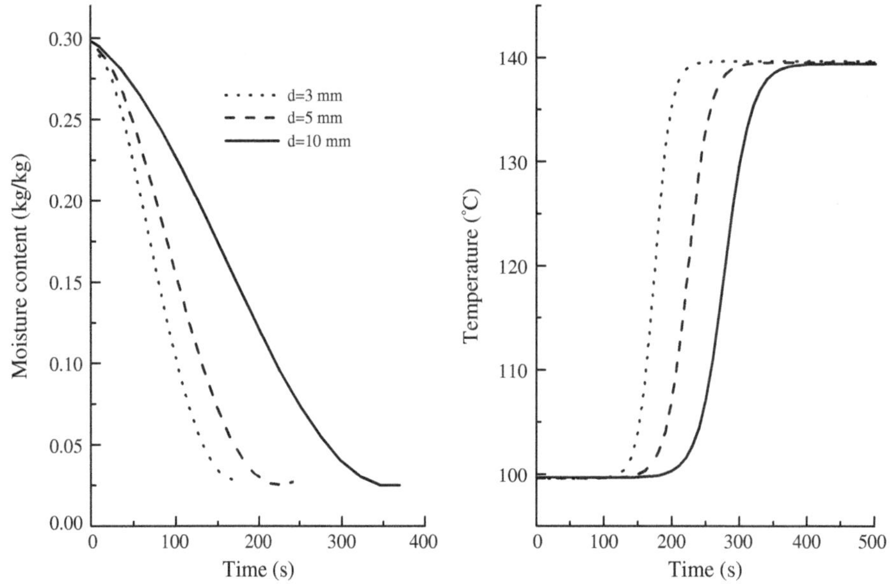

Figure 5. Effect of particle diameter on bed moisture content and temperature during superheated steam drying: $V_g = 1.5$ m s^{-1}, $T_g = 140$ °C.

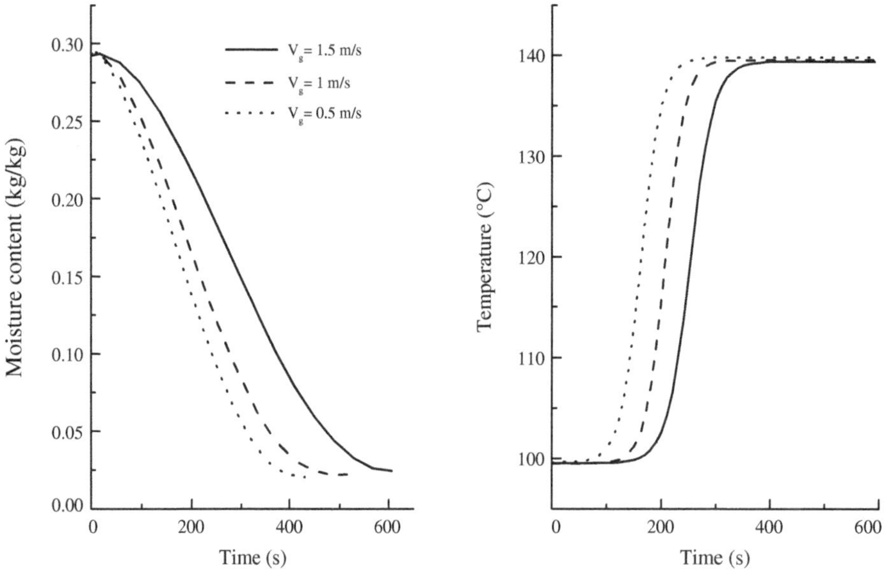

Figure 6. Effect of gas velocity on bed moisture content and temperature during superheated steam drying: $d = 10$ mm, $T_g = 140$ °C.

Comparison of 1D and 2D models

2D model is relevant in understanding real heat and mass transfer phenomena. Compared to 1D model, 2D model is used to study the transverse processes of heat and mass transfer. This model has the advantage of being more realistic, but some control mechanisms are retained. Figure 10 shows a comparison of these two models. We note from the curves of bed moisture content and temperature that the effect of taking into account 2D modeling, appears especially at the end of drying which show the influence of the internal process. Because the initial temperature is equal to that of steam saturation, the condensation phase is absent for the two models. The disadvantage of the condensation phase is the increase of the product moisture content which emphasizes the importance of preheating the product to be dried. This procedure is also used in the experimental study of Hager et al. [28]. Furthermore, the equilibrium moisture content for the 2D model is lower than that of 1D model which accentuates the importance of taking into account the lateral transfer for modeling packed bed drying. Figures 11 and 12 illustrate a comparison of the time space evolution of the moisture content and bed temperature for both models. We immediately note the existence of drying front for 2D model. The time-space evolution of 1D model shows the absence of a drying front. This is just a distribution of the moisture content and temperature within the medium. Indeed, the elevation of the

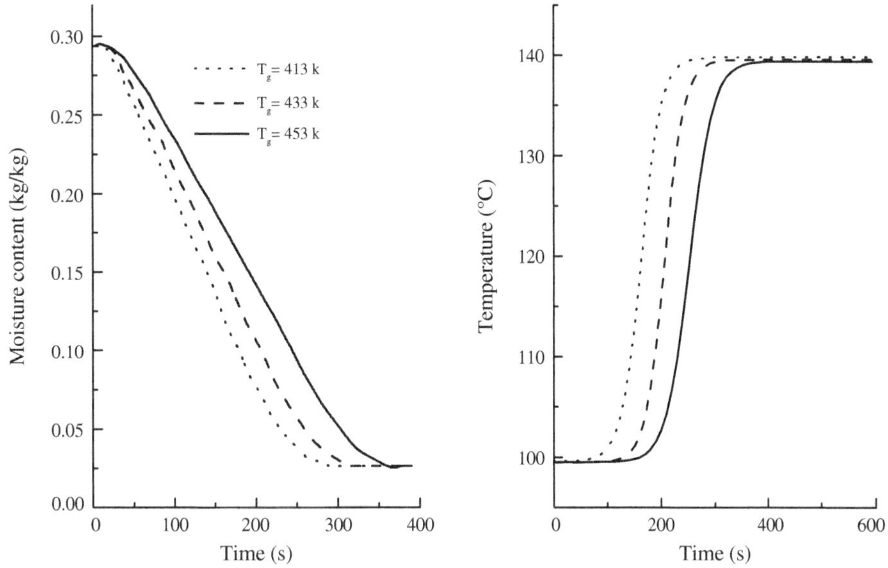

Figure 7. Effect of gas temperature on bed moisture content and temperature during superheated steam drying: d = 10 mm, V_g = 1.5 m s^{-1}

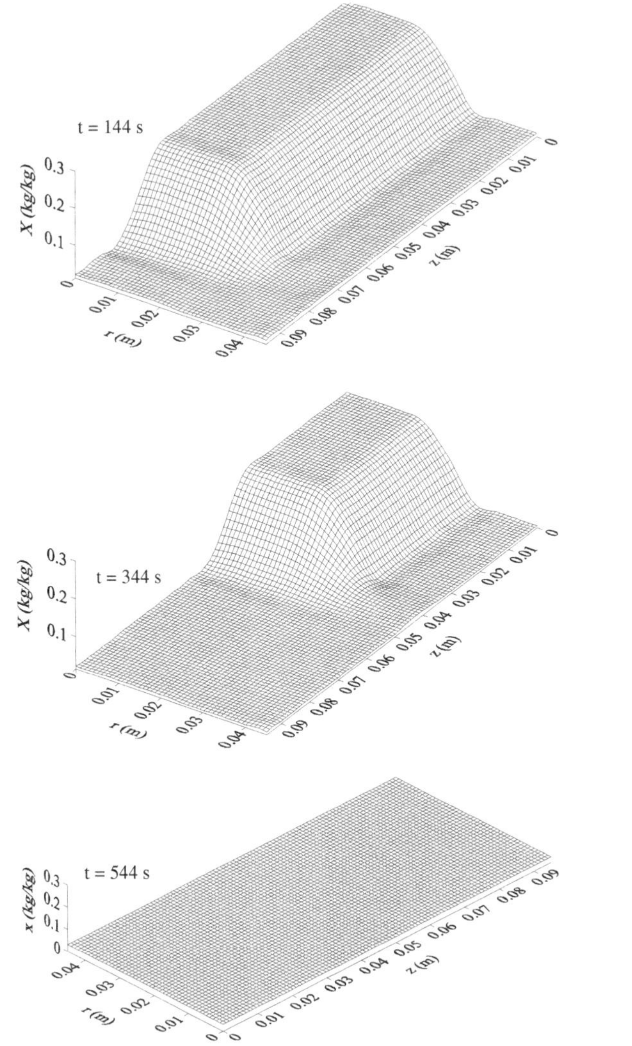

Figure 8. Time-space evolution of bed moisture content during superheated steam drying: d = 10 mm, T_g = 140 °C and V_g = 1.5 m s^{-1}.

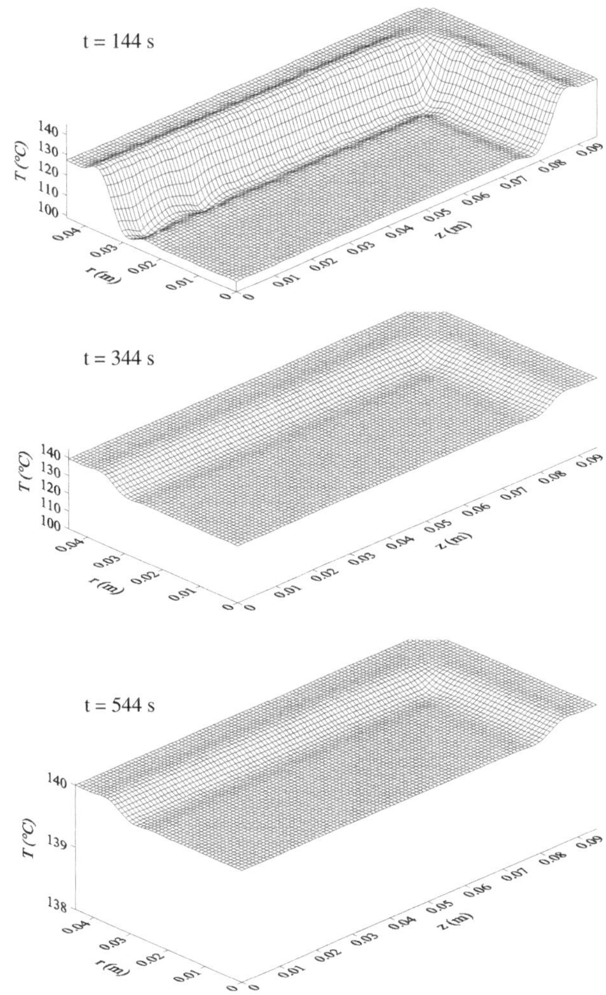

Figure 9. Time-space evolution of bed temperature during superheated steam drying: d = 10 mm, T_g = 140 °C and V_g = 1.5 m s^{-1}.

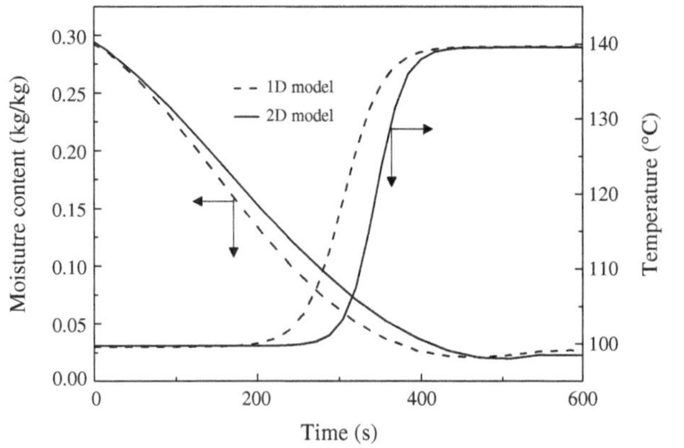

Figure 10. Comparison of moisture content and bed temperature for 1D and 2D models: d = 10 mm, T_g = 140 °C and V_g = 1.5 m s^{-1}

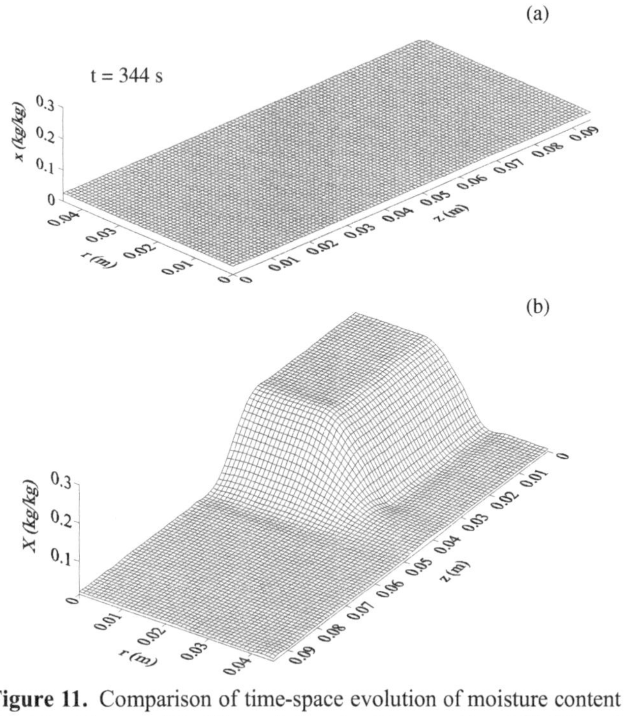

Figure 11. Comparison of time-space evolution of moisture content for 1D (a) and 2D (b) models: d = 10 mm, T_g = 140 °C and V_g = 1.5 m s^{-1}

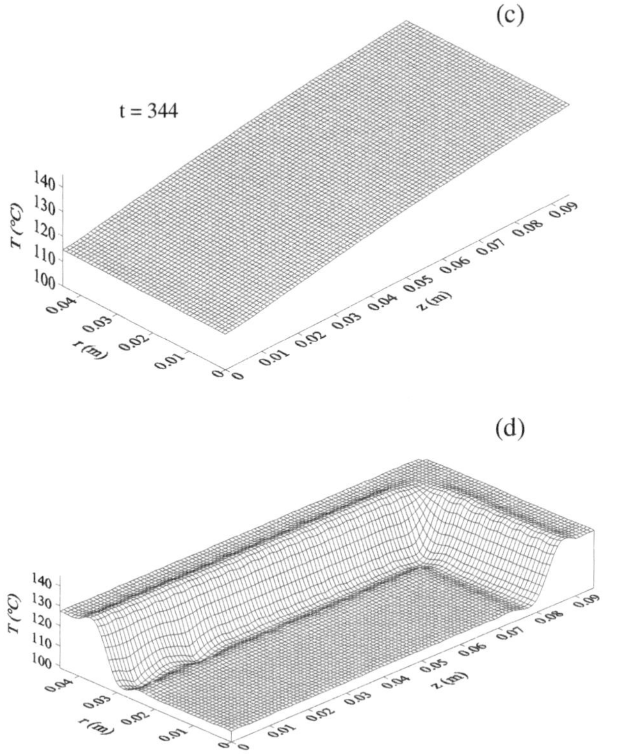

Figure 12. Comparison of time-space evolution of bed temperature for 1D (a) and 2D (b) models: d = 10 mm, T_g = 140 °C and V_g = 1.5 m s^{-1}.

evaporation rate leads to an increase of bed temperature. Consequently, the moisture content decreases. At the beginning, we note that the bed temperature is more important near the cylinder inlet and the wall in regard to its value at the cylinder middle (Figure 12). This is due to the drying agent (superheated steam) effects. In fact, superheated steam temperature is greater than that of the granular medium. Moreover, this temperature increase can be attributed to the wall heat flux application. Next the temperature decreases from the cylinder inlet to the outlet and from the wall to the center. Figure 11 presents the time-space evolution of the bed moisture content. In fact, the front movement follows the drying fluid blowing direction. Therefore the drying front moves from $Z = H$ to $Z = 0$.

Conclusion

A comprehensive heat and mass transfer model was developed for a packed bed drying process with superheated steam as drying agent. This model is based on the averaging volume approach. Taking into account the non validity of the thermal equilibrium hypothesis, a two temperatures model is presented: particle temperature (solid phase) and the gas temperature. The mass transfer is introduced by drying kinetics deduced from the particle model. The correlations of mass flux and moisture content, deduced from the numerical resolution of the single particle model, were incorporated in the bed model. The predicted results compared to the experimental data, indicated good agreements despite few discrepancies. The constructed numerical tool enabled us to predict the time-space variations of the bed temperature and the moisture content. We have also compared 1D and 2D models. Results show that the 2D model is more realistic. Indeed, time-space evolution of temperature and moisture content illustrates the existence of drying front for 2D model.

References

1. Thompson TL. 1970. Simulation for optimal grain dryer design. Transaction of the ASAE, 1(10), 844–848.
2. Riede Th, Schlünder EU. 1991. Selective through-drying in fixed beds. Chemical Engineering and Processing: Process Intensification, 29, 9–26.

3. Saastamoinen J, Impola R. 1997. Drying of biomass particles in fixed and moving beds. Drying Technology – International Journal, 5, 1919–1929.
4. Khan JA, Beasley DE, Alatas B. 1991. Evaporation from a packed bed of porous particles into superheated vapor. International Journal of Heat and Mass Transfer, 1, 267–280.
5. Hager J, Wimmerstedt R, Whitaker S. 2000. Steam drying a bed of porous spheres: theory and experiment. Chemical Engineering Science, 55, 1675–1698.
6. Chen Z, Agarwal PK, Agnew JB. 2001. Steam drying of coal, Part 2. Modeling the operation of fluidized bed drying unit. Fuel, 80, 209–223.
7. Tang Z, Cenkowski S, Muir WE. 2004. Modeling the superheated-steam drying of a fixed bed of brewers' spent grain. Biosystems Engineering, 87, 67–77.
8. Arnaud G, Fohr JP. 1988. Slow drying simulation in thick layers of granular products. International Journal of Heat and Mass Transfer, 31, 2517–2526.
9. Sözen M, Vafai K. 1990. Analysis of the nonthermal equilibrium condensing flow of gas through a packed bed. International Journal of Heat and Mass Transfer, 33, 1247–1261.
10. Mhimid A, Fohr JP, Ben Nasrallah S. 1999. Heat and mass transfer during drying of granular products by combined convection and conduction. Drying Technology – International Journal, 17, 1043–1063.
11. Ratti C, Mujumdar AS. 1995. Simulation of packed bed drying of foodstuffs with airflow reversal. Journal of Food Engineering, 26, 259–271.
12. Wang ZH, Chen G. 1999. Heat and mass transfer in fixed-bed drying. Chemical Engineering Science, 54, 4233–4243.
13. Kato K, Ohmura S, Taneda D, Onozawa I, Shimura K, Iijima A. 1981. Drying characteristics in a packed fluidized bed dryer. Journal Chemical Engineering Japan, 14, 365–371.
14. Izadifar M, Baik O-D, Simonson CJ. 2006. Modeling of the packed bed drying of paddy rice using the local volume averaging (LVA) approach. Food Research International, 39(6), 712–720.
15. Stakić M, Banjac M, Urošević T. 2011. Numerical study on hygroscopic material drying in packed bed. Brazilian Journal of Chemical Engineering, 28(2), 273–284.
16. Zili L, Ben Nasrallah S. 1999. Heat and mass transfer during drying in cylindrical packed beds. Numerical Heat Transfer, Part A, 36, 201.
17. Sitompul JP, Istadi I, Widiasa IN. 2001. Modeling and simulation of deep-bed grain dryers. Drying Technology – International Journal, 19(2), 269–280.
18. Istadi I, Sitompul JP. 2002. A comprehensive mathematical and numerical modeling of deep-bed grain drying. Drying Technology International Journal, 20(6), 1123–1142.
19. Sitompul JP, Istadi I, Sumardiono S. 2003. Modeling and simulation of momentum, heat and mass transfer in a deep-bed grain dryers. Drying Technology – International Journal, 21(2), 217–229.
20. Boukadida N, Ben Nasrallah S, Perre P. 2000. Mechanism of two-dimensional heat and mass transfer during convective drying of porous media under different drying conditions. Drying Technology : An International Journal, 18, 1367–1388.
21. Kumar P, Mujumdar AS. 1990. Superheated steam drying: a bibliography. Drying Technology – International Journal, 8(1), 195–205.
22. Pronyk C, Cenkowski S. 2003. Superheated steam drying technology. Paper number RRV03-0014, an ASAE Meeting Presentation.
23. Devahastin S, Suvarnakuta P, Soponronnarit S, Mujumdar AS. 2004. A comparative study of low-pressure superheated steam and vacuum drying of a heat-sensitive material. Drying Technology – International Journal, 22(8), 1845–1867.
24. Iyota H, Nishimura N, Onuma T, Nomura T. 2001. Drying of sliced raw potatoes in superheated steam and hot air. Drying Technology – International Journal, 19, 1411–1424.
25. Whitaker S. 1999. The Method of Volume Averaging. Kluwer: Dordrecht, The Netherlands.
26. Whitaker S. 1984. Moisture Transport mechanisms during the drying of granular porous media. Proceeding of the Fourth International Drying Symposium. Edited by Mujumdar AS, p. 31–42.
27. Whitaker S. 1991. Improved constraints for the principle of local thermal equilibrium. Industrial and Engineering Chemistry Research, 30, 983–997.
28. Hager J, Wimmerstedt R, Whitaker S. 2000. Steam drying a bed of porous spheres: theory and experiment. Chemical Engineering Science, 55, 1675.
29. Messai S, Sghaier J, Lecomte D, Belghith A. 2007. Drying kinetics of porous spherical particle and the inversion temperature. Drying Technology – International Journal, 26, 157.
30. Sghaier J, Jomaa W, Belghith A. 2008. Superheated steam of a spherical porous particle. Journal of Porous Media, 11, 633–646.
31. Patankar SV. 1980. Numerical Heat Transfer and Fluid Flow. Hemisphere Publishing Corporation, Mc Graw Hill: Washington.
32. Moyne C, Stemmelen D, Degiovanni A. 1990. Asymmetric drying of porous materials at high temperature: theoretical analysis and experiments. International Journal of Chemical Engineering, 30, 654.
33. Elustondo DM, Mujumdar AS, Urbicain MJ. 2002. Optimum operating conditions in drying foodstuffs with superheated steam. Drying Technology – International Journal, 20(2), 381–402.

5

Evolutionary numerical simulation approach for design optimization of gas turbine blade cooling channels

Narasimha R. Nagaiah* and Christopher D. Geiger

Department of Industrial Engineering and Management Systems, University of Central Florida, Orlando, Florida 32816, USA

Abstract – Gas turbine blade cooling system design is a multidisciplinary, iterative and often tedious task involving complex relationships among multiple design objectives. Typical blade design requires a broad range of expertise in the materials, structural, heat transfer, and cost optimization disciplines. The multiple objectives involved are often conflicting and must be solved simultaneously with equal importance. The traditional approaches researchers scalarize the multiple objectives into a single objective using a weight vector, thus transforming the original multiple objective problem into a single objective problem. This research addresses the shortcomings of existing traditional approaches of the optimization of blade cooling configuration design.

Key words: Multiobjective, Optimization, Evolutionary, Numerical simulation, Gas turbine, Blade cooling channel, Ribs.

1 Introduction

Electricity is expected to remain the fastest growing form of worldwide end-use energy through 2030, as it has been for several decades. Gas turbines have been considered energy workhorses and are at the heart of almost all of the world's electricity generating systems. Gas turbines are also used extensively for aircraft propulsion and other industrial applications. A gas turbine, also called a combustion turbine, is a type of combustion engine with a complex operating mechanism, and is known to operate in harsh environments that include extremely high temperature and pressure levels. The design of the modern gas turbine requires interdisciplinary knowledge of aerodynamics, thermodynamics, heat transfer, mechanical systems, material science and manufacturing technology. In addition, a designer of turbine engines must balance the conflicting requirements of heat transfer against those of aerodynamics, materials, stress, manufacturability and costs. In general they are subjected to significant thermo-mechanical stresses that affect the durability and reliability of its components and, in turn, the turbine itself. Achieving high reliability and thermal efficiency of gas turbines is of continuing concern and motivates engineers and researchers to study and improve turbine efficiency and turbine reliability. Increased thermal efficiency and power output require that turbines operate at high inlet temperatures. However, the increase in operating temperature requires high level of cooling system design, specifically blade internal cooling system design.

In this paper a new evolutionary numerical simulation approach for multiobjective design optimization is proposed and studied. To overcome the drawback of subjectively selecting weights to transform the multiple objective design problems into a single objective problem. A two design performance objectives and a set of design decision variables that influence the performance objectives are studied. This proposed approach automatically generates the most appropriate blade cooling channel design specifications that simultaneously optimize the two design objectives.

1.1 The working principle of gas turbines

The simplest and most common gas turbine is an in-line axial flow turbine, as shown in Figure 1. A gas turbine operates by guiding incoming air into the compressor, which in turn, compresses and delivers highly-pressurized air into the combustor section of the engine. This is the mainstream air flow. The combustor burns the injected fuel using the compressed air delivered from the compressor. The mainstream air flow (referred to as hot gas) is a combustion mix of air, fuel and unburned hydrocarbons, and it can reach temperatures as high as 1700 °C and can produce high pressure variations [1, 2]. The hot gas enters a series of turbine stages, where a stage is composed of a set of vanes and a set of blades. The hot gas expands towards atmospheric (or ambient) pressure in each

*e-mail: raju@ucf.edu

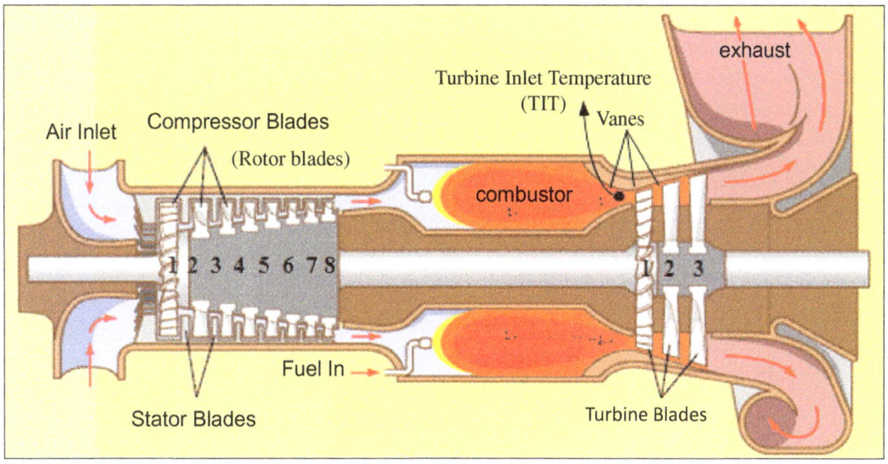

Figure 1. Cutaway view of typical industrial gas turbine engine (obtained from Britannica Encyclopedia, 1999).

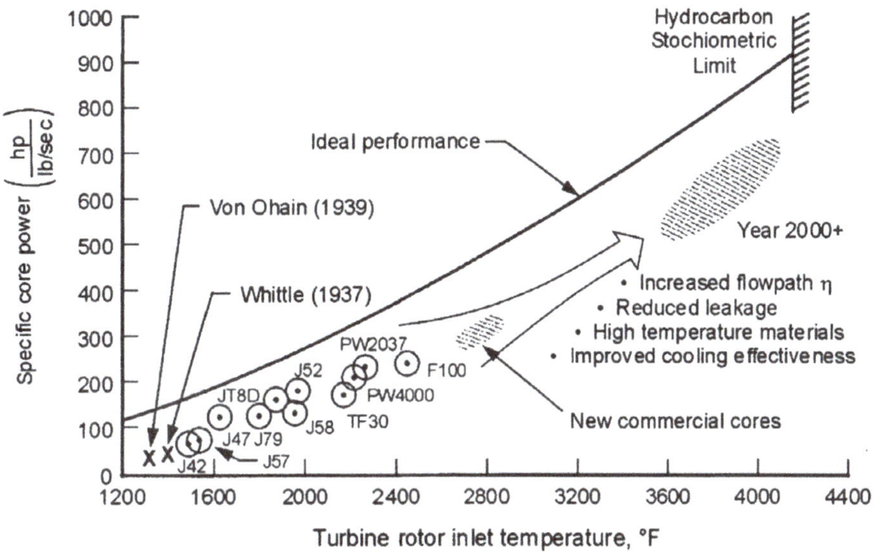

Figure 2. Historical trend of improving the core performance by increasing turbine rotor inlet temperature (Koff 1991; Reprinted with permission of the AIAA)

stage, and this gas expansion runs the turbine to generate output shaft power. This shaft power is used to drive the compressor, and it is also used to power the generator. In case of aircraft gas turbine the shaft power runs propeller to generate thrust and exhaust is diverted to nozzle to generate secondary thrust required to propel aircraft.

1.2 The gas turbine cooling system

The turbine engine parameter of greatest influence on core power and thermal efficiency is the turbine inlet temperature (TIT). To meet the demand requirements of power plants, such as increased thermal efficiency and increased power output, the owners of the gas turbines operate them at high inlet temperatures. During the last 30 years, the turbine inlet temperature for gas turbines has steadily increased (1200 °C–1700 °C) in order to improve thermal efficiency of the turbine. On the other hand, the allowable material temperature for the gas turbine hot gas path components has increased on a much slower rate. The historic increase in TIT, as shown in Figure 2, is a result of an attempt by gas turbine manufacturers to simultaneously increase the thermal efficiency and the specific core power per unit mass of air flow. The ideal Brayton Cycle curve, which is the performance theoretically obtainable with ideal components throughout the gas turbine engine, indicates a steady increase in specific core power until the Hydrocarbon Stoichiometric Limit is reached, which is the maximum temperature limit attained by burning fuel 100%. All existing gas turbine engine systems fall below this ideal curve. However, they follow the same general trend as this curve, from the very first gas turbine engines designed by Von Ohain (1939) and Whittle (1937) to more recent developments. Over this time span, there has been a several-fold increase in efficiency. However, there have also been very large increases in turbine inlet temperature.

One of the critical turbine components of hot gas path in gas turbine is the set of first-stage blades mounted on the rotor in the turbine section, as shown in Figure 1 and in Section 4 of

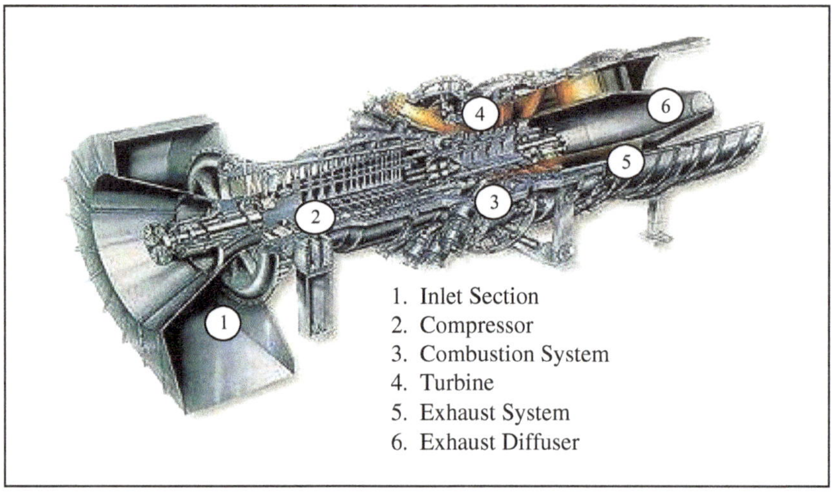

Figure 3. Sectional view of an industrial power plant gas turbine (Courtesy of Siemens).

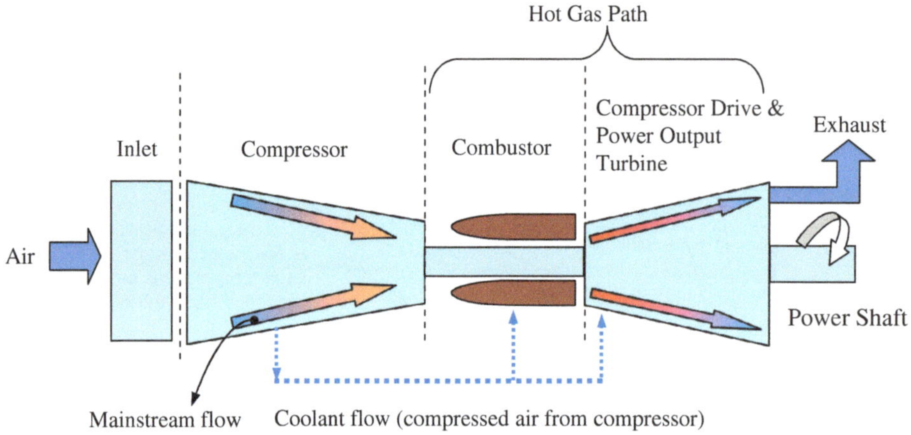

Figure 4. Schematic drawing of a gas turbine cooling supply system.

Figure 3. Turbine blades, in general, are subjected to high centrifugal forces due to their rotary motion and are vulnerable to high instances of failure compared to other turbine components in the hot gas path. Increasing the TIT is the primary contributor to creating the harsh operating environment for these critical components. In addition, as the TIT increases, the heat transferred to the turbine blade material also increases the probability of blade failure due to thermal stresses, load on the blades and ceramic thermal barrier coating that protects the blades.

In an effort to achieve the high levels of blade durability, the effective cooling of gas turbine components has received increasing attention in recent years. This has caused the rapid development of various methods to protect the gas turbine components. One such method is to cool these components using the highly-pressurized, compressed air of the mainstream flow that is diverted prior to the combustor section of the engine [3–5]. In essence, the turbine components in the hot gas path are cooled by a fraction of the compressed air (i.e., a secondary air flow) that is extracted by a cooling supply system from the compressor section of the engine (indicated by the dotted arrows in Figure 4). This secondary air flow of relatively cold and compressed air is called coolant, or cooling air. It is important to note that this extraction can reduce the thermal efficiency and power output of the turbine engine.

Figure 5 shows cooling techniques commonly used for first-stage turbine blades with three major internal cooling zones – the leading edge (Zone 1), pressure and suction surfaces (Zone 2), and the blade tip region (Zone 3). The leading edge is cooled by both jet impingement and film cooling. The middle portion of the blade is cooled by serpentine rib-roughened internal cooling passages. The rib-roughened internal cooling passages cause turbulence in the coolant as the coolant passes over and around the ribs (or turbulators). The turbulent air removes a fraction of the heat conducted in Zone 2 from the blade (see Figure 5). The cooling air exits through the cooling holes in Zones 1, 2 and 3 forming a thin, cool, insulating blanket along the external surface of the turbine blade. The cooling effectiveness (ϕ) of these turbine blade cooling techniques depends on two parameters – the mass flow rate ($\dot{m}$) and the flow velocity (u) of the coolant. These two parameters greatly depend on the rib configuration in the serpentine internal cooling passages.

The mass flow rate $\dot{m}$ of the coolant is usually measured as a percentage of the mainstream flow. Figure 6 shows that cooling effectiveness increases rapidly with a small percentage of

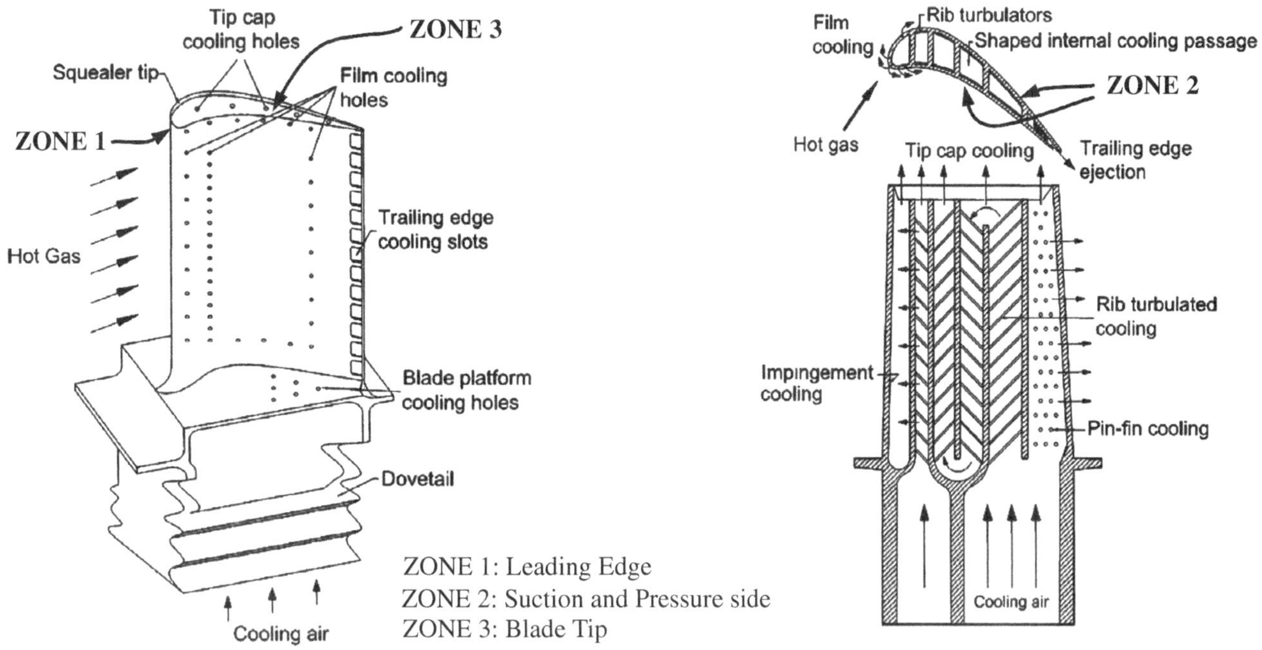

Figure 5. The schematic of a modern gas turbine blade with common cooling techniques [4].

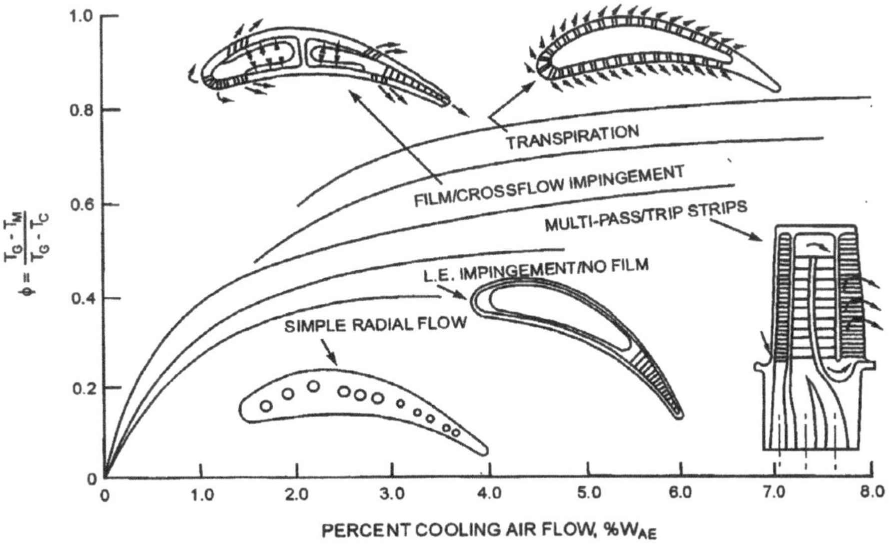

Figure 6. Cooling effectiveness of different blade cooling design vs. cooling air flow (Moustapha et al. [2]).

coolant, but then the growth of the cooling effectiveness slows. To increase further cooling, a large amount of cooling air and/or different cooling techniques must be used. Since the cooling air is extracted from the compressor section of the engine, it represents a direct loss of engine efficiency due to a reduced amount of mainstream flow [2]. Figure 6 also shows that improvement in blade design can significantly increase the cooling effectiveness of the coolant at the same percentage of coolant flow. Despite the recent developments in internal cooling system technology, it is difficult to cool the blades significantly beyond an average cooling effectiveness of 0.50 (as shown in Figure 6). Thus, there is a need to explore further cooling passage configurations that directly impact a blade's cooling effectiveness.

This research investigation focuses on optimizing the design configuration of the internal cooling passage of a turbine blade in order to enhance the cooling so that the desired cooling effectiveness is achieved. In general, the relationship among most of the critical components and internal subsystems in gas turbines are complicated, and the performance objectives of the turbine engine sometimes conflict. For example, the efficiency of gas turbines increases as TIT increases. However, operating at high temperatures decreases the life of the gas turbine engine and increases its operating costs. These conflicting objectives necessitate turbine blade design decisions that must consider trade-offs between the objectives. The optimization of the design configuration of the blade internal cooling passage gives rise to a multiobjective design optimization problem, where the two conflicting objectives are to: (1) maximize the cooling effectiveness, represented by the surrogate term heat transfer coefficient (h), to increase the blade life and reliability of the engine, and (2) minimize the pressure drop (Δp) in the cooling passage. The minimization of the pressure drop is important in that, enough pressure must be retained in the cooling passage for satisfactory ejection of the coolant flow. If there is insufficient pressure in the cooling air flow, the exit velocity of the coolant will be lower than that of the mainstream flow, and it will disturb the mainstream air flow. This disturbance is called mixing loss and can contribute to the loss of engine efficiency. Thus, minimization of pressure drop inside the cooling passage is an important objective for designers to consider.

2 Previous related literature

Over the last 50 years, there is a wide array of existing research concerning gas turbine blade cooling techniques. Researchers use analytical, computational and experimental methods to improve cooling techniques for gas turbine. Recent monographs focusing entirely on the gas turbine heat transfer and associated cooling technology is provided by Han et al. [3, 4] and Goldstein [6]. The use of multiobjective optimization in heat transfer problems is a relatively new area of focus and has been the point of interest only in recent years. In particular, the last few years have seen a sharp increase of heat transfer related optimization using evolutionary algorithms (EAs). Gosselin et al. [7] review multiobjective optimization using genetic algorithms, the more popular representative of the family of EAs, in heat transfer problems.

One of the well-known methods to improve the heat transfer (i.e., enhance material cooling) in a flow passage is to roughen the surfaces with turbulators (called ribs) in the blade's internal cooling passages so that the surface area increases and enhances cooling. Gas turbine researchers study different design configurations of ribs in blade internal cooling passages to enhance the cooling process [8]. However, the use of rib turbulators poses other risks such as a decrease in coolant air flow pressure and a decrease in the velocity of coolant flow. Thus, a design optimization process capable of addressing multiple design objectives simultaneously can be a suitable tool in such conditions. The application of multiobjective design optimization to internal cooling passages not only helps enhance the blade cooling, but it can eventually be used in other areas such as heat exchanger/heat sink design, where cooling passages need to be optimized.

The optimization of cooling passage is extensively studied by Kim and Kim [9], who consider the optimization of internal cooling passages with straight rectangular ribs [10], V-shaped ribs [11] and the angle of the ribs [13]. They identify the values of geometric design variables with the objective function defined as a linear function of heat transfer coefficient and friction drag coefficient (pressure drop). They suggest that using a numerical approach presents a reliable way of designing optimized heat transfer surfaces. It is important to note that the two objectives considered in their study are heat transfer coefficient and pressure drop. However, these two objectives are combined to form a composite function using a vector of weights. The selection of the weights is based on designer's experience, which could lead to errors in optimization if the factors are not carefully selected.

In summary, due to the complex nature of the flow and heat transfer phenomena involved in cooling passage design, only few attempts have been made in applying multiobjective optimization techniques to the design of turbine blade cooling passages. The limited studies consider two objective functions and convert the two objectives to a single composite objective function. However, no existing research simultaneously considers two objectives independently to optimize the design. The present work investigates a multiobjective design optimization method integrating evolutionary algorithms and numerical simulation to find a set of Pareto optima for the rib design in the internal cooling passage of the blade.

3 Proposed multiobjective design optimization framework

The overall goal of this research is to investigate and propose an approach that optimizes the gas turbine blade internal cooling channel design to enhance turbulent convective heat transfer while considering multiple design objectives simultaneously. The specific objectives of this research are to: (1) design a multiobjective procedure for the heat transfer optimization problem; (2) integrate a commercially-available simulation package used to build computational fluid dynamics (CFD) models for the analysis of the flow field and associated heat transfer of different design configurations of gas turbine blade cooling channels; and (3) automate the design optimization framework. Further the specific tasks aim of this research is to automate the process of finding the rib design for the cooling passage of a gas turbine blade that simultaneously maximizes the cooling effectiveness (i.e., the heat transfer coefficient h) increasing the blade life and reliability of the engine, and minimizes the pressure drop Δp increasing the engine efficiency. Due to symmetric nature of cooling passage and also to minimize the computational effort, only a periodic segment of cooling passage is selected in this study. Furthermore, this segment is simplified to a two-dimensional geometry with ribs on the bottom and top walls, as shown in Figure 7. It is the position and the shape of these ribs along the two walls that are to be optimized. The shape of the ribs depends on critical design variables. The radii (R_1 and R_2) of ribs 1 and 2 (in the periodic segment), and fillet radii (R_3, R_4, R_5 and R_6) between ribs and wall surface are considered as critical design variables that influence the values of objectives. Ribs induce separation and reattachment of flow to enhance the heat transfer by creating turbulent mixing. The heat transfer is greater at the reattachment locations, but it is low at the locations where flow separation takes place due to ribs. The flow separation and reattachment phenomenon is influenced by radii of the ribs.

The increase in radii R_1 and R_2 increases size of the ribs, surface area and turbulent mixing inside the cooling channel to enhance the heat transfer from blade to cooling air flow. However, an increase in rib size increases the drop in air flow pressure. Fillet radii decreases pressure drop and at the same

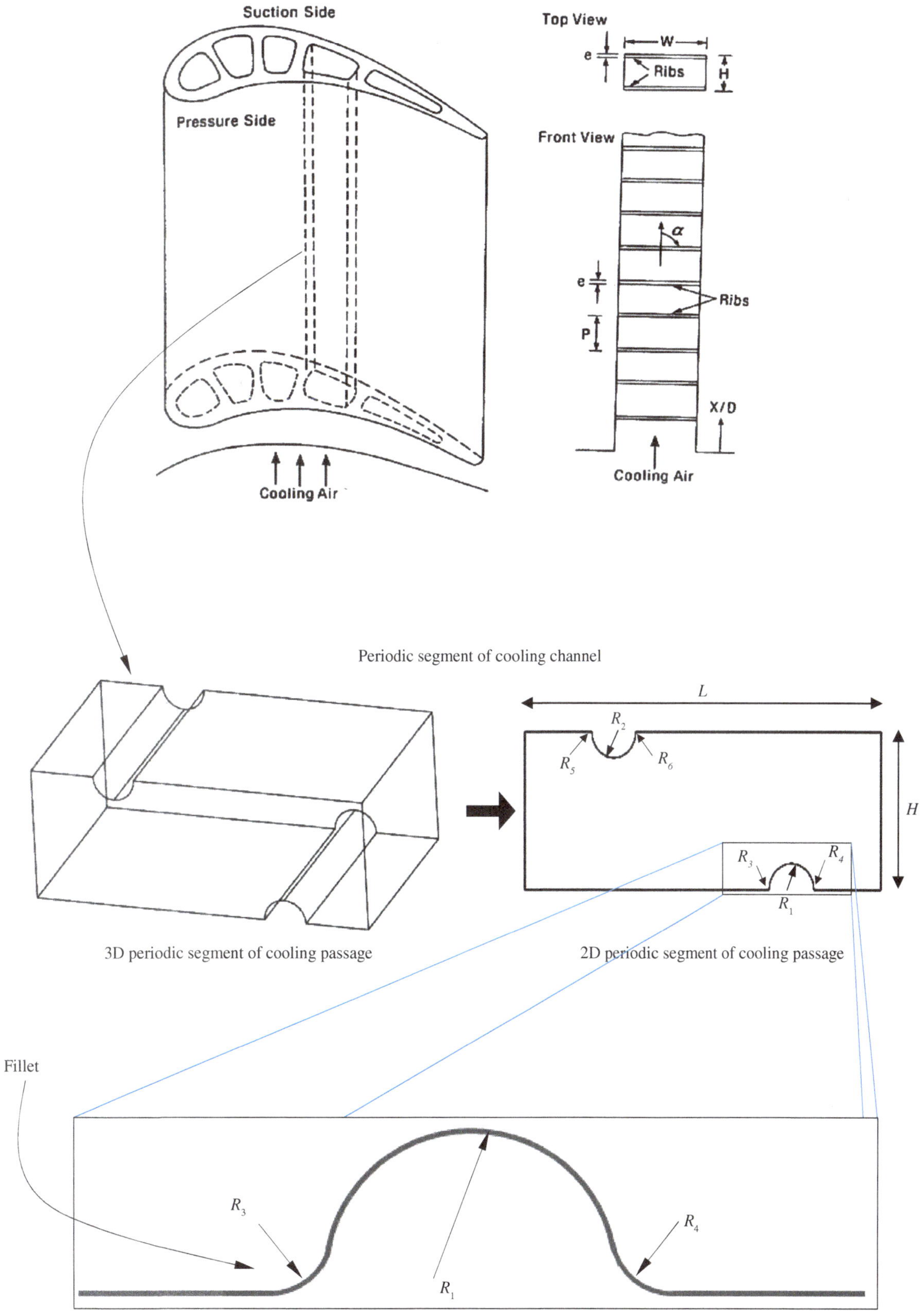

Figure 7. Selection of 2D periodic segment of the gas turbine blade cooling passage and design variables.

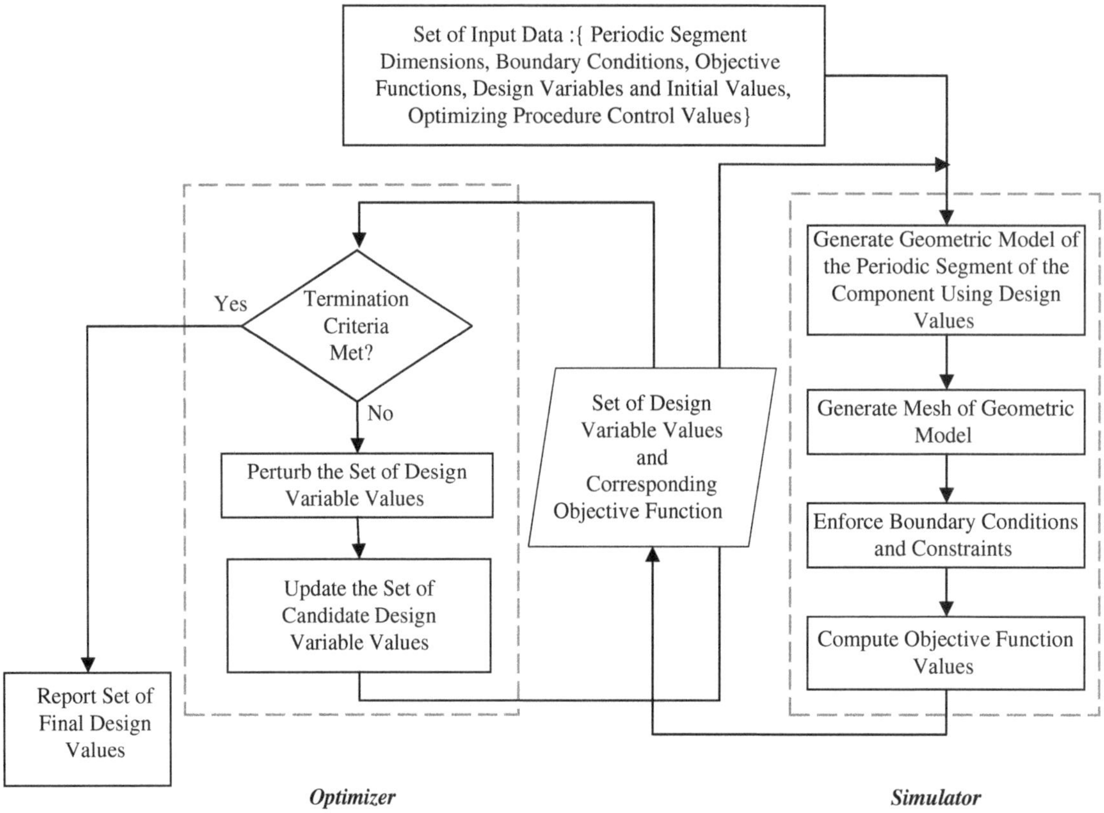

Figure 8. Overview of the proposed framework for mechanical component multiobjective design optimization.

time increases heat transfer rate by creating smooth surface contact between ribs and blade wall (refer to Figure 7 for an enlarged view of cooling channel ribs and fillets). Therefore, variations in these specifications can change the heat transfer coefficient h, and the pressure drop Δp.

The proposed optimization conceptual framework for gas turbine blade internal cooling passage design optimization is illustrated in Figure 8. In general, the framework is comprised mainly of an optimizer component and an evaluation component. The optimizer component iteratively generates multiple candidate rib configuration designs. Evaluation of the candidate designs are performed using numerical simulation, and the simulation can be viewed as a black box with: (1) an input interface that accepts and translates new design specifications and (2) an output interface that communicates objective function values to the optimizer component. The simulation component in this research uses computational fluid dynamics (CFD) numerical simulation. Based on the candidate design evaluation results, the optimizer utilizes these results to generate the next set of candidate designs through the perturbation of the evaluated designs. This cycle continues until the termination criteria are met. The next sections provide more details of the main design optimization procedure components therein.

3.1 Solution evaluation component

For the specific design problem under study, the solution evaluation component or simulator evaluates the objective function values, heat transfer coefficient h and pressure drop Δp for a set of candidate design variable values generated by optimizer. In the present study, the radii (R_1 and R_2) of ribs 1 and 2, and fillet radii (R_3, R_4, R_5 and R_6) between ribs and wall surface are considered as critical design variables (refer to Figure 7). To obtain the objective function values for a given rib configuration, the coolant flow and heat transfer problem is solved for the periodic segment by numerical simulation. To solve the turbulent flow and heat transfer problem and to obtain the values for the objective functions, four main steps are performed: (1) generate computational model of 2D periodic segment of cooling channel using design variable values, (2) generate the mesh, (3) enforce the initial and boundary conditions for the geometric model and (4) solve flow and heat transfer governing equations. Figure 9 shows a pictorial view of process involved while solving a computational model for objective functions.

3.2 Optimizer component

Evolutionary Algorithms (EAs) are appropriate choice for solving multiobjective optimization problems. The dramatic increase in computational power and at the same time decrease in computational cost showing upward trend of using EAs to solve multiobjective optimization problems [12]. EAs are population-based optimization algorithms capable of finding multiple solutions in its final population that optimizes each objective simultaneously. Briefly, an EA-based optimization process starts with a randomly-generated initial population of size N individual solutions. The representation of the individual

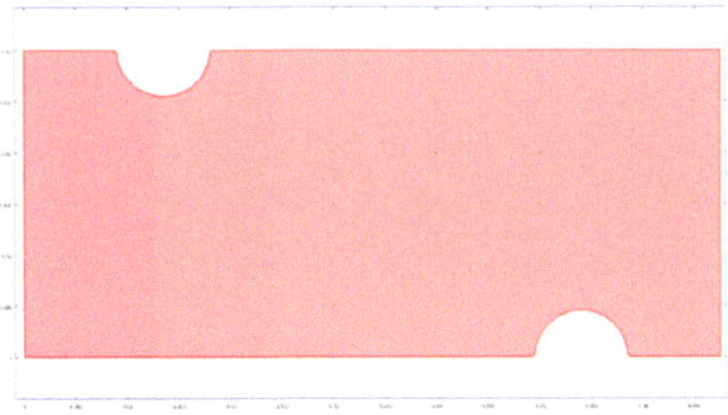

(a) Initial model

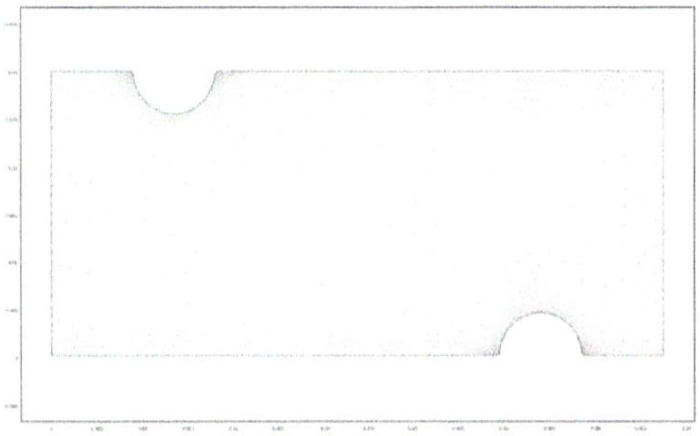

(b) Meshed model

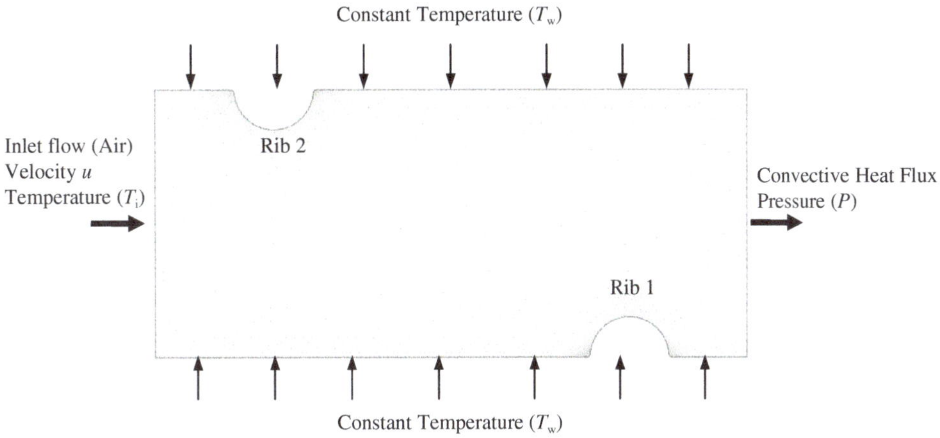

(c) Model after with boundary and initial conditions

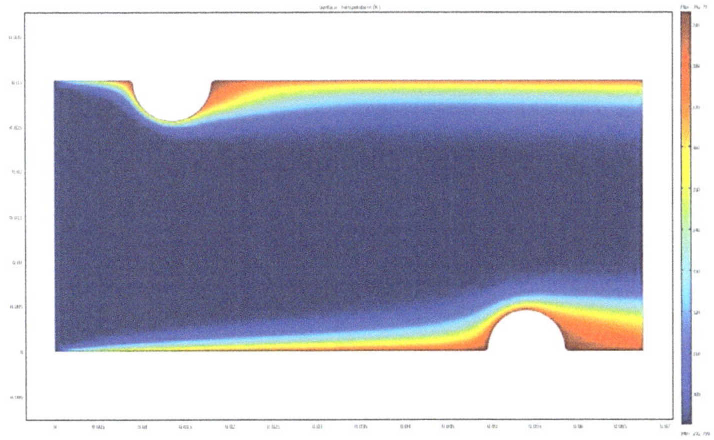

(d) A solved model of temperature distribution

Figure 9. Pictorial overview of the steps involved in solving computational model for objective function values.

solution is given by binary strings, where each string is the binary representation of design variable values. Each set of binary strings in the population of individuals is evaluated using the simulator to find the corresponding fitness value (i.e., objective function value). A new population of individuals is created using three genetic operators: reproduction, crossover and mutation. The new population of individuals is then passed to the Simulation component for evaluation. This procedure continues to cycle until a termination criterion is met (in this study, a maximum number of iterations, called generations). The elitist Non-Dominated Sorting Genetic Algorithm II (NSGA II) proposed by Deb et al. [14], is currently one of the most popular EAs used to solve complex and real-world multiobjective optimization problems. Some of the salient features of NSGA II are its fast elitist sorting method that involves a combined pool of both the parent and child populations and provides diverse population using an autonomous crowding distance method. NSGA II introduces elitism by comparing the current population of candidate solutions with the previously found best nondominated solutions. In NSGA II, the selection procedure uses two

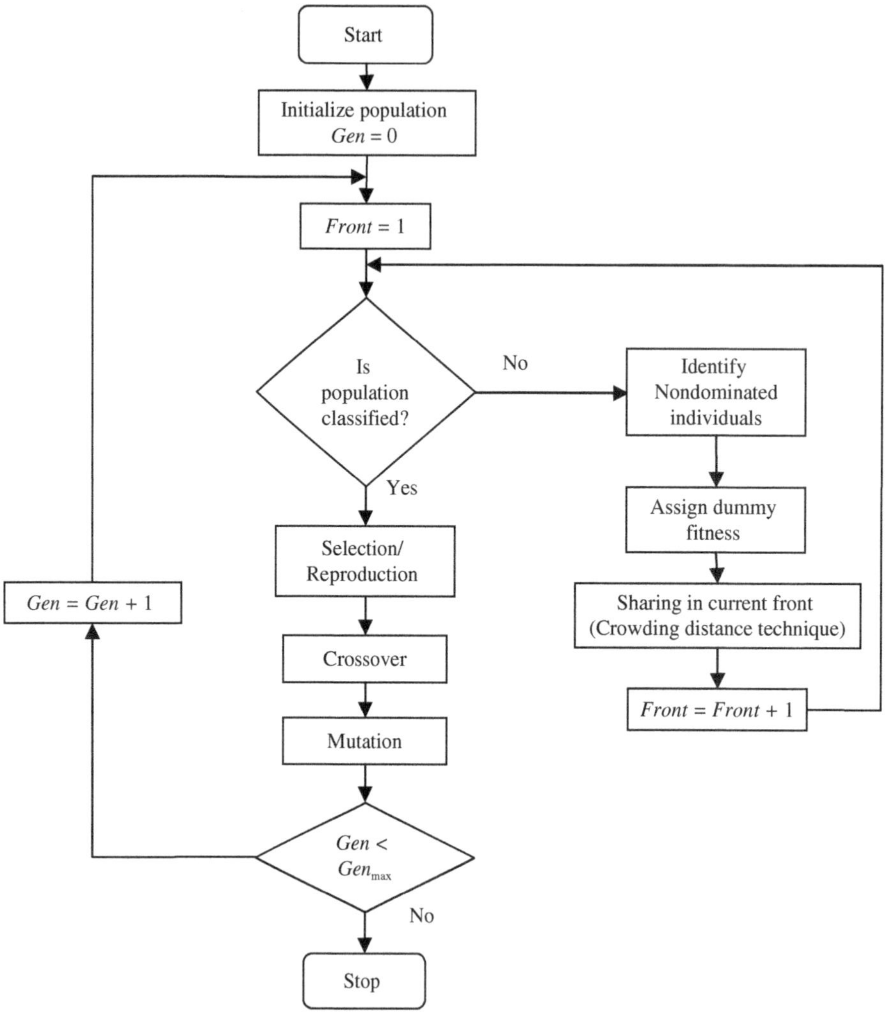

Figure 10. A flowchart of the working logic of NSGA II.

processes: (1) Nondominated ranking and (2) crowding distance assignment. Figure 10 shows process flowchart for NSGA II for better understanding of the crowding distance and nondomination rank assignment process. A more detailed description about evolutionary algorithms and NSGA II can be found in Deb et al. [14, 15].

3.2.1 Reproduction (or selection) operator

The primary objective of the reproduction operator is to copy the better performing candidate solutions and discard the poor performing solutions in the population, while maintaining the population size. This is achieved by performing the following tasks:

- Identify the better performing solutions in a population based on their fitness values.
- Make multiple copies of the better performing solutions to create the mating pool.
- Discard poor performing solutions from the population so that the copies of the better performing solutions can be placed in the population.

3.2.2 The crossover operator

A crossover operator, also referred to as the recombination operator, is applied next to the candidate solutions of the mating pool. The crossover operator exchanges information between selected solution pairs (called parent solutions) with a probability of occurrence *c*. The simulated binary crossover (referred to in the literature as SBX) operator introduced by Deb and Agarwal [16] is performed in this algorithm.

3.2.3 The mutation operator

The crossover operator is primarily responsible for the intensification of the search and the mutation operator allows for diversification of the search to prevent the search process from becoming trapped at a local optimum. After crossover, the newly-generated solutions undergo a mutation operation, where operator changes a 1 to 0, and vice versa, with a mutation probability of occurrence *m*. The polynomial mutation operator introduced by Deb and Goyal [17] is employed by NSGA II in which the probability distribution is polynomial.

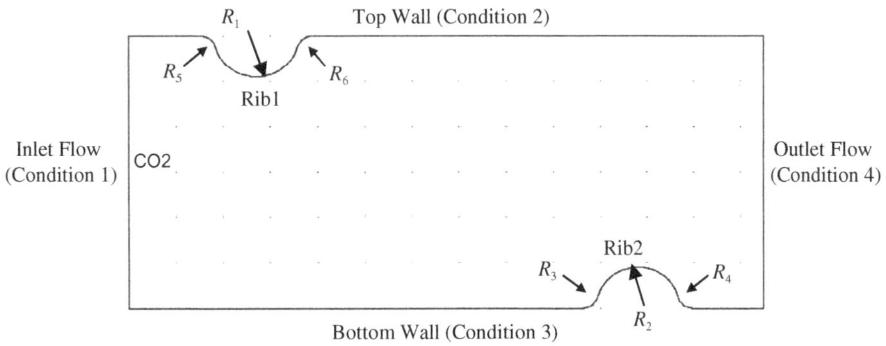

Figure 11. Cooling channel with design variables R_1, R_2, R_3, R_4, R_5, and R_6.

Table 1. Design variables and value ranges (meters).

Parameters	Radius of rib 1 (R_1)	Radius of rib 2 (R_2)	Fillet radius (R_3)	Fillet radius (R_4)	Fillet radius (R_5)	Fillet radius (R_6)
Lower bound	0.001	0.001	0.00001	0.00001	0.00001	0.00001
Upper bound	0.0055	0.0055	0.00004	0.00004	0.00004	0.00004

4 Optimization and results

In this research, the four steps of solution evaluation are performed using numerical Multiphysics modeling and simulation software COMSOL. In addition, the Non-dominated Sorting Genetic Algorithm II (NSGA II) proposed by Deb et al. [14] is used as the optimizer component. However, the impetus and success of this research investigation is not necessarily predicated upon using these specific approaches.

4.1 Design variables and input parameters

Recall that six design variables R_1 and R_2 (radii of Ribs 1 and 2, respectively) and fillet radii R_3, R_4, R_5, and R_6 are considered in this research investigation. Thus, results show only 2D rib shapes based on the radii of the ribs and fillet radii of the ribs in the cooling channel. Various parameters and variables are used in both the COMSOL numerical simulation and NSGA II. The parameters and range of initial values of the variables are now summarized. In COMSOL, the computational domain is selected and defined (Figure 11). The value ranges of the design variables R_1 through R_6 are given in Table 1. These ranges are approximated based on experimental results by Han et al. [3]. Next, Flow Physics is used to reflect the realistic conditions of internal cooling passage flow to correctly simulate the characteristics of the fluid flow. Here, compressed air (coolant) with non-isothermal and turbulent flow physics is used. Table 2 shows the material properties (density, dynamic viscosity) of air at atmospheric temperature and pressure, and these properties are imposed through the Graphical User Interface (GUI) of COMSOL. For simplicity the inlet boundary conditions are assumed to be constant boundary conditions. This study did not considered the effect on the inlet boundary conditions due to the change in ribs configuration at each simulation.

Table 3 summarizes initial boundary conditions used to solve objective functions in the COMSOL simulation environment. The cooling channel Inlet Flow (Condition 1) is subjected

Table 2. Initial subdomain conditions used for the COMSOL numerical simulation.

Fluid	Properties
Coolant air	Density (ρ) = 1.204 kg/m^3 Dynamic viscosity $\mu = 1.983 \times 10^{-5}$ kg/m s

Table 3. Initial boundary conditions used for the CFD simulation.

Boundary	Initial/boundary condition
Inlet flow (Condition 1)	Temperature (T) = 293 K Velocity (u) = 10 m/s; Reynolds number (Re) = 20,000
Wall (Conditions 2 and 3)	Temperature = 393 K Thermal wall function
Outlet flow (Condition 4)	Convective heat flux Pressure (p) = 0

to a temperature (T) and velocity (u) to create necessary turbulence in the flow. Conditions 2 and 3 are wall boundaries, where fluid velocity is zero, and it is maintained at constant temperature, larger than that of the coolant. At the Outlet Flow Condition 4 indicating fluid departure, typically a relative pressure and convective heat flux is imposed. The boundary conditions are used as initial conditions to solve the governing equations iteratively to predict approximate fluid flow and heat transfer properties inside the cooling passage.

A pilot study has been conducted to determine the effect of evolutionary algorithms control parameters on different problems. Table 4 lists suggested MOEA control parameters used in this research investigation. The population size, N, is the number of candidate design solutions at each generation. A small population size can limit the capability of exploration of the search space and inhibits the purpose of crossover operations. Conversely, use of large population size can be computationally-expensive. For this study, a population size of $N = 50$ is used. The maximum number of generations, Gen_{max}, denotes

Table 4. NSGA-II input control parameters.

MOEA parameters	Parameter values
Population size (N)	50
Maximum generations (Gen_{max})	100
Reproduction/selection	Tournament selection (Rank & crowding distance)
Crossover probability	c = 0.90 (or, 90%)
Mutation probability	m = 0.10 (or, 10%)

the number of generations (i.e., iterations) when to terminate the MOEA and to report the best set of design solutions so far. From the pilot study it is observed that convergence of solutions to optimal front at Gen_{max} = 100 generations. In EAs, the creation of new solution is carried out by crossover operator. The crossover probability, c, defines how often crossover is performed. A low crossover probability decreases the speed of convergence due to lower exploration rate. On the other hand, a high probability may contribute to premature convergence. In general, the recommended range of c is between 0.60 and 0.95. Likewise, the mutation probability, m, denotes how often parts of an individual solution undergoes random perturbations. It introduces diversity into the population and should be a small value to avoid the algorithm from becoming a random search. In general, the recommended range of m is between 0 and 0.20. From the pilot study c = 0.90 (or, 90%) and m = 0.10 (or, 10%) shown better convergence rate.

4.2 Discussion of the results

The heat transfer coefficient h is computed and summed along the top and bottom walls of periodic segment of the cooling passage using the boundary integration method. Pressure drop Δp is obtained by computing the difference between inlet and outlet air flow pressures using the boundary integration method. The optimization framework is executed for population size $N = 50$ and maximum generations $Gen_{max} = 100$ generations.

4.2.1 Single objective functions optimization

The goal in single objective optimization is to converge to the (global) optimum. Single-objective problems either minimize or maximize the objective function value depending upon the problem type attempting to reach single optimum value. In cooling channel design optimization, the heat transfer coefficient (h) is selected as an objective function because of its importance in blade cooling. The objective here is to maximize the value of h varying design variable values. The results of optimization of h for 2, 4 and 6 design variables are presented below.

Figure 12a shows graphical representation of single-objective optimization results for two design variables. The y-axis represents objective function, heat transfer coefficient (h), which needs to be maximized, whereas the x-axis represents number of generations (Gen_{max}). The objective function (h) value at each generation is average of 50 objective function values (Pop) in that generation and it is compared with highest value of h in the same generation (Figure 12a). It is observed that the convergence of objective function h to global optimal value is linear and rapidly converge within the first few generations (five generations). In other words, the convergence rate slows and remains almost constant after 5th generation. To save computational time, one could stop the optimization process just after the 5th generation and report results. The best value of h found in this case is 15.4253 W/m^2 K (over an average of 50 design specifications in the population). Similarly, the initial h value is 13.3949 W/m^2 K (over an average of 50 design specifications in the population). As a result, a 15.15% increase in heat transfer coefficient or cooling effectiveness (directly proportional to HTC) is achieved.

Next from Figure 12b, it is evident that the convergence rate in the case of four design variables is slow and appears to take more generations than the two design variable problem. The average global optimal value of h found in this case is 16.2793 W/m^2 K (over an average of 50 design specifications in the population). Similarly, the average initial h value is 13.7239 W/m^2 K (over an average of 50 design specifications in the population). A 18.62% increase in the heat transfer coefficient is achieved. By introducing the fillet radii as design variables, the four design variable problem resulted in 5.53% more cooling effectiveness than two design variable problem.

Similarly from Figure 12c, again it is evident that the convergence rate in case of six design variables further slows down compared to both the four and two design variable problem. The average global optimal value of h found in this case is 17.8476 W/m^2 K (over an average of 50 design specifications in the population). Similarly, the average initial h value is 13.9687 W/m^2 K (over an average of 50 design specifications in the population). An average a 27.75% increase in heat transfer coefficient is achieved. By introducing the additional fillet radii as design variables, the six design variable problem results in 15.70% more cooling effectiveness than the two design variable problem and 9.63% more cooling effectiveness than the four design variable problem.

4.2.2 Multiple (two) objective optimization

In the single-objective optimization problem, it is seen that the solution converges to one solution. Therefore, one can easily chose the final design specifications to use without ambiguity. But, when more than one objective functions are considered simultaneously for optimization and a Pareto-based optimization approach as proposed in this research, there exists a number of trade-off, or compromise, solutions. Without any further information, no solution from the set of compromise solutions can be said to be better than any other in the set. Thus, in multiobjective optimization, an effort must be made in finding the set of trade-off optimal solutions by considering all objectives to be equally important. Thus, it can be conjectured that there are two goals in a multiobjective optimization. First, set of solutions that is as close to the Pareto-optimal front as possible must be identified. Second, the set of solutions must be as diverse as possible. After a set of such trade-off solutions are found, a user can then use higher-level preference information to make a choice. The above such trade-off solutions are obtained a multiobjective optimization involves two search spaces instead of

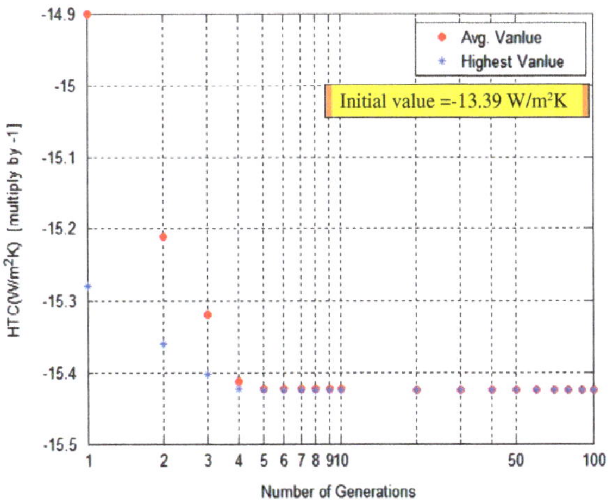

a) Single objective and two design variables

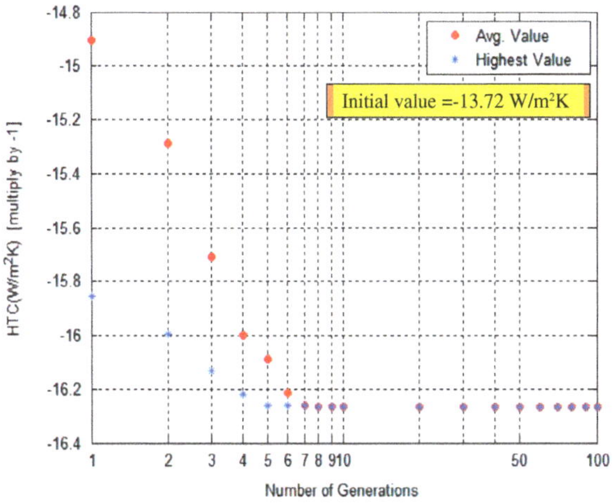

b) Single objective and four design variables

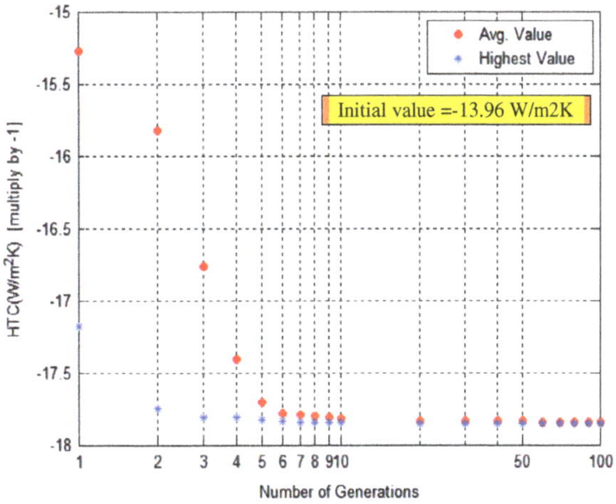

c) Single objective and six design variables

Figure 12. Convergence behavior of single objective to the (global) optimum.

one. In single-objective optimization, there is only one search space – the decision variable space. However, in multiobjective optimization, there exists an associated space called objective, or criteria, space. Section below presents results of cooling channel design considering two objectives – heat transfer coefficient (h) and coolant pressure drop (Δp).

Figure 13 shows Pareto optimal frontier for two objectives with two, four, and six design variables in Figures 13a, 13b, and 13c, respectively. For example, Figure 13a shows the initial solution (Gen_{max} = 0 generation) and the efficient frontier after 100 generations (Gen_{max} = 100). The x-axis represents pressure drop Δp, which is to be minimized. The y-axis represents heat transfer coefficient h, which is to be maximized. Note that the heat transfer coefficient objective function value is multiplied by -1 to convert it to an objective for minimization by the duality principle. After 100 generations, the solutions converge to the true Pareto optimal front with diverse set of solutions. The Pareto optimal front for the two objectives four design variables (Figure 13b) and two objectives six design variables (Figure 13c) suggest that the solutions are not converging to a smooth and uniform Pareto frontier as shown in the two-variable case. This is because the solution space of the problem increases exponentially with the increase in the number of decision variables. Therefore, more search iterations (i.e., generations) with an efficient search strategy (i.e., fine-tuning the search control parameters such as crossover c and mutation m) is required to explore more promising regions.

The interpretation of the turbine blade cooling passage designs along the Pareto front can be seen in Figure 14, which is the set of Pareto optima at N = 50 and Gen_{max} = 100. Figure 14 highlights three of the 50 design solutions along the Pareto frontier. Each of the three highlighted design solutions is shown with its corresponding temperature distribution color plot image of the periodic segment of the blade internal cooling passage. At the entrance (the left side boundary) of each plot, an inlet temperature of the coolant is applied. The top and bottom wall surfaces are maintained at a constant temperature. From the color plot, it can be concluded that the temperature of the incoming cool air (dark blue) increases because it is picking up heat from the surfaces (i.e., top and bottom) because of the turbulence caused by the ribs while flowing through the cooling passage. In Figure 14, it can be seen that tradeoff Solution 1

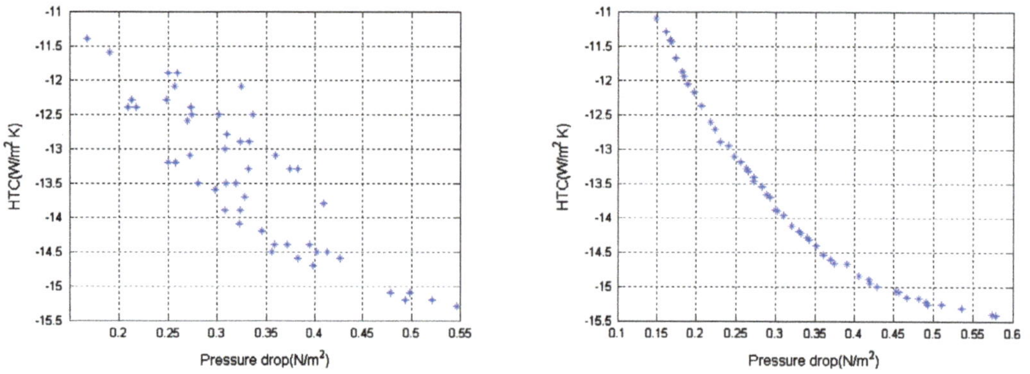

(a) Two objective and two design variables (left) Initial solutions Gen_{max}= 0, (right) Pareto optimal frontier after Gen_{max}= 100

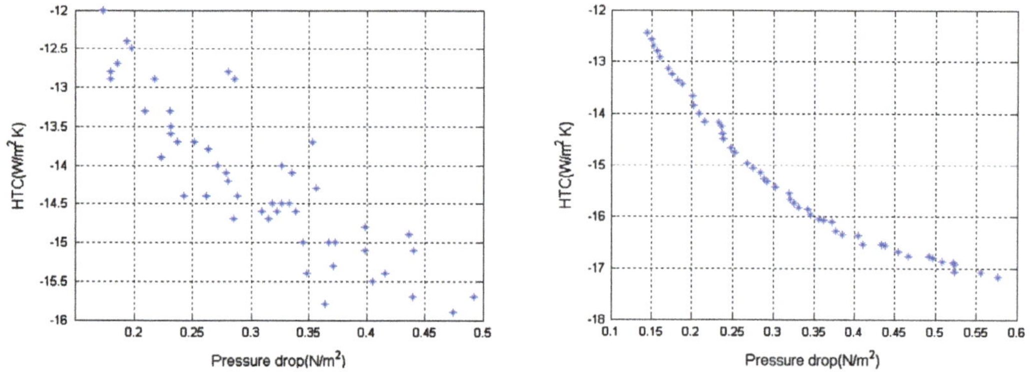

(b) Two objective and four design variables (left) Initial solutions Gen_{max}= 0, (right) Pareto optimal frontier after Gen_{max}= 100

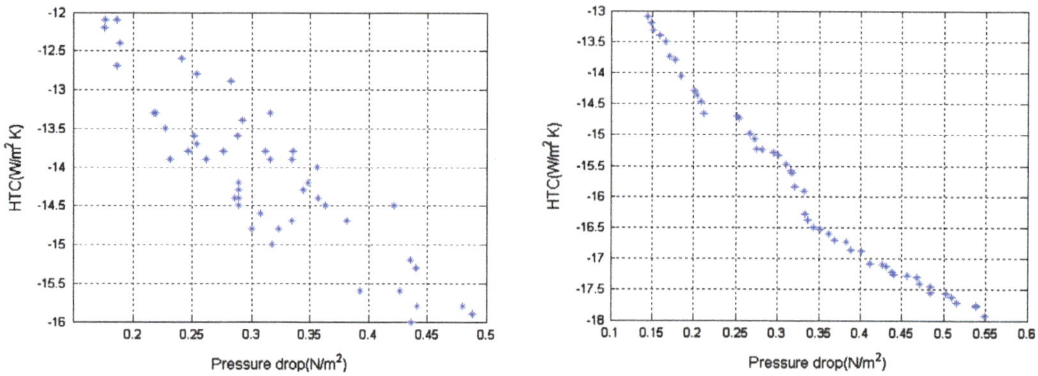

(c) Two objective and six design variables (left) Initial solutions Gen_{max}= 0, (right) Pareto optimal frontier after Gen_{max}= 100

Figure 13. Pareto optimal frontier for two objectives at the initial generation and at Gen_{max} = 100 generations.

(Design 1) has smaller rib radii seen in the temperature distribution color plot image. The values of the radii R_1 and R_2 design variables are given in the inside table (for illustration purpose two main design variables rib radii R_1 and R_2 selected here). Solution 1 has minimal pressure drop (Δp = 0.1485 N/m^2) and a low heat transfer coefficient (h = 11.09 W/m^2 K). The temperature distribution color plot image for tradeoff Solution 3 (Design 3) shows larger ribs with a high pressure drop (Δp = 0.5783 N/m^2) and a high heat transfer coefficient (h = 15.42 W/m^2 K). The tradeoff Solution 2 (Design 2) shows a combination of large and small ribs with moderate pressure drop (Δp = 0.295 N/m^2) and heat transfer coefficient (h = 13.82 W/m^2 K).

5 Summary and future work

The main objective of this research is to propose a multiobjective design optimization for mechanical component design, specifically gas turbine blade internal cooling passages. The proposed optimization framework is built by integrating multiobjective evolutionary algorithms and computational fluid dynamics numerical simulation. In this study, the MOEA NSGA-II and the multiphysics numerical simulation software COMSOL is used. The results from the proposed multiobjective design optimization show the promise of the proposed approach to be used for the automated optimization of the rib designs inside a 2D cooling channel of a gas turbine blade.

	Design 1	*Design 2*	*Design 3*
R1	0.001	0.001218	0.005493
R2	0.001	0.00437	0.0055
Δp	0.1485	0.2955	0.5783
h	−11.09	−13.82	−15.42

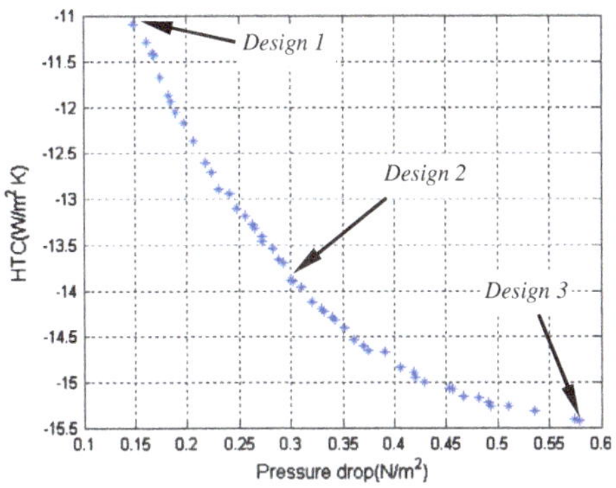

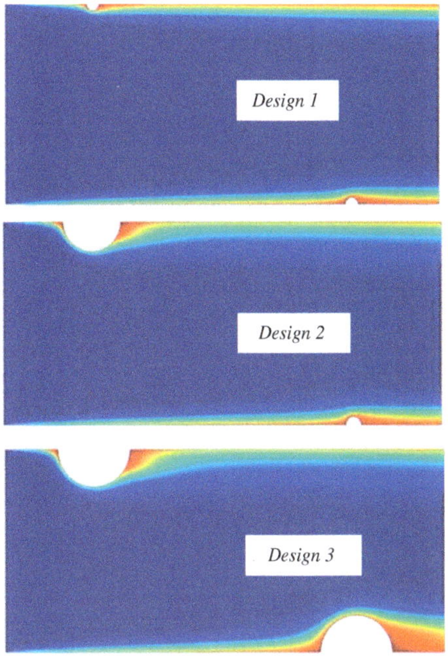

Figure 14. Specifications of gas turbine blade internal cooling channel ribs for three selected design solutions after 100 generations.

Even further, this research shows the promise of the proposed simulation multiobjective optimization approach to any mechanical component design optimization process.

The proposed approach is demonstrated for a 2D geometry with up to six design variables and two objectives considered. The next steps of this research effort include an enhancement on the work described in this paper. First, a third, relevant and very practical objective will be considered – the minimization of the material consumption in cooling passage design and blade manufacturing. Material consumption is a surrogate for material cost. Second, additional design variables that allow for more complex rib design geometries will be introduced. Finally, an intelligent approach to cluster Pareto optimal solutions can aid a decision-maker in reducing the design options thereby reducing the time to identify the best design option that satisfies the designer preferences.

References

1. Boyce MP. 2006. Gas Turbine Engineering Handbook, 3rd edn. Gulf Professional Publications: Boston.
2. Moustapha H, Zelesky MF, Baines NC, Japikse D. 2003. Axial and radial turbines. Concepts ETI Inc.
3. Han JC, Dutta S, Ekkad S. 2001. Gas turbine heat transfer and cooling technology, 1st edn. Taylor & Francis: New York.
4. Han JC. 2004. Recent studies in turbine blade cooling. International Journal of Rotating Machinery, 10(6), 443–457.
5. Logan E. 2003. Handbook of Turbo Machinery, 2nd edn. CRC Press: New York.
6. Goldstein RJ. 2001. Heat Transfer in Gas Turbine Systems. Annals of the New York Academy of Sciences: New York.
7. Gosselin L, Maxime G, François MP. 2009. Review of utilization of genetic algorithms in heat transfer problems. International Journal of Heat and Mass Transfer, 52, 2169–2188.
8. Ligrani PM, Oliveira MM, Blaskovich T. 2003. Comparison of heat transfer augmentation techniques. AIAA Journal, 41, 337–362.
9. Kim KY, Kim HM. 2002. Shape optimization of rib-roughened surface to enhance turbulent heat transfer. International Journal of Heat & Mass Transfer, 45, 2719–2727.
10. Kim HM, Kim KY. 2004. Design optimization of rib-roughened channel to enhance turbulent heat transfer. International Journal of Heat and Mass Transfer, 47, 5159–5168.
11. Kim KY, Lee YM. 2007. Design optimization of internal cooling passage with V-shaped ribs. Numerical Heat Transfer: Part A: Applications, 51, 1103–1118.
12. Deb K. 2007. Current trends in evolutionary multiobjective optimization. International Journal for Simulation and Multidisciplinary Design Optimization, 1(1), 1–8.
13. Kim KY, Kim HM. 2004. Optimization of the three-dimensional angled ribs with RANS analysis of turbulent heat transfer. Proceedings from the ASME Turbo Expo: Vienna, Austria.
14. Deb K, Agrawal, S, Pratap, A, Meyarivan, T. 2002. A fast elitist non-dominated sorting genetic algorithm for multi-objective optimization: NSGA-II, IEEE Transactions on Evolutionary Computation, 6(2), 182–197.
15. Deb K. 2001. Multi-Objective Optimization using Evolutionary Algorithms, 1st edn. John Wiley & Sons: New York.
16. Deb K, Agarwal, RB. 1995. Simulated binary crossover for continuous search space. Complex System, 9, 115–148.
17. Deb K, Goyal M. 1996. A combined genetic adaptive search (GeneAS) for engineering design. Computer Science and Informatics, 26(4), 30–45.

6

Mathematical methodology for optimization of the clamping forces accounting for workpiece vibratory behaviour

Riadh Chaari*, Moez Abennadher, Jamel Louati, and Mohamed Haddar

Mechanics Modelling and Production Research Unit (U2MP), University of Sfax, National Engineering School of Sfax (E.N.I.S.), B.P. 3038 Sfax, Tunisia

Abstract – This paper addresses the problem of determining the minimum clamping forces that ensure the dynamic fixturing stability. The clamping force optimization problem is formulated as a bi-level nonlinear programming problem and solved using a computational intelligence technique called particle swarm optimization (PSO). Indeed, we present an innovative simulation methodology that is able to study the effects of fixture-workpiece system dynamics and the continuously change due to material removal on fixturing stability and the minimum required clamping forces during machining. The dynamic behaviour of the fixtured workpiece subjected to time-and space-varying machining loads is simulated using a forced vibration model based on the regenerative vibrations of the cutter and workpiece excited by the dynamic cutting forces. Indeed, Material removal significantly affects the fixture-workpiece system dynamics and subsequently the minimum clamping forces required for achieving fixturing dynamic stability.

Key words: Machining, Fixturing, Stability, Vibration, Clamping, Optimization.

1 Introduction

In reality, machining processes such as milling are characterized by periodic forces. When the cutting frequency is near one of the natural frequencies of the fixture-workpiece system, consideration of dynamic effects in fixturing analysis is crucial. Therefore, the workpiece must remain stable in the fixture during machining in order to achieve operational safety and desired machining accuracy. In general, fixturing stability refers to the ability of the fixture to fully restrain the workpiece during the entire cutting process. As such, loss of contact and gross sliding during machining are considered to be indicators of an unstable workpiece and an inappropriate fixture design. These instabilities should be eliminated through correct design of the fixture and appropriate clamped forces.

This paper presents a dynamic model that study the effects of fixture-workpiece system dynamics and the continuously change due to material removal on fixturing stability and the minimum required clamping forces during machining. However, the majority of prior work treats the fixture-workpiece system as quasi-static and ignores the system dynamics. For example, De Meter et al. [1] considered the system deformation to be static for both clamping and machining and developed a linear programming (LP) model to compute the minimum required clamp pre-loads to prevent workpiece slip at the fixture-workpiece contacts during machining. Kang et al. [2] found the minimum clamping forces with the help of a contact stability index sensitivity matrix that is essentially a variation of the friction cone concept. Meyer and Liou [3] also developed an LP model in which the machining loads were varying and discretized over time but the system dynamics was ignored. Goals other than fixturing stability can be achieved through clamping force optimization. Hurtado and Melkote [4] formulated a multi-objective optimization model that found the minimum clamping loads to achieve workpiece shape conformability and fixture stiffness goals for a workpiece subjected to quasi-static machining forces. Chen et al. [5] presented a deformation control approach based on fixture layout design and clamping force optimization. In his approach, workpiece deformation measured accounting for chip removal effect, but the model contact is modelled by a linear spring based on Hertz contact theory. Zheng et al. [6, 7] presented a study of fixture stiffness. In first time Zheng et al. [6] developed a systematic FEA model to predict the fixture unit stiffness. In his model the contact problem is based on the penalty function method. In second time Zheng et al. [7] extended their study by an experimental study to identify the contact stiffness parameter. The normal and tangential contact stiffness is estimated by static and dynamic methods.

*e-mail: rchaari@yahoo.fr

Gui et al. [8] and Li and Melkote [9] examined the impact of clamping forces on workpiece location accuracy. Vishnupriyan et al. [10] proposed a genetic algorithm (GA) for optimizing the machining fixture layout for minimum machining error satisfying the tolerance requirements. A number of researchers (e.g., Wang et al. [11], Tao et al. [12], Liu and Strong [13]) proposed the idea of dynamic clamping because the minimum clamping loads required to stabilize a workpiece are dependent on the position of the cutting tool. They selected sampled toolpath points where the peak machining force was assumed to act and the clamping forces were adjusted using static linear or nonlinear optimization models.

Ignoring the material removal effect and thus the continuous change of system geometry and inertia can, as shown later in this paper, lead to erroneous the prediction of the required clamping forces for fixturing stability especially when a large percentage of material is removed. Several researchers considered system dynamics in their work. Mittal et al. [14] modelled a fixture-workpiece system using the dynamic analysis and design system (DADS) software and modelled the fixture-workpiece contact as a lumped spring-damper-actuator element. Liao and Hu [15] extended Mittal et al.'s work by including the workpiece compliance effect and contact friction through combined use of the finite element (FE) method and DADS. In both studies, separation between the fixture elements and the workpiece during machining was checked for a given set of clamping forces, but macro-slip was not considered. Li and Melkote [16] modelled the change of the workpiece gravity during machining. Kaya and Ozturk [17] applied an FE-based element death technique to simulate the chip removal process for fixture layout verification. However, both papers treat the fixture-workpiece system as quasi-static. On the other hand, Deng and Melkote [18] developed a model capable to determine the minimum required clamping forces for dynamically stable fixturing. This developed model takes account to the material removal effect and the fixture-workpiece system dynamics on the fixturing stability. But they didn't take account to the regenerative vibrations effect.

This paper addresses the problem of determining the minimum required clamping forces that ensure the fixture-workpiece dynamic stability during machining. The determination of the minimum required clamping forces takes account to the material removal effect and the fixture-workpiece system dynamics on the fixturing stability. The dynamic behaviour of the fixtured workpiece subjected to time-and space-varying machining loads is simulated using a forced vibration model based on the regenerative vibrations of the cutter and workpiece excited by the dynamic cutting forces. The clamping force optimization problem is formulated as a bi-level nonlinear programming problem and solved using a computational intelligence technique called particle swarm optimization (PSO). Through a simulation example, insight into the effects of fixture-workpiece system dynamics and its continuous change due to material removal on fixturing stability and the minimum required clamping forces is obtained.

2 Problem formulation and approach

This study focuses on machining workpieces surrounded by L locators and C clamps as shown schematically in Figure 1.

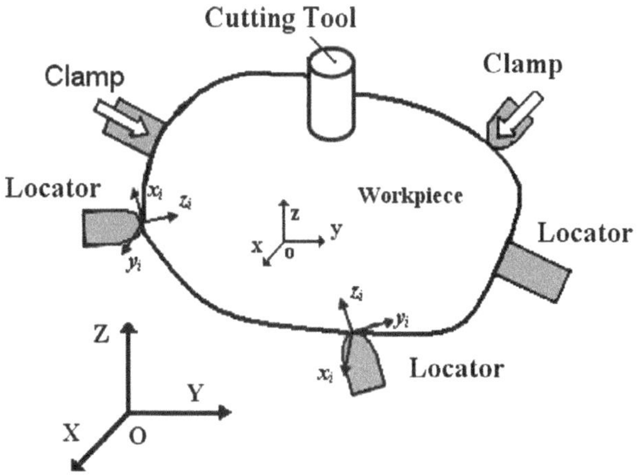

Figure 1. An arbitrary machining fixture-workpiece system [18].

Three coordinate systems are used to describe the position and orientation of the workpiece (see Figure 1): global coordinate system (*XYZ*), workpiece coordinate system (*xyz*), which is fixed to the mass center of the workpiece with its coordinate axes coinciding with the principal inertia axes of the initial workpiece shape, and local coordinate system (x_i y_i z_i), which is fixed at each fixture-workpiece contact. Matrices are represented in the workpiece coordinate system (*xyz*) unless noted otherwise. The workpiece undergoes dynamic and static motion during clamping and machining, slip, and/or lift-off at the contact regions. From the standpoint of operational safety and machining accuracy, lift-off and macro-slip at the fixture-workpiece contacts during machining are undesirable because they result in loss of total restraint of the workpiece in the fixture. To eliminate these two types of fixturing instabilities, the dynamic stability of the fixtured workpiece during machining must be properly assured. An important design variable is the clamping forces applied to the workpiece prior to the cutting operation. The clamping forces must be neither too large nor too small to prevent excessive workpiece distortion or instability during machining.

Therefore, the problem is to optimize the clamping forces. A procedure for checking the dynamic stability of the fixtured workpiece during machining needs to be established first. This procedure consists of a dynamic model for simulating the vibratory behaviour of the fixtured workpiece subjected periodic machining loads and computing the dynamic displacements at the fixture-workpiece contacts due to machining, and a static model for calculating the static deformation at the fixture-workpiece contacts due to clamping. In order to produce more accurate machining dynamics models for predicting the dynamic behaviour of the fixture-workpiece system, a developed FEM method are used and the continuous change of the system geometry and inertia during machining due to the material removal effect, must be taken account. Based on the stability check procedure, an optimization model (detailed by Figure 2) is formulated to search for the minimum set of clamping forces needed to stabilize the workpiece dynamics generated by the machining operation.

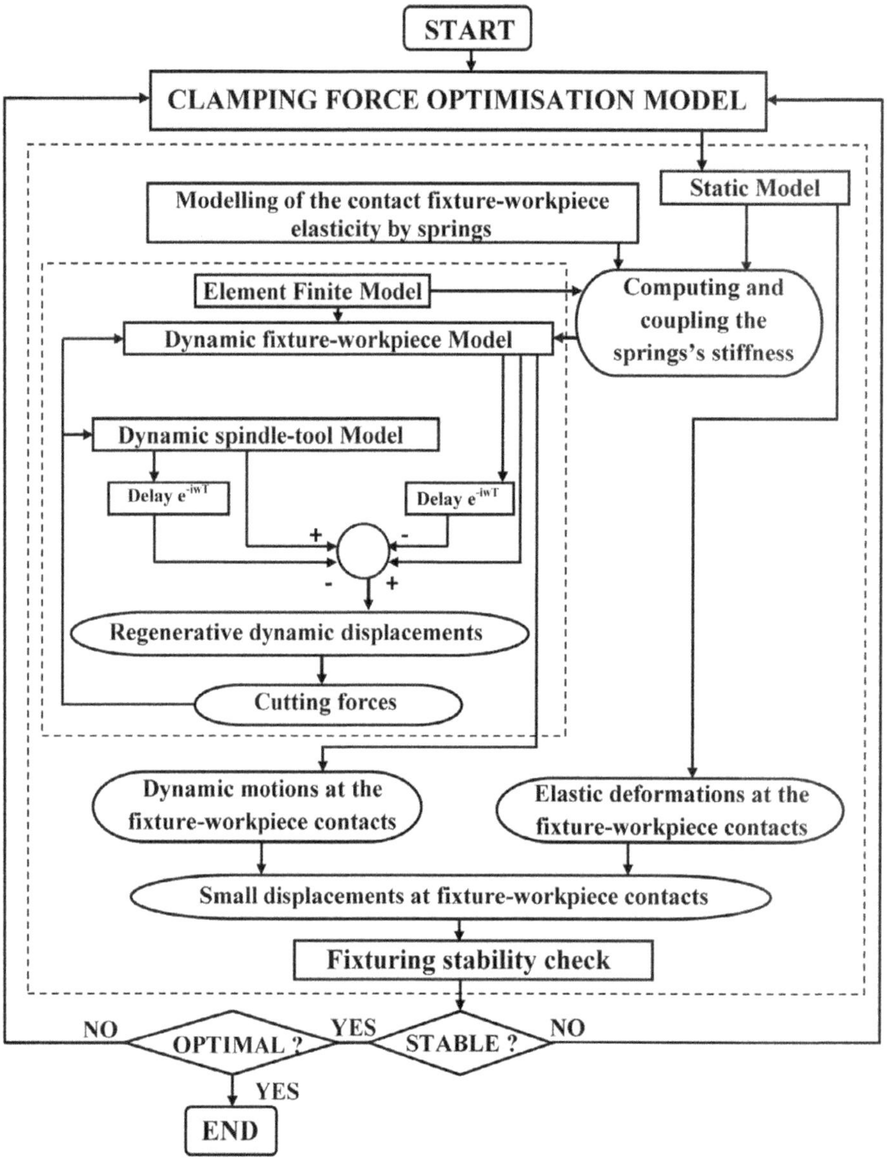

Figure 2. Procedure for optimization of the clamping forces.

2.1 Dynamic modelling

A multi-tooth machining operation such as milling is characterized by periodic cutting forces whose frequency of variation equals the tooth passing frequency. Therefore, the fixture-workpiece system is subjected to forced vibration caused by the dynamic cutting forces. When the excitation frequency is near the natural frequency of the fixture-workpiece system, the workpiece undergoes large motion that cannot be correctly predicted by a quasi-static analysis. Therefore, a dynamic model that is capable of predicting the vibrational behaviour of the fixture-workpiece system during cutting is required to analyze the fixturing stability in high-speed machining.

The regenerative chatter vibration often takes place in a milling process. In peripheral milling, due to the periodically interrupted cutting process, the dynamic cutting force oscillates periodically and excites vibrations between the cutter and workpiece.

A machining dynamics model is presented in this paper based on the regenerative vibrations of the cutter and workpiece excited by the dynamic cutting forces.

2.1.1 Assumptions

This developed procedure presented by Figure 2 is based on the following assumptions:

- The Hertzian model holds for all contacts.
- Coulomb friction is present at all contacts.
- The coupling between the vibration of the tool and the workpiece is done by the cutting force in a node.
- The materials of workpiece, the fixtures and the clamps are considered perfectly elastic and isotropic.
- All deformations are linearly elastic and small.
- The thermal error effects are considered negligible.

- The spindle model is simplified as a second-order spring-damper vibratory system.

2.1.2 Workpiece model

A workpiece FEM is established. The FEM is composed of parallelepiped elements with eight nodes and three degrees of freedom per node. In order to integrate the workpiece model to the spindle-tool set model at the interaction point representative of milling interaction, the dynamic equations of the part have to be established. Refer to [21], dynamic equations are obtained using Lagrange formulation associated with a finite element method:

$$[M_w]\ddot{q}_{wN} + [C_w]\dot{q}_{wN} + [K_w]q_{wN} = F_{\frac{c}{w}}(t). \tag{1}$$

The damping model used draws on Rayleigh viscous equivalent damping, which makes it possible to regard the damping matrix C_w as a linear combination of the mass matrix M_w, the spindle rigidity matrix K_w.

Where, $q_{wN} = [x \quad y \quad z] \in R^3$ the motion vector;

$[M_w] = \begin{bmatrix} m_w(t) & 0 & 0 \\ 0 & m_w(t) & 0 \\ 0 & 0 & m_w(t) \end{bmatrix} \in R^{3\times3}$ is the inertia matrix;

$[C_w] = \begin{bmatrix} C_{xw} & 0 & 0 \\ 0 & C_{yw} & 0 \\ 0 & 0 & C_{zw} \end{bmatrix}$ is Rayleigh viscous equivalent damping; $[K_w]$ is the intrinsic stiffness matrix of the workpiece;

$F_{c/w}(t) = \begin{Bmatrix} F_{X_{c/W}}(t) \\ F_{Y_{c/W}}(t) \\ F_{Z_{c/W}}(t) \end{Bmatrix} \in R^3$ is the vector of external cutting loads.

2.1.3 Spindle model

The spindle model is simplified as a second-order spring-damper vibratory system. The dynamic motion equations are obtained using Lagrange formulation.

$$[M_c]\ddot{q}_c + [C_c]\dot{q}_c + [K_c]q_c = F_{w/c}(t), \tag{2}$$

where, $q_c = [X \quad Y \quad Z] \in R^3$ motion vector;

$[M_c] = \begin{bmatrix} m_c & 0 & 0 \\ 0 & m_c & 0 \\ 0 & 0 & m_c \end{bmatrix} \in R^{3\times3}$ is the inertia matrix;

$[C_c] = \begin{bmatrix} C_{xc} & 0 & 0 \\ 0 & C_{yc} & 0 \\ 0 & 0 & C_{zc} \end{bmatrix}$; $[K_c]$ is the intrinsic stiffness matrix of the spindle model.

The damping model used draws on Rayleigh viscous equivalent damping, which makes it possible to regard the damping matrix C_c as a linear combination of the mass matrix M_c, the spindle rigidity matrix K_c

2.2 Dynamic cutting force model

A dynamic cutting force model of peripheral milling has been developed [22] and its brief mathematical formulation related to this paper is summarized in this section. The cutting forces excite vibrations of the cutter and workpiece simultaneously in the normal (*X*) and feed (*Y*) directions, causing dynamic displacements x_c, y_c and x_w, y_w of the cutter and workpiece, respectively. Therefore, the dynamic displacements are carried to the undeformed chip thickness direction v with the coordinate transformation as follows [24, 25]:

$$\begin{aligned} v_{ic} &= -x_c\cos(\varphi_i) - y_c\sin(\varphi_i) \\ v_{iw} &= -x_w\cos(\varphi_i) - y_w\sin(\varphi_i) \end{aligned}. \tag{3}$$

Based on the cutting force model proposed by the authors [22, 23], for down milling the total cutting force applied on the *i*th tooth of the cutter is given by:

$$\begin{cases} F_{ix} \approx -K'_S f_t a_{ixy} - K'_S a_{ixx}(\Delta x_c - \Delta x_w) - K'_S a_{ixy}(\Delta y_c - \Delta y_w) \\ \qquad -K'_S \partial t_i(j_i) a_{ix} \\ F_{iy} \approx K'_S f_t a_{iyy} + K'_S a_{iyx}(\Delta x_c - \Delta x_w) + K'_S a_{iyy}(\Delta y_c - \Delta y_w) \\ \qquad +K'_S \partial t_i(j_i) a_{iy} \end{cases}, \tag{4}$$

where K'_S is the cutting force coefficient [22, 23], and:

$$\begin{aligned} \Delta x_c &= x_c - x_c(-T),\ \Delta y_c = y_c - y_c(-T), \\ \Delta x_w &= x_w - x_w(-T),\ \Delta y_w = y_w - y_w(-T), \end{aligned} \tag{5}$$

$$\begin{aligned} a_{ixx} &= \left[0.5556(\sin(j_i))^{1.8} + 0.5c_1 j_i + 0.25c_1\sin(2j_i)\right]_{j_s}^{j_e}, \\ a_{ixy} &= \left[0.5j_i - 0.25\sin(2j_i) + 0.5556c_1(\sin(j_i))^{1.8}\right]_{j_s}^{j_e}, \end{aligned} \tag{6}$$

$$\begin{aligned} a_{iyx} &= \left[0.5\varphi_i + 0.25\sin(2\varphi_i) - 0.5c_1\varphi_i + 0.25c_1\sin(2\varphi_i)\right]_{\varphi_s}^{\varphi_e}, \\ a_{iyy} &= \left[0.5556(\sin(\varphi_i))^{1.8} - 0.5c_1\varphi_i + 0.25c_1\sin(2\varphi_i)\right]_{\varphi_s}^{\varphi_e}, \end{aligned} \tag{7}$$

$$\begin{aligned} a_{ix} &= \left[-\cos(j_i) + 1.25c_1(\sin(j_i))^{0.8}\right]_{j_s}^{j_e}; \\ a_{iy} &= \left[1.25(\sin(j_i))^{0.8} + c_1\cos(j_i)\right]_{j_s}^{j_e}, \end{aligned} \tag{8}$$

Because $0 \le \varphi \le \psi$, $\varphi_i = \varphi - \omega t_{ime} + (i-1)(2\pi/m)$ and $0 \le \varphi_i \le \Omega$ gives the extreme values of the parametric angle φ_i as:

$$\begin{aligned} \varphi_s &= \max(0, -\omega t_{ime} + (i-1)(2\pi/m)) \\ \varphi_e &= \min(\Omega, \psi - \omega t_{ime} + (i-1)(2\pi/m)) \end{aligned}, \tag{9}$$

$$\begin{aligned} \begin{Bmatrix} F_x(t_{ime}) \\ F_y(t_{ime}) \end{Bmatrix} &= \begin{bmatrix} -a_{xx}(t_{ime}) & -a_{xy}(t_{ime}) \\ a_{yx}(t_{ime}) & a_{yy}(t_{ime}) \end{bmatrix} \begin{bmatrix} \Delta x(t_{ime}) \\ \Delta y(t_{ime}) \end{bmatrix} \\ &+ \begin{bmatrix} -a_{xy}(t_{ime}) \\ a_{yy}(t_{ime}) \end{bmatrix} f_t + \begin{bmatrix} -a_x(t_{ime}) \\ a_y(t_{ime}) \end{bmatrix} \delta t_i(\varphi_i) \end{aligned}, \tag{10}$$

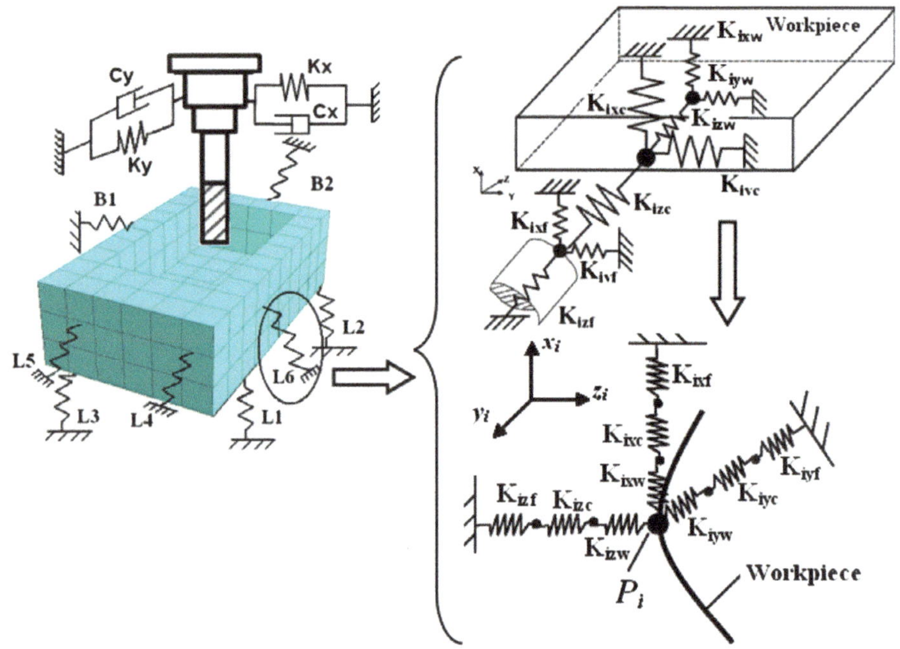

Figure 3. The mass-spring vibratory model of the coupled "spindle-workpiece-fixture" system.

where

$$\begin{gathered} a_{xx} = K'_s \sum_{i=1}^{m} a_{ixx},\ a_{xy} = K'_s \sum_{i=1}^{m} a_{ixy},\ a_{yx} = K'_s \sum_{i=1}^{m} a_{iyx} \\ a_{yy} = K'_s \sum_{i=1}^{m} a_{iyy},\ a_x = K'_s \sum_{i=1}^{m} a_{ix},\ a_y = K'_s \sum_{i=1}^{m} a_{iy} \end{gathered}. \quad (11)$$

The regenerative dynamic displacements of the cutter and workpiece can be expressed in the frequency domain as a product of the frequency domain as a product of the transfer function of the structure and the transfer function of the structure and the dynamic cutting forces [24, 25],

$$\begin{gathered} \begin{Bmatrix} \Delta x_c(i\omega) \\ \Delta y_c(i\omega) \end{Bmatrix} = (1 - e^{-iwT}) \begin{bmatrix} G^c_{xx}(i\omega) & 0 \\ 0 & G^c_{yy}(i\omega) \end{bmatrix} \begin{Bmatrix} F_x(i\omega) \\ F_y(i\omega) \end{Bmatrix} \\ \begin{Bmatrix} \Delta x_w(i\omega) \\ \Delta y_w(i\omega) \end{Bmatrix} = (1 - e^{-iwT}) \begin{bmatrix} G^w_{xx}(i\omega) & 0 \\ 0 & G^w_{yy}(i\omega) \end{bmatrix} \begin{Bmatrix} F_x(i\omega) \\ F_y(i\omega) \end{Bmatrix} \end{gathered} \quad (12)$$

Experimental results shown by [24, 25, 26] reveal that the measured cross transfer functions $G^c_{xy}(iw) \approx G^c_{yx}(iw) \approx G^w_{xy}(iw) \approx G^w_{yx}(iw) \approx 0$ were negligible. Where T is the tooth passing period, $G^c_{xx}(iw)$, $G^c_{yy}(iw)$ and $G^w_{xx}(iw)$, $G^w_{yy}(iw)$ are the direct transfer functions, respectively, of the cutter and the fixture workpiece-system in the X (normal) and Y (feed) directions.

2.3 Modelling of the contact fixture-workpiece elasticity by springs

At each fixture-workpiece interface, the overall compliance comes from three sources-fixture element, contact, and workpiece. Assuming that each source of compliance can be modelled as three linear springs in the x_i, y_i and z_i directions of the local frame, respectively. The torsional compliance from all sources is considered to be negligible. Therefore, as shown in Figure 3, the composite stiffness at the ith fixture-workpiece contact, p_i, is the summation in series of three stiffness components, k_{ijf}, k_{ijc} and k_{ijw} ($j = x$, y and z), representing the stiffness of the fixture element, contact and workpiece, respectively.

As shown in Figure 4, the workpiece structural compliance at the fixture-workpiece contact is approximated by three linear springs and is incorporated into the overall of the system by placing them in series with other springs representing the contact and fixture element compliances.

2.4 Calculation of the fixture-workpiece stiffness

The calculation of the local stiffness of the fixture-workpiece system is given by this equation [27, 30]:

$$k_{ij} = \left[(k_{ijf})^{-1} + (k_{ijc})^{-1} + (k_{ijw})^{-1}\right]^{-1}. \quad (13)$$

The spring constants for the workpiece structural stiffness (k_{ixW}, k_{iyW} and k_{izW} for $i = 1$–8) are obtained using a finite element (FE) method. The fixture-workpiece contact stiffness, k_{ijc} can be obtained by this equation [27]:

$$\begin{gathered} k_{zc} = \frac{3}{2}\left(\frac{16R(E^*)^2}{9}\right)^{1/3} P^{1/3} \\ k_{jc} = 8a\left(\frac{2-\upsilon_w}{G_w} + \frac{2-\upsilon_w}{G_w}\right)^{-1} \quad \text{for, } j = x \ \ or \ \ y \end{gathered} \quad (14)$$

where $\frac{1}{E^*} = \frac{(1-\upsilon_W^2)}{E_W} + \frac{(1-\upsilon_f^2)}{E_f}$, υ, E and G represent the Poisson's ratio, Young's modulus, and shear modulus of the material, respectively.

A fixture element (locator or clamp) is modelled as a short or long cylindrical cantilever, depending on its length-to-

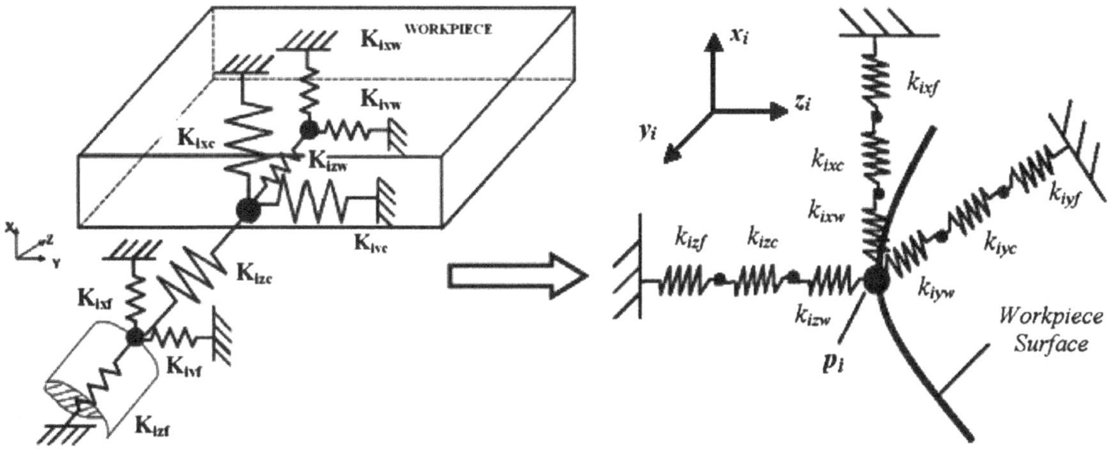

Figure 4. Composite stiffness at the *i*th fixture-workpiece contact.

diameter ratio. The structural stiffness of a fixture element is calculated by the following expression [27]:

$$k_{ixf} = k_{iyf} = \begin{cases} (3E_{if}I_{if})/L_{if}^3, & (L_{if}/d_{if}) \geq 3 \\ (3G_{if}A_{if})/4L_{if}, & (L_{if}/d_{if}) < 3 \end{cases}, \quad k_{izf} = \frac{E_{if}A_{if}}{l_{if}} \tag{15}$$

where E_{if}, G_{if}, L_{if}, A_{if} and I_{if} are the Young's modulus, shear modulus, length, cross-sectional and polar moment of inertia of the *i*th fixture element, respectively.

2.5 Static model

The principle of minimum complementary energy [19, 20] is utilized to develop the static model so that it can deal with an arbitrarily configured fixture-workpiece system without suffering from static indeterminacy of problems. The static model is then essentially a constrained nonlinear optimization model. The objective function of the model is the total complementary energy of the fixture-workpiece system subjected to clamping forces, and the design variables are the unknown contact forces of the system. Since the structural compliance of the workpiece and fixture elements are considered to be negligible relative to the contact compliance, the total complementary energy of the system is given by the sum of the stress energy of the compressed nonlinear virtual springs used to model the contact stiffness. Note that only constant clamping forces generated by force-controlled fixture are considered in this study. Therefore, denoting P_i, Q_{xi}, and Q_{yi} as the contact forces due to clamping loads in the z_i, x_i, and y_i directions, the static model can be written as:

Minimise Π_c

$$(P_i, Q_{xi}, Q_{yi}) = \sum_{i=1}^{(L+c)} \left(\int_0^{P_i} \delta z_i dP_i + \int_0^{Qx_i} \delta x_i dQ_{xi} + \int_0^{Qy_i} \delta y_i dQy_i \right), \tag{16}$$

$$\sum \bar{F} = \bar{0}, \sum \bar{M} = \bar{0}, \tag{17}$$

$$P_j = F_{cj} \text{ For } j = (L+1), \ldots, (L+C), \tag{18}$$

$$\sqrt{(Q_{xi})^2 + (Q_{yi})^2} - \mu_s^i P_i \leq 0 \text{ For } i = 1, \ldots, (L+c), \tag{19}$$

$$P_i \rangle 0 \text{ For } i = 1, \ldots, (L+c), \tag{20}$$

$$P_i - S_y(\Pi a_i^2) \leq 0 \text{ For } i = 1, \ldots, (L+c). \tag{21}$$

The constraint (17) represents the static equilibrium condition; (18) results from the fact that constant clamping forces are used; (19) represents the Coulomb friction law; (20) comes from the unilateral nature of a fixture-workpiece contact; (21) represents the non-yielding constraint on the contact stress where S_y is the yield strength of the workpiece material while a_i is the radius of the *i*th contact region and is a function of the normal force P_i. The forces are then substituted into the contact force-displacement relationship to calculate the contact elastic deformation due to clamping. The contact force-displacement relationship for a spherical-tipped fixture element pressed against a curved workpiece surface is given as follows [27]:

$$\delta_z = \left[\frac{9P^2}{16R(E^*)^2}\right], \delta_j = \frac{Q_j}{8a}\left(\frac{2-\upsilon_w}{G_w} + \frac{2-\upsilon_f}{G_f}\right). \tag{22}$$

For $j = x, y$.

Where, $R = \left(\frac{1}{R_w} + \frac{1}{R_f}\right)^{-1}$ is the relative curvature at the contact with R_w being the local radius of the workpiece surface and R_f being the tip radius of the fixture element; $\frac{1}{E^*} = \frac{(1-\upsilon_W^2)}{E_W} + \frac{(1-\upsilon_f^2)}{E_f}$, υ, E and G represent the Poisson's ratio, Young's modulus, and shear modulus of the material, respectively; the subscripts w and f refer to the workpiece and fixture elements, respectively; $a = \left(\frac{3PR}{4E^*}\right)$ is the radius of the contact region.

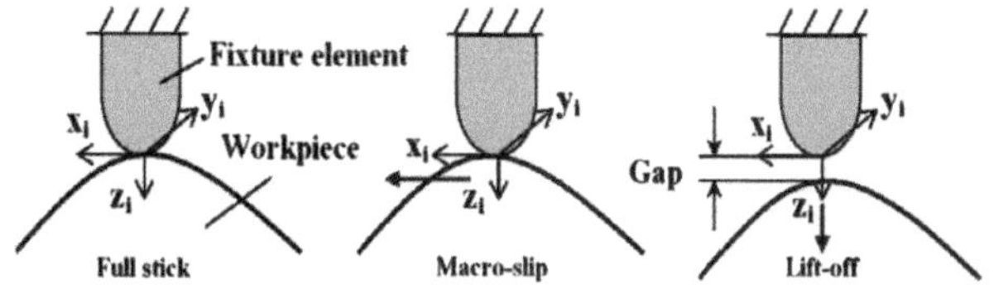

Figure 5. Dynamic contact interaction between workpiece and fixture element.

2.6 Fixturing stability criteria

The dynamic stability of the fixtured workpiece during machining can be analyzed by determining the status of the interaction between the workpiece and fixture elements (i.e., locators and clamps) at each contact. Prior to machining, the clamps are actuated and this causes localized elastic deformation at the fixture-workpiece contacts. Consequently, the workpiece assumes a static equilibrium position, from which it undergoes dynamic motion during machining. Superposition of the static elastic deformation due to clamping and the dynamic displacement due to machining at a fixture-workpiece contact gives the total motion of that contact. As shown in Figure 5, three types of contact status are possible: full stick, macro-slip and lift-off. Since total restraint of the workpiece by the fixture must be satisfied throughout the machining operation, lift-off of the workpiece from any fixture element and macro-slip of the workpiece at any contact at any instant are indicators of an unstable workpiece. In the local coordinate system (x_i y_i z_i) for the ith fixture-workpiece contact, lift-off is equivalent to a positive displacement of the workpiece in the z_i direction while macro-slip indicates that the Coulomb friction law is violated. Therefore, the two fixturing stability criteria can be stated mathematically as in equation (23), which must be satisfied for i = 1 to (L + C) during machining.

$$\max\{\Delta_{zi}(t)\} \le 0$$

$$\max\left\{\sqrt{[k_{xi}\Delta_{xi}(t)]^2 + [k_{yi}\Delta_{yi}(t)]^2} - \mu_s^i[k_{zi}|\Delta_{zi}(t)|]\right\} \le 0, \tag{23}$$

where, t is the machining time; $\Delta x_i(t)$, $\Delta y_i(t)$ and $\Delta z_i(t)$ are the superposed displacements of the ith fixture-workpiece contact in the x_i, y_i, and z_i directions, respectively; μ_s^i is the static friction coefficient at the ith contact; k_{xi}, k_{yi}, and k_{zi} are the local contact stiffness's in the x_i, y_i, and z_i directions, respectively.

The total motion of the ith is obtained as follows:

$$\Delta_{ij}(t) = d_{ij}(t) - \delta_{ij} \quad \text{For}\, j = x,\, y,\, z, \tag{24}$$

where, symbols d and δ represent the dynamic and static displacements of the workpiece, respectively.

2.7 Optimization clamping forces methodology

The overall approach used in this article is to determine the minimum required clamping forces during machining is summarized in the flowchart shown in Figure 2. It is seen that the approach is based on the fixturing dynamic stability analysis procedure and an optimization method to perform the task of searching for the minimum set of clamping forces. Development of the clamping force optimization model is presented in the next section.

2.7.1 Nonlinear optimization model

The objective function to be minimized is the 2-norm of the clamping force vector F_c. The number of components of this vector depends on the number of clamps C used in the fixture and thus $F_c \in R_c$. The task here is to search in a C-dimensional hyperbox for F_c with minimum length that satisfies the stability conditions given by equation (23). This clamping optimization task is then formulated as a bi-level nonlinear programming problem as follows:

$$\text{Minimize}, \theta(F_c) = \|F_c\|_2$$

$$\text{Subject to}, \Delta_{ji}(t) = d_{ji}(t) - \delta_{ji} \quad \text{for} \quad j = x, y, z, \tag{25a}$$

$$\max_t\{\Delta_{zi}(t)\} \le 0, \tag{25b}$$

$$\max_t\left\{\begin{matrix}\sqrt{[k_{xi}\Delta_{xi}(t)]^2 + [k_{yi}\Delta_{yi}(t)]^2} - \\ \mu_s^i[k_{zi}|\Delta_{zi}(t)|]\end{matrix}\right\} \le 0, \tag{25c}$$

$$L_b \le F_c \le U_b, \tag{25d}$$

where i = 1 to (L + C), L_b and U_b are the upper and lower bounds of F_c, respectively; δ is computed using equation (22) and the static optimization model given by equation (16).

This model includes two optimization tasks at two levels. The upper level is the minimization of the 2-norm of the clamping force vector and the lower level is the minimization of the total complementary energy of the fixture-workpiece system subjected to clamping forces. The constraint regions of the two optimization problems are implicitly dependent on each another.

2.7.2 Presentation of the PSO optimization model

PSO is an algorithm for finding optimal regions of complex search spaces through the interaction of individuals in a population of particles [28]. It was developed by Kennedy and Eberhart [29] based on the social behaviour metaphor.

This heuristic search algorithm is therefore used to solve equation (25)). The algorithm searches a solution space by adjusting the trajectories of individual vectors, called "particles" as they are conceptualized as moving points in multidimensional space. Each particle is assigned a randomized velocity. These particles are assigned velocities and are "flown" through the problem hyperspace. At the kth iteration, the best solution (pbestx) and the corresponding objective function value (pbest) that each particle has achieved so far are stored. Also stored is the index (gbest) of the particle that has the best

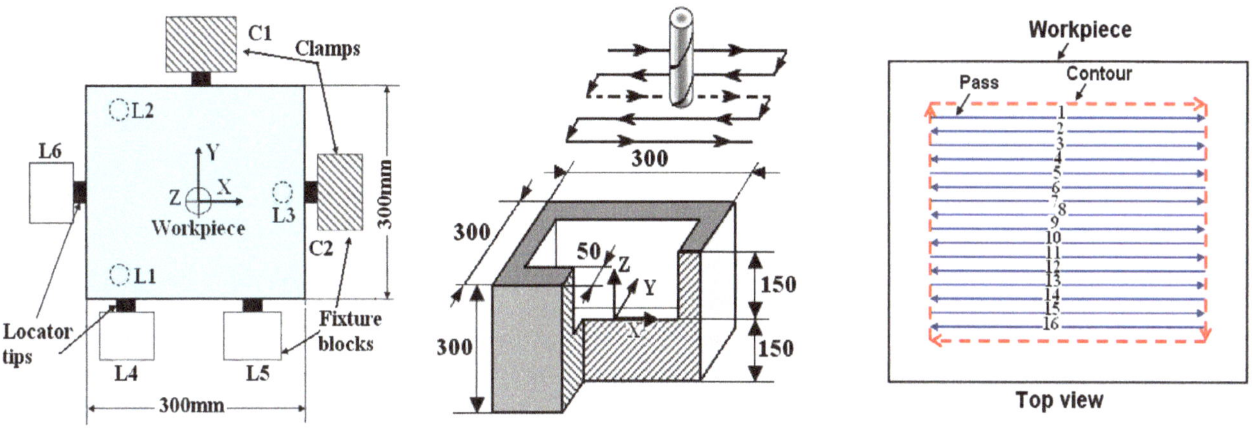

a) The original workpiece and fixture layout
b) Final part
c) The tool path

Figure 6. Final part and fixture layout (L1–L6: locators; C1–C2: clamps).

performance in the group up to that iteration. Equations (26) and (27) are used to update the velocity (v) and the position (x) of the ith particle at the $(k+1)$th iteration.

$$\overline{v}_i^{k+1} = w_i^{k+1}.\overline{v}_i^k + c_1\text{rand}(\,)(\text{pbest } x_i^k - \overline{x}_i^k) + c_2\text{Rand}(\,)(\text{pbest } x_{\text{gbest}}^k - \overline{x}_i^k) \quad (26)$$

where w is the inertia weight; c_1 and c_2 are positive constants; rand and Rand are generators of random numbers between 0 and 1.

$$\overline{x}_i^{k+1} = \overline{x}_i^k + \overline{v}_i^{k+1}. \quad (27)$$

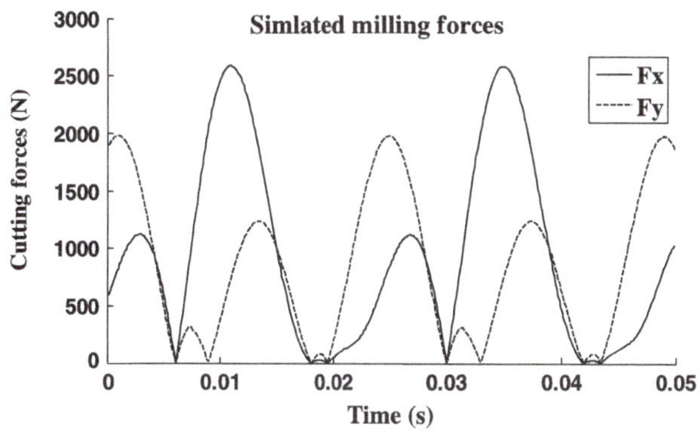

Figure 7. Simulated milling forces.

3. Simulation example

A pocketing end milling simulation example is used to illustrate the approach developed in this paper for to verify the dynamic stability of a fixtured workpiece and to search the minimum set of clamping forces needed during machining. This example considers an end milling operation. As illustrated in Figure 6, the original workpiece is a solid block of aluminium 7075 and the operation involves milling a step cut on the top surface of the workpiece. The operation consists of 30 depth levels along the Z axis with 16 tool passes along the X axis at each level. The removed volume is about 23% of the total volume of the original workpiece.

Cutting conditions used in this example are given by the following Table 1.

The spatial coordinates of the fixture-workpiece contacts in the (xyz) frame are listed in Table 2. All fixture (clamps) elements are identical with a cylindrical body (radius = 20 mm and length = 30 mm) and a spherical tip (radius = 19.8 mm).

The material properties of the workpiece and the fixture elements are given in Table 3.

The instantaneous machining forces, shown in Figure 7 (during two tool revolutions), are obtained from a milling force model presented in this paper (derived from [22, 23]). Note that the effect of the helix angle on the cutting forces is neglected in

Table 1. Cutting conditions (tool: end mill, Ø 25.4 mm, 4-flute, and 30º helix).

Feed rate (mm/s)	Axial depth (mm)	Radial depth (mm)	Spindle speed (rpm)
100	5	12.5	2500

the force model, resulting in a zero force in the Z direction. In general, the cutting force in the Z (axial) direction in milling is small especially for the small axial depth of cut (5 mm) and relatively small helix angle (30º) employed in the current example.

A modal analysis of the fixture-workpiece system was performed by the use of FEM for the initial and final workpiece geometries, and the first four natural frequencies are presented in Table 4. It can also be seen that the natural frequencies increase (become higher in the end of the milling operation) as the workpiece loses material.

A spindle speed of 2500 rpm corresponds to tooth passing frequency 166.7, which is near the lowest natural frequency of the initial fixture-workpiece system 206, was finally selected for the example so as to disqualify a quasi-static analysis but not in the immediate vicinity of the peak areas to avoid resonance.

Figure 8 shows the convergence of the PSO search for the example machining operation. The two plots are for the first

Table 2. Coordinates of fixture-workpiece contacts

Locator	Coordinate (x, y, z) (mm)	Clamp	Coordinate (x, y, z) (mm)
L1	(−100, −100, −150)	C1	(0, 150, −50)
L2	(−100, 100, −150)	C2	(150, 0, −50)
L3	(100, 0, −150)		
L4	(−100, −150, −50)		
L5	(100, −150, −50)		
L6	(−150, 0, −50)		

Table 3. Material properties.

Parameter	Workpiece	Fixture elements
Material	Aluminium 7075-T6	Steel 1018
Density(kg/m^3)	2700	
Young's modulus (GPa)	70.3	201
Poisson's ratio	0.354	0.296
Yield strength (MP)	500	
Static coefficient of friction	0.35	

Table 4. Natural frequencies.

	Natural frequency (Hz)			
	1	2	3	4
Begening	206	244	311	378
End	252	288	355	441
The spindle speeds which correspond to natural frequencies (rpm)				
	1	2	3	4
Begening	3090	3660	4665	5670
End	3780	4320	5325	6615

and last tool passes, respectively. The optimal clamping forces found for the two passes are given below the plots. The first observation confirms the search power of the PSO algorithm and the appropriateness of the values of design parameters listed in Table 5. The second observation indicates the significance of the material removal effect on the fixture-workpiece dynamics and consequently the minimum required clamping forces. It is suggests the need for time-varying clamping forces during machining.

Then the Figures 9 and 10 shows the fixturing dynamic stability of the system operating under the minimum required clamping forces simulated at the first and last pass. The horizontal axis of both plots in Figures 9 and 10 represents the fixture-workpiece contact indices which range from 1 to 8 corresponding to L1, L2, L3, L4, L5, L6, C1, and C2 (see Figure 6a), respectively. The vertical axes of plots in Figures 9 and 10 stand for the left hand sides of the two fixturing dynamic stability criteria (given in equation 23), respectively. Therefore, a stem above the zero horizontal line in Figures 9 and 10 is an indicator of lift-off or macro-slip at the corresponding fixture-workpiece contact. The height of a stem represents the degree of fixturing dynamic stability (if below zero) or instability (if above zero) of the contact. The Figures 9 and 10 illustrates that the two clamping forces simulated in the first and last pass are able to stabilize the workpiece during machining.

Note that the results of the convergence of the PSO search algorithm are:

- For the first pass: ${}^{1}{}_{1}F_{c}^{\text{opt}} = [6830 \quad 4330](N)$, i.e., $Fc1$ = 6830 N and $Fc2$=4330N.
- For the first pass: ${}^{2}{}_{2}F_{c}^{\text{opt}} = [3843 \quad 2180](N)$, i.e., Fc1 = 3843 N and $Fc2$ = 2180N.

Figures 11–18 compare the dynamic motions of the fixture-workpiece system at the contact points during the first and last passes when their corresponding "minimum" clamping forces (see Figure 8) are used. It is seen that the dynamic motion of the fixtured workpiece (at the contact points) during the first tool pass is larger than during the last pass. This says that the workpiece experiences smaller motion as it loses material. The reason for this may be found by examining the change in the system natural frequencies as the material is cut away. The lowest natural frequency of the system is 206 Hz during the first pass and 252 Hz during the last pass. Although the change is not very large, the system inertia decreases significantly. The small change in the natural frequency is because the system stiffness also decreases as lower "minimum" clamping forces are used for the last pass. The dominant excitation frequency (i.e., tooth passing frequency) for this simulation example is 166.7 Hz, which is near the lowest natural frequency of the initial fixture-workpiece system. As the workpiece loses material, the machining operation moves away from resonance and thus produces smaller system dynamic motion.

Note that workpiece motions (at the contact points) during only two tool revolutions are shown in each plot. As seen from these plots (Figures 11–18), the workpiece vibrations change significantly during the last pass in which 23% material has been removed. From the results, the key findings are as follows:

- The fixture-workpiece system during the end milling operation presents significant dynamics when certain spindle speeds are used such that the excitation frequency is in the vicinity of a natural frequency of the system. In this scenario, consideration of the fixture-workpiece system dynamics is critical for an accurate analysis of the system.
- Material removal in machining continuously changes the properties of the fixture-workpiece system, e.g., inertia and geometry. When a large portion of material is removed from the workpiece, the fixture-workpiece system behaves quite differently (in the example, the amplitudes of workpiece vibrations (at contact points) were found to be higher during the first tool pass than during the last pass). As a result, higher clamping forces are required to stabilize the workpiece during the first pass than during the last pass.
- Because of the material removal effect, dynamic clamping is an option to achieve the best possible performance of a machining fixture-workpiece system. In addition, allowing different forces at different clamps with a good combination of the clamping forces can improve the overall fixture performance.

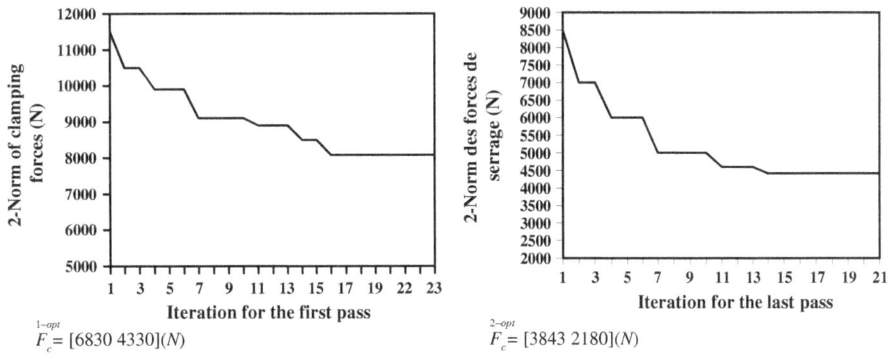

Figure 8. Convergence of the PSO search algorithm.

Table 5. Parameter values used in the PSO model.

Parameter	Value	Comments
Dimension of the solution space	2	Two clamps
Size of solution space	Lower bound: 1000 N	Specified by the user
	Upper bound: 8000 N	Can be updated during search
Population size	30	Depending on the size of the solution space
Inertia weight w	0.4–0.9	Case specific
Coefficients c_1, c_2	1;1	Case specific
Stop criteria	maxIt = 50	maxIt-maximun number of iterations
	$m = 8$	m is the number of consecutive iterations without considerable improvement in the solution

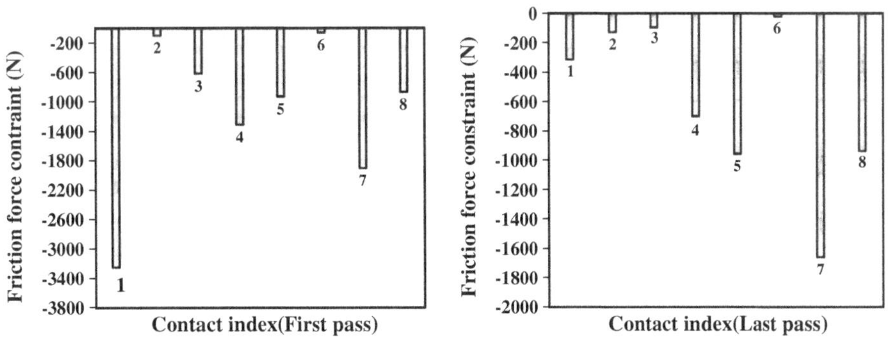

a) Lift-off check at the first pass b) Lift-off check at the last pass

Figure 9. Lift-off check.

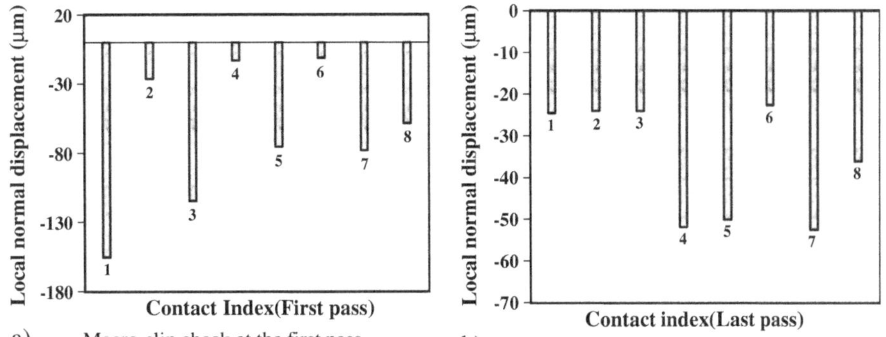

a) Macro-slip check at the first pass b) Macro-slip check at the last pass

Figure 10. Macro-slip check.

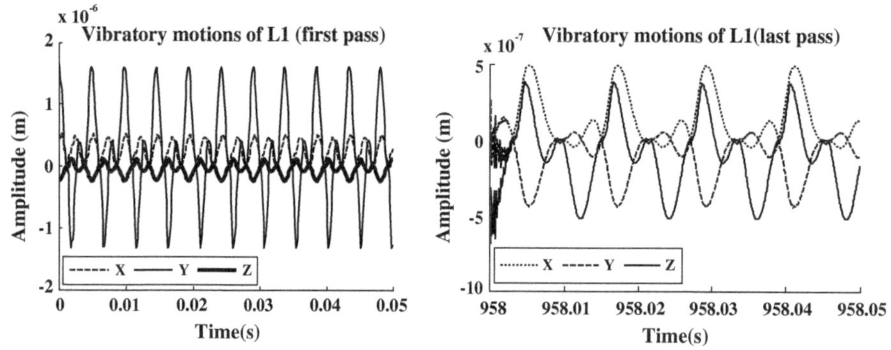

Figure 11. Dynamic motion at contact point L1 during the first and last depth level along *Z*.

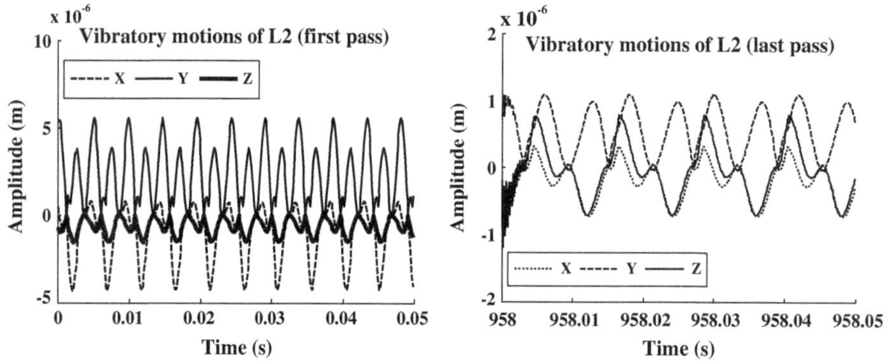

Figure 12. Dynamic motion at contact point L2 during the first and last depth level along *Z*.

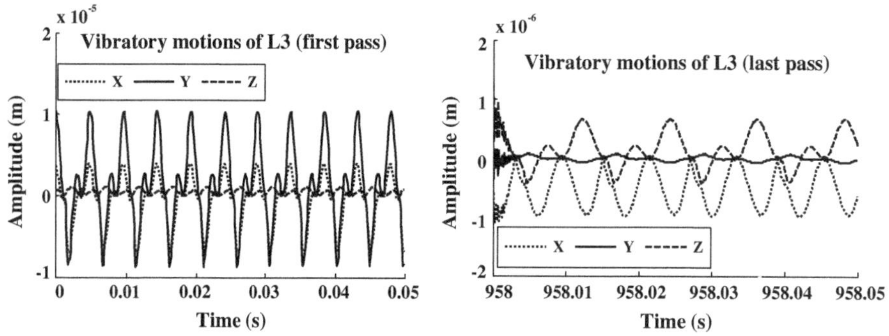

Figure 13. Dynamic motion at contact point L3 during the first and last depth level along *Z*.

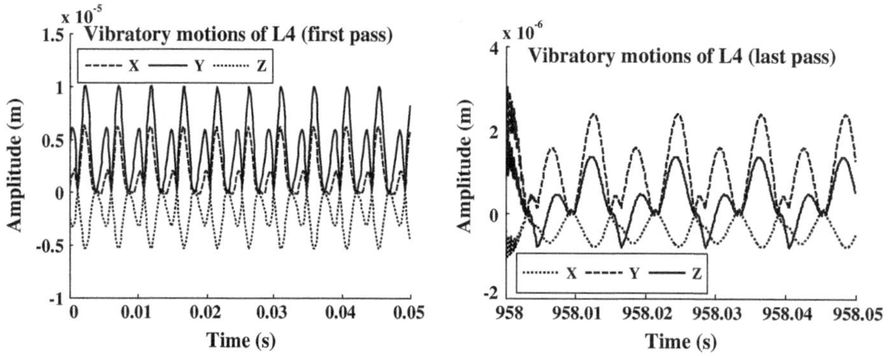

Figure 14. Dynamic motion at contact point L4 during the first and last depth level along *Z*.

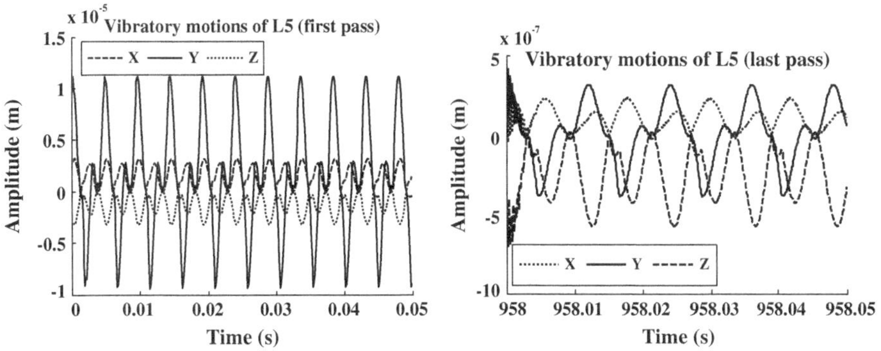

Figure 15. Dynamic motion at contact point L5 during the first and last depth level along *Z*.

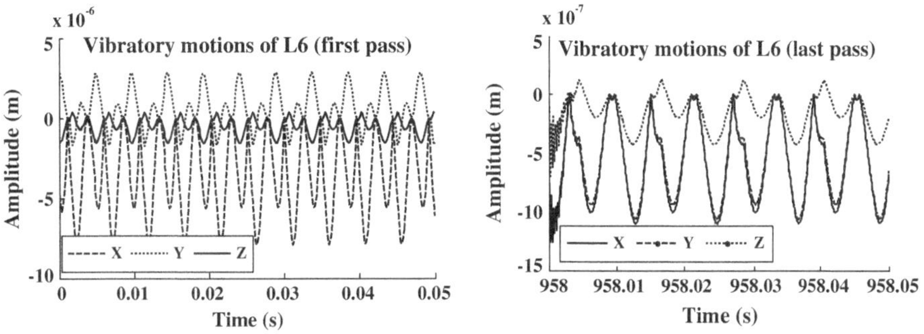

Figure 16. Dynamic motion at contact point L6 during the first and last depth level along *Z*.

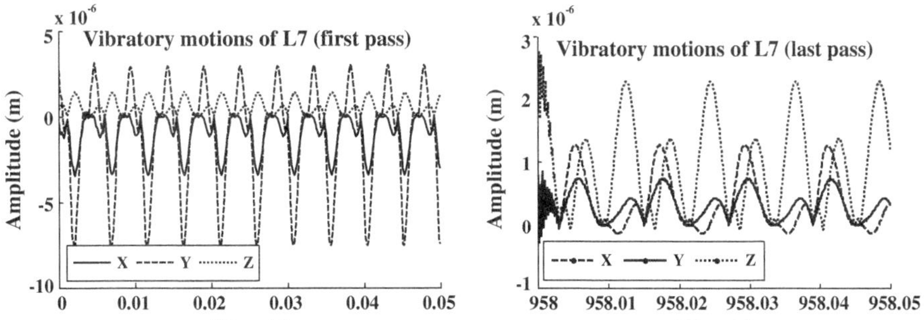

Figure 17. Dynamic motion at contact point L7 during the first and last depth level along *Z*.

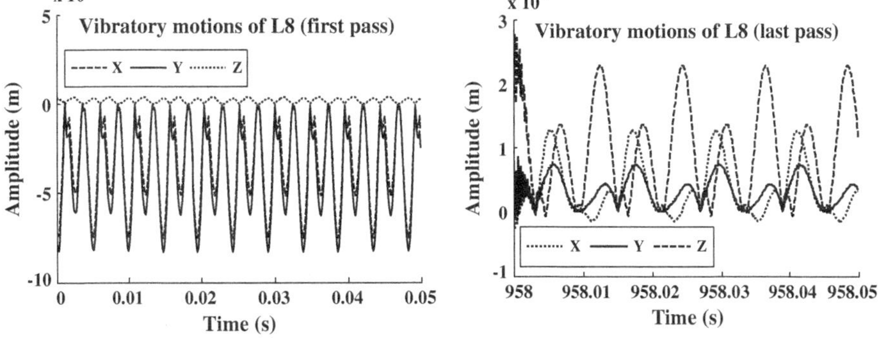

Figure 18. Dynamic motion at contact point L8 during the first and last depth level along *Z*.

4. Conclusion

A systematic procedure for optimization of the clamped forces of a fixture-workpiece system in machining has been established in this paper. Through this study, it is seen that:

- In determination of the minimum required clamping forces to ensure the fixturing dynamic stability of the fixture-workpiece system in the example milling operation, consideration of fixture-workpiece system dynamics is found to be crucial when the excitation frequency is in the vicinity of the natural frequency of the system.
- Material removal significantly affects the fixture-workpiece system dynamics and subsequently the minimum clamping forces required for achieving fixturing dynamic stability. The minimum required clamping forces are found to decrease as material is removed from the workpiece.
- Satisfaction of fixturing stability criteria requires good combinations of clamping forces, which can be determined from the model and approach developed in this paper.

Future research will focus on the influence of the structure compliance (for relatively flexible parts) on the stability of the fixtured workpiece.

References

1. De Meter EC, Xie W, Choudhuri S, Vallapuzha S, Trethewey MW. 2001. A model to predict minimum required clamp preloads in light of fixture-workpiece compliance. International Journal of Machine Tools and Manufacture, 41, 1031–1054.
2. Kang Y, Rong Y, Yang JA. 2003. Geometric and kinetic model based computer-aided fixture design verification. Journal of Computing and Information Science in Engineering, Transactions of the ASME, 3, 187–199.
3. Meyer RT, Liou FW. 1997. Fixture analysis under dynamic machining. International Journal of Production Research, 5, 1471–1489.
4. Hurtado JF, Melkote SN. 2002. A model for synthesis of the fixturing configuration in pin-array type flexible machining fixtures. International Journal of Machine Tools and Manufacture, 42, 837–849.
5. Chen W, Ni L, Xue J. 2007. Deformation control through fixture layout design and clamping force optimization. International Journal of Advanced Manufacturing Technology, 38, 9–10, 860–867.
6. Zheng Y, Rong Y, Hou Z. 2008. The study of fixture stiffness-part I: a finite element analysis for stiffness of fixture units. International Journal of Advanced Manufacturing Technology, 36, 865–876, DOI:10.1007/s00170-006-0908-5.
7. Zheng Y, Hou Z, Rong Y. 2008. The study of fixture stiffness-Part II: contact stiffness identification between fixture components. International Journal of Advanced Manufacturing Technology, 38, 19–31. DOI:10.1007/s00170-007-1077-x.
8. Gui X, Fuh JYH, Nee AYC. 1996. Modeling of frictional elastic fixture-workpiece system for improving location accuracy. IIE Transactions, 28, 821–827.
9. Li B, Melkote SN. 2001. Fixture clamping force optimization and its impact on workpiece location accuracy. The International Journal of Advanced Manufacturing Technology, 17, 104–113.
10. Vishnupriyan S, Majumder MC, Ramachandran KP. 2010. Optimization of machining fixture layout for tolerance requirements under the influence of locating errors. International Journal of Engineering, Science and Technology, 2(1), 152–162.
11. Wang YF, Wong YS, Fuh JYH. 1999. Off-line modeling and planning of optimal clamping forces for an intelligent fixturing system. International Journal of Machine Tools and Manufacture, 39, 253–271.
12. Tao ZJ, Kumar AS, Nee AYC. 1999. Automatic Generation of Dynamic Clamping Forces for Machining Fixtures. International Journal Production Research, 37(12), 2755–2776.
13. Liu JJ-X, Strong DR. 2003. Machining fixture verification for nonlinear fixture systems. International Journal of Advanced Manufacturing Technology, 21, 426–437.
14. Mittal RO, Cohen PH, Gilmore BJ. 1991. Dynamic modeling of the fixture-workpiece system. Robotics & Computer-Integrated Manufacturing, 8(4), 201–217.
15. Liao YG, Hu SJ. 2000. Flexible multibody dynamics based fixture-workpiece analysis model for fixturing stability. International Journal of Machine Tools & Manufacture, 40, 343–362.
16. Li B, Melkote SN. 2001. Optimal fixture design accounting for the effect of workpiece dynamics. The International Journal of Advanced Manufacturing Technology, 18, 701–707.
17. Kaya N, Öztürk F. 2003. The application of chip removal and frictional contact analysis for workpiece-fixture layout verification. International Journal of Advanced Manufacturing Technology, 21, 411–419.
18. Deng H, Melkote SN. 2006. Determination of minimum clamping forces for dynamically stable fixturing. International Journal of Machine Tools & Manufacture, 46, 847–857.
19. Bauchau OA. 2004. Aerospace Structural Analysis – Chapter 7: Variational and Energy Principles. Georgia Institute of Technology: Atlanta, USA.
20. Li B, Melkote SN. 1999. Elastic contact model for the prediction of workpiece-fixture contact forces in clamping. Journal of Manufacturing Science and Engineering, Transactions of the ASME, 121(3), 485–493.
21. Mané I, Gagnol V, Bouzgarrou BC, Ray P. 2008. Stability-based spindle speed control during flexible workpiece high-speed. International Journal of Machine Tools & Manufacture, 48, 184–194.
22. Liu X, Cheng K, Webb D, Luo XC. 2002. Improved dynamic cutting force model in peripheral milling-Part 1: theoretical model and simulation. International Journal of Advanced Manufacturing and Technology, 20, 631–638.
23. Liu X, Cheng K, Webb D, Longstaff AP, Widiyarto HM, Jiang X-Q, Blunt L, Ford D. 2004. Improved dynamic cutting force model in peripheral milling-Part 2: Experimental verification and prediction. International Journal of Advanced Manufacturing and Technology, 24, 794–805.
24. Altintas Y, Budak E. 1995. Analytical prediction of stability lobes in milling. Annals CIRP, 44(1), 357–362.
25. Altintas Y. 2000. Manufacturing Automation: Metal Cutting Mechanics, Machine Tool Vibrations, and CNC Design. Cambridge University Press.
26. Budak E, Altintas Y. 1998. Analytical prediction of chatter stability in milling; Part 1: General formulation, Part 2: Application of the general formulation to common milling systems. Journal of Dynamic Systems Measurement and Control, 120(1), 22–36.

27. Johnson KL. 1985. Contact Mechanics. Cambridge University Press: UK.
28. Clerc M, Kennedy J. 2002. The particle swarm-explosion, stability, and convergence in a multidimensional complex space. IEEE Transactions on Evolutionary Computation, 6, 58–73.
29. Kennedy J, Eberhart RC. 1995. Particle swarm optimization, in Proceedings of the IEEE International Conference on Neural Networks, vol. 4, Perth, Australia, IEEE Service Center, Piscataway, NJ. p. 1942–1948.
30. Jacobson B, Kalker JJ. 2000. Rolling Contact Phenomena. Springer: Wien, New York. p. 19–57.

7

Efficient supersonic air vehicle design using the Service-Oriented Computing Environment (SORCER)

Scott A. Burton[1,*], Edward J. Alyanak[2], and Raymond M. Kolonay[2]

[1] American Optimization LLC, Springboro, Ohio 45066, USA
[2] Aerospace Systems Directorate Air Force Research Laboratory, Wright-Patterson Air Force Base, Ohio 45433, USA

Abstract – The Air Force Research Lab's Multidisciplinary Science and Technology Center is investigating conceptual design processes and computing frameworks that could significantly impact the design of the next generation efficient supersonic air vehicle (ESAV). The ESAV conceptual design process must accommodate appropriate fidelity multidisciplinary engineering analyses (MDAs) to assess the impact of new air vehicle technologies. These analyses may be coupled and computationally expensive, posing a challenge due to the large number of air vehicle configurations analyzed during conceptual design. In light of these observations, a design process using the Service-Oriented Computing Environment (SORCER) software is implemented to combine propulsion, structures, aerodynamics, aeroelasticity, and performance in an integrated MDA. The SORCER software provides the automation and tight integration to grid computing resources necessary to achieve the volume of appropriate fidelity analyses required. Two design studies are performed using a gradient-based optimization method to produce long and short range ESAV wing designs. The studies demonstrate the capability of the ESAV MDA, the optimization algorithm, and the computational scalability and reliability of the SORCER software.

Key words: Optimization, Air vehicle design, SORCER.

1 Introduction

The ability of a new aircraft design to meet various requirements is largely dictated by decisions made during the conceptual design phase [1, 2]. Specifically, it is the conceptual design that determines the majority of the aircraft's life cycle costs [3]. It is therefore advantageous to assess as many air vehicle designs with the highest fidelity analyses possible given the available resources (e.g., human, schedule, and computational) This tradeoff between number of designs analyzed and available resources has typically favored the use of low fidelity closed-form equations and statistical models in conceptual design. These computationally inexpensive tools facilitate the assessment of large numbers of designs, but typically lack the fidelity to uncover significant engineering risks or are unable to assess the impact of new technologies.

In recent years, advances in computer processing speed and the prevalence of grid computing resources in industry have opened the door to more computationally expensive analyses during conceptual design [1, 4]. The conceptual design process described herein is based on the availability of such resources and incorporates analyses from several engineering disciplines. The efficient supersonic air vehicle (ESAV) multidisciplinary analysis (MDA) is used in a design example that combines techniques from traditional conceptual and preliminary design processes. Details of the ESAV MDA are discussed in Section 2.

While the existence of grid computing facilities is a necessary condition for performing large-scale conceptual design studies, it is not sufficient. Software to integrate the different engineering analysis codes and coordinate the execution of thousands of analyses remains an area of research. The study herein uses the Service-Oriented Computing Environment (SORCER) software to perform these functions [5]. Salient features of the SORCER ESAV application are discussed in Section 3. The ESAV MDA is exercised on two optimization problems described in Section 4. Lastly, results of the ESAV design studies and conclusions are presented in Sections 5 and 6, respectively.

2 Multidisciplinary analysis

The MDA described herein is a blend of conceptual and preliminary analysis methods from geometry, propulsion, structures, aerodynamics, aeroelasticity, and performance

*e-mail: scottaburton@amopti.com

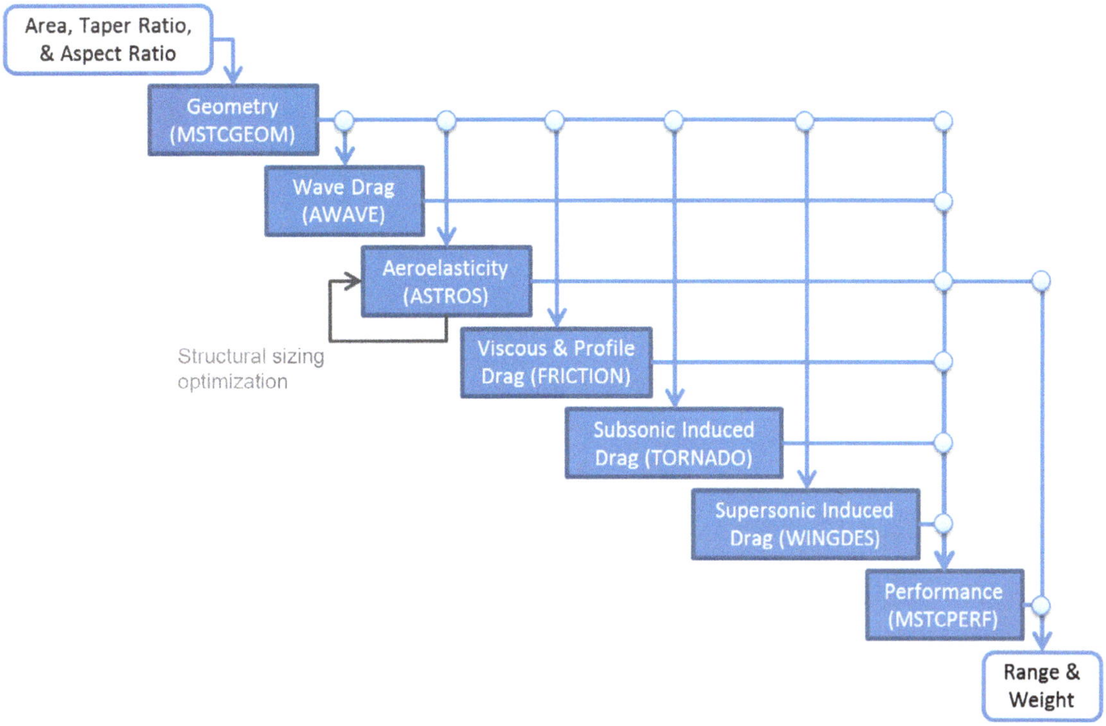

Figure 1. The ESAV MDA N^2 diagram includes geometry generation, an aerodynamic analysis, an aeroelastic analysis with structural sizing optimization, and a performance analysis.

disciplines. The analysis process and data flow is shown in the ESAV N^2 diagram in Figure 1.

The process begins by parametrically generating geometry, mesh, and input data suitable for several different analyses at varying fidelities. The geometry is used as input to compute several figures of merit of the air vehicle, which include the air vehicle drag polars, design mass, range, and aeroelastic performance. The different responses are evaluated for several flight conditions and maneuvers. These responses are then used to construct the objective and constraints of the multidisciplinary optimization (MDO) problem.

2.1 Airframe geometry

The purpose of the airframe geometry application, MSTCGEOM, is to create geometry data for the physics-based analyses in the MDA. MSTCGEOM is compiled Matlab™ code and is considered medium fidelity [6]. The application does not replace or replicate the extensive functionality of a commercial CAD package (i.e., a high fidelity geometry application); however, it does go beyond the functionality of a typical conceptual design tool. MSTCGEOM parametrically creates mid-plane panel aerodynamic models and FEM analysis and design models. In the case of ASTROS or NASTRAN, these FEM models may include static and dynamic aeroelastic load cases. In addition, composite structural elements and distributed masses to represent sub-systems may be a part of the FEM models. The geometric features of the aircraft modeled include the fuselage and all of the lifting aircraft components (wings, vertical stabilizer, and horizontal stabilizers). The physics-based analyses that use the MSTCGEOM output include ASTROS, FRICTION, TORNADO, AWAVE, WINGDES, and MSTCPERF [7–9].

MSTCGEOM is computationally inexpensive to execute and produces the required fidelity output for the ESAV analyses described herein. In this manner, MSTCGEOM is a bridge between conceptual and preliminary design fidelities. MSTCGEOM may be used to make both parametric and topological changes to the airframe wing during the MDO. Figure 2 shows an example of the MSTCGEOM application output for the baseline ESAV.

MSTCGEOM works by building up a series of major components such as cross sections, wing sections, point masses, and material definitions into assemblies such as wing components and a fuselage. Each of these components and assemblies are objects that contain methods that return information to support analysis model generation for a specific fidelity or type of model. For example, a cross section object may return mid-plane information to support the generation of a linear aerodynamic panel model. Likewise, the same cross section object can also return a completely discretized representation of itself for use in a FEM (i.e., the logic to produce both models exists in the same object). Any aerodynamic or structural mesh that is produced may be refined based on the user's requirements. The level of refinement used in the ESAV analyses is accurate enough to produce the correct physics-based global trends, but is not sufficient to identify stress concentrations that may be used in fatigue life prediction.

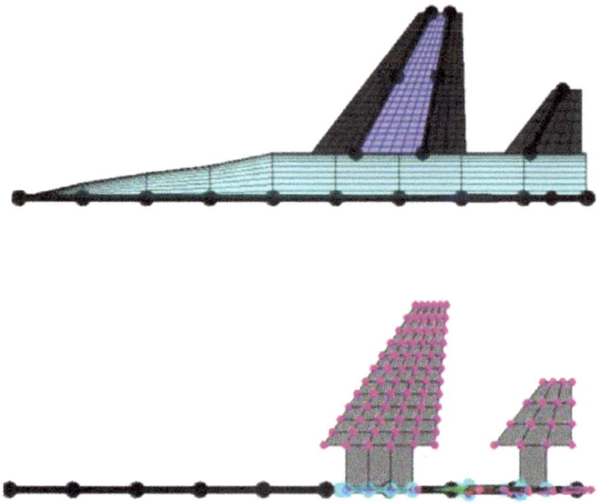

Figure 2. The MSTCGEOM application creates discrete geometry for ASTROS, FRICTION, TORNADO, AWAVE, WINGDES, and MSTCPERF applications (baseline ESAV half-span shown).

2.2 Aeroelasticity

Stress analyses and aeroelastic sizing for the ESAV are performed for four static aeroelastic maneuver cases at an altitude of 10,000 ft. Specifically, they are the following: (1) a 9.0 g pull-up at Mach 0.9; (2) a 7.2 g pull-up at Mach 1.2; (3) a negative 3.0 g push-over at Mach 1.2; and (3) an anti-symmetric 100 deg s^{-1} aileron roll at Mach 1.2. ASTROS is used to perform the aeroelastic analysis and structural sizing optimization that results in an optimized structural weight subject to stress constraints. The structural sizing optimization is considered a sub-optimization in regards to the overall optimization problem discussed in Section 4.1. To that end, the structural sizing optimization is an integral part of the MDA and is performed for each configuration analyzed. The structural sizing sub-optimization problem has 57 design variables that are distributed over the wing and tail surface spar, rib, and skin thicknesses. The ESAV is modeled with aluminum and 70 ksi is used as the maximum allowable stress.

2.3 Aerodynamics

To assess the aero performance of a configuration, estimations of zero-lift drag and induced drag are required. The zero lift drag components estimated are viscous and profile drag. They are calculated using the FRICTION code [8]. Solutions for zero lift drag are computed at a number of altitude and Mach numbers. Specifically, the altitudes are: 0; 10,000; 30,000 and 60,000 ft. The Mach numbers are: 0.0; 0.1; 0.5; 0.9; 1.0; 1.1; 1.5; and 2.0.

The induced drag is estimated using the vortex lattice code TORNADO [9]. A TORNADO solution is computed over a range of angles of attack and Mach numbers and combined with the FRICTION results to develop a set of drag polars. The resulting set of drag polars are used in the mission performance calculations for each configuration.

For the structural sizing problem described in the previous section, the aerodynamic analysis is internal to the ASTROS application. Specifically, USSAERO is the aerodynamic solver used in the version of ASTROS for the study herein. The USSAERO aerodynamic model contains all the lifting and control surfaces.

2.4 Propulsion

The aircraft engine performance is modeled using TERMAP (Turbine Engine Reverse Modeling Aid Program). TERMAP is an engine cycle performance program developed by Allison Engineering that uses a building block approach to predict both the steady state and transient performance of gas turbine engines [10]. The program is FORTRAN-based and requires a user to define engine component characteristics. TERMAP uses this information to estimate flow path information at station numbers throughout the engine.

To improve the computational efficiency of the N^2 process in Figure 1, TERMAP is used to build an engine deck approximation of thrust and specific fuel consumption over all flight conditions of interest. The engine deck is used during the optimization to calculate range and optimal altitude trajectory for the flight leg of interest.

2.5 Mission performance

The mission performance assessment involves calculating the aircraft range for a given Mach number. This is accomplished by using the design weight from ASTROS. This FEM-based weight calculation is higher fidelity than an empirically-based calculation typically done in conceptual design. The design weight includes non-structural masses and internal fuel load, such that the design weight is the gross takeoff weight for the vehicle. (Since a half-span model is used, the weight from ASTROS is half the takeoff gross weight.) The design weight along with the drag polars and the propulsion system performance assessment are used to calculate the range for the vehicle.

The range calculation assumes that the air vehicle is flying at an optimal altitude for its weight. The altitude is allowed to increase as fuel is burned, thus returning a range equal to the distance traveled over the optimal altitude trajectory. For the ESAV studies herein, the range calculation is done for a 6000 lb fuel burn at Mach 0.8.

3 Multidisciplinary analysis implementation

The practical large-scale MDO of an ESAV requires the MDA behind the objective and constraint functions be implemented such that the following hold true: (1) the evaluation of constraints and objectives is automatic and does not require user intervention; (2) the evaluation of constraints and objectives is reliable, repeatable, and accurate to the desired fidelity; (3) constraint and objective evaluations that encounter errors or exceptions are communicated to the MDO algorithm gracefully and the MDO algorithm handles them appropriately;

and (4) the evaluation of objective and constraint functions for different designs is done in parallel using the computational resources available.

With these considerations in mind, the Java-based Service-Oriented Computing Environment (SORCER) software is used to implement the ESAV MDA [5]. The MDA is then driven by an optimization algorithm implemented in Matlab via the SORCER *ModelClient* class.

3.1 SORCER fundamentals

SORCER is a Java-based, network-centric computing platform that enables engineers to perform analyses and design studies in a very flexible, robust, and distributed computing environment. At the foundation of an engineering analysis within SORCER is the concept of a service provider. A service provider – or simply "provider" – is Java code implemented in accordance with SORCER standards that makes a number of Java methods available to remote clients over a computer network [11]. These methods are referred to as the provider's service operations.

Providers typically provide service operations that leverage existing domain-specific analysis codes (e.g., CFD or FEM codes). These providers are referred to as analysis providers. The implementation of a given service operation in an analysis provider usually involves wrapping an engineering analysis executable with Java code. The executable is generally platform dependent and performs the bulk of the engineering-specific computations for a given service operation. An analysis provider's service operation may also enhance the functionality of the underlying executable with Java code or simply pass data to-and-from it.

Two common approaches to wrapping executables are to: (1) use the file system in conjunction with a system call from the analysis provider's service operation; or (2) use more sophisticated technology such as Java Native Interface to tightly couple the analysis provider's service operation to the executable. In the first case, the service operation writes an input file for the executable to read, invokes the executable via a system call, and concludes by reading the output file produced by the executable. The service operation may return the raw output files produced by the executable via a URL object or post-process the file and pass higher-level Java objects to the remote client. If errors or exceptions occur in an analysis provider's service operation, the caller is notified gracefully and may (or may not) take action to correct the problem. Once an analysis provider has been implemented to provide a number of service operations, it may be published from a computer on the network using SORCER. Remote clients may then access the analysis provider's service operations via a small Java program called a *service requester*, or use simple SORCER command-line tools to access service operations directly.

For example, the service operation used in the ESAV study to create input files for ASTROS is called *doMstcGeomService*. This service operation is implemented by a provider called Engineering Air Vehicle Provider. The underlying executable for the *doMstcGeomService* is the compiled Matlab program MSTCGEOM. MSTCGEOM creates all of the input decks required for the individual analysis tools in the ESAV MDA. The argument for the Java method *doMstcGeomService* is an object that contains all the information necessary to define the MSTCGEOM input deck, including the ESAV wing design variables (area, aspect ratio, and taper ratio). Using the input object, the Engineering Air Vehicle Provider's *doMstcGeomService* service operation writes an input deck for the MSTCGEOM Matlab executable, invokes MSTCGEOM via system call, and returns the URL of the ASTROS input file produced by MSTCGEOM to the remote client. In addition to the ASTROS input file, the MSTCGEOM executable also writes input files required by the *doTornadoService* and *executeFriction* service operations. Both of these service operations are also provided by the Engineering Air Vehicle Provider.

Like most service operations, the *doMstcGeomService* service operation returns multiple outputs that may be used by multiple downstream service operations to form a complete MDA. The returned objects from a service operation may be used directly or post-processed on the client-side (i.e., at the user's location) such that the output may be meaningful for design optimization. In the previous example, the output URL object pointing to the ASTROS input file is used directly by a downstream service operation. If a scalar-value was desired – e.g., to assign a value to an objective function during an optimization – the URL could be opened and parsed to yield a number. The means of reducing output from service operations in this manner is accomplished with the SORCER *Filter* subclasses, discussed later.

To provide a layer of abstraction from a specific implementation of a service operation (i.e., Java method), providers implement Java interfaces to identify the service operations they provide. (Java interfaces are defined by a list of method signatures, which are composed of a method name, argument types, and return type.) These Java interfaces are referred to as service types. In this manner, a remote-client is generally only concerned about finding a provider that implements the service type containing the service operation of interest. The specific provider usually does not matter, since it is the service operation that the client desires. Many providers on the network may have different implementations of the same service operation in which case they would all implement the same interface (i.e., service type). If a client desires to select a specific provider for a given service operation, the client may specify the name of the provider in addition to the service type.

To use SORCER providers that are published on a network, a client must specify the name of the service operation, the service type that contains the service operation of interest, and the arguments for the service operation. The client does not have to specify the hostname, port, or IP address of the computer that is running the service operation process. The underlying architecture, Jini Connection Technology, supports dynamic discovery and join on both the provider- and client-side, thus relieving the client from knowing where to find service operations.

The arguments for all SORCER service operations are instances of the *Context* class. A Context object – or "context"

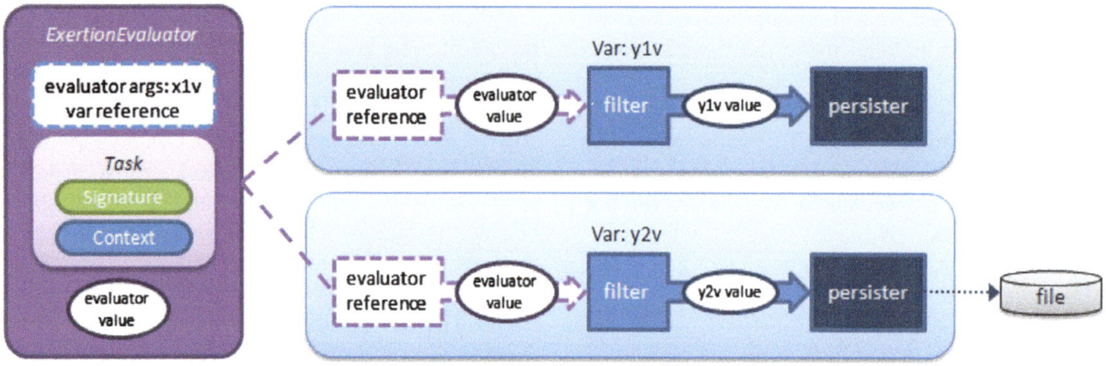

Figure 3. A single Evaluator object may be referenced by multiple Var objects employing different filters to reduce aggregate data to scalar values; the values may be then written to a file (or object field) using a var's persister.

– is a generic container that encapsulates name-value pairs. In this manner, a context possesses its own namespace for identifying argument objects for a service operation. The service operation name and service type are encapsulated in the SORCER *ServiceSignature* class. (Instances of *ServiceSignature* are referred to as "signatures".)

The combination of the service operation name, service type, and a context are encapsulated in an instance of the SORCER *Task* class. A *Task* instance – or "task" – represents the basic unit of work that service providers operate on. Since tasks involve using remote service operations, a layer of abstraction exists to encapsulate both remote and locally executed units of work. The *Evaluator* class and its subclasses are the entities responsible for doing computational work via their *evaluate* method from both remote and local resources in SORCER.

An *Evaluator* instance – or "evaluator" – that encapsulates a task is called an *exertion* evaluator. The word "exertion" implies the *remote* use of service providers. This is in contrast to other subclasses of *Evaluator* that are capable of doing simple calculations locally when the heavy lifting nature of an engineering analysis service operation is not required. For example, the *ExpressionEvaluator* class enables a client to combine *Var* instances in algebraic equations without using a remote service operation (the *Var* class is discussed later). In this light, evaluators are the entities responsible for either defining the process of how data is produced via remote service operations or producing the data themselves locally via their own logic.

Var instances – or "vars" – are used to form both independent and dependent variables in SORCER. In the case of independent variables, vars have an instance of *IndependentEvaluator* that contains the value of the independent variable. The *IndependentEvaluator* object acts as a simple container to store a value and does not perform any calculations or use service operations on the network.

Dependent vars are typically created to implement scalar-valued mathematical functions, such as objective, constraint, or other mathematical functions for use in analysis or optimization studies. They may also be used to form composite functions. Vars contain the objects necessary to produce – rather than store – the values they represent, which may include *Evaluator*, *Filter*, and *Persister* instances. The three types of objects are used in a specific sequence when the method *getValue* is invoked on a dependent var object.

When the *getValue* method is called on a dependent var, the evaluator's *getValue* method is subsequently called. The evaluator's *getValue* method then, in turn, calls *getValue* on its argument vars to ensure their respective values are current before proceeding. (If a var represents a function dependent on other vars, i.e., argument vars, the evaluator of the var representing the function holds references to those argument vars.) The evaluator then checks to see if it is out of date due to argument vars changing since the last invocation of *evaluate*. If the argument vars have not changed, the evaluator returns its current value without further processing. In this manner, all calls to *evaluate* are demand-driven, i.e., the *evaluate* method is invoked only if the evaluator's arguments have changed since the last invocation of *evaluate*. As described earlier, the *evaluate* method may use a remote service operation implemented by a provider on the network to perform the necessary calculations, or the calculations may be done locally via its own logic. Once the evaluator's *evaluate* and *getValue* methods complete and return a new value to the var, an instance of a *Filter* subclass may be employed.

Since exertion evaluators typically return aggregate forms of data, reducing the output from them is often necessary. This data reduction is accomplished using the *Filter* subclasses. Using a *Filter* subclass object – or "filter" – on the returned objects from an evaluator allows clients to reduce or transform the aggregate data produced by evaluators to yield new objects. The result of filtering is often a scalar value of a mathematical function. In the case where an evaluator produces aggregate data, the evaluator may be referenced by a number of different vars with different filters to yield different values, see Figure 3. A series of filters may be combined to form a filter pipeline, such that downstream filters operate on the output of upstream filters. The resulting value returning from the filter (or pipeline filter) is the value of the var, which is then be passed to the var's *Persister* instance if it exists.

The *Persister* instance – or "persister" – uses the value from the filter as an argument to its *setValue* method. Depending on the specific subclass of *Persister* used, the persist method may write the var value to a file or set a field on an object. Once the persister is done executing, the var value is returned to the caller of the var's *getValue* method.

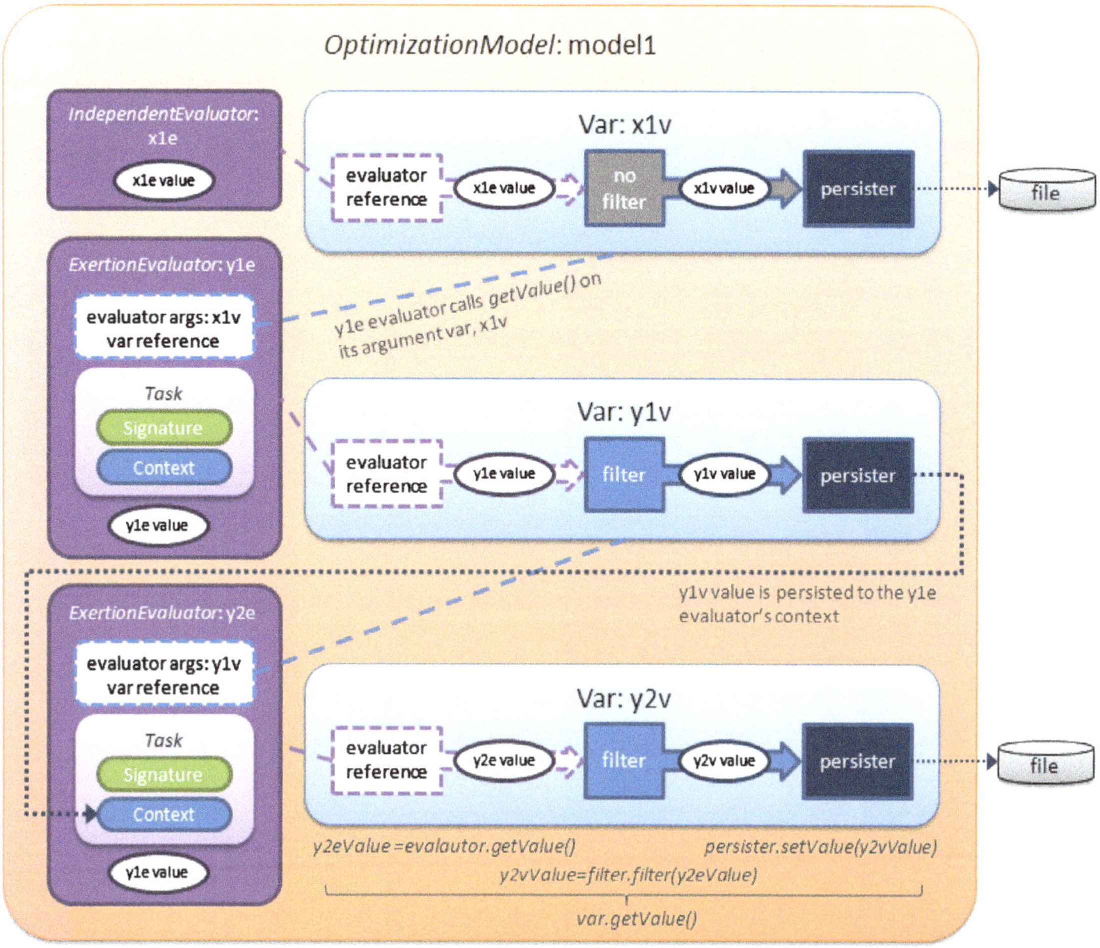

Figure 4. An OptimizationModel is a collection of Var objects; each var references an evaluator and may be referenced by other vars to form composite functions.

An *OptimizationModel* instance – or "model" – is a collection of any number of vars that define a specific optimization problem. Figure 4 illustrates a simple example of an *OptimizationModel* called *model1*. The model contains three vars; one independent and two dependent. The var *y1v* is a function of *x1v*, which is the independent var. The var *y2v* is a function of *y1v*, thus it is a composite function.

The vars are the implementation of the design variables, optimization objective function, and constraint functions. Additional vars may be included for monitoring other figures-of-merit. Once a model is created, it may be published as a provider for clients to access remotely. This type of provider is referred to as a model provider to distinguish it from analysis providers. The model provider behaves like shared memory that is accessible to remote clients over the network. There is only one state of the model, which is subject to manipulation by remote clients. There are two techniques for clients to interact with a published model provider: (1) via a single model query; or (2) via a table model query.

A single model query is done from a requestor or SORCER command-line tool. In both cases, a query object is constructed containing the design variable var names and values and the var names of the objective and constraint functions that the client wishes to calculate. The query also contains the name of the model provider, so that it may be found on the network.

In contrast to analysis providers, accessing model providers over the network does not require clients to specify a service type and service operation name. Since all *OptimizationModel* instances have only one service type with a single service operation, the information is implied. This does, however, necessitate that the client specify the model provider name when multiple models are published.

Once the query is constructed, it is executed by the client. The published model provider receives the query object and proceeds to invoke *setValue* on all the design variables. Once the *setValue* calls are completed, the model begins calling *getValue* on the client-specified objective and constraint functions. After all the *getValue* calls are complete, the query object is returned to the remote client with the updated var values.

The second technique of using models is via a table query. Rather than passing a single set of design var values to the model provider, a client may construct a table of runs containing the names of the design vars and their values for each run. The table is added to the query with the name of the model provider and the var names of the objective and constraint vars the client wishes to calculate. Since a table of runs is being sent

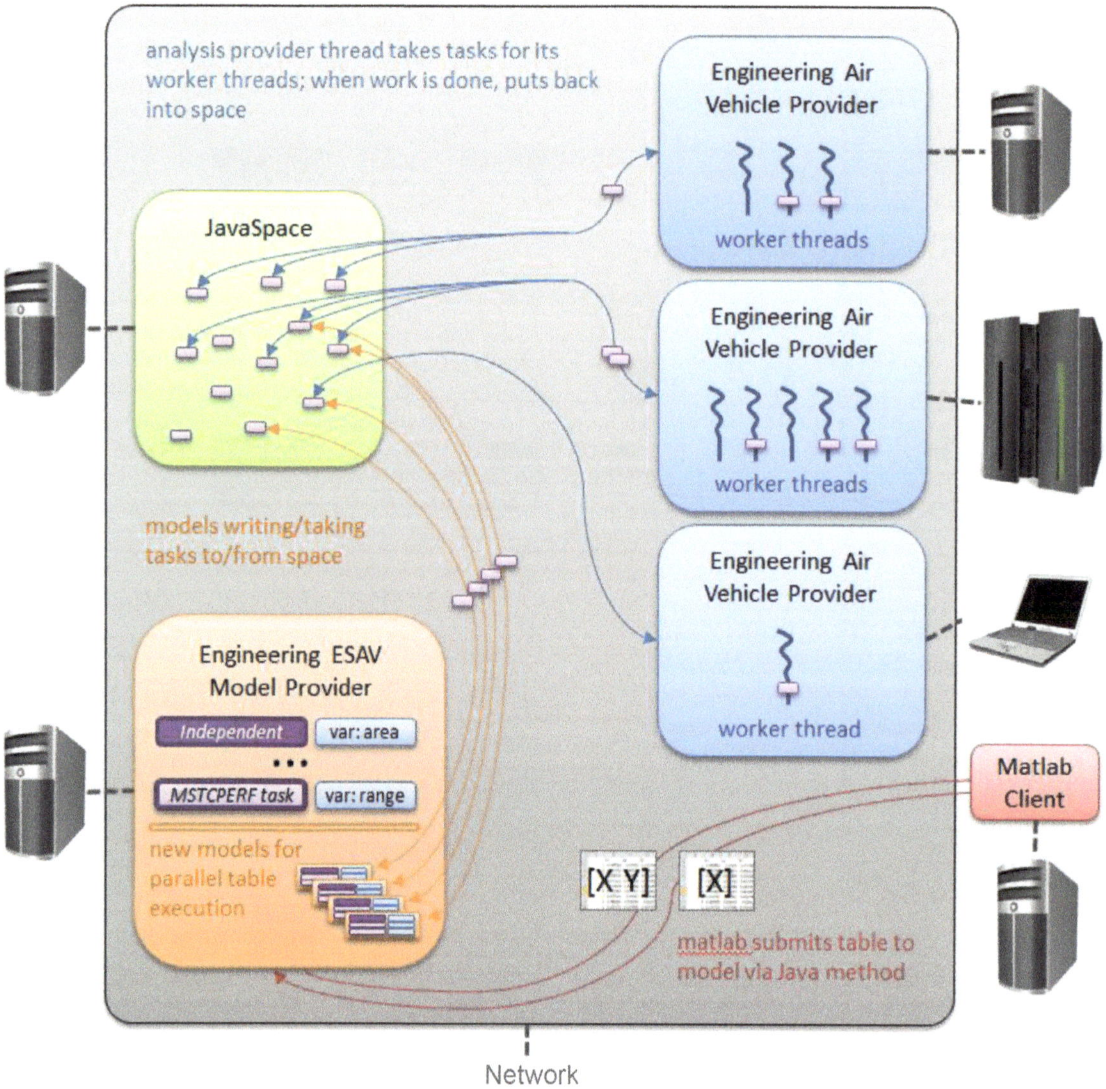

Figure 5. SORCER uses JavaSpaces technology to provide a flexible, dynamic grid computing facility for ESAV optimization studies.

to the model provider, the client may specify the number of runs the model provider is to execute in parallel.

Once the table query is constructed and sent to the model provider, the model provider creates new child instances of itself for parallel execution of the table. (The child instances are newly constructed objects, such that the states of their respective vars are not inherited from the model provider.) The model provider then creates a thread for each child model and begins to *setValue* and *getValue* on the vars as in the single model query. Once all the threads complete, the var values for each run are returned to the caller in a table object.

In contrast to the single model query, the table query does not affect the state of the parent model provider's vars. The child models are isolated from one another and the parent model, and are used exclusively to perform the client-requested calculations. Once the calculations are complete, the child models are discarded and the states of the parent model vars remain unaffected.

3.2 SORCER space computing

Large numbers of engineering analyses are generally required to perform conceptual-level aircraft design. This significant computational burden is addressed at the AFRL MSTC by using the SORCER software. The network-centric approach of SORCER enables the use of heterogeneous computing resources, including a variety of operating systems, hardware, and software. Specifically, ESAV studies performed at the AFRL MSTC use SORCER in conjunction with a mix of Linux-based cluster computers, desktop Linux-based PCs, Windows PCs, and Macintosh PCs. The ability of SORCER to leverage these resources is significant to MDO applications in two ways: (1) it supports platform-specific executables that may be required by an MDA; and (2) it enables a variety of computing resources to be used as one entity (including stand-alone PCs, computing clusters, and high-performance computing facilities). The main requirements for using a

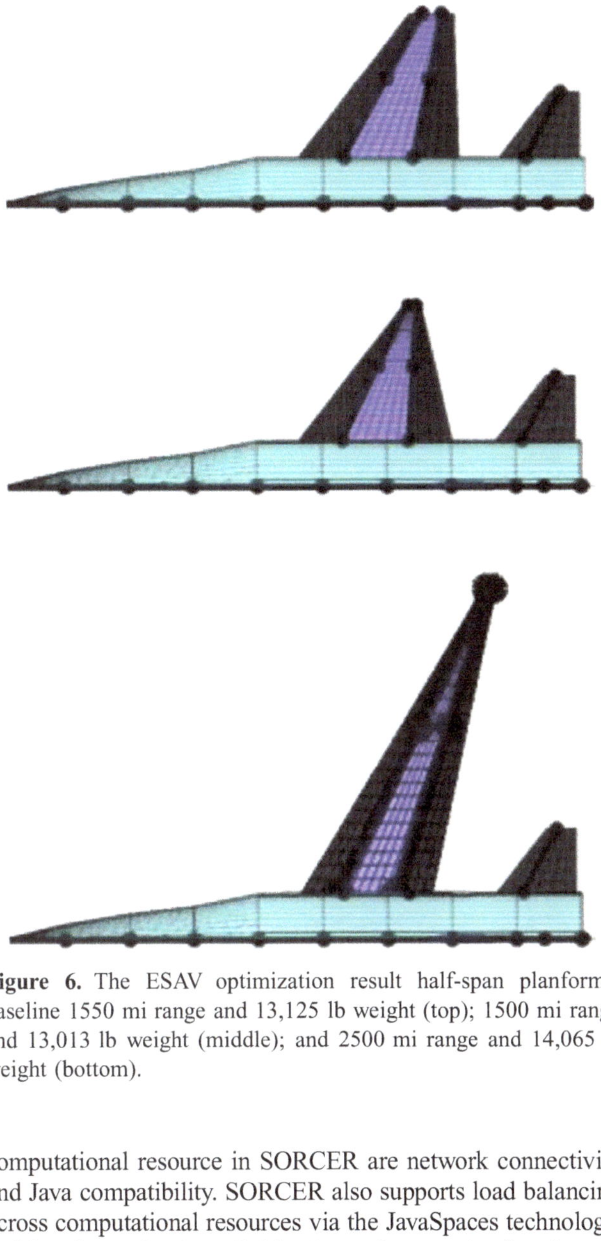

Figure 6. The ESAV optimization result half-span planforms: baseline 1550 mi range and 13,125 lb weight (top); 1500 mi range and 13,013 lb weight (middle); and 2500 mi range and 14,065 lb weight (bottom).

computational resource in SORCER are network connectivity and Java compatibility. SORCER also supports load balancing across computational resources via the JavaSpaces technology, making the evaluation of objective and constraint functions in parallel a simple and a dynamically scalable process [12].

SORCER employs JavaSpaces technology to implement a loosely-coupled distributed computing system. JavaSpaces enables different processes on different computers to communicate asynchronously in a reliable manner [12]. Using this technology, SORCER implements a self-load balancing grid computing system that can dynamically grow and shrink during the course of an optimization study. The self-load balancing quality of SORCER is necessary to ensure efficient use of computational resources given the varying computational expense of different tasks executing simultaneously during a design study [13].

A JavaSpace – or "space" – is Java code running on a computer that is accessible on the network in the same manner as service providers, see Figure 5. The space provides a type of shared memory where exertion evaluators can put tasks they wish to be processed by service providers. Service providers, in turn, use Jini discovery mechanisms upon startup to find spaces on the network and monitor them for tasks with their service type. If a service provider sees a task it can operate on in a space, and the task has a flag indicating it has not been processed, the provider takes the task from the space. The provider then executes the appropriate service operation and returns the task to the space with a flag indicating the task has been processed. Once the task has been returned to the space, the process that initially wrote the task to the space detects the returned task and checks to see if it has been processed. If the task indicates it has been processed, the evaluator removes the task from the space.

To achieve the load balancing across multiple computers, a service provider may be configured to have a fixed number of worker threads. The number of worker threads determines the number of tasks a provider can process in parallel. By configuring the number of worker threads for a specific service provider on a specific computer, the provider can self-load balance the computer it is hosted on (assuming that it is the only service provider operating on the host computer).

4 Design studies

The design of a supersonic air vehicle involves assessing many figures of merit and the tradeoffs between them. In this study, a preliminary exploration of the design space is performed through univariate parametric sweeps of the design variables. In this manner, each design variable is independently swept from its lower bound to its upper bound in equal steps. This process is costly, but it serves three important purposes: (1) the sweeps provide a set of results that engineers may use to assess the validity of the MDA implementation; (2) the results may be used to screen design variables and responses for significance, which may result in a smaller MDO problem; and (3) the sweep results are used to calculate the sensitivities for the first iteration of the gradient-based optimization process. Once the preliminary study is completed, the MDO problem is solved using a robust successive linear programming (SLP) technique.

4.1 Optimization problem statement

A single objective function is identified along with nonlinear and side constraint functions to exercise the ESAV model in SORCER. The optimization problem takes the general form:

$$\left\{ \begin{array}{c} \text{Minimize } f(\boldsymbol{x}) \\ \text{subject to } \boldsymbol{g}(\boldsymbol{x}) \leq \boldsymbol{0} \\ \text{and } \boldsymbol{x}^{\text{lb}} \leq \boldsymbol{x} \leq \boldsymbol{x}^{\text{ub}} \\ \text{where } \boldsymbol{g} \in \mathbb{R}^{ncon} \text{ and } \boldsymbol{x} \in \mathbb{R}^{ndv}. \end{array} \right. \tag{1}$$

The objective is to minimize the structural weight of the ESAV, which includes the fuselage and lifting surface weights. In this manner, the life cycle cost of the aircraft is kept down due to its correlation with takeoff gross weight [14]. Each evaluation of the objective function, $f(\boldsymbol{x})$, involves a structural sizing sub-optimization of the wing and tail surface spar, rib, and

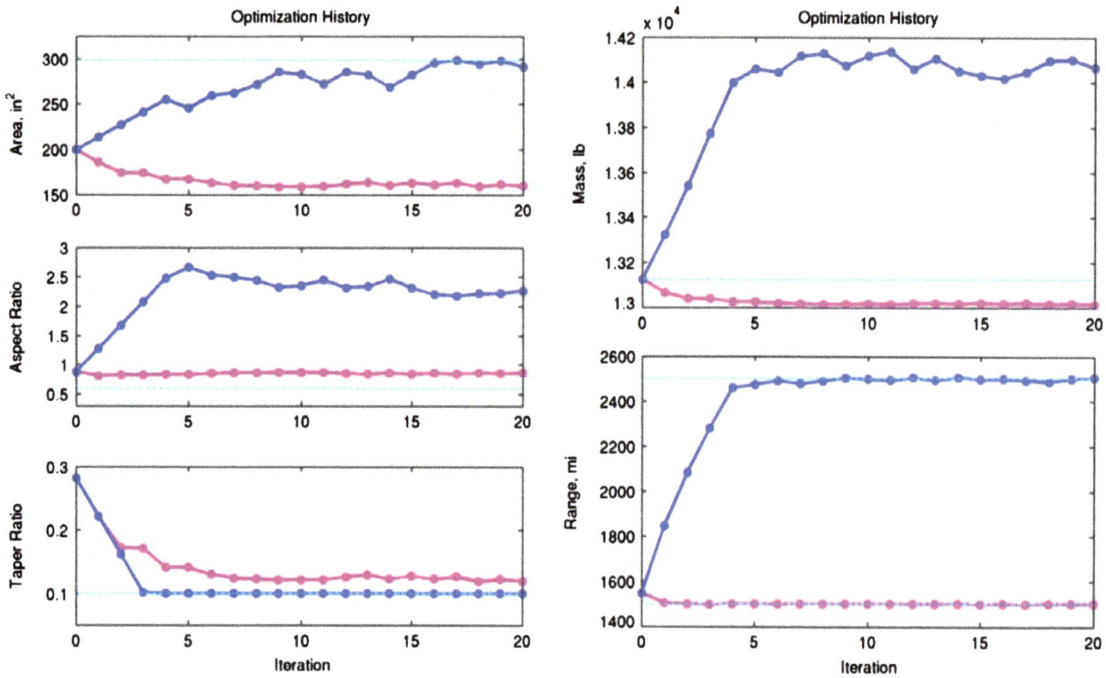

Figure 7. Results from ESAV optimization illustrate how the design variables, objective and constraint functions are manipulated by the SLP technique (1500 mi range constraint, magenta; 2500 mi range constraint, blue).

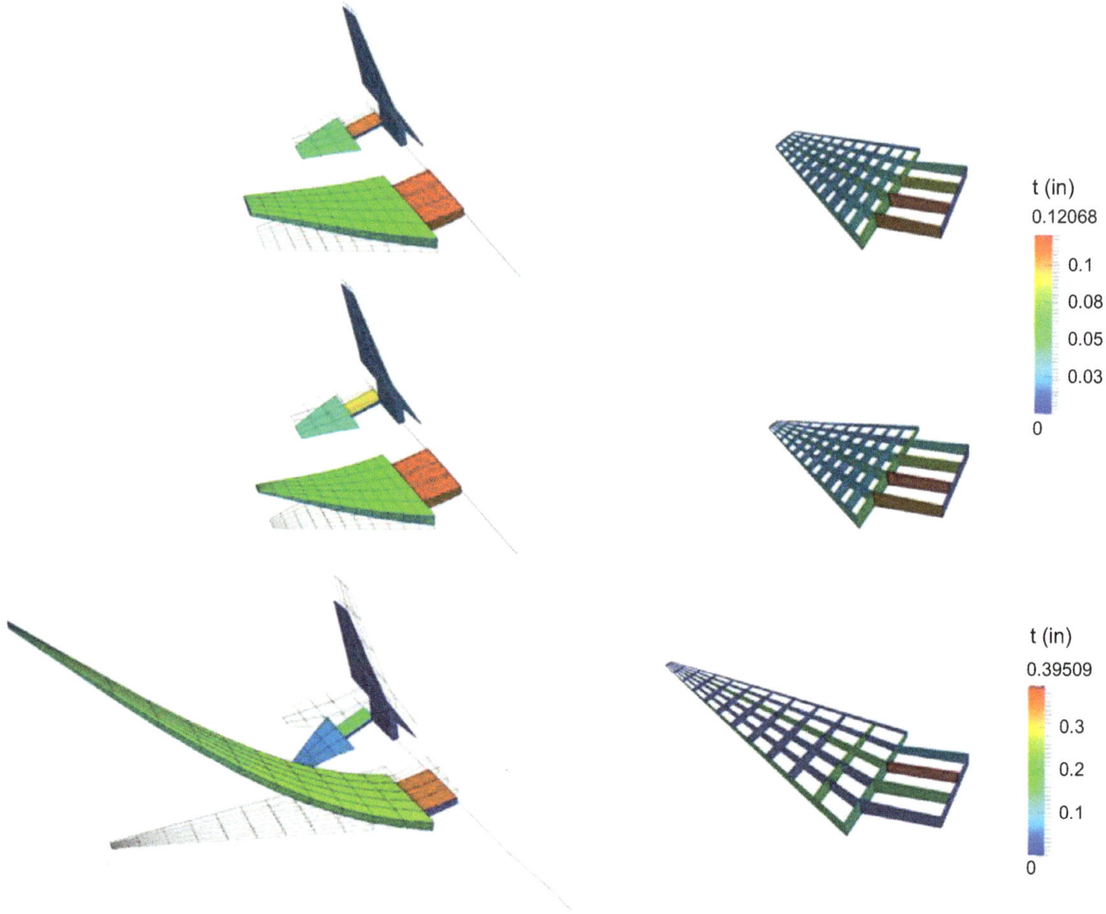

Figure 8. The ESAV planform and structural design thicknesses from the ASTROS structural sizing sub-optimization problem are shown above for the 1550 mi range baseline design (top row), the 1500 mi range design (middle row), and the 2500 mi range design (bottom row) optimal designs. (The left column shows the thicknesses for the wing skins and the right column shows the thicknesses of the ribs and spars.)

skin thicknesses. This sub-optimization process is performed by ASTROS and is integral to the MDA, see Section 2.2 for details.

Two optimization problems are solved with a single range constraint (*ncon* equal to one). In the first case, the cruise leg range, $g(\boldsymbol{x})$, is constrained to be above 1500 mi at Mach 0.8 for 6000 lb of fuel. In the second case, the range is increased to 2500 mi. The design variables, $\boldsymbol{x}$, are the wing area, taper ratio, and aspect ratio (*ndv* equal to three). The lower and upper bounds on the design variables ($\boldsymbol{x}^{lb}$ and $\boldsymbol{x}^{ub}$) are: 100 and 300 ft^2 for wing area, 0.6 and 5.0 for aspect ratio, and 0.1 and 1.0 for taper ratio, respectively.

4.2 Optimization algorithm

The optimization technique applied is an adaptation of the Frank-Wolf implementation of the successive linear programming (SLP) method [15]. It is a gradient-based method that uses a finite-difference approach to compute sensitivities. Issues of computational efficiency and robustness are addressed using parallel computations with SORCER and high-order/adaptive finite-differencing techniques.

The SLP method used herein is tailored for parallel computing on a large number of CPU cores such that gradient and line search calculations are executed in parallel [16]. The algorithm is tolerant to failed runs in both the gradient and line search phases of an optimization iteration. This robustness comes with the computational cost associated from the extra analyses required over that of a typical first-order finite-difference technique.

To ensure a sensitivity calculation is successful, four objective and constraint function evaluations are executed in parallel for each design variable at each SLP iteration. (The number of points used in the sensitivity calculation is based on the computational resources available and the experience of the designer.) The objective and constraint evaluations are used to calculate the highest-order finite-difference sensitivity possible given the number of successful function calls. If all of the function calls fail, the robust SLP algorithm attempts to adapt the step-size and recalculate the sensitivity. If this attempt fails, the user is prompted to take corrective action.

The optimization begins by calculating the gradient of the objective and constraint functions at the initial design point. The gradients are then used to form a constrained linear programming (LP) problem where the optimum is found using an LP solver. In traditional implementations of gradient-based methods, a line search follows that involves sequentially searching for a minimum along a line from the initial design to the estimated optimum. This approach is inefficient given the computing resources available to most engineers today.

Since many CPU cores are available on a single computer, cluster, or high-performance computing facility, the line search is converted to a partial enumeration of the line. The number of designs evaluated on the line depends on the computational resources available, the anticipated nonlinearity of the search, the computational expense of the analysis, and the schedule of the MDO project. Once the line is partially enumerated, the local optimum may be found via a simple curve fit to a quadratic polynomial. Using the local optimum determined from the partial line enumeration, a subsequent iteration begins with a gradient evaluation at this point. The SLP process continues based on optimization convergence criteria or schedule considerations.

The SLP algorithm is implemented using Matlab and uses the *lingprog* LP solver to determine search directions. The responses required to calculate the objective and the constraints are supplied via the SORCER Engineering ESAV Model Provider. The communication between Matlab and the Engineering ESAV Model Provider is done by instantiating the SORCER *ModelClient* class within a Matlab session. The *ModelClient* class provides a simple interface for setting design variable values, running the MDA, and obtaining the responses necessary to form the objective and constraint functions.

5 Results

Figure 6 shows the planforms for the baseline (approximately 1550 mi range and 13,125 lb half-span weight), the 1500 mi range design (13,013 lb half-span weight), and the 2500 mi range design (14,065 lb half-span weight). The optimization iteration history is shown in Figure 7 for the design variables, constraint (range), and objective (weight). Figure 8 plots the results of the ASTROS structural sizing sub-optimization performed on the three ESAV designs. The figure shows the optimal thicknesses of the skins, ribs, and spars for the three configurations.

In the first case with the 1500 mi range constraint, the wing area and taper ratio were reduced to minimize weight. Since the baseline range was slightly greater than 1500 mi, the optimizer had some design freedom to reduce the objective function in this manner. The overall reduction in weight from the baseline was approximately 0.9%. The structural sizing sub-optimization was driven in this case by two of the four maneuvers: the 9.0 g pull-up at Mach 0.9; and the 7.2 g pull-up at Mach 1.2.

The second optimization problem solved was for the 2500 mi range constraint. The planform of the wing in this case became long and slender with a high aspect ratio, high wing area, and low taper ratio. Since the initial design was infeasible, a considerable amount of weight was added via increases in wing area and aspect ratio to achieve the 2500 mi range. The weight was increased approximately 7.2% from the baseline design. The structural sizing sub-optimization was driven by the 9.0 g pull-up at Mach 0.9 maneuver.

6 Conclusions

The SORCER engineering software was used successfully in the design of two prototype ESAVs. The results from both optimization cases exhibited the correct trends consistent with historical aircraft design. These results provide a degree of validation of the implementation of the Matlab SLP code, the SORCER ESAV model, the SORCER providers, and the JavaSpaces technology.

The use of the JavaSpaces technology for parallel distributed computing proved reliable and efficient. It was a straightforward process to add computers to the SORCER grid as needed during the course of the two optimization studies. This flexibility proved valuable as the number of computers available varied from day-to-day.

Lastly, the application of a gradient-based SLP optimization technique capable of handling failed design points was demonstrated successfully. The robust nature of the SLP technique and its ability to perform gradient and line search functions in parallel made the ESAV optimization studies possible in a short period of time with minimal user intervention.

Although the SLP technique did not necessarily determine a global optimum, it did: (1) find a feasible design when started from an infeasible one; (2) improve the objective function when started from a feasible design; and (3) operate autonomously while providing the user with feedback on the optimization process. The results demonstrate the practical utility of the robust SLP approach. Furthermore, the algorithm was straightforward to implement, troubleshoot, and its successive nature (i.e., design improvements with each iteration) provided re-start capability. These qualities make it a sound approach for design optimization, especially when working with newly implemented MDAs where the univariate sweeps executed for the finite-difference sensitivities provide valuable feedback to the user.

Future work will include using design space exploration techniques and emergent optimization methods, such as genetic algorithms. Also, varying the fidelity of several of the analyses in the MDA in a variable complexity optimization process may be examined.

References

1. Integration of tools and processes for affordable vehicles, Tech. Report RTO-TR-AVT-093 AC/323(AVT-093)TP/102, NATO Research and Technology Organisation, 2006.
2. Pre-Milestone A and Early-Phase Systems Engineering: A Retrospective Review and Benefits for Future Air Force Acquisition, Committee on Pre-Milestone A and Early-Phase Systems Engineering: A Retrospective Review and Benefits for Future Air Force Acquisition, National Research Council, ISBN: 0-309-11476-4, 2008.
3. Fantini P, Balachandran LK, Guenov MD. 2008. Computational system for multi disciplinary optimization at conceptual design stage. International Journal of Simulation and Multidisciplinary Design Optimization, 2, 177–185.
4. Schumacher G, Daoud F, Petersson O, Wagner M. 2012. Multidisciplinary airframe design optimization, 28th International Congress of the Aeronautical Sciences.
5. Kolonay RM, Sobolewski M. 2011. Service ORiented Computing EnviRonment (SORCER) for large scale, distributed, dynamic fidelity aeroelastic analysis & optimization, International Forum on Aeroelasticity and Structural Dynamics, Paris.
6. Alyanak E, Kolonay R. 2012. Efficient supersonic air vehicle structural modeling for conceptual design, 12th AIAA Aviation Technology, Integration, and Operations (ATIO) Conference and 14th AIAA/ISSM, Indianapolis, DOI: 10.2514/6.2012-5519
7. Neill D, Herendeen D. 1995. ASTROS Enhancements: Volume I – Astros User's Manual, Wright Laboratory WL-TR-96-3004.
8. Mason WH. 2002. Software for Aerodynamics and Aircraft Design. Retrieved 2012, from Skin Friction/Form Factor Drag estimation: http://www.dept.aoe.vt.edu/~mason/Mason_f/MRsoft.html#SkinFriction
9. Melin T. 2000. A Vortex Lattice MATLAB Implementation for Linear Aerodynamic Wing Applications, Royal Institute of Technology, http://www.redhammer.se/tornado
10. Macinnis DV. 1994. A Simplified Cycle-Matching Transient Simulation of a Turbofan Engine, AIAA Joint Propulsion Conference, Indianapolis.
11. Newmarch J. 2006. Foundations of Jini 2 Programming, Apress, Inc., Chap. 1.
12. Freeman E, Hupfer S, Arnold K. 1999. JavaSpaces Principles, Patterns, and Practice, Addison Wesley Longman Inc., Chap. 1.
13. Argod V, Belegundu AD, Aziz A, Agrawala V, Jain R, Rajan SD. 2008. MPI-enabled shape optimization of panels subjected to air blast loading. International Journal of Simulation and Multidisciplinary Design Optimization, 2, 273–282.
14. Raymer D. 1992. Aircraft Design: A Conceptual Approach, 2nd ed., AIAA Education Series, Chap. 18.
15. Reklaitis GV, Ravindran A, Ragsdell KM. 1983. Engineering Optimization: Methods and Applications, John Wiley & Sons, Inc., New York, Chap. 8.
16. Burton SA, Prakash C, Machnaim J. 2007. Multistage Low Pressure Turbine Airfoil Shape Optimization using the C^3 Process, 3rd AIAA Multidisciplinary Design Optimization Specialist Conference, Honolulu.

8

Numerical study of water and heat transfer in unsaturated clay-loam soil

Nacer Lamrous*, Said Makhlouf, and Nora Belkaid

Mouloud Mammeri University, L.M.S.E. Laboratory, PO Box 17, RP 15000, Tizi-Ouzou, Algeria

Abstract – This present study is the numerical estimation of the temperature distribution and the water content distribution underground soil under the Mediterranean climate type. We use as input data of ambient temperature, air humidity and solar radiation, average values during 10 years estimated from data supplied by the local meteorological station (Tizi-Ouzou, Northern Algeria, 36°47′59″, North latitude and 4°1′59″, East longitude). A theoretical model, based on heat and water transfer equations, was established for an unsaturated soil submitted to the climatic conditions of this site. The mathematical model established in mono dimensional type, for a semi infinite transfer model, is based on Whitaker theory of heat and mass transfers in unsaturated porous medium (Withaker 1977, 1980) with the hypothesis that air pressure into soil porosity is equal to atmospheric pressure. The equations were discretized according to the finite volume method, which is more adapted for this type of problem, and were solved by the Newton-Raphson iterative method in the environment of Matlab software. The simulations have been done for two typical days (January 15 and May 15). Curves of temperature and water content evolutions in term of depth and time were obtained.

Key words: Heat and mass transfer, Soil temperature, Unsaturated porous soil, Simulation.

1. Introduction

Theoretical prediction of mass and heat transfer in an unsaturated porous medium is still complex. This is due to the discontinuity of these media and the diversity of the phenomena taking part in this process. It is also due to the difficulty of predicting the phenomenological coefficients to be introduced in the calculation programs, as reported by Moyne [7]. In this theoretical study, the mathematical model is based on Whitaker's approach [12,13], and the technique of the representative elementary volume (REV, see Figures 1 and 2).

This allows us to assimilate the porous medium to a continued soil. The air and water mass conservation equations in both liquid and steam forms and the heat conservation equation are written down at the scale of pores, and then integrated in the domain of REV. We have developed a code of calculation and simulated the behaviour of the soil. In this program, the thermo physical characteristics such as density, heat capacity, absolute permeability, thermal conductivity and porosity have been measured in the LCTP (Central Laboratory of Public Works, Tizi-Ouzou). The meteorological conditions of the site are taken out of files where temperature and humidity, solar radiation and wind velocity during the period of ten years are consigned by meteorological station of Tizi-Ouzou.

*e-mail: `lamrous_nacer@yahoo.fr`

2. Basic mathematical model

The variables that we have retained for the study of this problem are the moisture content X, the temperature T along with the pressure P of the gas phase constituted of a mixture of water steam and dry air. In order to be confined to the dominating phenomena and simplify a certain number of calculations, we adopt some main hypothesis such as:

- The deformations of the solid phase are neglected; thus its speed is zero;
- The local temperature is identical for the three phases;
- The liquid phase is incompressible;
- The constituents of the gas phase are expected to behave like perfect gas.

2.1 Pore scale equations

Before developing the conservation equations, we must introduce certain variables which intervene in this problem. In fact, the moisture content X can be written as follows:

$$X = \frac{m_w}{m_s} = \frac{\rho_w V_w + \rho_v V_g}{\rho_s V_s}. \tag{1}$$

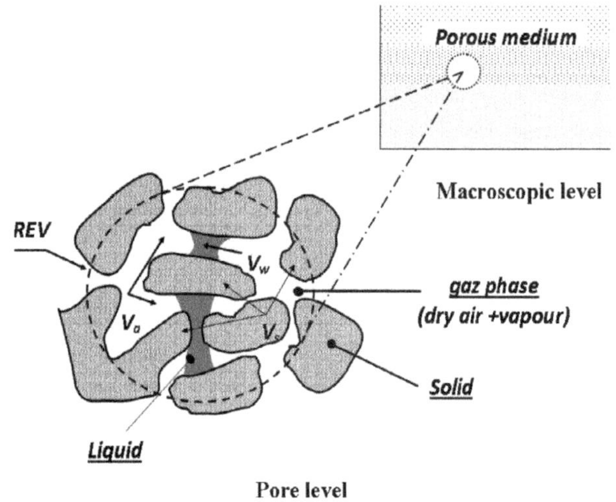

Figure 1. Illustration of representative volume element (REV).

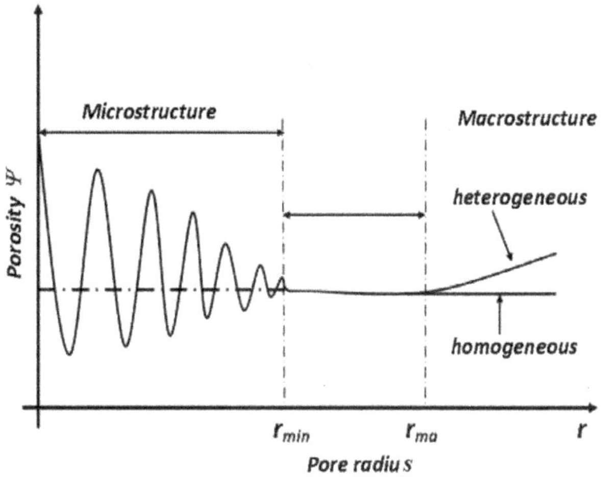

Figure 2. Porosity variation in porous medium at different scales.

We define also the volume fractions of water, gas and solid in the porous medium as follows:

$$\varepsilon_w = \frac{V_w}{V};\ \varepsilon_g = \frac{V_g}{V};\ \varepsilon_s = \frac{Vs}{V}. \tag{2}$$

We obtain thus:

$$X = \frac{\rho_w \varepsilon_w + \rho_v \varepsilon_g}{\rho_s \varepsilon_s}. \tag{3}$$

Saturation S is the volume fraction of void space filled by liquid:

$$S = \frac{V_w}{V_w + V_g} = \frac{V_w}{V - V_s}. \tag{4}$$

The porosity ψ of a porous material is the ratio of the total void or volume pore to the total volume of the material:

$$\Psi = \frac{V_w + V_g}{V}. \tag{5}$$

The relationship between moisture content and saturation is:

$$X = \frac{\Psi \rho_w}{1 - \Psi} S. \tag{6}$$

In these relations, m, ρ, and V represent respectively mass, density and volume and the subscripts a, g, s, v and w are used to designate respectively dry air, gas, solid, vapour or water components.

The conservation equations of water, dry air and energy are written as follows:

$$\frac{\partial}{\partial t}(\rho_w \varepsilon_w + \rho_v \varepsilon_g) + \nabla \cdot (\rho_w v_w + \rho_v v_g) = \nabla \cdot \left[\rho_g D_{\text{eff}} \cdot \nabla\left(\frac{\rho_v}{\rho_g}\right)\right], \tag{7}$$

$$\frac{\partial}{\partial t}(\varepsilon_g \rho_a) + \nabla \cdot (\rho_a v_g) = \nabla \cdot \left[\rho_g D_{\text{eff}} \cdot \nabla\left(\frac{\rho_a}{\rho_g}\right)\right], \tag{8}$$

$$\begin{aligned}&\tfrac{\partial}{\partial t}(\varepsilon_s \rho_s h_s + \varepsilon_w \rho_w h_w + \varepsilon_g \rho_v h_v + \varepsilon_g \rho_a h_a) \\ &+\nabla \cdot \left[\rho_w h_w v_w + (\rho_v h_v + \rho_a h_a) v_g\right] = \nabla \cdot \left[\rho_g h_a D_{\text{eff}} \cdot \nabla \frac{\rho_a}{\rho_g}\right] \\ &+\nabla \cdot \left[\rho_g h_v D_{\text{eff}} \cdot \nabla \frac{\rho_v}{\rho_g}\right] + \nabla \cdot (\lambda_{\text{eff}} \nabla T).\end{aligned} \tag{9}$$

In addition to above conservation equations, gas and water velocities must be evaluated. Their expressions, given by the generalized law of Darcy [2] are written by neglecting the gravitational terms for gas and liquid phases, as follows:

$$v_w = -\frac{K \cdot k_w}{\mu_w} \cdot \nabla P_w \text{ and } v_g = -\frac{K \cdot k_g}{\mu_g} \cdot \nabla P_g. \tag{10}$$

Here v, P, ε, h and T are respectively the velocity, pressure, volume fraction, mass enthalpy and temperature of a considered porous media component. D_{eff} and λ_{eff} are the effective vapour diffusivity and effective thermal conductivity. They are obtained from the process of up scaling from pore scale transport equations to macroscopic equations [12]. K and k are the absolute and relative permeability respectively and μ is the dynamic viscosity.

2.2. Boundary conditions

In order to complete the conservation equations listed beforehand, we associate the following boundary conditions:

The temperature T_s, on the soil surface, is calculated by thermal balance on the assumption that the dominating exchanges are done within the atmosphere (conductive exchanges in the soil are neglected):

$$(1-\alpha)R_g - \varepsilon\sigma_B\left(T_s^4 - (T_a - 6)^4\right) - h_c(T_s - T_a) = L_v J_w, \tag{11}$$

Table 1. Average meteorological parameters of Tizi-Ouzou site, 2011.

	Jan	Feb	Mar	Apr	Ma	Jun	Jul	Au	Sep	Oct	Nov	Dec
Solar radiation (w m^{-2})	472	541	624	671	633	640	620	641	586	506	453	438
Relative humidity (%)	84.8	65.5	64.5	60.8	54.5	47.8	42.0	46.2	53.9	56.9	63.1	65.1

Table 2. Physical and thermal soil parameters.

	Density ρ_s (kg m^{-3})	Thermal conductivity λ_s (W m^{-1} K^{-1})	Mass thermal capacity $C_{p,s}$ (kJ kg^{-1} K^{-1})	Porosity Ψ (%)	Absolute permeability Ka (m^2)
Value	1515	1	1270	40	10^{-14}

where T_a is the ambient temperature, L_v is the latent heat of evaporation, α is the soil albedo and R_g is the global solar flux.

J_w, the flux of water mass is described for convective drying by the boundary layer theory with Stefan correction:

$$J_w = \frac{K_E M_v}{RT_a}\left(h_r P_v^{\text{sat}}(T_s) - P_v(T_a)\right). \qquad (12)$$

Here, K_E is the mass transfer coefficient, M_v is the vapour molar mass, h_r is relative humidity and R is the universal gas constant. The superscript sat designate saturation state. The conductive heat flux is zero because the temperature undergoes very slight variations at a certain depth, whereas the moisture content of the soil is maximal and equal to the moisture content of the saturated soil.

2.3 Parameters of simulation

To study the behaviour of soil submitted to climatic conditions of Tizi-Ouzou, we rely on meteorological data collected at the local station on typical days of a 12-month year. As initial conditions, we have taken the measured values of temperatures and for moisture content, we choose a linear profile. Here are included the curves of annual evolution of the monthly average of temperatures of both air and soil noted down in Tizi-Ouzou. Table 1 shows the average monthly values of global solar radiation and relative humidity measured at meteorological local station.

To simulate heat and water transfers in soil, we use the thermo physical characteristics provided by the LCTP laboratory and shown in Table 2.

It is known that the effective conductivity coefficient is independent of pore size distribution. As heat conduction occurs in all phases in parallel, the heat flux or thermal conductivity contributions must be weighted according to their respective volume fractions of the phases. The serial model of conductivities of different components is appropriate in these conditions. Thus, if the contribution of gas is neglected, the effective thermal conductivity can be computed as:

$$\lambda_{\text{eff}}(S,T) = (1-\psi)\,\lambda_S + S\psi\lambda_w(T), \qquad (13)$$

Ψ is the porosity of soil, λ_s is soil thermal conductivity, S is saturation and λ_w is the water conductivity. According to Van Genuchten, cited by Lefebvre [6], the relative permeability of liquid water is:

$$k_w(S) = \left(1-\left(1-S^{1/m}\right)^m\right)^2\sqrt{S}. \qquad (14)$$

For the capillary pressure, we retained the expression of Van Genuchten (1980), written in terms of the effective saturation S_{eff} and which introduces two empirical parameters, β and m.

$$P_c(S_{\text{eff}}) = \left((S_{\text{eff}})^{(-1/m)} - 1\right)^{(1-m)}/\beta \text{ with } S_{\text{eff}} = (\theta-\theta_r)/(\theta_m-\theta_r). \qquad (15)$$

For a silty soil, we took the values recommended by Calvet [1]: $\beta = 0.0115$ Pa^{-1}; $m = 0.5169$; $\theta_m = 0.52$ kg m^{-3} et $\theta_r = 0.218$ kg m^{-3}.

For the relative humidity of air inside the soil, we will take again the expression obtained from fitting curves of the sorption isotherm [10], reported by Hong Vu [4], considering:

$$H_r = \frac{P_v}{P_v^{\text{sat}}(T)} = \begin{cases} 1 & \text{if } X > X_{\text{irr}} \\ \dfrac{X\left(2-\frac{X}{X_{\text{irr}}}\right)}{X_{\text{irr}}} & \text{if } X \le X_{\text{irr}} \end{cases}, \qquad (16)$$

X_{irr} is the irreducible water content defined as the value of moisture content above which the water is free in the pores.

The saturating vapour pressure is calculated according to the relation given by [11]:

$$P_v^{\text{sat}}(T) = 133.32\exp\left[18.584 - 3984.2/(233.426+T)\right]. \qquad (17)$$

The binary diffusion coefficient of vapour in air is calculated from equation given by Schirmer [5]:

$$\delta_{va}(T,P) = 2.26\,10^{-5}(T/T_R)^{1.81}\,P_R/P_g, \qquad (18)$$

where, T_R and P_R are reference temperature and pressure, respectively.

The surface tension in porous medium is expressed according to [11]:

$$\sigma(T) = -1.3\,10^{-7}T^2 - 1.58\,10^{-4}T + 0.07606. \qquad (19)$$

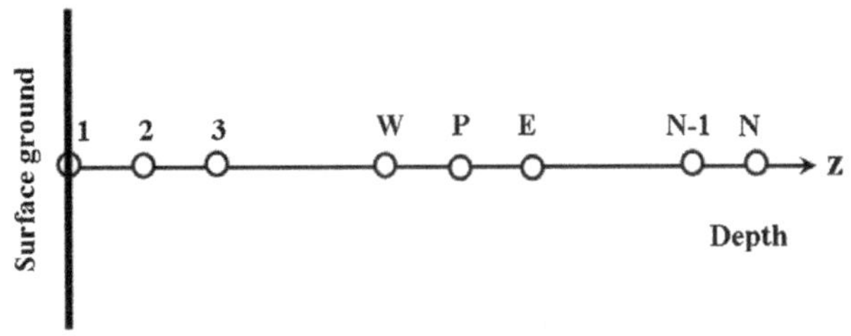

Figure 3. Control volume mesh.

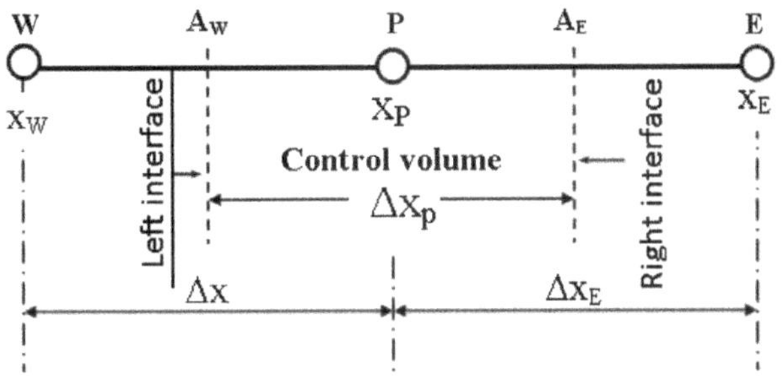

Figure 4. Normal control volume element for one-dimensional problems.

2.4 Discretization of the equations

For the discretization of these equations, the volume control method was used as recommended by Patankar [9]. Integrating the conservation equation of water over the control volume element and time duration, we obtain for each node P as shown in Figures 3 and 4:

$$\int_t^{t+\Delta t}\int_{x_W}^{x_E}\left[\varepsilon_s\rho_s\frac{\partial X}{\partial t}+\frac{\partial}{\partial t}(\varepsilon_g\rho_v)\right]\mathrm{d}V\mathrm{d}t-\int_t^{t+\Delta t}\times\int_{x_W}^{x_E}\nabla J_w\mathrm{d}V\mathrm{d}t=0. \tag{20}$$

Then:

$$\frac{(\Delta x_W+\Delta x_E)}{2\Delta t}\left[(\varepsilon_s\rho_s X+\varepsilon_g\rho_v)_P^{t+\Delta t}-(\varepsilon_s\rho_s X+\varepsilon_g\rho_v)_P^{t}\right]+\left[(J_w)_E^{t+\Delta t}-(J_w)_W^{t+\Delta t}\right]=0. \tag{21}$$

The water flux J_w from east and west boundary surfaces of control volume can be expressed as:

$$(J_w)_e^{t+\Delta t}=\left(\rho_w\frac{Kk_w}{\mu_w}\right)_{A_E}^{t+\Delta t}\frac{(P_w)_E^{t+\Delta t}-(P_w)_P^{t+\Delta t}}{\Delta x_E}+(\rho_g D_{\text{eff}})_{A_E}^{t+\Delta t}\frac{(y_v)_E^{t+\Delta t}-(y_v)_P^{t+\Delta t}}{\Delta x_E}, \tag{22}$$

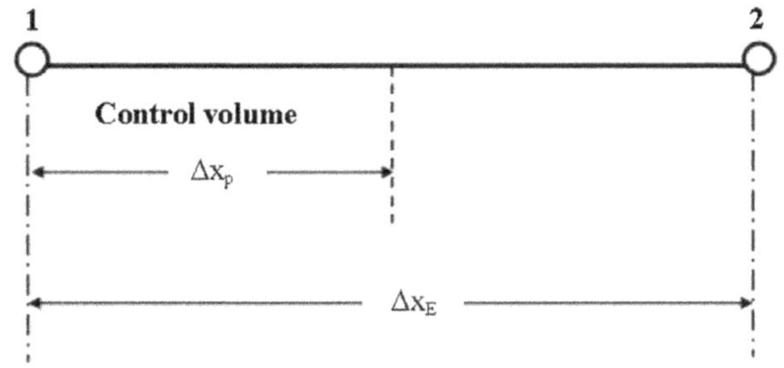

Figure 5. Control volume for the 1st element.

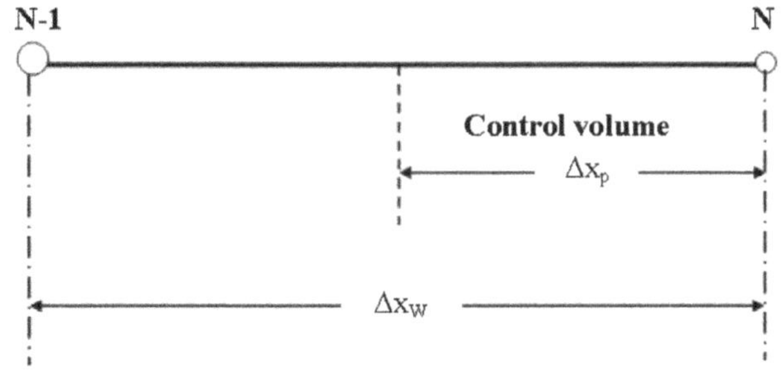

Figure 6. Control volume for the nth element.

$$(J_w)_w^{t+\Delta t}=\left(\rho_w\frac{Kk_w}{\mu_w}\right)_{A_W}^{t+\Delta t}\frac{(P_w)_P^{t+\Delta t}-(P_w)_W^{t+\Delta t}}{\Delta x_W}+(\rho_g D_{\text{eff}})_{A_W}^{t+\Delta t}\frac{(y_v)_P^{t+\Delta t}-(y_v)_W^{t+\Delta t}}{\Delta x_W}. \tag{23}$$

For the 1st and the Nth element (cf. Figures 5 and 6), the discrete version of water conservation writes:

$$\frac{(\Delta x_E)}{2\Delta t}\left[(\varepsilon_s\rho_s X+\varepsilon_g\rho_v)_1^{t+\Delta t}-(\varepsilon_s\rho_s X+\varepsilon_g\rho_v)_1^{t}\right]+\left[(J_w)_{A_E}^{t+\Delta t}-(J_w)_1^{t+\Delta t}\right]=0, \tag{24}$$

$$\frac{(\Delta x_W)}{2\Delta t}\left[(\varepsilon_s\rho_s X+\varepsilon_g\rho_v)_N^{t+\Delta t}-(\varepsilon_s\rho_s X+\varepsilon_g\rho_v)_N^{t}\right]+\left[(J_w)_N^{t+\Delta t}-(J_w)_{A_W}^{t+\Delta t}\right]=0. \tag{25}$$

The conservation equation of energy is discretized in a similar way as that of water:

$$\frac{(\Delta x_W+\Delta x_E)}{2\Delta t}\left[\Psi_P^{t+\Delta t}-\Psi_P^t\right]+\left[(J_e)_{A_E}^{t+\Delta t}-(J_e)_{A_W}^{t+\Delta t}\right]=0, \tag{26}$$

where:

$$\Psi_P^{t+\Delta t}=\left(\overline{\rho C_p}(T-T_R)+\varepsilon_g\rho_v\Delta h_{v0}\right)_P^{t+\Delta t}\ \text{et}\ \Psi_P^t=\left(\overline{\rho C_p}(T-T_R)+\varepsilon_g\rho_v\Delta h_{v0}\right)_P^{t}, \tag{27}$$

$$
\begin{aligned}
(J_e)_{A_E}^{t+\Delta t} &= (T_{A_E}^{t+\Delta t} - T_{\text{ref}}) \Bigg[(C_a \rho_g D_{\text{eff}})_{A_E}^{t+\Delta t} \\
&\frac{(y_a)_E^{t+\Delta t} - (y_a)_P^{t+\Delta t}}{\Delta x_E} + (C_v \rho_g D_{\text{eff}})_{A_E}^{t+\Delta t} \\
&\times \frac{(y_v)_E^{t+\Delta t} - (y_v)_P^{t+\Delta t}}{\Delta x_E} + \left(\rho_w C_w \frac{K k_w}{\mu_w} \right)_{A_E}^{t+\Delta t} \\
&\times \frac{(P_w)_E^{t+\Delta t} - (P_w)_P^{t+\Delta t}}{\Delta x_E} \Bigg] + (\Delta h_{v0} \rho_g D_{\text{eff}})_{A_E}^{t+\Delta t} \\
&\times \frac{(y_v)_E^{t+\Delta t} - (y_v)_P^{t+\Delta t}}{\Delta x_E} + (\lambda_{\text{eff}})_{A_E}^{t+\Delta t} \frac{(T)_E^{t+\Delta t} - (T)_P^{t+\Delta t}}{\Delta x_E}, \qquad (28)
\end{aligned}
$$

$$
\begin{aligned}
(J_W)_{A_W}^{t+\Delta t} &= (T_{A_W}^{t+\Delta t} - T_{\text{ref}}) \Bigg[(C_a \rho_g D_{\text{eff}})_{A_W}^{t+\Delta t} \\
&\frac{(y_a)_P^{t+\Delta t} - (y_a)_W^{t+\Delta t}}{\Delta x_W} + (C_v \rho_g D_{\text{eff}})_{A_E}^{t+\Delta t} \\
&\times \frac{(y_v)_P^{t+\Delta t} - (y_v)_W^{t+\Delta t}}{\Delta x_W} + \left(\rho_w C_w \frac{K k_w}{\mu_w} \right)_{A_W}^{t+\Delta t} \\
&\times \frac{(P_w)_P^{t+\Delta t} - (P_w)_W^{t+\Delta t}}{\Delta x_W} \Bigg] + (\Delta h_{v0} \rho_g D_{\text{eff}})_{AW}^{t+\Delta t} \\
&\times \frac{(y_v)_P^{t+\Delta t} - (y_v)_W^{t+\Delta t}}{\Delta x_W} + (\lambda_{\text{eff}})_{A_W}^{t+\Delta t} \frac{(T)_P^{t+\Delta t} - (T)_W^{t+\Delta t}}{\Delta x_W} \qquad (29)
\end{aligned}
$$

For element 1:

$$
\frac{\Delta x_E}{2\Delta t}\left[\Psi_1^{t+\Delta t} - \Psi_1^{t}\right] + \left[(J_e)_{A_E}^{t+\Delta t} - (J_e)_1^{t+\Delta t}\right] = 0. \qquad (30)
$$

For element N:

$$
\frac{\Delta x_E}{2\Delta t}\left[\Psi_N^{t+\Delta t} - \Psi_N^{t}\right] + \left[(J_e)_N^{t+\Delta t} - (J_e)_{A_W}^{t+\Delta t}\right] = 0. \qquad (31)
$$

The mass conservation equation of the dry air is not used because the total pressure is considered as constant in the soil since the dry air pressure can be deduced from the partial pressure of the vapour.

3. Numerical results and interpretations

We represent here the curves of time evolution of temperature and moisture levels and their evolution as a function of depth, starting first at the free surface of the soil, then deeper up to 3 m. Figure 7 shows the evolution of average month temperature measured at ground surface and in ambient air under shedder.

Figures 8 and 9 show the evolution of the soil temperature at free surface level and at different depths: 0.0; 0.1; 0.3 and 0.5 m. Both curves reveal clearly that the variations of the outdoor climatic conditions during the day affect only the upper layers of the soil, nearly down to 50 cm in the case of the characteristics of the soil on which we carried out the study.

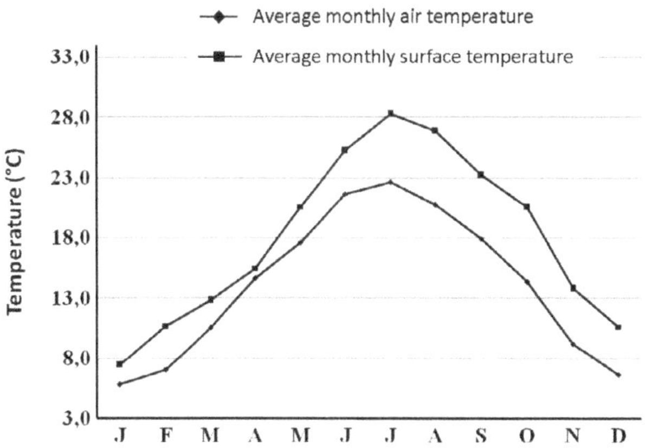

Figure 7. Average monthly air and surface soil temperature.

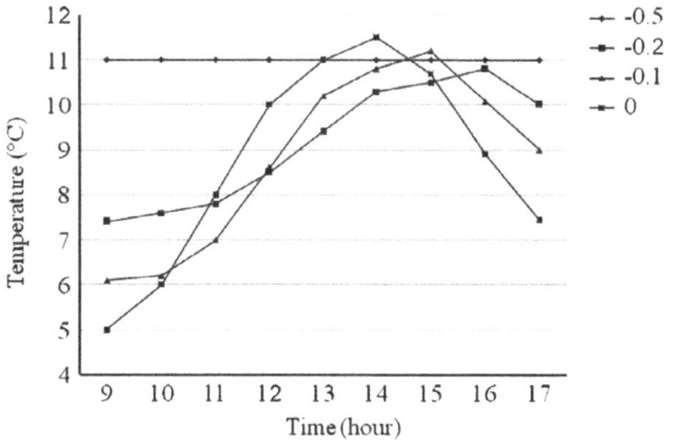

Figure 8. Horary evolution of the soil temperature (January 15).

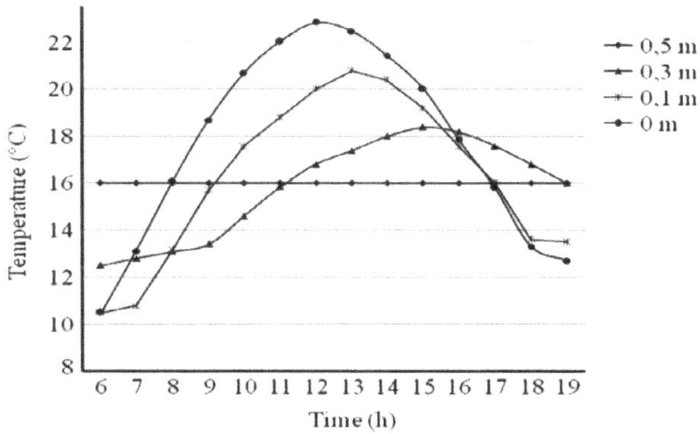

Figure 9. Horary evolution of the soil temperature (May 15).

The surface temperature is attenuated in depth and is affected by a progressive dephasing, associated with the inertia of the soil. Thus, during a typical day of May, the maximum temperature reaches 23 °C at the free surface level, at 12 h 30 mn, and decreases to 18.2 °C at 30 cm depth, with a gap of 3 h. This phenomenon is also reproduced in January with lower temperature maxima and a more important time difference in the afternoon. These results as well as the evolution of the temperature within this free surface – the top layer – are confirmed by previous works on this theme [3,8]. At the lower layers, the

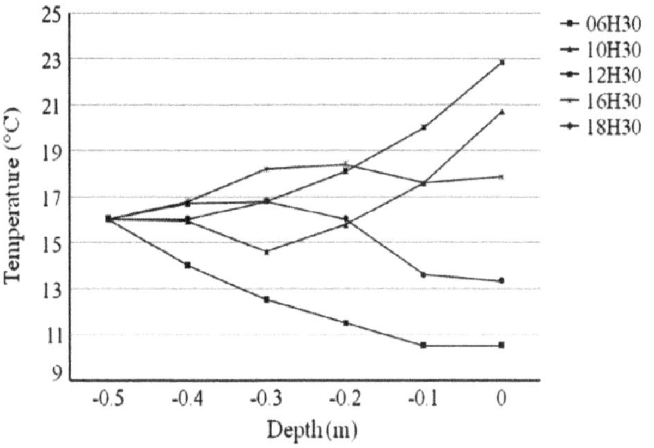

Figure 10. Temperature evolution in terms of depth (May 15).

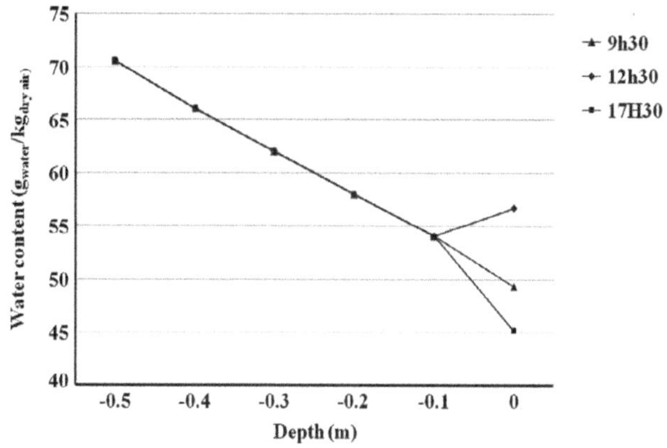

Figure 12. Water content evolution in terms of depth (May 15).

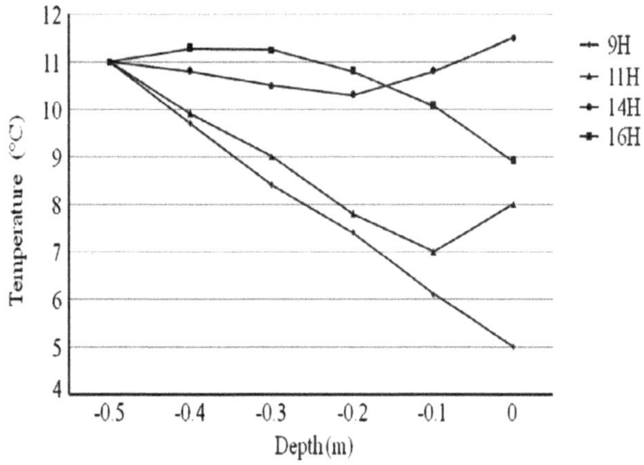

Figure 11. Temperature evolution in terms of depth (January 15).

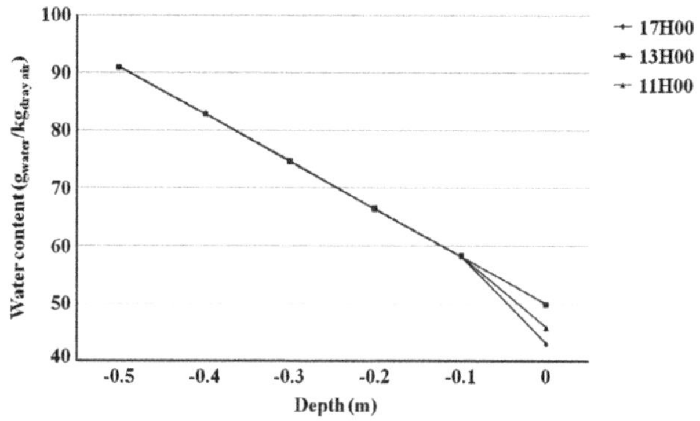

Figure 13. Water content evolution in terms of depth (January 15).

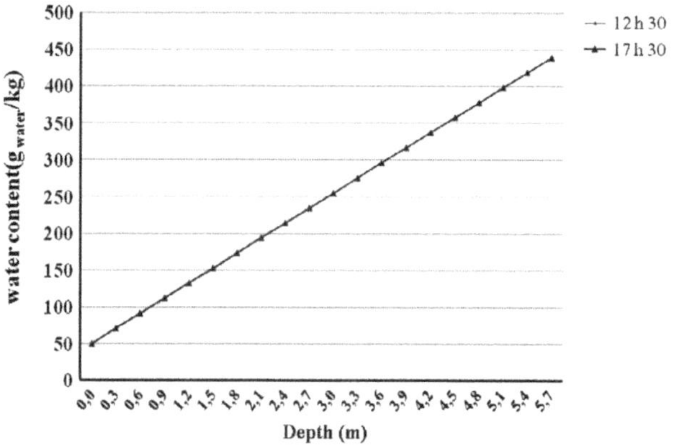

Figure 14. Water content in soil depth.

temperatures are constant during the day (we have noticed this in some measures carried out in situ). As we go deeper over 3 m, the temperature goes down to the monthly average soil temperature (17 °C in May).

Figure 10 (for May 15) and Figure 11 (for January 15) show the evolution of the temperature in depth at different moments of the day. Thus for May 15 at 6 h 30, the temperature is lower on the surface and increases as we go deeper until 0.5 m where it stabilises at 16 °C. On the other hand, at 12 h 30, the temperature is higher on the surface and decreases until 0.5 m of depth where it stabilises again at the same value of 16 °C. For January 15 (see Figure 11), the temperature is greater on the surface for 11 h and 14 h. It decreases in the first centimeters and increases again until it reaches 11 °C at 50 cm depth.

Figures 12 and 13 show the evolution of water content down to 50 cm of the upper ground layer. This parameter is slightly affected by the climatic variation during the day. However, we notice a slight rising of humidity on the surface at the hottest moment of the day. But this phenomenon is observed only in the first 10 cm.

When we are interested by the water content evolution in greater depths, we observe that it does not change during all the day time. Figure 14 shows that water content raises until it reaches the saturation at 3 m depth while keeping the initial linear profile (condition is set up at this limit). The study of this phenomenon must be conducted over a longer duration, such a complete season.

4 Conclusion

This theoretical study allows the reconstitution of the thermal and hydrous behaviour of unsaturated silty soil in real conditions at the scale of a day and as a function of depth. The results obtained by theoretical model are satisfactory and confirmed by the referred literature. However, the validation via experiment is necessary to validate again the value of the numerous parameters and the thermo physical characteristics which intervene in this problem. The computer code developed here can be a valuable tool in several domains such as agricultural science to help choose the plants and depth of the root system and the bioclimatic engineering and facilities such as underground heat exchangers in Canadian wells. With soil temperature and water content data, we can evaluate the heat flux and water flux through the soil layers. Thus, the water flux by capillary suction can be estimated for soil at known porosity and saturation.

References

1. Calvet R. 2003. Le sol: propriétés et fonction. Volume 2. Phénomènes physiques et chimiques. Applications agronomiques et environnementales. France agricole. Dunod 2003.
2. De Vries DA, Kruger AJ. 1966. On the value of the diffusion coefficient of water vapour in air. Proceeding du Colloque International du CNRS n°160: Phénomènes de transport avec changement de phase dans les milieux poreux ou colloïdaux. CNRS éditeur, 18 au 20 Avril 1966. p. 61–72.
3. Gao Z, Bian L, Hu Y, Wang L, Fan J. 2007. Determination of soil temperature in an arid region. Journal on Arid Environments, 71, 157–168.
4. Hong Vu T. 2006. Influence of pore size distribution on drying behaviour of porous media by a continous model, PhD-engineer. Otto Von Guericke University of Magdeburg, Germany.
5. Krischer O, Kast W. 1992. Die wissenschaftlichen Grundlagen der Trocknungstechnik, erster Band, dritte Auflage. Berlin: Springer.
6. Lefevbre R. 2003. Ecoulement Multiphasique en Milieux Poreux. Hiver 2003.
7. Moyne C. 1987. Transferts couplés chaleur-masse lors du séchage: prise en compte du mouvement de la phase gazeuse. Doctorat thesis of Polytechnical National Institut of Lorraine, France.
8. Nebbali R. 2008. Modélisation de la dynamique du climat interne distribué dans une serre de culture. Doctorat thesis of Franche-Comté University, France.
9. Patankar SV. 1980. Numerical Heat Transfer and Fluid Flow. Hemisphere Publ. Co: New York.
10. Perre P, Turner IW. 1999. A 3-D version of transPore: a comprehensive heat and mass transfer computational model for simulating the drying of porous media. International Journal of Heat and Mass Transfer, 42, 4501–4521.
11. Vargaftik NB, Vinogradov JK, Yargin VS. 1996. Handbook of Physical Properties of Liquids and Gases: Pure Substances and Mixtures, 3rd edn. Begell House: New York.
12. Whitaker S. 1977. Simultaneous heat, mass, and momentum transfer in porous media: a theory of drying. Advances in Heat Transfer, 13, 119–203.
13. Whitaker S. 1980. Heat and mass transfer in granular porous media, in Advances in Dryng 1. Mujumdar AS, Editor. Hemisphere Publ. Co: Washington, DC. p. 23–61.

9

Modified Covariance Matrix Adaptation – Evolution Strategy algorithm for constrained optimization under uncertainty, application to rocket design

Rudy Chocat[1], Loïc Brevault[2,3], Mathieu Balesdent[2,*], and Sébastien Defoort[2]

[1] IFMA, EA3867, Laboratoires de Mécanique et Ingénieries, Clermont Université, CP 104488, 63000 Clermont-Ferrand, France
[2] Onera – The French Aerospace Lab, BP 80100, 91123 Palaiseau Cedex, France
[3] CNES – Launchers Directorate, 52 rue Jacques Hillairet, 75612 Paris, France

Abstract – The design of complex systems often induces a constrained optimization problem under uncertainty. An adaptation of CMA-ES(λ, μ) optimization algorithm is proposed in order to efficiently handle the constraints in the presence of noise. The update mechanisms of the parametrized distribution used to generate the candidate solutions are modified. The constraint handling method allows to reduce the semi-principal axes of the probable research ellipsoid in the directions violating the constraints. The proposed approach is compared to existing approaches on three analytic optimization problems to highlight the efficiency and the robustness of the algorithm. The proposed method is used to design a two stage solid propulsion launch vehicle.

Key words: Evolutionary Strategy, Covariance Matrix Adaptation, CMA-ES, Uncertainty, Constrained optimization, Rocket design.

1 Introduction

The design of complex systems, such as aerospace vehicles, can be expressed as a non linear constrained optimization problem solving. The determination of the optimal system architecture requires a global exploration of the design space involving repeated evaluations of computationally expensive black box functions used to model the different disciplines (e.g. in aerospace: structure, propulsion, aerodynamics). Moreover, in the early design phases, uncertainties arise due to lack of knowledge of the system characteristics and the environment (e.g. propellant combustion rate, wind gust) and to the use of low fidelity analyses (for instance analytic code instead of numerically costly finite element analysis). Therefore, the design of a complex system at the conceptual design phase entails a constrained optimization problem under uncertainty that can be formulated as:

$$\begin{array}{ll} \text{Min} & \Xi[F(\mathbf{z},\mathbf{U})] \\ \text{wrt} & \mathbf{z} \end{array} \tag{1}$$

$$\text{st} \quad \mathbf{K}[\mathbf{g}(\mathbf{z},\mathbf{U})] \leq 0 \tag{2}$$

$$\mathbf{z}_{\min} \leq \mathbf{z} \leq \mathbf{z}_{\max} \tag{3}$$

with:

- $\mathbf{z}$: the vector of design variables $\mathbf{z} \in \mathbb{R}^n$ (e.g. rocket diameter, propellant masses),
- $\mathbf{U}$: the vector of the uncertain variables $\mathbf{U} \in \mathbb{R}^p$ (e.g. wind gust during the rocket launch, material characteristics). In the paper, the probability theory is used to model uncertainty.
- F: $\mathbb{R}^n \times \mathbb{R}^p \rightarrow \mathbb{R}$: the performance function (e.g. propulsive speed increment ΔV, Gross Lift-Off Weight),
- Ξ: a measure of uncertainty for the performance function F (e.g. the expected value, a linear combination of expected value and standard deviation),
- $\mathbf{g}$: the vector of the inequality constraint functions, g_i, $i \in \{1, \ldots, m\}$ (e.g. maximal allowed stress in the structures, maximal tolerance on orbit injection accuracy),
- $\mathbf{K}$: a vector of measures of uncertainty for the inequality constraint functions (e.g. linear combination of expected value and standard deviation, probability of failure),
- $\mathbf{z}_{\min}$ and $\mathbf{z}_{\max}$ the lower and upper bounds for the design variables.

Being able to handle uncertainty at early design phases is essential to efficiently characterize the optimal system design

*e-mail: mathieu.balesdent@onera.fr

and its performances. It can reduce time and cost of the next design phases by avoiding redesign process [18]. However, solving the optimization problem (Eqs. (1)–(3)) is challenging due to the presence of both the uncertainty and inequality constraints. A brief overview of the existing methods to handle uncertainty and inequality constraints is provided in the next paragraphs.

The presence of noise in the optimization problem (Eqs. (1)–(3)) results from the estimation of the measures of uncertainty (Ξ and $\mathbf{K}$). Several methods exist to compute these measures and the most classical approach is a numerical approximation by the Crude Monte Carlo (CMC) method. The CMC estimator of the uncertainty measure is also a random variable. In order to numerically handle the presence of noise in optimization, several approaches have been proposed:

- Re-sampling [3]: the re-sampling method consists of repeated evaluations of the objective and the constraint uncertainty measures for the same design variable value $\mathbf{z}$. Then, a statistics of the repeated samples is used instead of the single evaluation of the objective and the constraint measures to decrease the impact of noise. The main drawback of this approach is the increase of the computational cost due to the repeated evaluations of expensive functions.
- Surrogate model [3]: surrogate models of the objective function measure Ξ (and/or the constraint measures $\mathbf{K}$) are built from the evaluated measures. In general, surrogates smooth the noisy functions and decrease the impact of uncertainty in the optimization. The main drawback lies in the difficulty to build accurate surrogate models in high dimensions.
- Population based algorithms [10]: to address uncertainty, optimization algorithms relying on a population of candidates allow to increase the size of the population to enlarge its spread and to obtain more information and smooth the noise.

Moreover, the presence of uncertainty makes the use of classical optimization algorithms, such as gradient based optimization algorithms, not suited for optimization. Indeed, the gradients of the objective uncertainty measure or the constraint uncertainty measures are noisy resulting in possible erroneous descent directions. Diverse derivative-free algorithms have been proposed in the literature to solve optimization under uncertainty problems [15]. Among these, the population based optimization algorithms seem promising [6]. Swarm Intelligence (Particle Swarm Optimization [7], Artificial Bee Colonies [13], etc.), Differential Evolution [17], Evolutionary Algorithms (Genetic Algorithm [12]) or Evolution Strategies (Covariance Matrix Adaptation – Evolution Strategy [10]) have been investigated to solve noisy optimization problems. One of the current issue with the derivative-free algorithms is the constraint handling which relies mainly on heuristic approaches and is problem dependent [16]. A comprehensive overview of constraint handling in derivative-free algorithms is presented in the survey [16]. The most commonly applied methods to handle the constraints are:

- Death penalty [19]: it is the simplest method to handle constraints. The solution that does not satisfy the constraints is rejected and another potential solution is re-evaluated until one candidate solution satisfies the constraints. The advantage of the approach is that it does not modify the optimization algorithm but the method is very expensive because no information is learned from an unfeasible solution (a solution which does not satisfy the constraints) to characterize the non feasible space. Furthermore, if the feasible space is restricted compared to the design space, the computational cost becomes prohibitive because a high number of samples has to be generated to obtain feasible solutions.
- Penalization [5, 6]: this approach consists in replacing the objective function $\Xi\ [F(\mathbf{z}, \mathbf{U})]$ by a combination of the objective function and a penalization function Π such as: $\Xi\ [F(\mathbf{z}, \mathbf{U})] + \Pi\ (\mathbf{K}[\mathbf{g}(\mathbf{z}, \mathbf{U})])$. The penalization function can be fixed or can change as a function of the number of iterations. When a solution violates the constraints, the objective function is deteriorated by a factor proportional to the penalization function and the value of the constraints. Despite its simplicity, the main drawback of this approach lies in the determination of a suitable penalization function which depends on the objective function and the constraints and is thus problem dependent.
- Multi-objective [4]: this method transforms the optimization problem into a multi-objective optimization problem by considering the minimization of the violation of the constraints as an objective. Dedicated multi-objective optimization algorithms can be used, however, it often results in an increase of the computational cost [16].
- Surrogate model [14]: this approach builds a surrogate model based on the unfeasible solutions in order to approximate the non feasible zones. However, this approach requires enough unfeasible solutions to construct accurate surrogate models. Moreover, it can be difficult to build the surrogates in high dimension or if the constraints are highly non linear.

Among the derivative-free algorithms, the Covariance Matrix Adaptation – Evolution Strategy (CMA-ES) [10] is particularly competitive for real-valued black box functions as highlighted in several extensive benchmarks [8, 9]. Moreover, a treatment of uncertainty has been proposed for CMA-ES and has been successfully tested in a benchmark of optimization under uncertainty problems [8]. However, the test problems used to evaluate the performances of CMA-ES are unconstrained optimization problems and only few studies focus on the application of CMA-ES to constrained optimization problems [2, 6]. An adaptive penalty function has been proposed to update the penalty coefficient as a function of the sum of the violated constraint values [6]. Arnold and Hansen [1] proposed a new approach to handle the constraints for a simplified version of CMA-ES: (1+1)-CMA-ES which involves one offspring candidate generated from one parent. Modified (1+1)-CMA-ES approximates the normal vector directions of the

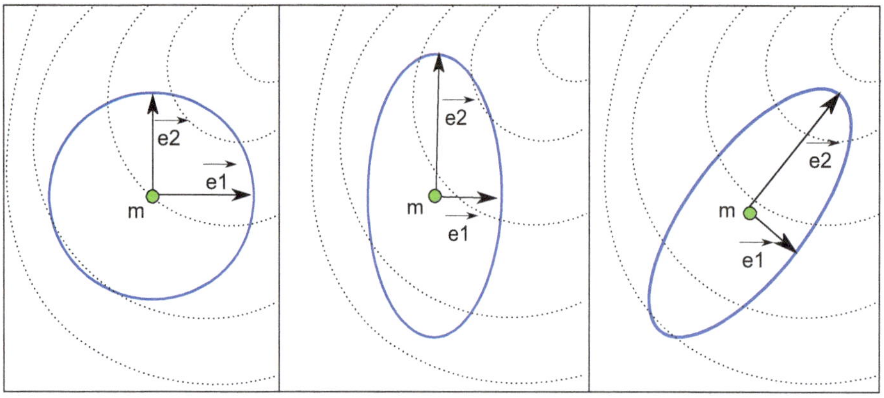

Figure 1. Three ellipsoids, depicting three different normal distributions, where **I** is the identity matrix, **D** is a diagonal matrix, and **C** is a positive definite covariance matrix. Thin dot lines depict objective function contour lines.

constraint boundaries in the vicinity of the current candidate solution. A control of the covariance of the distribution of the offspring candidate allows to get closer to the boundary of the feasible regions without violating them. The approach provides interesting results but is limited to (1+1)-CMA-ES which becomes inefficient in high dimensions (local convergence, large number of calls to the functions) [1].

The main objective of this paper is to adapt CMA-ES(λ, μ), which involves λ offspring candidates generated from μ parents, to efficiently solve constrained optimization problem under uncertainty. The proposed approach is based on a control of the covariance of the distribution of the offspring candidates through a modification of the covariance matrix that characterizes the probable research hypervolume in order to generate candidates without violating the constraints. The proposed adaptations take into account the specificities of the update and the selection mechanisms of CMA-ES(λ, μ) that differ from (1+1)-CMA-ES due to the presence of a population instead of a single candidate. The remainder of the paper is organized as follows. Section 2 provides an overview of the original CMA-ES(λ, μ) and the adapted (1+1)-CMA-ES to handle constraints. Section 3 presents the modifications of CMA-ES(λ, μ) to handle constraints in optimization problems. Section 4 evaluates the algorithm efficiency on three analytic test functions and on the design of a two stage solid rocket. A comparison with penalized CMA-ES(λ, μ) and modified (1+1)-CMA-ES is provided.

2 Description of CMA-ES(λ, μ) and modified (1+1)-CMA-ES

2.1 CMA-ES(λ, μ) algorithm

The Covariance Matrix Adaptation – Evolution Strategy (CMA-ES) introduced by Hansen et al. [10] belongs to the Evolution Strategy algorithm family. A brief overview of CMA-ES(λ, μ) is provided in the section in order to understand the proposed modifications for the constraint handling, for more information on the algorithm see [10]. CMA-ES(λ, μ) is used to solve unconstrained optimization problems. CMA-ES relies on a distribution model of a candidate population (parametrized multivariate normal distribution) in order to explore the design space. It is based on a selection and an adaptation process of the candidate population. In CMA-ES(λ, μ), at each generation, λ offspring candidates are generated from μ parents. At the next generation, to select the new parents from the offspring candidates, a (λ, μ)-selection is used, the μ best offspring candidates are chosen (with respect to their ranking according to the objective function). The multivariate normal distribution has an infinite support, but an iso-probability contour (for instance at ±3 standard deviation of the mean) is characterized by an ellipsoid delimiting a probable research hypervolume (Figure 1). Throughout the generations, the research hypervolume is updated in order to converge and to shrink around the global optimum. CMA-ES(λ, μ) generates the population by sampling a multivariate normal distribution:

$$\mathbf{z}_t^{(k+1)} \sim \mathbf{m}^{(k)} + \sigma^{(k)}\mathcal{N}\left(0, \mathbf{C}^{(k)}\right), \quad \text{for } t = 1, \ldots, \lambda \tag{4}$$

with $\mathbf{z}_t^{(k+1)} \in \mathbb{R}^n$ an offspring candidate generated from a mean vector $\mathbf{m}^{(k)}$, a step size $\sigma^{(k)}$ and a multivariate normal distribution $\mathcal{N}(0, \mathbf{C}^{(k)})$ with zero mean and a covariance matrix $\mathbf{C}^{(k)} \in \mathbb{R}^n \times \mathbb{R}^n$. λ is the size of the population generated at each iteration (k). The normal distribution is characterized by a positive definite covariance matrix $\mathbf{C}^{(k)}$ in order to allow homothetic transformations and rotations of the probable research hypervolume (Figure 1). The update of the covariance matrix incorporates dependence between the past generations and between the μ best candidates from the previous generation [10]. The mean vector characterizes the center of the next population and is determined by a combination process through the weighting of the μ best candidates: $\mathbf{m}^{(k+1)} = \sum_{i=1}^{\mu} w_i \mathbf{z}_{(i)}^{(k+1)}$ with $\sum_{i=1}^{\mu} w_i = 1$, $w_1 > w_2 > \cdots > w_\mu > 0$ the weighting coefficients and $(\mathbf{z}_{(1)}, \ldots, \mathbf{z}_{(\mu)})$ the best candidates among the offspring ranked according to the fitness value. The weighting coefficients are determined based on the number of μ best candidates according to [10]. A simplified version of CMA-ES(λ, μ) is described in Algorithm 1. A detailed description of the selection and update mechanisms can be found in [10].

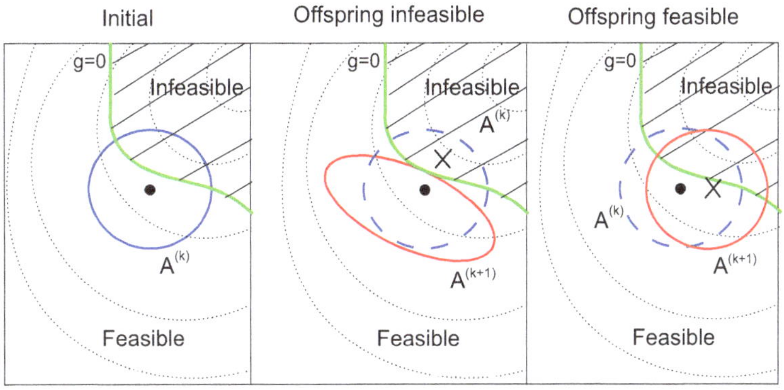

Figure 2. The dot is the parent, the cross is the generated offspring, the solid line circle is characterized by $\mathbf{A}^{(k)}$ defining the ellipsoid delimiting an iso-probability research hypervolume. At the center, the red ellipsoid represents the update of $\mathbf{A}^{(k+1)}$ in order to take into account the constraint violation by the offspring. On the right figure, the offspring does not violate the constraint resulting in a standard covariance matrix update.

Algorithm 1 CMA-ES(λ, μ) [10]

1) Initialize the covariance matrix $\mathbf{C}^{(0)} = \mathbf{I}$, the step size $\sigma^{(0)}$ and the selection parameters [10]
2) Initialize the mean vector $\mathbf{m}^{(0)}$ to a random candidate, $k \leftarrow 0$

while CMA-ES convergence criteria are not reached **do**

3-1) Generate λ new offspring candidates according to: $\mathbf{z}_t^{(k+1)} \sim \mathbf{m}^{(k)} + \sigma^{(k)}\mathcal{N}(0, \mathbf{C}^{(k)})$, $t \in \{1, \ldots, \lambda\}$

3-2) Evaluate candidates and rank them based on the objective function

3-3) Determine the mean vector given the weighting coefficients of the μ best candidates: $\mathbf{m}^{(k+1)} = \sum_{i=1}^{\mu} w_i \mathbf{z}_{(i)}^{(k+1)}$

3-4) Update covariance matrix $\mathbf{C}^{(k+1)}$ and the step size $\sigma^{(k+1)}$ according to [10], $k \leftarrow k + 1$

end while

4) **return** best candidate $\mathbf{z}_{\mathbf{best}}$

The convergence criteria can either be based on the maximum number of iterations (function evaluations), the fitness value, the standard deviation of the current population smaller than a given tolerance, or the covariance matrix $\mathbf{C}$ which becomes numerically not positive definite. In the next paragraph, an adaptation of CMA-ES for uncertainty handling is described.

2.2 CMA-ES(λ, μ) algorithm for optimization under uncertainty

Several features ensure the CMA-ES robustness with respect to the presence of uncertainty in the objective function: the population-based approach, the weighted averaging in the recombination process, the rank based and the non-elitist selection (not based on the best offspring candidate). However, if the noise is too high compared to the objective function (signal to noise ratio too low) it perturbs the algorithm convergence. An appropriate handling of uncertainty has been proposed by Hansen et al. [11] to overcome this issue. Modified selection and update mechanisms are performed when the noise is above a given threshold. It is based on a re-sampling approach and involves re-evaluation of the objective function. Because CMA-ES(λ, μ) is only based on the rank of the candidates, the effective noise is evaluated by monitoring changes or stability of the offspring candidate ranking. If the offspring candidate ranking is changed after the re-evaluation of the objective function, the ranking change of the offspring candidates is aggregated into a metric quantifying the uncertainty level [11]. If the noise is higher than a given uncertainty level threshold, the step size σ is increased. The increase of σ ensures that despite the noise, sufficient selection information is available [11].

A benchmark of algorithm dealing with optimization under uncertainty has been performed and the treatment of uncertainty with CMA-ES(λ, μ) allows to obtain accurate results [8]. However, CMA-ES(λ, μ) is not able to efficiently handle the constraints. Indeed, only penalization [5, 6] or surrogate based methods [14] have been proposed. These approaches can be effective but are problem dependent and require accurate tuning of hyper parameters. A simplified version of CMA-ES(λ, μ) called (1+1)-CMA-ES has been adapted to handle constraints and is detailed in the next section.

2.3 (1+1)-CMA-ES with constraint handling

(1+1)-CMA-ES is a simplified version [1] of CMA-ES(λ, μ) with only one offspring generated from one parent, "+" means that the selection is done between the parent and the offspring. As in CMA-ES(λ, μ), the offspring candidate solution is generated as:

$$\mathbf{z}^{(k+1)} \sim \mathbf{z}^{(k)} + \sigma^{(k)}\mathcal{N}\left(0, \mathbf{C}^{(k)}\right) \tag{5}$$

(1+1)-CMA-ES is easier to implement as only one offspring $\mathbf{z}^{(k+1)}$ is generated from one parent $\mathbf{z}^{(k)}$ at each generation and the selection is between the parent and the offspring. The update mechanisms for (1+1)-CMA-ES are detailed in [1].

To incorporate the handling of constraints, Arnold and Hansen [1] proposed to reduce the covariance of the distribution of the offspring candidate in the approximated directions of the normal vectors of the constraint boundaries in the vicinity of the current parent candidate solution. For that purpose, the matrix $\mathbf{A}^{(k)}$, which is the Cholesky decomposition of $\mathbf{C}^{(k)} : \mathbf{A}^{(k)}\mathbf{A}^{(k)^T} = \mathbf{C}^{(k)}$, is updated in case of constraint violations in order to avoid generating candidates in the next generations that will violate the constraints. $\mathbf{A}^{(k)}$ is used as it is easier to compute its inverse than for $\mathbf{C}^{(k)}$. A vector characterizing the constraints $\mathbf{v}_j^{(k)} \in \mathbb{R}^n$ is defined, initialized to be zero, and updated according to:

$$\mathbf{v}_j^{(k)} \leftarrow (1 - c_c)\mathbf{v}_j^{(k)} + c_c\mathbf{A}^{(k)}\mathbf{z}^{(k)} \quad \forall j \in \{1, \ldots, m\} \tag{6}$$

where $\mathbf{v}_j^{(k)}$ is an exponentially fading record of steps that have violated the constraints and c_c a parameter characterizing how fast the information present in $\mathbf{v}_j^{(k)}$ fades. In the generations in which the offspring candidate is unfeasible, the Cholesky matrix is updated according to:

$$\mathbf{A}^{(k)} \leftarrow \mathbf{A}^{(k)} - \frac{\beta}{\sum_{j=1}^{m} \mathbb{1}_{g_j(\mathbf{z}^{(k)})>0}} \sum_{j=1}^{m} \mathbb{1}_{g_j(\mathbf{z}^{(k)})>0} \frac{\mathbf{v}_j^{(k)}\mathbf{w}_j^{(k)^T}}{\mathbf{w}_j^{(k)^T}\mathbf{w}_j^{(k)}} \tag{7}$$

with $\mathbf{w}_j^{(k)} = \mathbf{A}^{(k)^{-1}}\mathbf{v}_j^{(k)}$, the indicator function associated to the constraint g_j: $\mathbb{1}_{g_j(\mathbf{z}^{(k)})>0}$ and β a parameter controlling the reduction of the covariance of the distribution. For $\beta = 0$, the algorithm is identical to the standard (1+1)-CMA-ES. The update of the matrix $\mathbf{A}^{(k)}$ allows to modify the scale and the orientation of the research hypervolume in order to be tangential to the constraints and to avoid its violation (Figure 2). Modified (1+1)-CMA-ES for constraint handling is interesting because it is not problem dependent. Experimental evaluations have been performed highlighting its efficiency for unimodal constrained optimization problems. However, as (1+1)-CMA-ES, it is not able to optimize multimodal functions and becomes inefficient in high dimensions [1].

To overcome these drawbacks, in the next section, an adaptation of CMA-ES(λ, μ) for constraint handling is proposed inspired from the (1+1)-CMA-ES approach.

3 Proposed adaptation of CMA-ES(λ, μ) for constraint handling

The proposed approach of CMA-ES(λ, μ) for constraint handling is based on the same approach as modified (1+1)-CMA-ES. However, it is necessary to adapt it to take into account the specificities of CMA-ES(λ, μ). Indeed, CMA-ES(λ, μ) generates a population instead of a single offspring candidate. Thus, each offspring candidate can potentially violate one or several constraints. Moreover, the selection of the μ best candidates is based on the rank of the objective function. However, these best candidates can also violate the constraints. Depending if the μ best candidates are feasible or not, or if only a fraction of them is feasible, or on the number of violated constraints, the covariance matrix used to generate the offspring candidates has to be modified in order to avoid the generation of unfeasible offspring candidates. The research hypervolume engendered by an iso-probability contour of the multivariate normal distribution $\mathcal{N}(0, \mathbf{C})$ can be represented by a n-dimensional ellipsoid. In the proposed approach, the constraint handling method allows to reduce the semi-principal axes of the research ellipsoid in

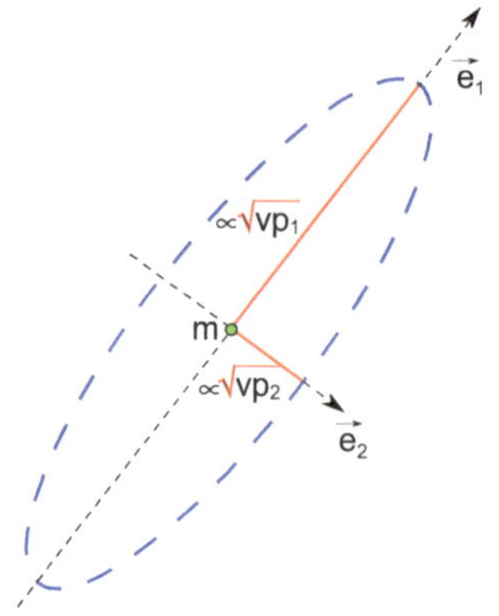

Figure 3. Parametrization of the ellipsoid defining the probable research hypervolume.

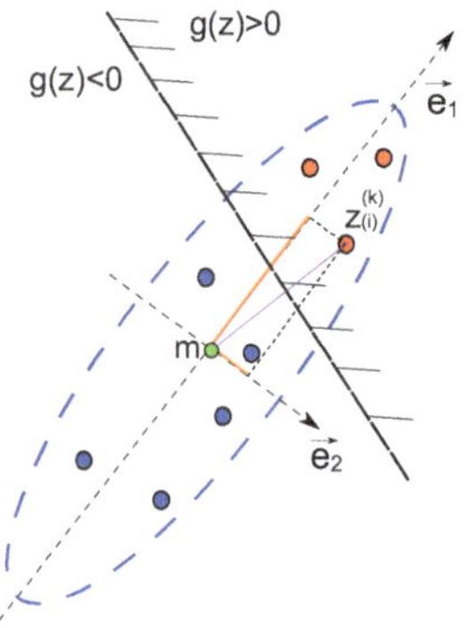

Figure 4. Violation of the constraint and projection over the eigenvectors, blue = feasible, red = unfeasible candidates.

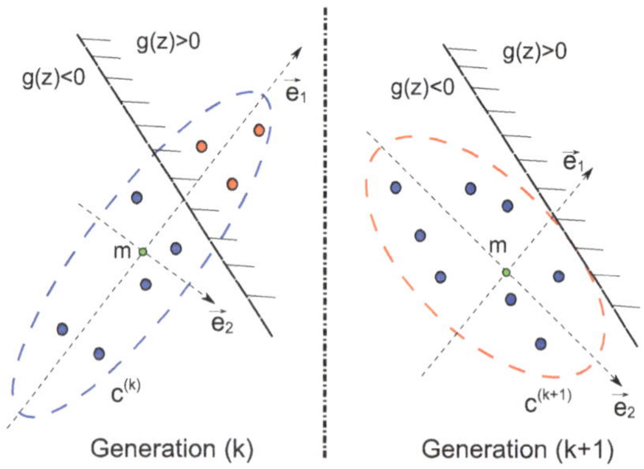

Figure 5. Evolution of the covariance matrix due to the constraint violation, blue = feasible, red = unfeasible candidates.

the directions violating the constraints. The eigenvalues of the covariance matrix **C** control the length of the semi-principal axes. The decrease of the eigenvalues reduces the semi-principal axis lengths. The covariance matrix, which is symmetric positive definite, can be decomposed according to:

$$\mathbf{C} = \mathbf{P}\mathbf{D}\mathbf{D}\mathbf{P}^T = \mathbf{P}\mathbf{D}^2\mathbf{P}^T \tag{8}$$

where $\mathbf{P}$ is an orthogonal matrix such that: $\mathbf{P}\mathbf{P}^T = \mathbf{P}^T\mathbf{P} = \mathbf{I}$. The columns of $\mathbf{P}$ form an orthogonal basis of eigenvectors of $\mathbf{C}$. $\mathbf{D} = \mathrm{diag}(\sqrt{vp_1}, \ldots, \sqrt{vp_n})$ is a diagonal matrix with the square roots of eigenvalues of $\mathbf{C}$. As illustrated in Figure 3, the square roots of the covariance matrix eigenvalues $\sqrt{vp_i}$ are proportional to the semi-principal axis lengths of the ellipsoid defining the sampling hypervolume.

At the generation (k), between the step 3-3) and 3-4) of Algorithm 1, if any of the m constraints is violated by any of the μ best offspring candidates, the covariance matrix is modified according to Algorithm 2.

Algorithm 2 Proposed CMA-ES(λ, μ) covariance matrix modificiation

$$3.3.1)\ \text{Diagonalize}\ \mathbf{C}^{(k)} \text{such that}\ \mathbf{P}\mathbf{D}^{(k)^2}\mathbf{P}^T = \mathbf{C}^{(k)} \quad (9)$$

$$3.3.2)\ \mathbf{S}^{(k)} \leftarrow \mathbf{P}\mathbf{D}^{(k)^2}\mathbf{P}^T - \gamma \mathbf{P}\mathrm{diag}\left(\sqrt{vp_{m_1}^{(k)}}, \ldots, \sqrt{vp_{m_n}^{(k)}}\right)^2 \mathbf{P}^T \quad (10)$$

$$3.3.3)\ \mathbf{C}^{(k)} \leftarrow \left[\frac{\det(\mathbf{C}^{(k)})}{\det(\mathbf{S}^{(k)})}\right]^{1/n} \mathbf{S}^{(k)} \quad (11)$$

with:

$$\sqrt{vp_{m_i}^{(k)}} = \left(\sqrt{vp_i^{(k)}} \sum_{j=1}^{m} \frac{\sum_{l=1}^{\mu} \mathbb{1}_{g_j(\mathbf{z}_{(l)}^{(k)})>0} w_{lj} \mathrm{Proj}_{\vec{e}_i}\left[\mathbf{z}_{(l)}^{(k)} - \mathbf{m}^{(k)}\right]}{\sum_{t=1}^{\mu} \mathbb{1}_{g_j(\mathbf{z}_{(t)}^{(k)})>0} w_{tj}}\right) \quad (12)$$

The covariance matrix is diagonalized, equation (9) and the eigenvalues $vp_i^{(k)}$ of the covariance matrix $\mathbf{C}^{(k)}$ are modified. The new eigenvalues are the former eigenvalues minus a term $vp_{m_i}^{(k)}$ taking into account the violation of the constraints. $\sqrt{vp_{m_i}^{(k)}}$, equation (12), is a function of the former eigenvalues $\sqrt{vp_i^{(k)}}$, of the indicator function of the constraint $\mathbb{1}_{g_j(\mathbf{z}_{(l)}^{(k)})>0}$, of the weighting coefficients w_{ij} and of the projection $\mathrm{Proj}_{\vec{e}_i}\left[\mathbf{z}_{(l)}^{(k)} - \mathbf{m}^{(k)}\right]$ of the distance between an ordered candidate violating the constraints $\mathbf{z}_{(l)}^{(k)}$ and the mean point $\mathbf{m}^{(k)}$ in the direction of the eigenvector $\vec{e}_i$ corresponding to the eigenvalue $vp_i^{(k)}$. γ is a parameter similar to β in (1+1)-CMA-ES. For $\gamma = 0$ the proposed algorithm is similar to the classical CMA-ES(λ, μ). For each constraint g_j, the μ_{cj} candidates among the μ best candidates that violate the constraint are ranked according to the constraint value. The weighting coefficients w_{ij} for each constraint g_j are defined according to the same rule as for the recombination process for the calculation of $\mathbf{m}^{(k)}$:

$$w_{ij} = \frac{\ln(\mu_{cj} + 1) - \ln(i)}{\mu_{cj}\ln(\mu_{cj} + 1) - \sum_{k=1}^{n} \ln(k)} \quad (13)$$

and $\sum_{i=1}^{\mu} w_{ij} = 1$ where: $w_{1j} \geq \cdots \geq w_{\mu j} \geq 0$. w_{1j} is associated to the candidate that violates the most the constraint g_j and $w_{1\mu_c}$ with the candidate that violates the less the constraint. For the candidates among the μ best that do not violate the constraint, the indicator function is equal to zero and therefore these candidates do not participate in the modification of the covariance matrix. The projection of the violation distance along the eigenvector (Figure 4) allows to reduce the covariance matrix in the direction orthogonal to the constraint violation.

Equation (11) allows to keep the hypervolume of the ellipsoid constant before and after the modification of the covariance matrix in order to avoid premature convergence. The volume of the ellipsoid is reduced in the direction orthogonal to the constraints but is increased in the direction tangential to the constraints (Figure 5). The modified CMA-ES(λ, μ) algorithm for constraint handling is detailed in Algorithm 3.

Algorithm 3 Proposed modified CMA-ES(λ, μ) for constraint handling

1) Initialize the covariance matrix $\mathbf{C}^{(0)} = \mathbf{I}$, the step size $\sigma^{(0)}$ and the selection parameters [10]
2) Initialize the mean vector $\mathbf{m}^{(0)}$ to a random candidate, $k \leftarrow 0$

while CMA-ES convergence criterion is not reached **do**
 3-1) Generate λ new offspring candidates according to: $\mathbf{z}_t^{(k+1)} \sim \mathbf{m}^{(k)} + \sigma^{(k)}\mathcal{N}(0, \mathbf{C}^{(k)}),\ t \in \{1, \ldots, \lambda\}$
 3-2) Evaluate candidates and sort them based on the objective function
 if all the μ best candidates are infeasible **then**
  Modify the covariance matrix according to Algorithm 2.
  Return to step 3.1)
 else
  If all the μ best candidates are feasible
  Determine the mean vector given the weightings of the μ best candidates: $\mathbf{m}^{(k+1)} = \sum_{i=1}^{\mu} w_i \mathbf{z}_{(i)}^{(k+1)}$
  Update covariance matrix $\mathbf{C}^{(k+1)}$ according to [10]
  end if
 else
  If at least one of the μ best candidates is infeasible and at least one is feasible
  Modify the covariance matrix according to Algorithm 2.
  Use the feasible candidates to determine the mean vector $\mathbf{m}^{(k+1)}$
  Use the feasible candidates to update covariance matrix $\mathbf{C}^{(k+1)}$ according to [10] **end if**
 end if
 3-3) Update the step size $\sigma^{(k+1)}$ according to [10], $k \leftarrow k + 1$
end while
4) **return** best candidate $\mathbf{z}_{best}$

The evolution of the ellipsoid between the generations (k) and $(k + 1)$ if one of the μ best candidates violates a constraint is illustrated in Figure 5. The modification of $\mathbf{C}^{(k)}$ allows homothetic transformations in order to avoid to generate candidates in the non feasible zone.

If the mean vector $\mathbf{m}^{(k)}$ after the combination process is not feasible, instead of reducing the covariance matrix, the ellipsoid hypervolume is increased in order to generate candidates

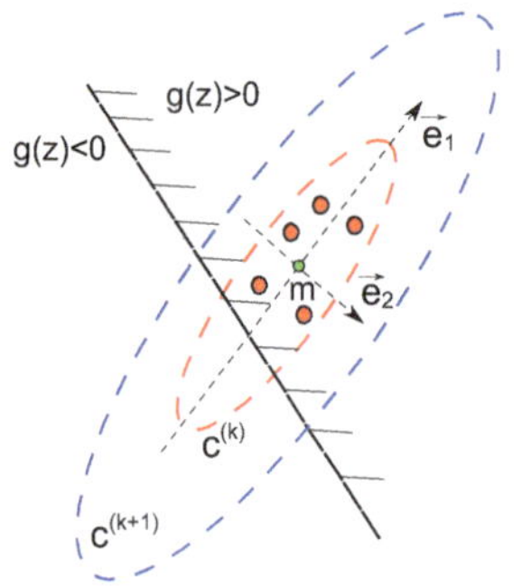

Figure 6. Modification of the covariance matrix due to the mean vector constraint violation, blue = feasible, red = unfeasible candidates.

in the feasible zone. Therefore, the ellipsoid hypervolume is increased according to:

$$\mathbf{S}^{(k)} \leftarrow \mathbf{PD}^{(k)^2}\mathbf{P}^T + \gamma\mathbf{P}\mathrm{diag}\left(\sqrt{vp_{m_1}^{(k)}}, \ldots, \sqrt{vp_{m_n}^{(k)}}\right)^2\mathbf{P}^T \quad (14)$$

The mean vector is displaced to the best feasible candidate generated at the next generation (Figure 6).

The modified CMA-ES(λ, μ) allows to take into account the constraints without degrading the objective function by penalization and avoids to tune the penalization parameters. Moreover, the proposed algorithm relies on the same update and selection mechanisms of the original CMA-ES(λ, μ) adapted for constraint handling and it keeps the invariance and unbiased design principles of CMA-ES(λ, μ) [10]. In the next sections, the proposed algorithm is tested on a benchmark of analytic functions and on the design of a two stage rocket in order to evaluate its performances.

4 Benchmark on analytic optimization problems

The proposed modified CMA-ES(λ, μ) is tested and compared to a penalized version of CMA-ES(λ, μ) with a constant penalization function, to the death penalty applied to CMA-ES(λ, μ) and to the modified (1+1)-CMA-ES on a benchmark of three analytic functions. The benchmark consists of a modified Six Hump Camel problem in 2 dimensions, the G04 optimization problem [8] in 5 dimensions and a modified Rosenbrock problem in 20 dimensions. These optimization problems are used in order to evaluate the proposed algorithm for different design space dimensions and different types of and numbers of inequality constraints (linear, non linear). In the following, the benchmark problems are introduced with the results. A discussion and a synthesis of the results for all the tests are provided in Section 4.4.

In the three problem formulations, the expected value is computed by Crude Monte Carlo method (CMC). A sample of 1000 points is used to estimate the expected value of the objective function. For each method (Modified CMA-ES(λ, μ), Death Penalty CMA-ES(λ, μ), Penalization CMA-ES(λ, μ), Modified (1+1)-CMA-ES) the optimization

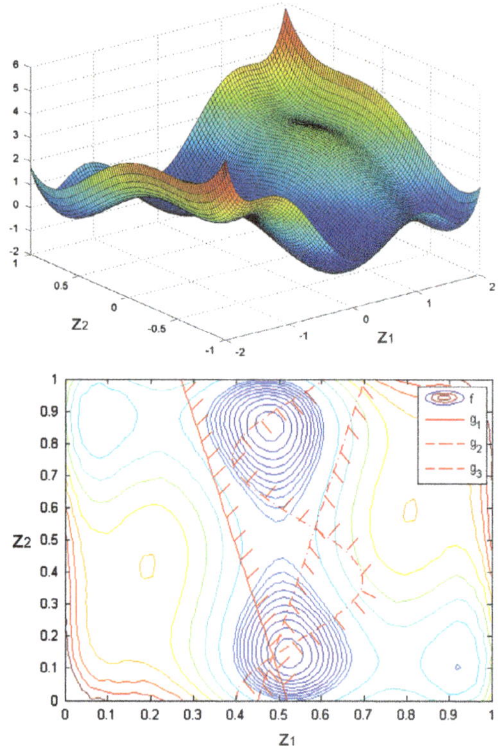

Figure 7. Modified Six Hump Camel function and constraints.

is repeated 50 times. The initialization is chosen randomly in the design space and the same initialization and the same random number seed are used for the four optimization algorithms. The same stopping criterion is used for all the algorithms: the distance in the design space between the mean vector and the best point found $\| \mathbf{m}^{(k)} - \mathbf{z_{best}} \|^2 < 10^{-3}$ must be lower than a tolerance 20 times in a row.

4.1 Modified Six Hump Camel problem

A modified version of the Six Hump Camel problem is used in order to introduce uncertainty and three inequality constraints. The formulation of the problem is the following:

$$\text{Min} \quad \mathbb{E}\big[f_{6\text{-hump}}(z_1, z_2) + f_{6\text{-hump}}(z_1\cos(U) + z_2\sin(U), - z_1\sin(U) + z_2\cos(U))\big] \quad (15)$$

$$\text{wrt} \quad \mathbf{z} = [z_1, z_2]$$

$$\text{st} \quad g_1(z_1, z_2) = z_1 + z_2/4 - 0.52 \le 0 \quad (16)$$

$$g_2(z_1, z_2) = z_1 + 0.01z_2 - 0.7 + 0.30\cos(60z_2^2/6) \le 0 \quad (17)$$

$$g_3(z_1, z_2) = z_1 - z_2/4 - 0.45 \le 0 \quad (18)$$

$$\mathbf{z}_{\min} \le \mathbf{z} \le \mathbf{z}_{\max} \quad (19)$$

Table 1. Results of modified Six Hump Camel problem. Average over 50 optimizations (in parenthesis the Relative Standard Deviation (RSD) – $\sigma/\mathbb{E}$).

Results	Modified CMA-ES(λ, μ)	Death penalty CMA-ES(λ, μ)	Penalization CMA-ES(λ, μ)	Modified (1+1)-CMA-ES
Objective function (global optimum[1])	−1.875 (0.80%)	−1.875 (0.89%)	−1.874 (0.88%)	−1.865 (0.88%)
Objective function (local optimum[2])	−0.347 (0.75%)	−0.347 (0.84%)	−0.347 (0.79%)	−0.348 (0.85%)
Constraint functions	-4.72×10^{-4} (96%)	-2.57×10^{-4} (75%)	-6.43×10^{-4} (68%)	-4.86×10^{-4} (127%)
Number of objective function + constraint evaluations	1211 (13%)	1395 (12%)	1618 (12%)	781 (16%)
Percentage of convergence to the global minimum	37%	22%	33%	48%
Percentage of convergence to the local minimum	63%	78%	67%	52%

[1] Statistics based on the optimizations that converged to the global optimum.
[2] Statistics based on the optimizations that converged to the local optimum.

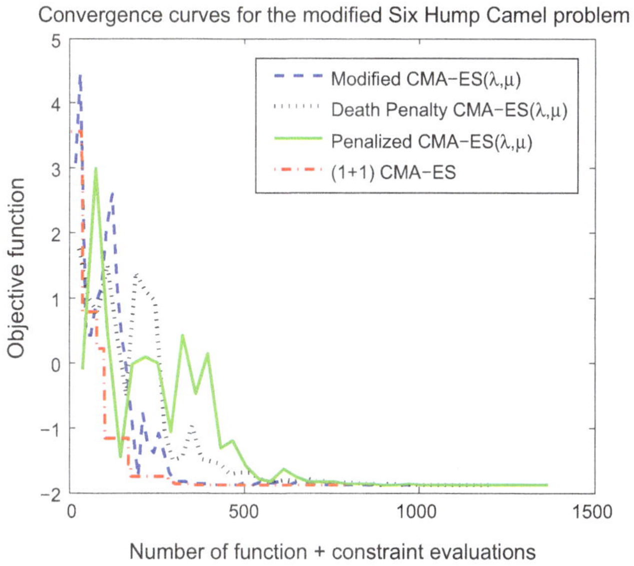

Figure 8. Convergence curves of the Six Hump Camel problem in 2 dimensions, based on one optimization run.

with $\mathbf{z} \in [-3, \ 3] \times [-2, \ 2]$, $f_{6-\text{hump}}(z_1, z_2) = (4-2.1z_1^2 + z_1^4/3) + z_1 z_2 + (4z_2^2 - 4)z_2^2$ and U a random variable distributed according to a normal distribution $U \sim \mathcal{N}(0, 0.05)$.

Representations of the function and the constraints are provided in Figure 7. The problem has one local optimum and one global optimum. The results are presented in Table 1 and the convergence curves for one optimization are given in Figure 8.

4.2 G04 optimization problem [8]

The G04 optimization problem involves 6 inequality constraints and is defined as following:

$$\text{Min} \quad G_{04}(\mathbf{z}) = 5.3578547 \times z_3^2 + 0.8356891 \times z_1 \times z_5 + 37.293239 \times z_1 - 40792.141 \tag{20}$$

$$\text{wrt} \quad z = [z_1, z_2, z_3, z_4, z_5]$$

$$\text{st} \quad g_1(\mathbf{z}) = u(\mathbf{z}) - 92 \leq 0 \tag{21}$$

$$g_2(\mathbf{z}) = -u(\mathbf{z}) \leq 0 \tag{22}$$

$$g_3(\mathbf{z}) = v(\mathbf{z}) - 110 \leq 0 \tag{23}$$

$$g_4(\mathbf{z}) = -v(\mathbf{z}) + 90 \leq 0 \tag{24}$$

$$g_5(\mathbf{z}) = w(\mathbf{z}) - 25 \leq 0 \tag{25}$$

$$g_6(\mathbf{z}) = -w(\mathbf{z}) + 20 \leq 0 \tag{26}$$

$$\mathbf{z}_{\min} \leq \mathbf{z} \leq \mathbf{z}_{\max} \tag{27}$$

with $\mathbf{z} \in \mathbb{R}^5$, $\mathbf{z}_{\min}$ = [78, 33, 27, 27, 27], $\mathbf{z}_{\max}$ = [102, 45, 45, 45, 45] and:

$$u(\mathbf{z}) = 85.334407 + 0.0056858 \times z_2 \times z_5 + 0.0006262 \times z_1 \times z_4 - 0.0022053 \times z_3 \times z_5 \tag{28}$$

$$v(\mathbf{z}) = 80.51249 + 0.0071317 \times z_2 \times z_5 + 0.0029955 \times z_1 \times z_2 + 0.0021813 \times z_3^2 \tag{29}$$

$$w(\mathbf{z}) = 9.300961 + 0.0047026 \times z_3 \times z_5 + 0.0012547 \times z_1 \times z_3 + 0.0019085 \times z_3 \times z_4 \tag{30}$$

The results are presented in Table 2 and the convergence curves for one optimization are given in Figure 9.

Table 2. Results of G04 optimization problem. Average over 50 optimizations (in parenthesis the RSD – $\sigma/\mathbb{E}$).

Results	Modified CMA-ES(λ, μ)	Death penalty CMA-ES(λ, μ)	Penalization CMA-ES(λ, μ)	Modified (1+1)-CMA-ES
Objective function	−30 487 (0.48%)	−28 105 (6.10%)	−28 758 (12%)	−30 452 (0.46%)
Number of objective function + constraint evaluations	1618 (33%)	9120 (40%)	13 139 (107%)	7048 (93%)

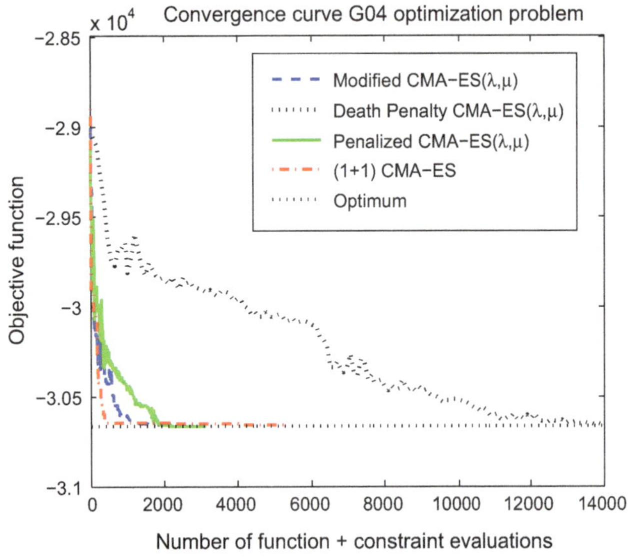

Figure 9. Convergence curves of the G04 optimization problem, based on one optimization run.

4.3 Modified Rosenbrock problem

The Rosenbrock optimization problem has been modified in order to incorporate uncertainty and an inequality constraint (Figure 10). The problem is formulated as following:

$$\text{Min} \quad \mathbb{E}\left[100\sum_{i=1}^{n-1}\left(z_{i+1}-z_i^2\right)^2+\sum_{i=1}^{n-1}(1-z_i)^2+U\right] \tag{31}$$

$$\text{wrt} \quad z=[z_1,\ldots,z_{20}]$$

$$\text{st} \quad g(\mathbf{z})=2-\prod_{i=1}^{n}z_i\leq 0 \tag{32}$$

with $n = 20$, $\mathbf{z}\in\mathbb{R}^{20}$ and U a random variable distributed according to $U\sim\mathcal{U}(-0.1, 0.0)$ a uniform distribution.

The results are presented in Table 3 and the convergence curves for one optimization are given in Figure 11.

4.4 Result and synthesis

The analytic test cases involve different dimensions (2, 5 and 20) and different number of constraints (1, 3 and 6) in order to evaluate the efficiency of the proposed modified CMA-ES(λ, μ) on various optimization problems. A qualitative synthesis of the obtained results is given in Figure 12. For all

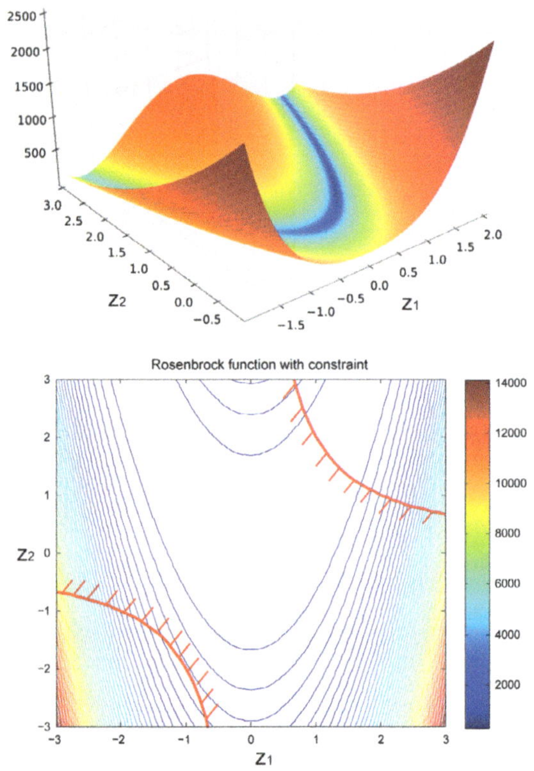

Figure 10. Modified Rosenbrock function and the constraints in 2 dimensions.

the three criteria (number of evaluations, robustness to initialization and value of the optimum), the lower value the better the quality of the method for the given criterion.

The Six Hump Camel problem has one local optimum and one global optimum. All the optimization algorithms converge either to the local or the global optimum. It illustrates the robustness property of the algorithms with respect to the initialization (relative standard deviation ~0.85% for all the algorithms). The found optima are all feasible. Modified (1+1)-CMA-ES converges in 48% of the optimization runs to the global optimum and the proposed modified CMA-ES (λ, μ) in 37% of the cases. The penalization and the death penality approaches converge only in 33% and 22% of the optimization runs to the global optimum. The number of calls to the objective function and the constraints is in increasing order: Modified (1+1)-CMA-ES (781), modified CMA-ES(λ, μ) (1211), Death Penalty CMA-ES(λ, μ) (1395) and Penalization

Table 3. Results of constrained Rosenbrock problem. Average over 50 optimizations (in parenthesis the RSD – $\sigma/\mathbb{E}$).

Results	Modified CMA-ES(λ, μ)	Death penalty CMA-ES(λ, μ)	Penalization CMA-ES(λ, μ)	Modified (1+1)-CMA-ES
Objective function	0.046 (0.12%)	0.109 (42.77%)	0.096 (26%)	0.0841 (53.85%)
Constraint functions	-5.05×10^{-4} (96%)	−0.74 (75%)	1.45 (68%)	−0.4294 (127%)
Number of objective function + constraint evaluations	12 798 (11%)	2708 (33%)	2689 (49%)	3481 (28%)

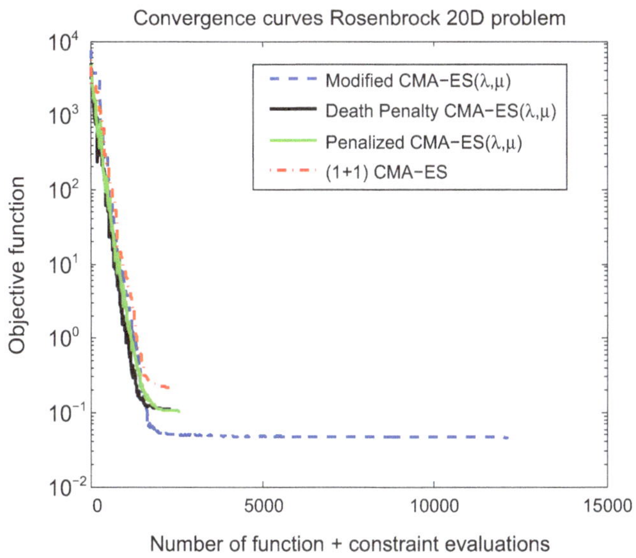

Figure 11. Convergence curves of the modified Rosenbrock problem in 20 dimensions, based on one optimization run.

CMA-ES(λ, μ) (1618). Modified (1+1)-CMA-ES is more efficient in this test case due to the low dimension and the simplicity of the optimization problem. The proposed modified CMA-ES(λ, μ) provides better results than the penalization and the death penalty approaches.

In the G04 problem, only the proposed modified CMA-ES(λ, μ) and the modified (1+1)-CMA-ES converge to the global minimum (with sufficient robustness with respect to the initialization). The number of calls to the objective function and the constraints is lower in the proposed algorithm (1618) compared to modified (1+1)-CMA-ES (7048) and the relative standard deviation is lower in the proposed approach. Moreover, the proposed approach converges efficiently to the global optimum. The Death Penalty and the penalization approaches do not succeed to reach the global optimum and are not robust to the initialization.

In the modified Rosenbrock problem, only the proposed modified CMA-ES(λ, μ) reaches the global optimum (with sufficient robustness (RSD: 0.12%) to the initialization). All the other algorithms are not robust to the initialization and do not converge to the global optimum. The number of calls to the objective function and the constraints is larger for the proposed approach (12 798) compared the other algorithms.

Consequently, from the benchmark, in small dimensions (<8), the modified (1+1)-CMA-ES provides good results in terms of convergence to the global optimum and robustness with respect to the initialization, however, as expected, in large dimensions, it presents issues to converge to the global optimum. The proposed modified CMA-ES(λ, μ) succeeds in small and large dimensions to find the global optimum. Moreover, this algorithm appears as robust to the initialization. In the next section, the proposed algorithm is used to design a two stage solid rocket and is compared to the existing CMA-ES based optimization algorithms.

5 Two stage solid propulsion rocket design

A multidisciplinary design problem consisting in maximizing the propulsive speed increment ΔV provided by a two stage rocket under geometrical and physical feasibility constraints is solved. The conceptual design models use simplified analysis of a two stage cylindrical solid propellant rocket motor. The multidisciplinary analysis involves four disciplines: the propulsion, the mass and sizing, the structure and the performance and constraint assessment (Figure 13). At the early design phase, model uncertainties exist and are taken into account. Two uncertainties are considered: the density of the propellant ρ and the ultimate strength σ for the rocket case material (Figure 14).

The problem is formulated as follows:

$$\text{Max} \quad \mathbb{E}[\Delta V(\mathbf{z}, \mathbf{U})] \tag{33}$$

$$\text{wrt} \quad \mathbf{z} = [D_{t1}, D_{s1}, P_{c1}, M_{p1}, D_{t2}, D_{s2}, P_{c2}, M_{p2}]$$

$$\text{st} \quad P_f[g_1(\mathbf{z}, \mathbf{U}) \geq 0] \leq 10^{-2} \tag{34}$$

$$P_f[g_2(\mathbf{z}, \mathbf{U}) \geq 0] \leq 10^{-2} \tag{35}$$

$$P_f[g_3(\mathbf{z}, \mathbf{U}) \geq 0] \leq 10^{-2} \tag{36}$$

$$\mathbf{z}_{\min} \leq \mathbf{z} \leq \mathbf{z}_{\max} \tag{37}$$

with: $\mathbf{U} = [U_1, U_2]$ with $U_1 \sim \mathcal{N}(1, 0.02)$ the uncertainty of the density of the propellant (ρ) and $U_2 \sim \mathcal{N}(1, 0.05)$ the uncertainty of the ultimate strength limit (σ) for the rocket case material. The design variables are described in Table 4. An overview of the disciplines is detailed in the next paragraphs.

Propulsion. The propulsion discipline computes for a given set of propellant characteristics (density ρ, combustion speed, flame temperature, heat capacity ratio), the thrust T,

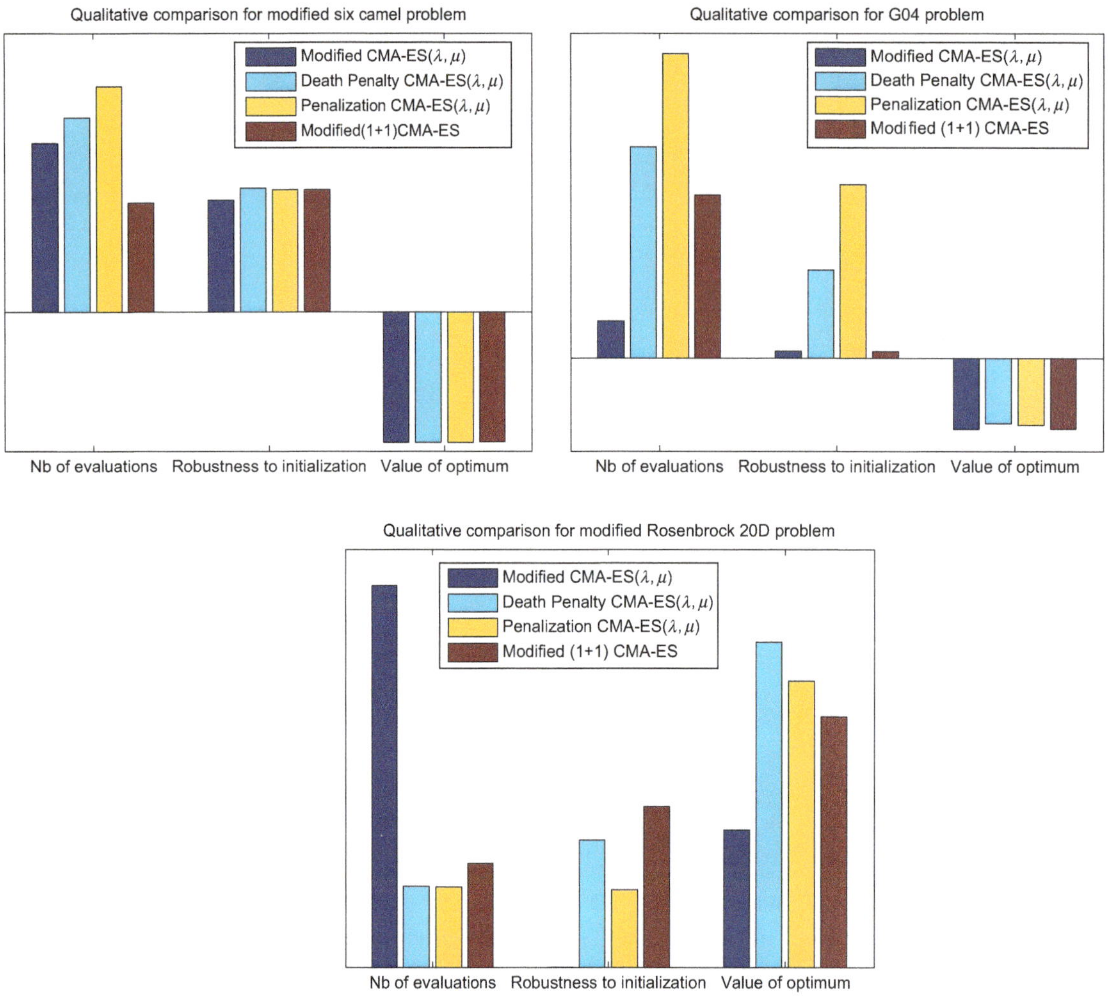

Figure 12. Qualitative results obtained for the different test cases.

the mass flow rate $\dot{m}$, the thrust coefficient c_T and the characteristic velocity c^*, under the assumption of constant thrust. The discipline takes nozzle shapes Dt, D_s and combustion pressure P_c as inputs. The used propellant is the Butargols with polybutadiene binder without aluminium additive.

Mass and Sizing. The mass and sizing discipline computes the dry mass m_d and the geometry of the two stage solid propulsion rocket. The dry mass involves the mass of the rocket case and the mass of the nozzle and the pyrotechnic igniter. The rocket geometry consists of the initial combustion area, the packaging ratio and the size of the central channel. The overall dimensions (rocket length L = 22 m and diameter D = 1.07 m) are considered as fixed.

Structure. The structure discipline computes the tank walls thickness (t) which is sized under the combustion pressure based on the material characteristics (yield strength, ultimate strength limit) and rocket geometry. Moreover, it computes the stress in the rocket case.

Performance and constraints. The performance is the propulsive speed increment ΔV and the expected value of ΔV is the objective function to be maximized. CMC based on 1000 samples is used to compute the expected value of the propulsive speed increment. The three constraints are: $g_1(.)$ which ensures packaging ratio (Propellant volume/Available volume) that has to be inferior to 87%, $g_2(.)$ which ensures that central channel diameter is 30% greater than the nozzle throat diameter and $g_3(.)$ which ensures that combustion area is greater than the minimum feasible (area of central channel walls). The probabilities of failure for the three constraints have to be inferior to 1%. The probabilities of failure are computed with a CMC of 10^4 samples, numerically corresponding to a relative standard error ($\sigma/\mathbb{E}$) of the probability estimation in the order of 5%.

5.1 Results

The optimization for each algorithm is repeated 10 times. All the optimizations start from the same baseline given in Table 4. The same stopping criterion is used for all the optimization algorithms: the distance in the design space between the mean vector and the best point found

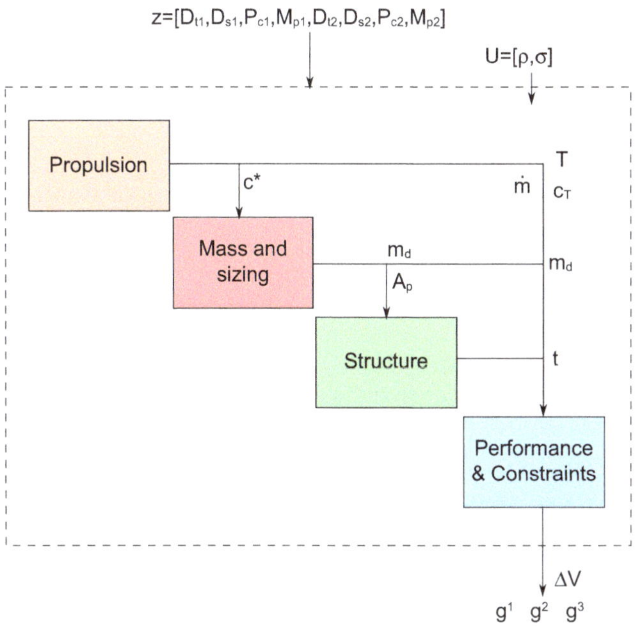

Figure 13. Design Structure Matrix for the two stage solid rocket.

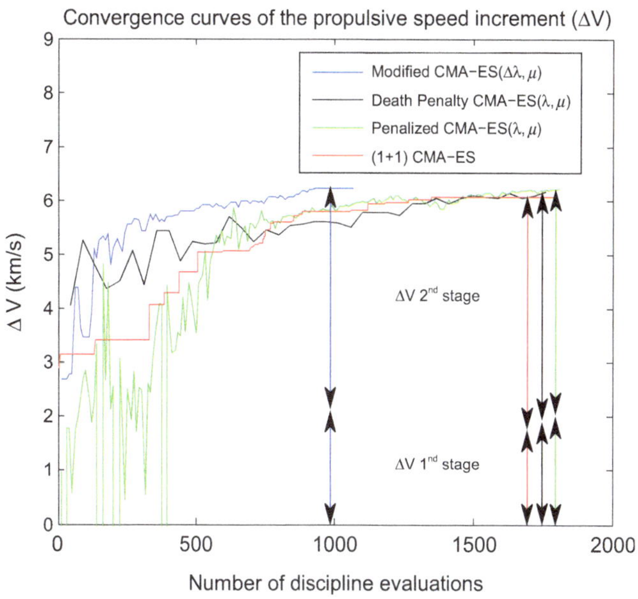

Figure 14. Convergence curves of the propulsive speed increment for the two stage solid propulsion rocket.

Table 4. Design variables for the two-stage rocket.

Variables	Symbol	Domain of definition	Baseline
1st stage nozzle throat diameter	D_{t1}	[0.05, 1] m	0.75 m
2nd stage nozzle throat diameter	D_{t2}	[0.05, 1] m	0.75 m
1st stage nozzle exit diameter	D_{s1}	[0.5, 1.4] m	1 m
2nd stage nozzle exit diameter	D_{s2}	[0.5, 1.4] m	1 m
1st stage combustion pressure	P_{c1}	[1, 500] bar	100 bar
2nd stage combustion pressure	P_{c2}	[1, 500] bar	100 bar
1st stage propellant mass	M_{p1}	[2000, 15 000] kg	8000 kg
2nd stage propellant mass	M_{p1}	[2000, 15 000] kg	8000 kg

Table 5. Results of the two stage rocket optimization. Average over 10 optimizations (in parenthesis the RSD – $\sigma/\mathbb{E}$).

Results	Modified CMA-ES(λ, μ)	Death penalty CMA-ES(λ, μ)	Penalization CMA-ES(λ, μ)	Modified (1+1)-CMA-ES
Design variables	[0.321, 0.9081, 205.46, 12 428.1, 0.637, 1.288, 21.08, 7911.2]	[0.449, 1.240, 64.12, 10 781.4, 0.535, 1.112, 32 785, 9726.6]	[0.283, 0.7131, 181.83, 12 462.5, 0.571, 1.39, 26.20, 9316.9]	[0.3558, 0.9961, 199.87, 11 750.2, 0.5476, 1.314, 31.78, 9723.6]
Objective function	6234.1 (0.95%)	6093.2 (1.55%)	6097.4 (1.43%)	6080.1 (1.19%)
Constraint functions	[−0.01, −0.0023,−0.0014, −0.01, −0.0023,−0.0032]	[−0.01, −0.025, −0.01, −0.01, −0.01, −0.0037]	[−0.01, −0.0079, −0.0053, −0.01, −0.0054, −0.0067]	[−0.01, −0.0086, −0.0037, −0.01, −0.0028, −0.0046]
Number of discipline evaluations	1127 (0.73%)	1758 (0.88%)	1813 (1.54%)	1675 (0.94%)

$||\mathbf{m}^{(k)} - \mathbf{z}_{\text{best}}|| < 10^{-3}$ must be under a tolerance 20 times in a row. The algorithms do not converge to the same optimum. Modified CMA-ES (λ, μ) provides a better optimum in terms of propulsive speed increment: 6234.1 m/s with a better robustness. The proposed algorithm converges on the average in 1127 discipline evaluations. The other optimization algorithms converge in the same order of number of discipline evaluations (~1750). Only four constraints are active at the optimum. The better optimum found by the proposed approach is essential as it has a better propulsive speed increment which could be used to increase the payload mass (Table 5).

6 Conclusion

The design of complex systems often induces a constrained optimization problem under uncertainty. In this paper, an adaptation of CMA-ES(λ, μ) has been proposed in order to efficiently handle the constraints. The probable research hypervolume engendered by an iso-probability contour of the multivariate normal distribution $\mathcal{N}(0, \mathbf{C})$ used to generate the candidate population is modified. The constraint handling method allows to reduce the semi-principal axes of the iso-probable research ellipsoid in the directions violating the constraints by decreasing the eigenvalues of the covariance matrix $\mathbf{C}$. The proposed approach has been tested with three analytic optimization problems highlighting the efficiency of the algorithm and the robustness with respect to the initialization. The proposed method has been used to design a two stage solid propulsion launch vehicle. A better optimum has been found with the proposed approach with respect to the existing CMA-ES based optimization algorithms resulting in a potential increase in the payload mass.

Acknowledgements. The work of L. Brevault is part of a CNES/ONERA PhD thesis.

References

1. Arnold DV, Hansen N. 2012. A (1+1)-CMA-ES for constrained optimisation, in Proceedings of the fourteenth international conference on Genetic and evolutionary computation conference, New York, NY, USA, p. 297–304, ACM.
2. Beyer H-G, Finck S. 2012. On the design of constraint Covariance Matrix Self-Adaptation Evolution Strategies including a cardinality constraint. IEEE Transactions on Evolutionary Computation, 16(4), 578–596.
3. Branke J, Schmidt C. 2003. Selection in the presence of noise, in Genetic and Evolutionary Computation Conference, Chicago, IL, USA.
4. Coello Coello CA. 2000. Constraint-handling using an evolutionary multiobjective optimization technique. Civil Engineering Systems, 17(4), 319–346.
5. Collange G, Reynaud S, Hansen N. 2010. Covariance Matrix Adaptation Evolution Strategy for multidisciplinary optimization of expendable launcher family, in 13th AIAA/ISSMO Multidisciplinary Analysis Optimization Conference, Fort-Worth, TX.
6. De Melo VV, Iacca G. 2014. A modified Covariance Matrix Adaptation Evolution Strategy with adaptive penalty function and restart for constrained optimization. Expert Systems with Applications, 41(16), 7077–7094.
7. Eberhart RC, Kennedy J. 1995. A new optimizer using particle swarm theory. Proceedings of the sixth international symposium on micro machine and human science, New York, NY, p. 39–43.
8. Hansen N. 2009. Benchmarking a BI-population CMA-ES on the BBOB-2009 noisy testbed, in Proceedings of the 11th Annual Conference Companion on Genetic and Evolutionary Computation Conference: Late Breaking Papers, Montreal, Qubec, Canada, p. 2397–2402, ACM.
9. Hansen N, Auger A, Ros R, Finck S, Pošík P. 2010. Comparing results of 31 algorithms from the black-box optimization benchmarking BBOB-2009, in Proceedings of the 12th annual conference companion on Genetic and evolutionary computation, p. 1689–1696, ACM.
10. Hansen N, Müller S, Koumoutsakos P. 2003. Reducing the time complexity of the derandomized Evolution Strategy with Covariance Matrix Adaptation (CMA-ES). Evolutionary Computation, 11(1), 1–18.
11. Hansen N, Niederberger AS, Guzzella L, Koumoutsakos P. 2009. A method for handling uncertainty in evolutionary optimization with an application to feedback control of combustion. IEEE Transactions on Evolutionary Computation, 13(1), 180–197.
12. Holland JH. 1975. Adaptation in natural and artificial systems: an introductory analysis with applications to biology, control, and artificial intelligence. University of Michigan, USA.
13. Karaboga D. 2005. An idea based on honey bee swarm for numerical optimization, Technical report, Technical report-tr06, Erciyes University, engineering faculty, computer engineering department.
14. Kramer O, Barthelmes A, Rudolph G. 2009. Surrogate constraint functions for CMA Evolution Strategies, in KI 2009: Advances in Artificial Intelligence, Springer, Verlag Berlin Heidelberg, p. 169–176.
15. Larson JM. 2012. Derivative free optimization of noisy functions. PhD thesis, University of Colorado.
16. Mezura-Montes E, Coello Coello CA. 2011. Constraint-handling in nature-inspired numerical optimization: past, present and future. Swarm and Evolutionary Computation, 1(4), 173–194.
17. Price K, Storn RM, Lampinen JA. 2006. Differential evolution: a practical approach to global optimization, Springer, Verlag Berlin Heidelberg.
18. Rowell LF, Korte JJ. 2003. Launch vehicle design and optimization methods and priority for the advanced engineering environment. NASA Technical Report NASA/TM-2003, 212654.
19. Schwefel H-P, Rudolph G. 1995. Contemporary evolution strategies, Springer, Verlag Berlin Heidelberg.

10

Structural optimization in ESAC: annals 2011

Ji-Hong Zhu* and Wei-Hong Zhang

Engineering Simulation and Aerospace Computing, Northwestern Polytechnical University, 710072, Xi'an, Shaanxi, P.R. China

Abstract – The purpose of this paper is to give an overall introduction of the structural optimization research works in ESAC group in 2011. Four main topics are involved, i.e. 1) topology optimization with multiphase materials, 2) integrated layout and topology optimization, 3) prediction of effective material properties and 4) composite design. More detailed techniques and some numerical results are also presented and discussed here.

Key words: ESAC annals, Topology optimization, Shape optimization, Composite.

1 Introduction

The Laboratory of Engineering Simulation and Aerospace Computing (ESAC) in Northwestern Polytechnical University is an active international research team in the field of structural optimization design. Since 1990s, we have been dedicated to the theoretical study of shape and topology optimization and their applications in aeronautics and aerospace industries.

In 2011, the research works and the published papers are mainly concentrated on the research of the following four topics, i.e., (1) topology optimization with multiphase materials, (2) integrated layout and topology optimization, (3) prediction of effective material properties and (4) composite design.

2 Topology optimization with multiphase materials

This work is focused on the topology optimization of lightweight structures consisting of multiphase materials [1, 2]. Instead of adopting the common idea of using volume constraint, a new problem formulation with mass constraint is firstly proposed, which is a significant breakthrough of the topology optimization with multiphase materials. For a discretized structure with n designable finite elements, the mass constraint is stated as

$$\sum_{i=1}^{n} \rho_i V_i \leq M, \tag{1}$$

where ρ_i denotes the material density related to element i, which has to be interpolated from the set of available material densities. M is the upper bound controlling the structural mass.

Meanwhile, to have a deep understanding of the mass constraint, recursive multiphase materials interpolation (RMMI) and uniform multiphase materials interpolation (UMMI) schemes are discussed and compared based on numerical tests and theoretical analysis.

RMMI interpolation scheme can be treated as an extension of the traditional SIMP/RAMP interpolation model reserved for the case of only one solid material phase and the void to the case of multiphase materials. This scheme is constructed through a recursive process. For a structure with m solid materials and n designable finite elements, the RMMI scheme interpolates Young's modulus of element i in the following manner

$$E_i = E(\mathbf{x}_i) = v_{im}(x_{im})E_i^{(1,m)}, \tag{2}$$

where

$$\begin{aligned} E_i^{(m,m)} &= E^{(m)}, \\ E_i^{(m-\xi,m)} &= v_{i\xi}(x_{i\xi})E_i^{(m-\xi+1,m)} + [1 - v_{i\xi}(x_{i\xi})] \\ & E_i^{(m-\xi)};\ (\xi = 1, 2, ..., m-1), \mathbf{x}_i \\ &= \{x_{ij}\};\ (i = 1, 2, ..., n; j = 1, 2, ..., m), \end{aligned} \tag{3}$$

where $E_{(j)}$ denotes Young's modulus of material phase j. $\boldsymbol{x}_i$ denotes the set of design variables associated with element i. Clearly, each element has m design variables that have different significations. $v_{ij}(x_{ij})$ is the weighting function related to element i.

It is indicated that the nonlinearity of the mass constraint introduced by RMMI brings numerical difficulties to attain the global optimum of the optimization problem. On the contrary, the UMMI-2 scheme makes it possible to formulate the mass constraint in a linear form with separable design variables.

*e-mail: jh.zhu@nwpu.edu.cn

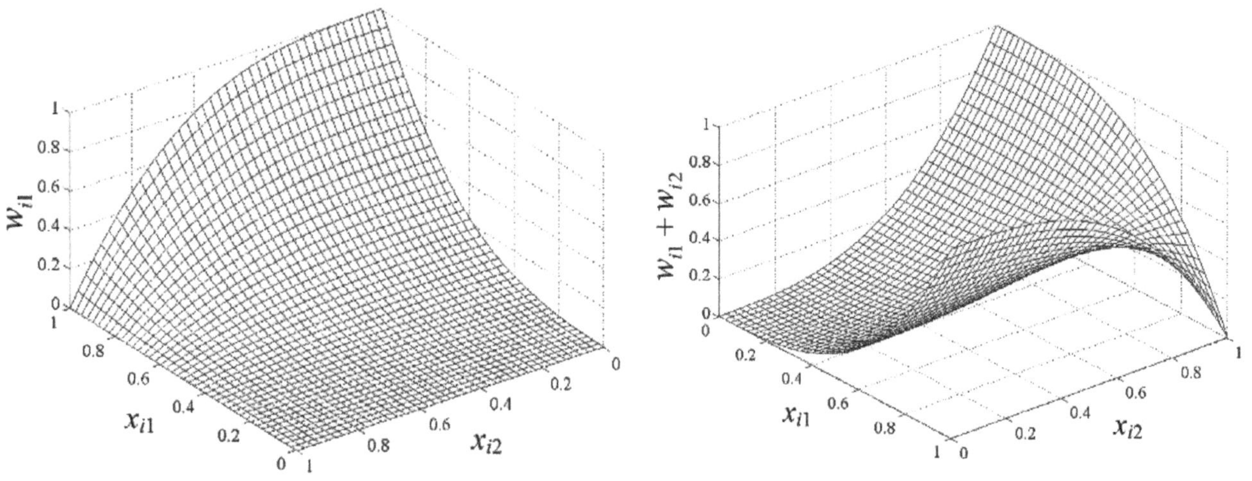

(a) w_{i1} (b) $w_{i1}+w_{i2}$

Figure 1. UMMI-2 interpolation model with two solid materials (m = 2, p = 3).

UMMI originated from the so-called DMO. It means that the continuous laminate design of fiber orientations was treated as a DMO problem. In this paper, the concept of DMO is generalized for structural topology optimization with multiphase materials. As the weighting function is uniform for each material phase, the interpolation scheme is, in fact, a uniform multiphase materials interpolation (UMMI). For a problem of m solid materials and n designable elements, each element will have m design variables. The UMMI interpolation of Young's modulus can be expressed as the weighted sum of all candidate material phases:

$$E_i = \sum_{j=1}^{m} w_{ij} E^{(j)}, \tag{4}$$

w_{ij} is the weighting function associated with the jth material phase which can be defined in UMMI-2 as

$$w_{ij} = x_{ij}^{p} \prod_{\substack{\xi = 1 \\ \xi \neq j}}^{m} (1 - x_{i\xi}^{p}). \tag{5}$$

As shown in Figure 1, one such formulation favors very much the problem resolution by means of mathematical programming approaches, especially the convex programming methods. Moreover, numerical analysis indicates that fully uniform initial weighting is beneficial to seek the global optimum when UMMI-2 scheme is used. Besides, the relationship between the volume constraint and mass constraint is theoretically revealed. The filtering technique is adapted to avoid the checkerboard pattern related to the problem with multiphase materials.

Numerical examples show that the UMMI-2 scheme with fully uniform initial weighting is reliable and efficient to deal with the structural topology optimization with multiphase materials and mass constraint. Meanwhile, the mass constraint formulation is evidently more significant than the volume constraint formulation.

As shown in Figures 2a and 2b, with the RMMI scheme and uniform x_{ij} or w_{ij}, the optimal structures have configurations consisting of both VM1 and VM2. Comparatively, the result shown in Figure 2c with only one single solid material (VM2) is found to be stiffer. This implies that the RMMI scheme cannot yield the global optimum at least even uniform initial weighting are used.

Using the UMMI-2 scheme with fully uniform initial variables, the optimal structure only consists of VM2, as shown in Figure 2d, even VM1 andVM2 are available. Although configurations given in Figures 2c and 2d are very similar, the structural compliances using UMMI-2 are smaller. Besides, partial uniform initial variables of Case B are adopted and the results are shown in Figures 2e and 2f. Obviously, different initial variables lead to totally different configurations. Note that some elements at the right-bottom corner in Figure 2f consist of "mixed" materials.

Another interesting thing is to compare the optimization results obtained with mass and volume constraints. To do this, suppose the structure mass is less than 25.92×10^3 kg. The upper bound of the volume constraint can be correspondingly determined for VM3 if the upper bound of the volume constraint of VM1 is fixed a priori. For different volume fractions of VM1, optimization results are plotted in Figure 3. The optimal solution with the mass constraint corresponds to the volume fractions of VM1 and VM3 being 0.463 and 0.068, respectively. It can be seen that this is the best solution among all possible ones using volume constraint in the sense of compliance minimization. In other words, the mass constraint always leads to a better solution than the volume constraint with the precondition of the same structure mass. Meanwhile, the percentage of each solid material phase will be automatically identified. In fact, the mass constraint has a more significant and important sense than the volume constraint in most engineering problems.

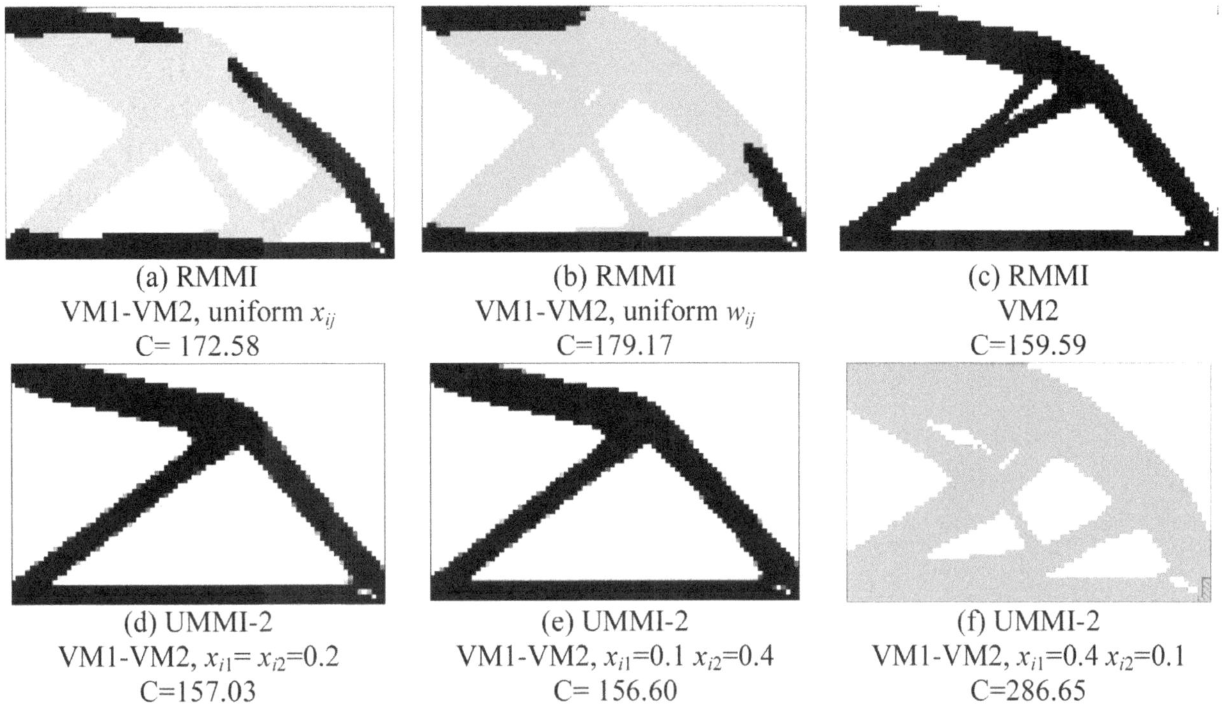

Figure 2. Optimization results of a cantilever beam.

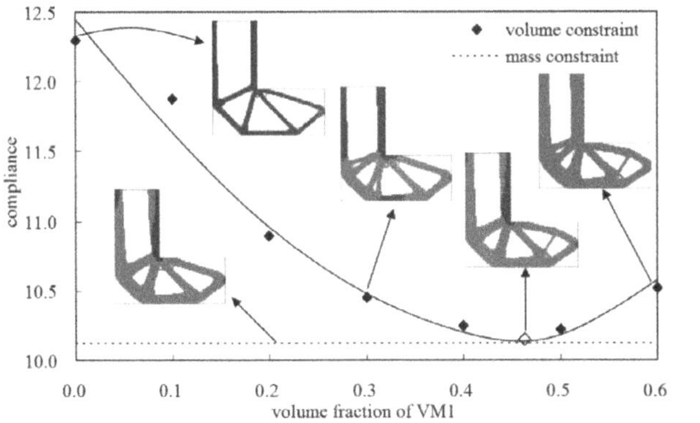

Figure 3. Comparisons of the optimization results under the mass and volume constraints with the equal structure mass.

From the development of this work, it is seen that the combination of UMMI-2 scheme together with fully uniform initial variables and the convex programming methods provides an efficient approach for the structural layout design of multiphase materials in static load case when the mass is controlled. Meanwhile, due to its generality, this approach may be extended to the structural design and microstructure design of multiphase materials in generalized load cases.

3 Integrated layout and topology optimization

In many engineering products, typical structural systems consists of components and structures with integrated interconnections inside a limited packing space. Global system performances significantly depend upon the layout of components and the topology of supporting structures. The design of multi-component structural systems is a relatively new area and is still not extensively researched. Since most real life engineering design problems involve multiple components or structures, it is a subject of great relevance

Recently, further researches are carried out in our group with significant improvements and systematical methodologies [3, 4]. Packing optimization is performed by means of the FCM (Finite Circle Method), and the supporting structure is designed by means of topology optimization techniques. Because movable components inevitably change the FE mesh and locally corrupt the updating of the material distribution, the traditional topology optimization method using a fixed FE mesh is no longer applicable. To circumvent this problem, a density point technique was proposed to associate the density variables with the centroids of a fixed background mesh. As shown in Figure 4, density points are a set of predefined points in the design domain to control the material distribution. No matter how the mesh changes, each element receives the pseudo-density value from its nearest density point.

In order to reduce the computing cost and simplify sensitivity analysis with respect to location design variables of each component, the super-element technique is adopted to model the movable components.

Similar to classic topology optimization, the sensitivity of C with respect to pseudo-density design variables can be analytically obtained by

$$\frac{\partial C}{\partial \eta_i} = \frac{\partial \mathbf{G}^T}{\partial \eta_i} \cdot \mathbf{u} - \frac{1}{2} \mathbf{u}^T \cdot \frac{\partial \mathbf{K}}{\partial \eta_i} \cdot \mathbf{u}. \tag{6}$$

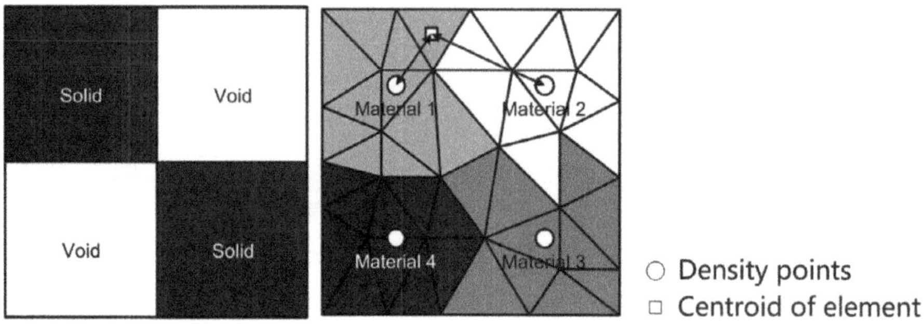

Figure 4. Element material properties received from the density points.

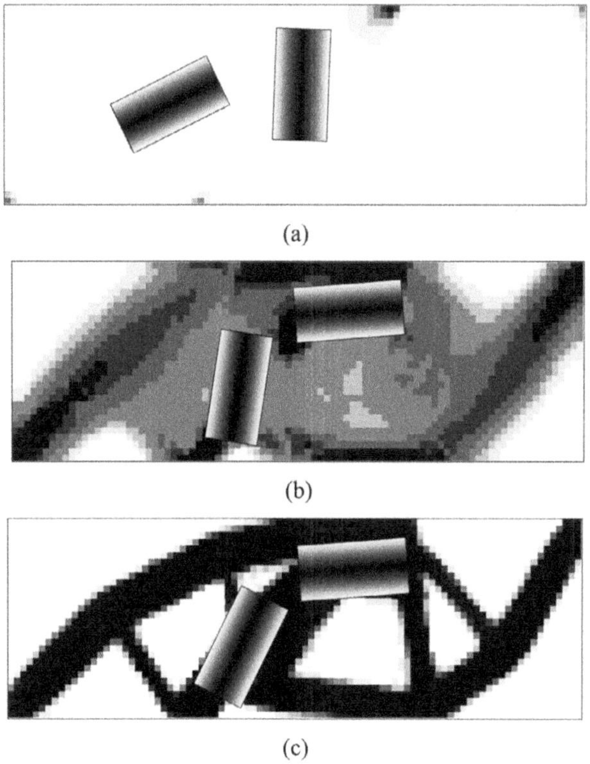

Figure 5. Integrated layout design with components of stiffer materials [40]: (a) iteration 5, (b) iteration 20, and (c) final design; final strain energy C50.074 J.

As to the sensitivity with respect to location design variables of components, a semi-analytical scheme combined with super-element formulation is applied in the following way. The global stiffness matrix of the system is partitioned into three parts: supporting structure, mesh transition region, and components. The derivative of the system compliance with respect to s_ε representing one of the three location parameters x_ε, y_ε, and θ_ε of component ε gives rise to

$$\frac{\partial C}{\partial s_\varepsilon} = \mathbf{u}^T \left[\frac{\partial \mathbf{G}^T}{\partial s_\varepsilon} - \frac{1}{2} \left(\frac{\partial \mathbf{K}_{(S)}}{\partial s_\varepsilon} + \sum_{i=1}^{n_c} \frac{\partial \mathbf{K}_{(T)l}}{\partial s_\varepsilon} + \sum_{i=1}^{n_c} \frac{\partial \mathbf{K}_{(C)l}}{\partial s_\varepsilon} \right) \mathbf{u} \right]. \tag{7}$$

A design example of simple structure system with two rectangular components is considered here. The global strain energy is the objective function to be minimized and the volume fraction is set to be 50%. Assume that all components are made of steel, and the supporting structure is made of aluminum. As shown in Figure 5, the components and the supporting structure are optimized simultaneously. Finally, the components are located in the main loading path of the structure due to their high stiffness.

Design of multicomponent systems can also be extended to support layout of bridge-like problems. Fixations are, here, defined as movable solid support components, which are partially embedded into the design domain. Positions of these support components are optimized together with the structural topology to obtain the maximum stiffness of the global system.

A typical example is illustrated in Figure 6. Materials used for the structure and support components are aluminum and steel, respectively. The road on the bridge is a non-designable area. Four identical support components are located symmetrically along the bottom boundary of the design domain. A uniform surface pressure is applied on the top of the bridge. The volume fraction for the design domain is set to be 40%. By optimization, it is found that a slight change of the support locations will lead to a significant variation of the structural topology.

It is possible to extend the methodology to the layout design of 3D multicomponent systems. However, the embedded meshing technique for 3D problems is much more complicated than in 2D situation. Meanwhile, the semi-analytical sensitivity analysis scheme for location design variables of components is time-consuming because each 3D component has now six location design variables. Recently, a Heaviside function-based approach is investigated for sensitivity analysis with respect to location design variables. The material discontinuity across the component boundary is approximately smoothed by means of the Heaviside function. When a location design variable is perturbed, sensitive elements attached to the component boundary are assumed to undertake a material property change, while the mesh remains unchanged. In this way, sensitivity analysis with respect to location design variables is carried out as easily as for pseudo-density variables. Figure 7 shows the optimization result of a 3D structure, in which tetrahedral elements are used and the involved components have a greater Young's modulus than that of the supporting structure.

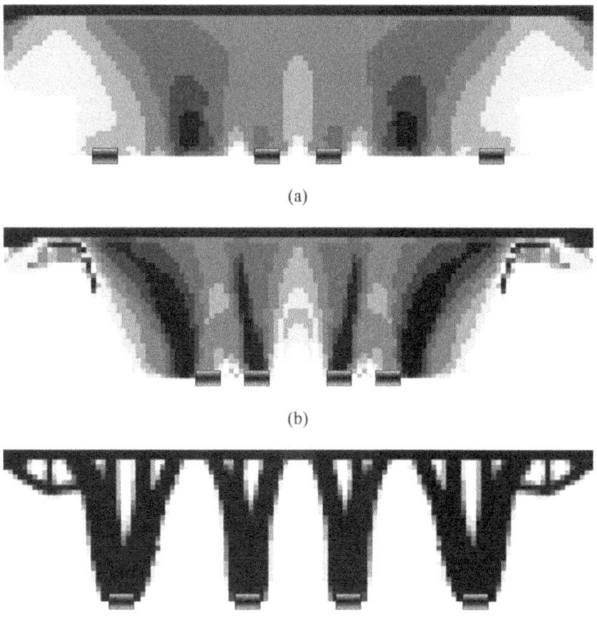

Figure 6. Integrated support and layout design of a bridge-like structure: (a) iteration 5, (b) iteration 10, and (c) final design, C50.1298 J.

4 Prediction of effective material properties

This work presents a microstructural modeling method considering the oxidation effect. Further studies on the prediction of material properties were carried for the multiphase and multilayer braided composite [5]. Regarding the multi-scale characteristics of the composite, the microstructure modeling is carried out sequentially from fiber to tow scale. The geometrical configuration of the microstructure is first analyzed, and mathematical relations among different geometrical parameters are derived on each scale. Second, effective elastic properties are obtained based on the sequential homogenization from fiber to tow scale. A strain energy based method is proposed to evaluate effective elastic properties with specific boundary conditions imposed on the microstructure. Numerical results obtained by the proposed method and the microstructure model show a good agreement with the results measured experimentally.

The present study aimed at the microstructure modeling and numerical prediction of elastic properties of 3D four-step MPML braided composites. First, the microstructure modeling is carried out sequentially from fiber to tow scale. Geometrical descriptions and finite element modeling of the microstructure are presented in detail.

The RVC (Representative Volume Cell) model of the 3D four-step MPML braided composite involves two scales: fiber and tow scales. The first scale concerns the modeling of RVC for MPML tows, and the second scale concerns the modeling of RVC for 3D four-step MPML braided composites as shown in Figure 8. Considering the multi-scale characteristics, the multi-scale analysis methods are needed for the analysis of MPML plain weave composites. The multi-scale analysis procedure is as follows. First, on the fiber scale, finite element model is built to obtain the effective moduli of the fiber scale RVC, and the results are used for the tows that were treated as a homogeneous transversely isotropic material. Second, based on the obtained elastic properties of tows, finite element model of the tow scale RVC is created to evaluate effective elastic properties of the composite. Here, RVC models of both scales are established using the ANSYS finite element software.

On the tow scale, the finite element model of RVC is depicted in Figure 9. Here, a 10-node tetrahedral solid element SOLID187 is used for the discretization, and the finite element models of both tows and matrices are shown respectively.

Second, a strain energy based method is applied to evaluate the effective elastic properties. Numerical results are then compared with the experimental results to validate the modeling.

In the elastic regime, the macroscopic behaviors of the RVC can be characterized by the effective stress tensor $\boldsymbol{\sigma}$ and strain tensor ε over the homogeneous equivalent model. They are interrelated by the effective, also termed homogenized, stiffness matrix $\boldsymbol{C}^H$

$$\sigma = \mathbf{C}^H \varepsilon. \tag{8}$$

The matrix coefficient $\mathbf{C}^H{}_{1111}$ can be derived

$$\mathbf{C}^H_{1111} = 2E^{(1)}/V. \tag{9}$$

In the same way, demonstrations can be made for other coefficients. The effective stiffness matrix can be written as

$$\begin{bmatrix} \frac{2E^{(1)}}{V} & \frac{E^{(7)}-E^{(2)}-E^{(1)}}{V} & \frac{E^{(9)}-E^{(3)}-E^{(1)}}{V} & 0 & 0 & 0 \\ & \frac{2E^{(2)}}{V} & \frac{E^{(8)}-E^{(3)}-E^{(2)}}{V} & 0 & 0 & 0 \\ & & \frac{2E^{(3)}}{V} & 0 & 0 & 0 \\ & & & \frac{2E^{(4)}}{V} & & \\ & & & & \frac{2E^{(5)}}{V} & \\ & \text{sym} & & & & \frac{2E^{(6)}}{V} \end{bmatrix}. \tag{10}$$

The elastic constants can be derived by inversing the above matrix. In practice, the considered RVC will be discretized as a finite element model on which the initial strain will be imposed to evaluate the strain energy. The comparison with the experimental results shows the validity and the rationality of the present method. Future studies will be focused upon the numerical prediction of thermal properties (the thermal expansion coefficient and thermal conductivity), progressive failure analysis and design optimization of the microstructure for 3D MPML braided composite.

Similar model incorporates the modeling of oxidized microstructure and computing of degraded elastic moduli can also be extended to simulate the oxidation behaviors of 2D C/SiC composites exposed to air oxidizing environments below 800 °C

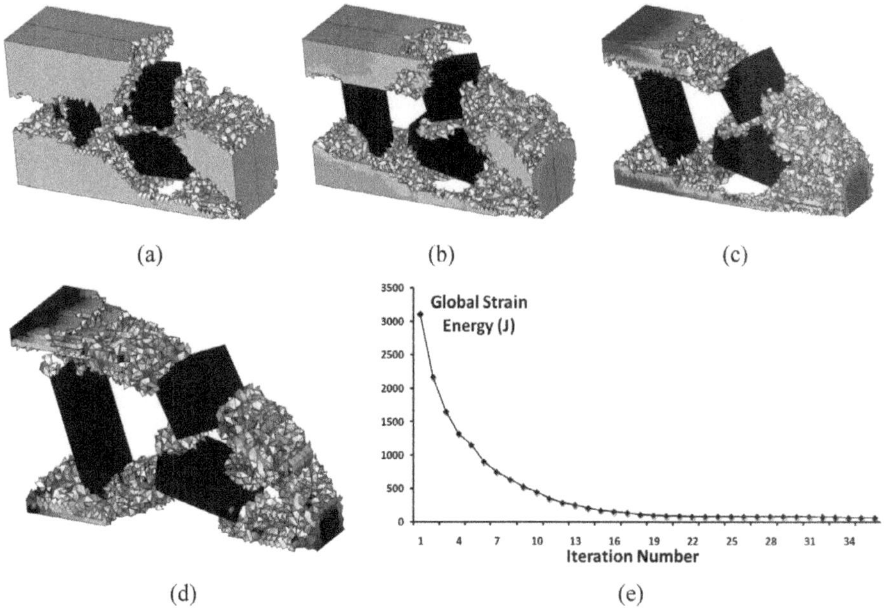

Figure 7. Integrated layout design of a 3D structure with three components: (a) iteration 2, (b) iteration 5, (c) iteration 10, (d) iteration 18, and (e) iteration history.

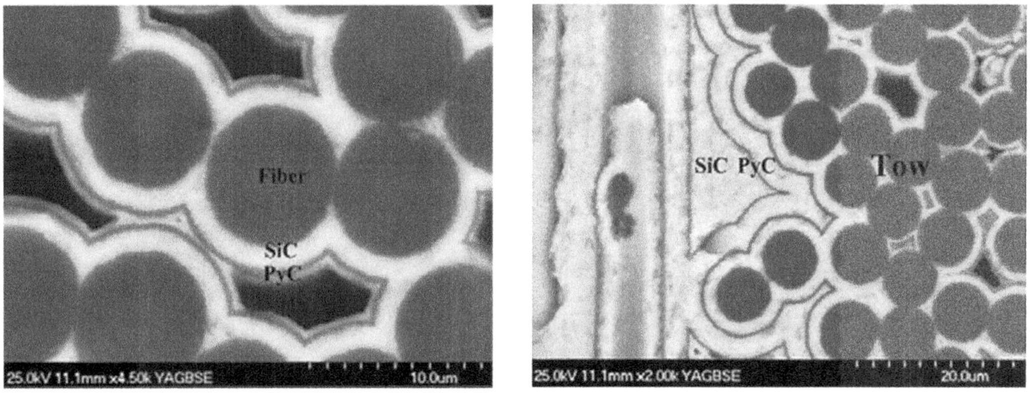

(a) fiber scale (b) tow scale

Figure 8. Images (SEM) of 3D braided MPML C/C-SiC.

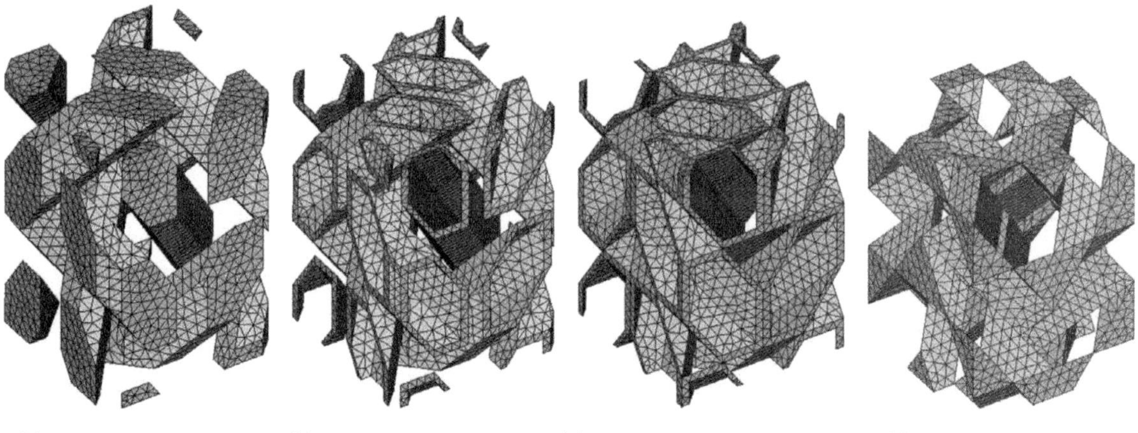

(a) (b) (c) (d)

(a) Tow (b) First layer of matrix (c) Second layer of matrix (d) Last layer of matrix

Figure 9. Finite element model of RVC on tow scale. (a) Tow, (b) First layer of matrix, (c) Second layer of matrix, and (d) Last layer of matrix.

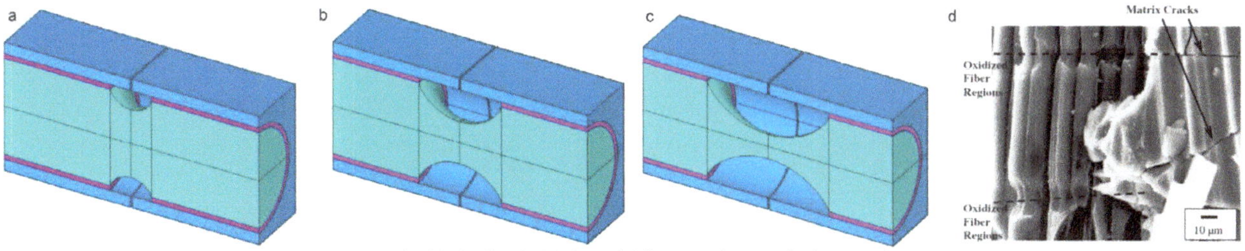

(a) 10 h (b) 20 h (c) 30 h and (d) experimental observation

Figure 10. Morphologies of microscopic RVC at different oxidation time. (a) 10 h, (b) 20 h, (c) 30 h, and (d) experimental observation.

[6]. Regarding the multi-scale characteristics of 2D C/SiC composite, the microstructure modeling is carried out on microscopic and macroscopic scale, respectively to compute the degraded elastic properties in terms of time duration, temperature and pressure, whose influences upon the oxidation microstructure morphology and degraded properties of 2D C/SiC composites are also investigated. The predicted microstructure morphology after oxidation is found in good agreement with experiment data as shown in Figure 10.

It is found that the computed in-plane extensional moduli after oxidation are in good agreement with experimental data over the oxidation time. In addition, numerical results show that the temperature and pressure have a profound influence upon the oxidation results. The ablation of carbon phases is aggravated with the increase of temperature and pressure and therefore the elastic modulus of composite decreases gradually. An immediate perspective is to address other composite architectures such as the 3D braided C/SiC composite and to extend the developed oxidation model to other oxidizing environments, like water or gas-water environments. Besides, it is worth to note that the oxidation simulation of 2D C/SiC composite above 800 °C is not taken into account in the current work because the oxidation scheme is non-uniform and much complicated.

5 Composite design

Composite design with fiber orientation optimization plays an important role in aircraft structure design. Earlier orientation optimization methods face the main problem of huge number of design variables. Recently, a patch concept was proposed to reduce the number of design variables as shown in Figure 11. The elements in one patch share identical definition of fiber orientation. However, the traditional stress-based method cannot deal with patch orientation optimization of composite structures. In this paper, we propose an extended stress-based method to deal with such problems. The considered problems are to minimize the mean compliance under multiple load cases or to maximize the eigenvalues of a composite structure. Four numerical examples are solved in order to demonstrate the efficiency of the new method. It is shown that the new method has the ability to deal with constraints on orientation angle, such as symmetric, anti-symmetric and discrete orientation angle constraints. The iteration is less time-consuming because no sensitivity analysis is needed and a quick convergence rate can be achieved.

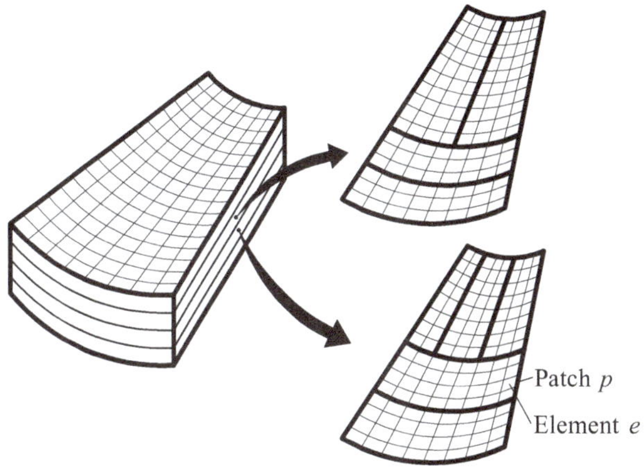

Figure 11. Definitions of layers, patches, and finite elements.

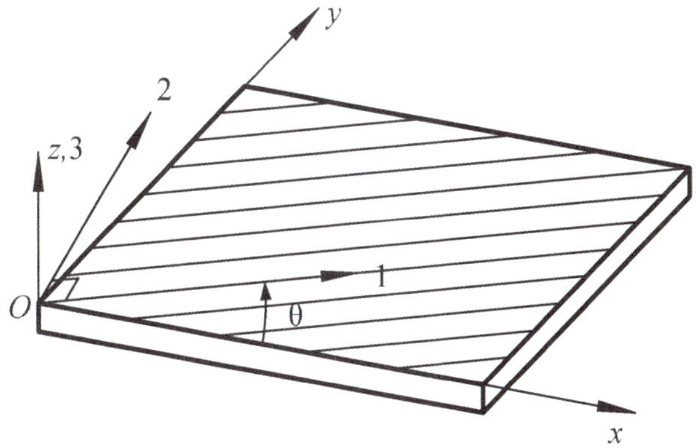

Figure 12. Orthotropic material.

In this work [7], the stress-based method is extended to deal with optimal patch orientation problems for composite structures under single and multiple load conditions. Some special constraints related to patch orientation angles are applied for the symmetric and anti-symmetric laminates in which patch orientation angles have to take the same and the opposite values, respectively. Besides, according to manufacturing considerations, investigations are made to the cases where the

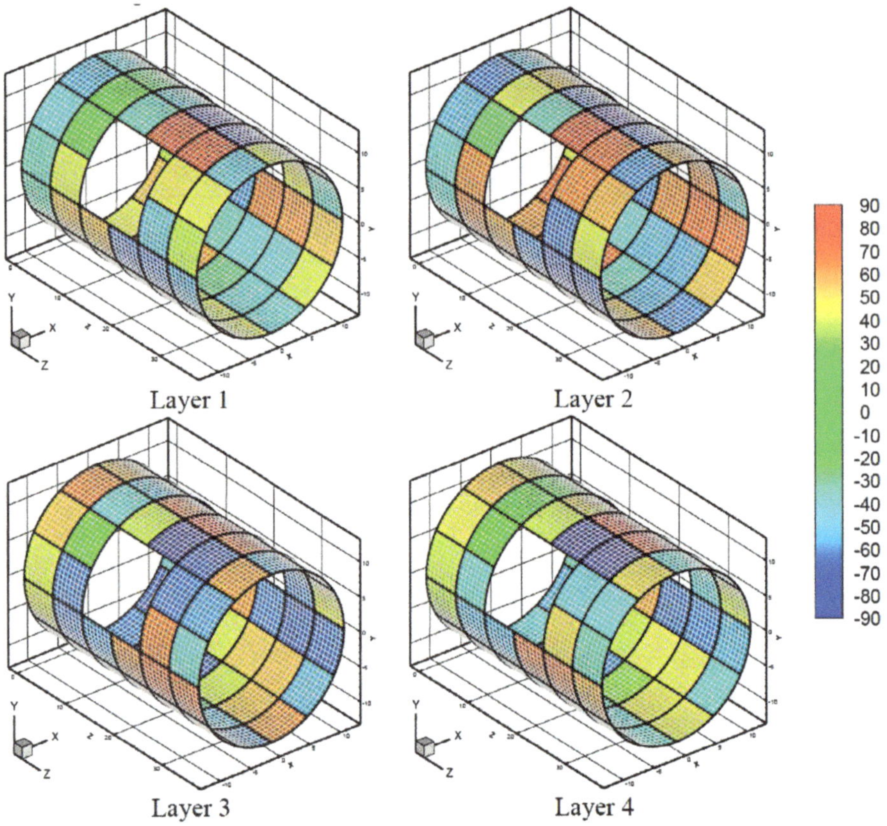

Figure 13. Optimal patch orientation angles with anti-symmetric laminate.

orientation angles are restricted to discrete values of 0°, 45°, −45°, and 90°. Several examples are finally given to show the validity of this method.

For thin laminated composite structures, the plane-stress condition may be applied due to the fact that out-of-plane stresses are often small compared with in-plane-stresses. Suppose each layer of the laminate is made of orthotropic materials. Figure 12 shows a patch over a layer where the stress-strain relationship of a point is

$$\begin{Bmatrix} \varepsilon_1 \\ \varepsilon_2 \\ \gamma_{12} \end{Bmatrix} = \begin{bmatrix} S_{11} & S_{12} & 0 \\ S_{12} & S_{22} & 0 \\ 0 & 0 & S_{66} \end{bmatrix} \begin{Bmatrix} \sigma_1 \\ \sigma_2 \\ \tau_{12} \end{Bmatrix} = \mathrm{S}_0 \begin{Bmatrix} \sigma_1 \\ \sigma_2 \\ \tau_{12} \end{Bmatrix} \quad (11)$$

where $\boldsymbol{\varepsilon} = \{\varepsilon_1, \varepsilon_2, \gamma_{12}\}^T$ and $\boldsymbol{\sigma} = \{\sigma_1, \sigma_2, \tau_{12}\}^{\mathrm{T}}$ represent the planar strains and stresses in the primed coordinate system (*O*-123), respectively; $\boldsymbol{S}_0$ is the un-rotated orthotropic compliance matrix.

Unlike the traditional stress-based method related to each element, the Extended stress-based method (ESBM) expands the integration over the patch domain so that

$$\frac{\partial \Phi}{\partial \theta_p} = \int_{\Omega_p} \sigma^T \frac{\partial \mathrm{S}}{\partial \theta_p} \sigma \mathrm{d}\Omega = 0, \quad (12)$$

where θ_p is the orientation variable related to patch *p*; Ω_p represents the spatial domain of patch *p*.

Let σ_x, σ_y, and τ_{xy} denote components of the stress tensor in the unprimed coordinate system (*O-xyz*), equation (12) can explicitly be developed as a function of the patch orientation variable θ_p

$$A \sin 2\theta_p + B \cos 2\theta_p + D \sin 4\theta_p + E \cos 4\theta_p = 0. \quad (13)$$

where coefficients *A*, *B*, *D* and *E* are computed as

$$\begin{aligned} A &= c \int_{\Omega_p} (\sigma_x^2 - \sigma_y^2) \mathrm{d}\Omega, \\ B &= 2c \int_{\Omega_p} \tau_{xy}(\sigma_x + \sigma_y) \mathrm{d}\Omega, \\ D &= d \int_{\Omega_p} [4(\tau_{xy})^2 - (\sigma_x - \sigma_y)^2] \mathrm{d}\Omega, \\ E &= 4d \int_{\Omega_p} \tau_{xy}(\sigma_x - \sigma_y) \mathrm{d}\Omega. \end{aligned} \quad (14)$$

Note that the numerical Gaussian integration can be applied for computing coefficients in equation (13). In practice, if the finite element mesh is dense enough, the stress in each element can be assumed to be a constant approximated by the value at the centroid of the element.

When $\theta_p \neq \pi/2$, roots of equation (13) can equivalently be obtained by solving the following fourth-order polynomial equation in terms of $\tan \theta_p$ within the interval $(-\pi/2, \pi/2)$.

$$\begin{aligned} (E - B)\tan^4\theta_p + (2A - 4D)\tan^3\theta_p - 6E\tan^2\theta_p \\ + (2A + 4E)\tan\theta_p + B + E = 0, \end{aligned} \quad (15)$$

By means of equation (13), the following evaluation function can be defined to determine the optimal value of θ_p selected from the real roots of equation (15) and $\theta_p = \pi/2$.

$$\begin{aligned} f(\theta_p) &= \int \frac{\partial \Phi}{\partial \theta_p} \mathrm{d}\theta_p \\ &= \int (A \sin 2\theta_p + B \cos 2\theta_p + D \sin 4\theta_p + E \cos 4\theta_p) \mathrm{d}\theta_p \\ &= \frac{1}{4}(-2A \cos 2\theta_p + 2B \sin 2\theta_p - D \cos 4\theta_p + E \sin 4\theta_p) \\ &\quad + \text{constant}. \end{aligned} \tag{16}$$

Note that if each patch has just one finite element, the ESBM is identical to the traditional stress-based method.

One example as shown in Figure 13 is to show the validity of the ESBM in dealing with constraints of patch orientation variables. The number of patches is 64 and each patch consists of 8×8 elements. The left side of the cylinder ($Z = 0$) is clamped and a bending moment Mx is applied on the right side ($Z = 36$) of the structure. Assume the laminated composite with four layers is used and they are numbered sequentially from the inside to the outside of the cylinder. Likewise, the objective function is to minimize the structural mean compliance.

Acknowledgements. This work is supported by National Natural Science Foundation for Distinguished Young Scholar (10925212), National Natural Science Foundation of China (90916027, 11002113, 51275424, 11172236), 973 Program (2011CB610304), the 111 Project (B07050), NPU Foundation for Fundamental Research (JC20120229).

References

1. Gao T, Zhang WH. 2011. A mass constraint formulation for structural topology optimization with multiphase materials. International Journal for Numerical Methods in Engineering, 88(8), 774–796.
2. Gao T, Zhang WH, Duysinx P. 2011. Topology optimization of structures designed with multiphase materials: volume constraint or mass constraint? Chinese Journal of Theoretical and Applied Mechanics, 43(2), 296–305.
3. Zhang WH, Xia L, Zhu JH, Zhang Q. 2011. Some recent advances in the integrated layout design of multicomponent systems. Journal of Mechanical Design, 133(10), 104503-1-15.
4. Xia L, Zhu JH, Zhang WH. 2011. New strategies for the efficient integrated layout design of multi-component system. Journal of Mechanical Engineering, 46(10), 148–155.
5. Xu YJ, Zhang WH, Domaszewski M. 2011. Microstructure modelling and prediction of effective elastic properties of 3D multiphase and multilayer braided composite. Material Science and Technology-London, 27(7), 1213–1221.
6. Xu YJ, Zhang WH. 2011. Numerical modelling of oxidized microstructure and degraded properties of 2D C/SiC composites in air oxidizing environments below 800 degrees C. Material Science and Engineering A, 528(27), 7974–7982.
7. Zhang J, Zhang WH, Zhu JH. 2011. An extended stress-based method for orientation angle optimization of laminated composite structures. Acta Mechanica Sinica, 27(6), 977–985.

11

A genetic algorithm based optimization framework to visualize, evaluate, and modify 3D space configurations in Desktop VR

Magesh Chandramouli[1,*] and Gary R. Bertoline[2]

[1] School of Technology, Purdue University, Calumet 46323, USA
[2] School of Technology, Purdue University, West Lafayette 47907, USA

Abstract – This paper presents the design and implementation of a Desktop VR (Virtual Reality) framework for generating and evaluating Pareto-optimal alternate 3D spatial configurations using GA (genetic algorithms). The 3-tier framework involves the generation of the Pareto-optimal plans using GA which are subsequently visualized first using a Java-based 2D Interface and finally in the form of a 3D VR scene. The search spaces (function domains) are extremely large in today's multifaceted interior design situations, and the optimization procedure involves conflicting objective functions, and limitations in the form of constraint functions. The interior space allocation problem is formulated and implemented as the "optimal configuration of artifacts". When using GAs, a group of Pareto-optimal solutions (Pareto set) are available for the planners and decision-makers, wherefrom one solution ought to be picked. Therefore, this study applies a tool to not only visually evaluate the plans, but also to interact with those plans to develop them further if needed. Besides enabling the optimal spatial configuration of the scene elements, this framework also facilitates evaluation and interaction via the 3D VR worlds. The framework aids the proactive exploration, analysis, and finalization of design aspects such as color, size, lighting, etc. of the various elements prior to the actual construction. The results demonstrate the robust performance of the GA and the final 3D VR environment with dynamic interactive capabilities. This final interface facilitates "GA-Compliant" transformations and scene modifications thereby facilitating the exploration and examination of alternative scene configurations.

Key words: Multiobjective optimization, Design optimization, Genetic algorithms, Desktop virtual reality.

1 Introduction

The optimal utilization of valuable interior spaces is impeded by the lack of efficient procedures to sift through the voluminous search spaces and support the generation/evaluation of alternate 3D spatial configurations. Only a very small fraction of the total search space is explored while solving interior space configuration problems involving MOO (Multiobjective Optimization). The design effort should pay critical attention to the notion of "the function" of the space i.e., the function this interior space is going to serve. From the seminal works of notable authors like Ballast, Pile, and Piotrowski [1–3] the importance of interior design becomes lucid and it is also evident that the interior space not only directly and indirectly influences our activities at work and home, but also influences several other daily activities such as shopping, dining, etc. One frequently overlooked, but fundamental element in the design of efficient interior spaces is the optimal arrangement of the artifacts. Vosinakis et al. [4] opine that effective interior space designs are typically purpose-driven, implying that the design is very mindful of the need of the clients. This innovative approach that integrates GA-based MOO with CG-based virtual scene rendering techniques facilitates design generation and evaluation. Systematic configuration of spaces and the arrangement of spatial components are absent or minimal either because it is considered trivial (which is incorrect) or due to the computational complexity involved in formulating and solving such MOO problems. Instead of following a logical procedure, the arrangement of artifacts in interior spaces continues to follow, by and large, a whimsical approach that has barely explored or evaluated the multitudes of options available.

There is a clear lack of the explicit articulation of objectives based on the "function" (purpose) of space. This is not to imply that planning is not at all involved in such interior spaces, but is not done in a methodical/systematic manner. While the service of interior designers is solicited for large-scale projects, the arrangement of the artifacts within the constructed spaces (e.g., libraries) rarely even tends to explore the "search space".

*e-mail: mchandr@purduecal.edu

The GA-based method exhaustively explores the search space to find the optimal solution. The rest of the paper is organized as follows: Section 2 is the literature review that addresses the related work. Section 3 explains the methodology involving the 3-tier framework for creating the Desktop VR scenes. Section 4 covers results and discussion and finally, Section 5 provides the conclusion.

2 Related work

Numerous examples of GA within computer graphics are found. Sims [5] applied evolutionary techniques to generate complex structures and textures. Notable works include those by Sims [5], Dawkins [6], Graf and Banzhaf [7], and Tood and Latham [8]. Chen et al. [9] employed GA to pick the best possible feature set for discerning computer graphics from digital photographic images. Yu et al. [10] used GA for automatically extracting features and processing stereo images. Nishino et al. [11] developed an interactive graphics optimizer using immune algorithms and continuing their work on using immune algorithms in Computer Graphics, Nishino et al. [12] developed an interactive 3D graphics modeler by simulating the human immune system.

Genetic algorithms were chosen for the MOO step in this study because of the following reasons:

(i) GAs have demonstrated higher probability of convergence within the parametric space;
(ii) GAs do not get trapped in the local optimum and try to find the global optimum;
(iii) GAs offer a diverse set of increasingly viable solutions (preservations of useful variations);
(iv) GAs have been efficiently employed for combinatorial optimization problems with extremely large search space of the order of 10^{100} or higher;
(v) GAs simultaneously evaluate many points in the parameter space, more likely to yield the global optimum GO;
(vi) By employing elitism, GAs try to retain the solutions with the highest fitness (so that the best solutions are not lost).

Stewart et al. [13] state that considering the inherent complex nature of modern design problems, the definitions of many objectives are not always nonlinear or additive. There are innumerable applications within the domain of construction graphics and interior design wherein MOO can be employed for solving problems involving combinatorial optimization [5]. Various MOO techniques including biologically inspired algorithms such as SA (Simulated Annealing), PSO (Particle Swarm Optimization), IWD (Intelligent Water Drops), CSS (Charged System Search), and ANT colony optimization were explored from the perspective of their suitability to this research. According to Balling et al. [14], one important reason that genetic algorithms have been exceptionally effective in solving design problems with multiple conflicting objectives and constraints is the contradictory nature of the two fundamental operations of crossover and mutation. Crossover intends to sustain the goodness in current generation, while, mutation aims to induce modifications or changes in the chromosomal structure (reverse effect).

Nevertheless, as GA commonly offers a pool of optimal Pareto plans or designs, advanced techniques are needed to evaluate and expedite the decision making process while reducing subjectivity, wherein visualization performs a vital role. Visualizing the candidate solutions of the design space is a lucid way to evaluate these optimal alternatives whilst minimizing abstraction. Therefore there is an imminent need for a tool to explore, analyze and evaluate the Pareto solutions. Chandramouli et al. [15] demonstrated that by using 3D visual scene renderings, planners who are experts in the fields of design planning can identify desirable or undesirable patterns. Karlen [16] states that "space planning is not a simple process involving a single category of information; rather, it is a complex dovetailing of several processes involving knowledge in many categories of information". Such information is of diverse nature ranging from Architectural Graphics applications by Ching [17] and Graphics Communication works by Bertoline et al. [18] to Kopec's work on Environmental Psychology [19], Kilmer's work on functional/visual design concepts [20], and Panero and Zelnik's work on human dimension [21].

When designing interior spaces, Rengel [22] precisely states that "... we claim space and subdivide in particular ways to suit the needs of the project". A judicious configuration of the artifacts is an essential component in the design of efficient interior spaces. According to Kubba [23], "an understanding of the use and allocation of space is the cornerstone in many design disciplines". Rengel [22] states that, "the number, size, and placement of large pieces of furniture, fixtures, and equipment, has a major impact on the interior space". Such earlier works have overlooked the "exploration and exploitation of search space" as evident from the detailed survey of software tools used by interior designers by Lok [24]. However, artifacts are largely arranged as and when the need arises and is dictated by the expertise of the person(s) involved in the process without necessarily following a standard protocol. This is in sharp contrast to the construction process which follows a standard procedure in accordance with a well laid-out set of rules, which must not be violated. Works by Maher et al. [25] and Whyte et al. [26] illustrate that computer based three-dimensional (3D) tools have evidently reduced the design time and facilitated concept evaluation.

3 Methodology

3.1 Research framework

There are three major components (Figure 1) in the design and implementation of the prototype for generating virtual representations for the alternate interior space configurations (Pareto plans). The first component in the 3-tier framework is the GA-based MOO process that results in the Pareto-optimal plans. The second part involves taking the numerical results of the GA-based computation and presenting them in the form of a Java-based 2D Graphic Interface. The third component

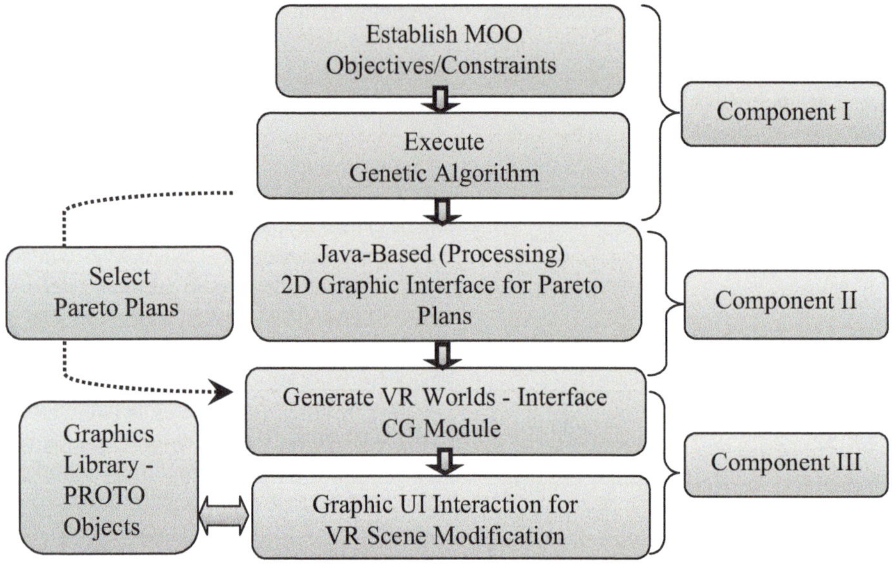

Figure 1. Framework for creating GA-based VR scenes.

involves the CG module wherein the virtual worlds are generated for viewing the plans in desktop Virtual Reality. The following sections provide a detailed description of the individual components of the aforementioned framework.

3.1.1 GA-based multiobjective optimization

Classically, a MOO is expressed as follows:

$$\text{Optimization } [f(x), or, y = (f_1(x), f_2(x) \cdots f_n(x))]$$
$$\text{Subject to } C(x) = \left(C_1(x), C_2(x), \cdots, C_p(x)\right)$$

n – Number of objective functions;
p – Number of constraints.

The set of values $(x_1, x_2, \ldots, x_n) \in X$ refers to the decision space and the set of values $(y_1, y_2, \ldots y_n) \in Y$ represents the objective space. The set of constraints i.e., $C(x) \leq 0$, limits the feasible set of solutions. GA-based MOO overcomes the limitations of traditional methods owing to its efficiency in solving non-linear, non-additive optimization problems without having to re-articulate or reformulate the design problem.

... C_{91}	... C_{92}	... C_{93}	... C_{94}	... C_{95}	... C_{96}	... C_{97}	... C_{98}	... C_{99}	... C_{100}
... C_{81}	... C_{82}	... C_{83}	... C_{84}	... C_{85}	... C_{86}	... C_{87}	... C_{88}	... C_{89}	... C_{90}
... C_{71}	... C_{72}	... C_{73}	... C_{74}	... C_{75}	... C_{76}	... C_{77}	... C_{78}	... C_{79}	... C_{80}
... C_{61}	... C_{62}	... C_{63}	... C_{64}	... C_{65}	... C_{66}	... C_{67}	... C_{68}	... C_{69}	... C_{70}
... C_{51}	... C_{52}	... C_{53}	... C_{54}	... C_{55}	... C_{56}	... C_{57}	... C_{58}	... C_{59}	... C_{60}
... C_{41}	... C_{42}	... C_{43}	... C_{44}	... C_{45}	... C_{46}	... C_{47}	... C_{48}	... C_{49}	... C_{50}
... C_{31}	... C_{32}	... C_{33}	... C_{34}	... C_{35}	... C_{36}	... C_{37}	... C_{38}	... C_{39}	... C_{40}
... C_{21}	... C_{22}	... C_{23}	... C_{24}	... C_{25}	... C_{26}	... C_{27}	... C_{28}	... C_{29}	... C_{30}
... C_{11}	... C_{12}	... C_{13}	... C_{14}	... C_{15}	... C_{16}	... C_{17}	... C_{18}	... C_{19}	... C_{20}
9 C_{01}	3 C_{02}	11 C_{03}	8 C_{04}	2 C_{05}	2 C_{06}	... C_{07}	... C_{08}	... C_{09}	... C_{10}

Figure 2. Grid representing physical layout.

3.1.2 GA formulation

The phenotype in this study is the physical layout of the library and the appropriate genotype has to be formulated (Figure 2) for this space. The study area considered here is the floor space corresponding to a library. Also, other interior spaces such as a restaurant, a space, and office space are demonstrated to illustrate other areas of application. The objective is to generate an optimal layout, which is represented as a raster grid of cells. These cells in the raster grid (layout) are divided into categories. Within this genetic framework depicting the study area, every variable (changeable) zone is denoted by a "gene" (integer value). A linear array of values is used to represent the layout.

Figure 2 illustrates the layout of the interior space in x and y dimensions. This is shown along with the genotypic representation using a single-dimensional array (Figure 3) The genes in the chromosome (G_1, G_2, …, G100) correspond to the cells in the grid representing the layout (C_1, …, C100). Index value represents the artifact element (Table 1). In both the genotype and phenotype in the (Figure 3) the "values" (integers representing the artifact type have been highlighted in green. It should be noted that the cells whose values are highlighted in yellow match (correspond between the linear integer array, the genotype and the layout, the phenotype). In the physical layout, the origin is at the bottom left.

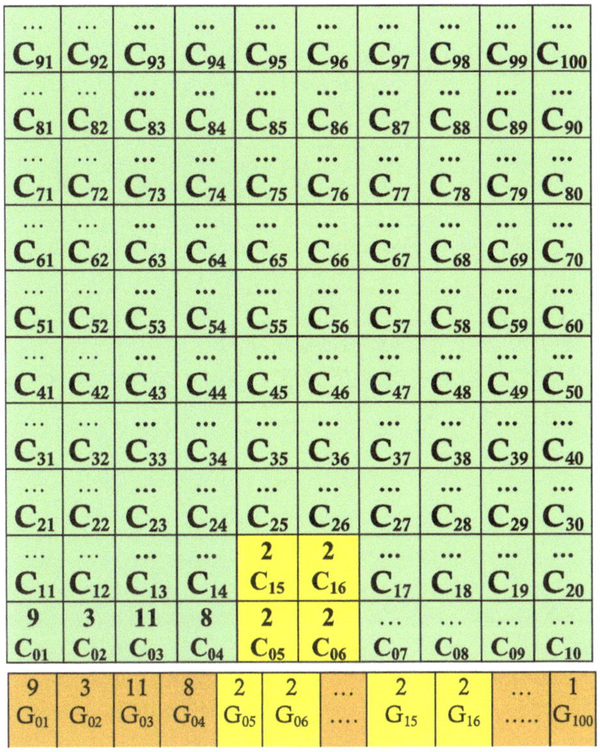

Figure 3. Encoding an artifact in a layout.

Table 1. Artifact elements and dimensions.

Index	Description	*X* dim.	*Y* dim.
1	IT lab	2	2
2	Washroom	2	2
3	Recreational space	2	2
4	Study Rm. Ind	1	1
5	Study Rm. Grp	3	3
6	Conf. room	4	4
7	Table/Ch. 1	1	1
8	Furniture 1	1	1
9	Furniture 2	1	1
10	Book-shelf	2	2
11	Ref. desk	1	1
12	Safety/emergency	4	4

The genotype conventions are as follows:

- G_n symbolizes the *n*th gene. Hence, G9 represents the 9th gene on a genotype with 100 genes.
- $G_{x\text{-}y}$ symbolizes the genotype component representing the *y*th occurrence of the artifact no. *x* (Index Value – *x*). Hence, $G_{1\text{-}2}$ symbolizes the 2nd occurrence of artifact number 1.
- $\min(G_{x:y})$ symbolizes the first gene of the genotype component representing the *y*th occurrence of the artifact no. *x* (Index Value – *x*). So, $G_{1\text{-}2}$ symbolizes the 2nd occurrence of artifact number 1.
- $\max(G_{x:y})$ symbolizes the last gene of the genotype component representing the *y*th occurrence of the artifact no. *x* (Index Value – *x*). Thus, $G_{1\text{-}2}$ symbolizes the 2nd occurrence of artifact number 1.
- C_n symbolizes the nth cell on the phenotype. Hence represents the 11th cell on the raster grid representing the actual layout of the interior space.

An artifact element occupying an area of 2 cells × 2 cells might seem continuous phenotypically, but on the genotype, they are not actually continuous (Figure 3).

3.1.3 Feasible set

Typically, for the first generation, each gene in the chromosome (representing the interior space) is designated a random value. One generation represents the size of the GA for a single run corresponding to 100 plans for the interior space layout. The suitability or "fitness" of each plan in the generation is calculated, using a fitness function. The plans that meet the constraints of the GA collectively constitute what is called the "feasible set". After the initial generation is obtained, the selection, recombination, and mutation processes are performed to create the subsequent generation. In this study, the cell location of the emergency facilities such as fire extinguisher and emergency door should not be changed. Hence, those artifacts continue to remain in their original location throughout the iterations. Also, every plan must have a minimum proportion of IT space and minimum amount of space for public utilities.

(a) Cells with index values 9 and 10 (emergency shelter and fire extinguisher) are not to be changed.
(b) The percentage of IT space and space for public utilities should have to be greater than 15% and 10%.

3.2 GA operations

3.2.1 Selection

The selection process signifies the choice of the individuals that are "fit-enough" for transmitting to the next generation and is based on the Darwinian notion of "survival of the fittest". Generally, this step is based on the "goodness" of a solution which is measured by the fitness of the individual. From the sorted pool of genotypes (Figure 4), the top 10 ($n = (\text{Sel}_{\text{Rate}} \times \text{Pop}_{\text{Gen}})$) highly ranked genotypes (with high fitness values) are selected for passing on the next generation. The Sel_{Rate} selected for the GA is 0.1 (10%). The remaining 90 genotypes ($n - (\text{Sel}_{\text{Rate}} \times \text{Pop}_{\text{Gen}})$) are created using the processes of crossover and mutation.

3.2.2 Crossover

The GA operation of recombination corresponds to the crossover (or recombination) process in the evolutionary cycle that involves combining the genetic information from two parent chromosomes (to produce the offspring. In this stage, "parent genotypes" are chosen and these are subjected to the crossover operation to generate children or "offspring" genotypes. To maintain the stochastic element in this process, a likelihood ratio documented in GA terminology as the "crossover probability" is employed. In this research, single point crossover is employed as explained here. The genetic material from

GA - SELECTION OPERATION															
Genotype (Plans sorted based on Fitness)														Index Value	Sorted Fitness
9	3	5	17	..	..	..	..	..	..	4	2	8	4	C_{11}	F_{11}
4	11	11	1	..	..	..	..	..	..	6	1	7	10	C_{27}	F_{27}
6	5	8	5	..	..	..	..	..	..	1	5	3	17	C_{33}	F_{33}
4	2	16	4	..	..	..	..	..	..	2	8	8	2	C_{42}	F_{42}
3	8	9	1	..	..	..	..	..	..	4	2	9	4	C_{12}	F_{12}
2	2	8	14	..	..	..	..	..	..	2	1	1	1	C_{72}	F_{72}
4	5	4	12	..	..	..	..	..	..	1	1	1	2	C_{54}	F_{54}

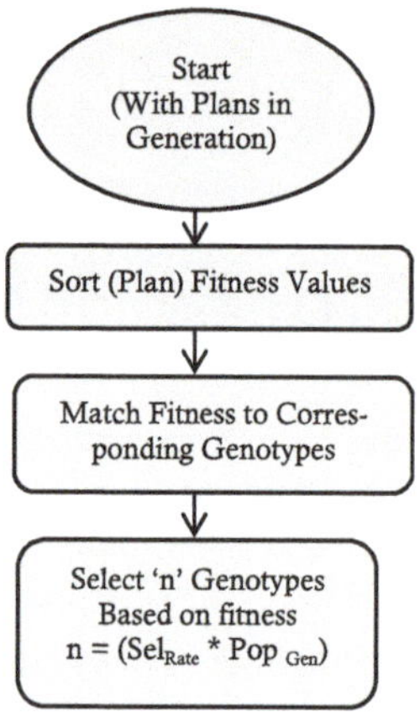

Figure 4. Fitness-based sorting of plans in a generation for selection.

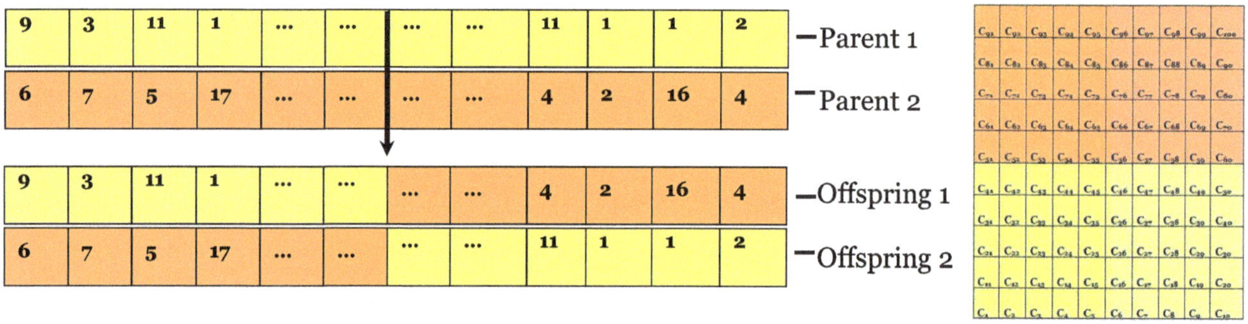

Figure 5. Crossover and resultant offspring genotype.

Parent 1 genotype (on the left of the crossover point) and from Parent 2 genotype (on the right of the crossover point) are joined to create Offspring1 (Figure 5).

The method of selection employed in this study for crossover is called tournament selection. In this type of selection, a small subsection of chromosomes from the parent generation is selected at first. From this the two chromosomes with the greatest fitness values are chosen. This method hence includes both a fitness based component as well as a non-fitness based component. Hence, it includes a stochastic element, while still considering the fitness of the parent chromosomes, which is why researchers consider this a blend of random as well as fitness-biased methods.

In the case of an artifact element occupying cells that are spread on more than a single row, the genotypic representation is not continuous. Hence, the most important consideration during the process of crossover is to prevent the splitting or breaking (of cells containing artifacts). Breaking refers to a single artifact element occurring at different locations, instead of being continuously represented at a single location. In order that unbroken artifact elements are obtained as a result of the crossover operation, other than performing the actual recombination between 50 and 51, care needs to be exercised in the GA formulation stage itself. For this reason, in the original chromosomes (representing the plans for the interior spaces) themselves, an artifact element ends before gene 50 or begins after 51 (Figure 6). Artifact element 6 (index value – 6) occupies four cells on the actual interior space layout (Table 1), the sub-section on the genotype representing this artifact element should occupy four genes.

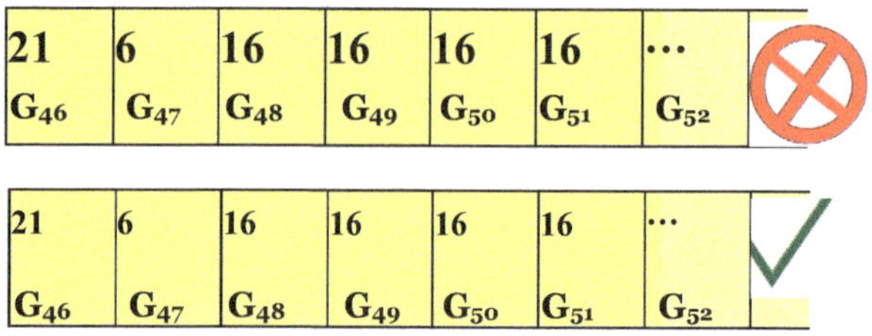

Figure.6. Correct and incorrect genotypes.

Let us consider the case when genes 48–51 represent this artifact element (G_{48}–G_{51}). During crossover, the other chromosomes 51st gene need not necessarily have the value 16. Consequently, the resultant plan of the interior space produced will have an area that is a mismatch for accommodating the corresponding artifact element. Thus, in the original chromosomes, by having an artifact element end before gene 50 or begin after 51. The general rule for any artifact element occupying more than one cell is as follows: $\max(Gx:y) < 50$ and $\min(Gx:y) > 51$, where x value can range from 1 to 12 (Index values of the elements)and the maximum value of y is the maximum number of times an artifact element occurs in the layout. In generic terms, the above can be stated as follows:

$$\min(Gx:y) < \frac{n}{2} \quad \text{and} \quad \max(Gx:y) > \frac{n}{2} + 1,$$

where x value can range from 1 to i_{max} (i_{max} – no. of artifacts) and the max. Value of y, max $(y) = \frac{n}{2}$, where n is the total no. of genes in the genotype.

3.2.3 Mutation

When the crossover operations are finished, the next generation has its population of chromosomes. At this point in time, random mutations are initiated in the population of chromosomes with a two-fold purpose: to prevent premature convergence and to enable the search process to diversify. These random mutations are induced at selected points (genes) located anywhere along the length of the chromosome. A value of 0.05 is selected as the probability for the mutation operation. A random value in the range of 0–1 is generated for each gene in the two offspring resulting in the crossover operation, the value of which determines whether mutation occurs or not. The integer value of the gene is subjected to change within the range of 1–12 if the mutation probability is higher than the random number so generated. Else, the gene is not subjected to mutation. The mechanism of the genetic algorithm rests on two fundamental operations namely selection and variation. Typically, when the GA proceeds with searching through the search space, it is desired to accrue the "good" traits or characteristics and prevent retaining the bad traits. In other words, it is required to preserve the good or "fit" individuals from generation to generation and eliminate those individuals or plans with poor fitness values (not suitable). Also, in order to facilitate obtaining diverse solutions, GA search process should be directed to explore the unexplored regions in the search space. All these are ensured by the GA operations of selection, crossover, and mutation.

3.3 Algorithms

The key algorithms (expressed as subroutines in the program) in the MOO framework are as follows:

3.3.1 Accidental re-allocation deterrent algorithm (ARDA)

During the process of creation of genotypes for a generation, a random function is used to generate integers in the range of 1–12 corresponding to the 12 artifact elements considered in this study. Following is the Matlab function used for this random generation of 100 chromosomes with 100 genes each: *randi* (12,100,100); However, as will be seen in the algorithms discussed subsequently here, there are other procedures that allot values to the cells based on the dimensions of the artifact and based on the constraints. This leads to the possibility that a cell whose value has been allocated (on a specific purpose) might be accidentally overwritten by a subsequent algorithm. In order to prevent this, the ARDA (Accidental Re-allocation Deterrent Algorithm) is implemented, which ensures that a cell whose value has been set on purpose by a specific function will not be accidently modified subsequently. From the earlier sections, it is known that every cell in the raster grid matches an artifact element which is represented in the genotype by the corresponding integer (index value).

Let us consider that for instance, an artifact element (index value x) occupies four cells (2×2), the genotypic depiction is not continuous as it is denoted by (G_6, G_7) and (G_{16}, G_{17}). ($G_6 = G_7 = G_{16} = G_{17} = x$). However, if later one of the cells preceding $G_6/G_7/G_{16}/G_{17}$ occupies more than one cell, then the subsequent corresponding cell will be overwritten with a value other than x. To prevent this from happening, besides the array containing the integer representing the index value for each cell in the genotype, another array of "dummy variables" is created. This dummy variable is given a default value of "1". When a gene value is changed "on purpose" by another function, say DVA explained below, then this dummy variable's value is changed to "2". Now, whenever a cell whose dummy variable value equals 2 is encountered it implies that this cell's value has been already set and should not be modified. In essence, the ARDA ensures the following: A cell's value can be changed only if dummy_{var} = "1".

3.3.2 Dimension validator algorithm (DVA)

The actual function of this DVA is evident from the name of this algorithm. The purpose of the DVA is to ensure that all the genes in the genotype/all the cells in the physical layout are allocated the values in accordance to their dimensional attributes mentioned in the dimension table. In other words, if artifact × has dimensions $m \times n$, then DVA's role is to ensure that the corresponding number of genes in the genotype are allotted in accordance with the positional requirements. Earlier, it was noted that "However, if later one of the cells preceding $G_6/G_7/G_{16}/G_{17}$ occupies more than one cell, then the subsequent corresponding cell will be overwritten with a value other than x". The vice-versa is what DVA2 is concerned about. If a gene value, say G_{15}, is found to be equal to an index value corresponding to an artifact occupying more than one cell and the genes following it are not available for allotment G_{16}/G_{17}, then the value has to be changed to one corresponding to single-celled artifacts.

3.3.3 Constraint validator algorithm (CVA)

Once again, as the name indicates this algorithm ensures that the GA constraints are imposed. It was noted that cells with index values 18 and 19 (Emergency Shelter and Fire Extinguisher) are not to be changed. Throughout the MOO process, this CVA ensures that the corresponding cells/genes always continue to have the same values as required by the constraints.

3.3.4 Optional GA modifier algorithm (GMA)

This is an optional algorithm that should be executed if the same interior space project is carried with slightly modified layout. For instance, for the same library interior space allocation project discussed earlier, if for various reasons, it is decided that book shelves are located on the northern corner of the layout, then cells 91–100 need not be actually included in the GA process. This is in a way a modified CVA algorithm that ensures that the MOO process is carried out according to this corresponding layout.

3.4 Fitness calculation and objectives

The following three objectives are considered:

1. Maximizing IT usability (labs, individual work stations),
2. Maximizing reading space (table/chairs, study rooms, and conference rooms), and
3. Maximizing public amenities (phones, ATMs, etc.).

Phenotypically, these objectives are innately conflicting since an enlargement in one space will lead to the shrinking of the space available for the other objectives. Also, to ensure that some minimum requirements are satisfied, the following constraints were imposed:

(a) Cells with index value 12 (emergency or safety features) are constant
(b) IT space and space for public utilities should be greater than 15% and 10%.

Balling's [27] Maximin function is used for calculating the fitness values of the genotypes. This function was originally applied for geospatial problems and has been employed in earlier studies involving spatial optimization [15, 28]. After taking into consideration several fitness evaluation strategies, this particular function described here has been found to be very suitable for GA studies involving a spatial component. The process of objective normalization is carried out using a basic scaling procedure. This is done using the formula shown in equation (1) for each objective k for a set of plans in a generation and then performing re-scaling subsequently. In equation (2), val_{ik} is the value of the objective k when assuming plan i, $\mathrm{val}_{\min,k}$ is the least value of objective k among all the plans in the current generation, and $\mathrm{val}_{\max,k}$ is the highest value of objective k among all the plans in a generation. The objectives of each plan were compared with other plans in the generation to find the fit ones in the current generation. Hence, when an objective of a plan i is compared with that of another plan j, plan j is better than plan i if the difference between j and i is positive (Eq. (2)).

$$\mathrm{Obj}_k(\mathrm{Plan}_i) = \frac{\mathrm{val}_{ik} - \mathrm{val}_{\min,k}}{\mathrm{val}_{\max,k} - \mathrm{val}_{\min,k}},\ k = 1, 2, 3\ i = 1, \ldots, 100$$

$$\mathrm{Obj}_k(\mathrm{Plan}_j) - \mathrm{Obj}_k(\mathrm{Plan}_i) = \frac{\mathrm{val}_{jk} - \mathrm{val}_{\min,k}}{\mathrm{range}_k} - \frac{\mathrm{val}_{ik} - \mathrm{val}_{\min,k}}{\mathrm{range}_k} = \frac{\mathrm{val}_{jk} - \mathrm{val}_{ik}}{\mathrm{range}_k}$$

$$f_i = 1 - \max\left(\min \times \left(\sum\left(\frac{\mathrm{Obj1}_j - \mathrm{Obj1}_i}{\mathrm{range}_1}, \frac{\mathrm{Obj2}_j - \mathrm{Obj2}_i}{\mathrm{range}_2}, \ldots, \frac{\mathrm{Obj}n_j - \mathrm{Obj}n_i}{\mathrm{range}_n}\right)\right)\right)^p.$$

The fitness of each plan in a generation is calculated with regard to the other plans in the same generation. Range corresponds to the difference between the maximum and the minimum values of that particular objective. Considering two plans $plan_j$ and $plan_i$ in a particular generation, $plan_i$ is dominated by $plan_j$ if *$plan_j$ surpasses the other plan in all the objectives*. The fitness is calculated as explained in equation (3). Based on the fitness formula described above, the Pareto-optimal plans were selected from a generation based on the fitness values obtained. While dominated plans had a fitness value between 0 and 1, Pareto-optimal plans had fitness values greater than 1 [14, 27]. Besides, in this study, a higher value for p is employed. This value was chosen to pursue Pareto-optimality more vigorously. The fitness of plans with fi more than 1 gets higher, and the fitness of those plans with fi values less than 1 gets further lower.

3.5 Pareto-optimality and visualization

A brief discussion of the notion of "Pareto-optimality", especially Pareto-optimal plans is very important from the perspective of this study on the whole. The GA iterations engenders a set of optimal solutions (Pareto-optimal set), which signifies the pool of solutions that denote the optimal trade-off [13] for the interior space problem with the three objectives subject to the constraints. For a plan to be considered as a Pareto plan it should not be inferior to or underperform any other plan from that generation. Stated otherwise, a plan may outclass the Pareto plan when considering one objective and a different plan may be better with regards to another objective; however, a "single plan" must not outdo a Pareto plan in all the objectives considered for the MOO. Balling, Taber, Brown, and Day [27] clearly state that the "*Pareto set is independent of the relative importance of the objectives*". The above discussion evinces that plans do not belong to the Pareto set (non-Pareto plans) are "dominated", as a Pareto plan that is better (or that which dominates) has been found. The fitness function used in this study is the one proposed by Balling et al. [14, 27].

However, the challenge now is to select one solution or plan for implementation. For any design optimization problem, a single satisfactory solution that can be implemented is required. As suggested by Chandramouli et al. [28], the use of genetic algorithms for design optimization entails meticulous scrutiny of the differences among the candidate solutions to obtain a better knowledge of the basic processes and the fulfillment of the objective functions. The process of choosing one single solution over others entails exhaustive domain knowledge. As noted by Seixas et al. [29], typically, many GA-based design optimization procedures make the final choice of the solution (from the Pareto set) based on some "higher level information".

However, when it comes to design and planning, it is not possible to implement all the conflicting solutions in the Pareto set (resulting from the GA process). Hence, a virtual reality based visualization framework is generated to explore and study the alternative solutions.

3.6 Visualizing GA results

As explained by the previous section, while using GAs for MOO, the procedure literally stops with the pool of the Pareto solutions and thereafter, one solution from the Pareto set is chosen based on some "higher-level or expert knowledge". Arguably, these involve several subjective measures and this seriously undermines the efforts to objectively select a plan without prejudice from the various stakeholders involved. Therefore there is an imminent need for a tool to explore, analyze and evaluate the Pareto solutions. Desktop VR-based graphic visualization can meet such need by facilitating not only presented information, but also enabling seeing and understanding of hidden information among datasets. Connolly [30] defines virtual reality as the application of an artificial environment generated by computer technology to simulate some targeted activity. However, directly creating virtual representations from the results of the GA process is a tedious task. Hence, a processing (Java-variant) 2D graphic interface is used to obtain the numerical array from the GA and output a 2D Graphic layout, which then serves as a reference for building the virtual worlds.

3.7 2D Graphic interface

In order to facilitate the transition from the numerical GA results to virtual representations, a 2D Graphic Interface is built in this study. For generating the 2D graphic interface, a Java-variant called Processing is employed. Processing, an open source programming language built by Reas and Fry [31], is largely used for graphic programming. The Pareto-optimal plan obtained from the GA-based iterations is a layout representing an optimal configuration of artifacts in the interior space. This is in the form an array of integers. This integer array contains 100 genes corresponding to the 100 cells in the raster grid representing the interior space layout (Figure 7). However, when constructing the 3D virtual scenarios corresponding to the Pareto-optimal plans, using this numerical array of integers can be extremely cumbersome.

Hence, the numerical array (Pareto plan) is provided as the input in the processing file (.pde), which produces an output in the form of a graphic layout. Figure 8 shows the iconized graphic interface for another Pareto plan. This iconized graphic interface can now be used as a reference for the 3D Virtual world scenarios depicting Pareto plans.

3.8 Virtual worlds for visualization

The tool used for visualizing the scenarios is the virtual reality modeling language (VRML/X3D) specifications laid out by Ames et al. [32]. The primary reasons for using this tool are the web advantage it offers and the programmable external

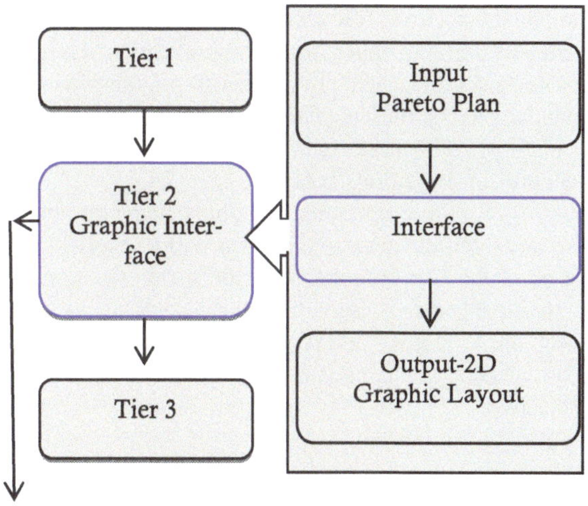

1	1	6	6	2	1	9	3	3	10
1	1	3	3	3	3	2	3	3	10
7	7	3	3	3	3	2	2	6	10
6	12	1	1	6	2	1	1	7	10
6	8	1	1	8	8	1	1	7	10
6	3	3	7	3	3	3	3	7	10
6	3	3	9	3	3	3	3	7	10
6	3	3	9	3	3	3	3	7	10
5	2	2	2	2	4	3	3	4	10
5	1	1	2	2	4	3	3	4	10

Figure 7. Tier II: Graphic interface.

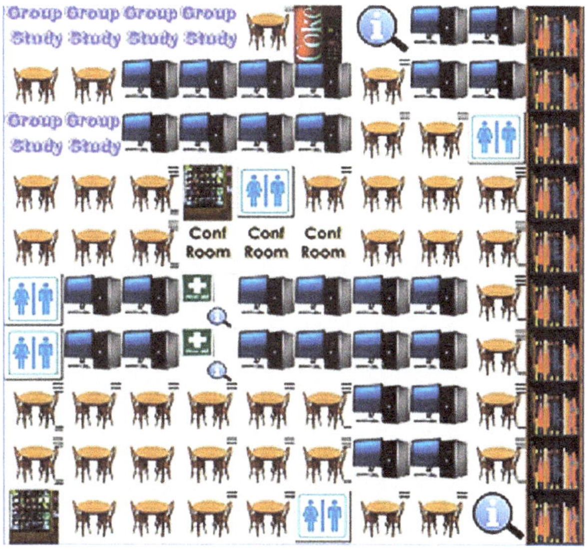

Figure 8. Tier II: Iconized graphic interface.

authoring interface (EAI). VR scenes thus created can be hosted online directly without much further modification. Users can view these desktop virtual environments on their computer using standard web browsers enabled with simple, open-source VRML plugins. Also, these are easily convertible into the latest web standard for 3D worlds, x3D.

Using a VR-based representation offers immense advantages. The users can navigate within these virtual worlds, move the objects in the worlds, rotate or scale them, and transform them in multiple ways. These virtual worlds facilitate user interaction with the 3D objects and provide a sense of immersion. In this study visualization is used as a tool to evaluate the Pareto plans objectively.

3.9 Virtual world generation

The scene is defined by "nodes" (X3D/VRML). A visualization scene can be considered to be composed of objects with properties. In the preceding sentence, the phrase "objects with properties" is highlighted or emphasized, since these properties determine how an objects looks and/or behaves. Considering a sample scene, say a library, as considered in this study. If further broken down into smaller fragments, the elements that result include tables, chairs, lamps, washrooms, etc. The furniture may be of a particular material, color, and dimensions. All these are the attributes of the furniture. Similarly, each element has its own characteristic features or attributes.

A scene can be viewed as being composed of elements or objects, each of which has its own properties or attributes. A parent object can include any number of children, which can be grouped or assembled to function as one single entity. This sort of hierarchical arrangement helps in the step-by-step design of the object and also understanding the framework at any later stage. A scene-tree construction is used in virtual scene renderings. The root or the parent object consists of whole scene grouped together and all the other components are grouped under this parent object using "parent-child" relationships. Individual scene elements corresponding to each floor type were created positioned according to their corresponding positions as per the Pareto plan obtained in the previous step.

For complex objects including multiple parts, various object parts are grouped to form parent objects, leading to complete objects that are combined and re-positioned to create the final 3D scene. Another advanced way of customizing the virtual objects is by the use of PROTO nodes, which the following section addresses.

3.10 Virtual objects and customizable nodes

Virtual world objects are described as shapes with geometry and appearance. All features such as buildings, rooms, artifacts, etc. can be modeled as shapes which can be grouped together and transformed (translated or rotated) within the coordinate system within which they are built. A vast number of VR objects, however complex they might be, are built using the fundamental shape node with the principal fields namely geometry and appearance. The geometry field is used to describe the geometric properties of the object and the appearance field is

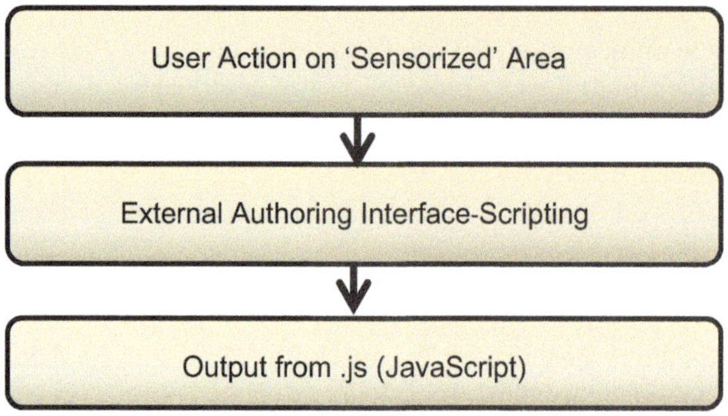

Figure 9. I/O sequence (event-response) framework.

used to describe how the object looks. In order to be able to develop a framework that can interface the GA results with a virtual environment, the objects in the virtual environment should be mutually compatible with the GA. This implies that the scene or the objects that are formulated in the GA should be in a form that can be represented and programmed within the virtual environment (GA-compatible). A significant step towards this goal involves generating customizable nodes, the PROTOS. PROTOS represents reusable scene objects and these can be referenced from within the file and externally (EXTERNPROTO). The PROTO library created as part of this study can serve as an extremely useful addition that can be extended into several other object oriented applications as well. The framework also provides functionalities to dynamically change graphical attributes such as diffuseColor /specularColor/ transparency to generate various appearances of the same object and also dynamically manipulate the size of the object.

3.11 Programmed interface: user interaction

One of the primary reasons for implementing the graphic library using PROTOS is to program the objects created. The exposedFields can be compared with Input and Output jacks and can be wired to create "ROUTES". Figure 9 below demonstrates the Input/Output flow corresponding to this. The key attributes of a graphics scene include geometry and appearance. There are numerous fields and exposed fields that cumulatively control the shape and appearance of objects.

By routing user events through appropriate Input and Output jacks, these fields can be controlled "dynamically" to manipulate geometry and appearance, such as these:

- Dynamically changing diffuseColor/specularColor/ transparency for alternate appearances of object(s);
- Dynamically changing the size of the object.

While these can be achieved using commercial software, these are numerous restrictions to what can be done within such IDEs. However, these programmed objects offer innumerable possibilities whereby their behavior and attributes can be controlled. By way of creating custom-made PROTOS and EXTERNPROTOS, the behavior of the scene and the constituent objects can be controlled very precisely to generate desired actions. Next, controlling the position and orientation by manipulating transformational attributes is discussed.

The virtual world/scene is divided into pixels/voxels. Objects are embedded within these grids (2D/3D) with built-in sensors to recognize user actions. When the user clicks within the perimeter/boundary of the pixel/voxel, the position (geometric center) is highlighted. The highlighted object always exists in the scene, however, initially in the un-activated state (visibility set to zero). Subsequently, using JavaScript functions, the user click event is used to activate the object (by modifying the material node's transparency value). In order to facilitate the user clicking, a buffer is provided whereby the user can clicking within a radius of the geometric center of the grid and the object is still selected. Ultimately, the crux of this interaction and manipulation involves building sensors and programming them to sense user actions.

Once the virtual scene environments have been built for the corresponding Pareto-optimal plans, further modifications without affecting the constraints can be performed as shown above wherein the decision makers can apply changes to attributes including position and appearance to "envision" alternative scenarios. This offers a tremendous capability whereby once a VR design is generated and the user wants to manipulate the final design. Some parts are programmed to be translated or rotated or scaled, while others are fixed or stationary. The GA constraints require them to be at that fixed position (for design criteria) and hence these cannot be changed.

Earlier sections covered in detail the methodology for the design and implementation of the GA-enabled visualization framework to generate design scenarios involving the arrangement of artifacts in interior spaces. The framework involves three major elements namely GA-based MOO, a 2D graphic interface, and finally a VR-based visualization environment. This discussion of the results of processes of MOO and visualization is especially from the perspective of the interior space application.

The GA-based MOO in conjunction with the VR-based visualization described previously was tested on a common interior space e.g., library. The interactive and navigable virtual worlds generated are discussed in detail here. Such visualizations can efficiently illustrate design scenario(s) than a conventional (paper-based) or PC-based 2D scene representations. However, unless the GA data is converted into the 3D format it is not of significant use to planners and decision makers, since analyzing voluminous numerical data and inferring from them is a cumbersome task, which may not ultimately yield the desired results. Visualization greatly facilitates understanding the design scenario and comprehending its function from diverse perspectives; this is especially important when multiple stakeholders with varying interests are involved.

4 Results and discussion

4.1 Important considerations while interpreting results

As mentioned in the earlier sections, the "primary focus" of this study is the amalgamation of the GA-based MOO with the Desktop VR based CG component to design and implement an innovative framework to generate alternative interior space configurations. However, the essential innovation involves the actual integration of GA with CG. The emphasis here is the demonstration of the usefulness of the visualization scenarios resulting from this integrated framework and the analysis of the environments as a tool to facilitate "informed decision-making". From this point of view, the GA metrics are not as significant as the process of interpreting the visual results. It was explained in the literature review that earlier research applying GA-based MOO to the domain of interior space design using a framework similar to that of this study could not be found; this is one of the major reasons justifying the innovative aspect of this research.

From the above discussion the need for extensive research in this area of "GA metrics for MOO in interior space planning" is evident. This in itself can be a subject matter deserving extensive research spanning over another graduate research. Hence, even though this section discusses the GA metrics, the focus is essentially to elucidate the visualization results and demonstrate the usefulness of the research outcomes in generating as well as evaluating alternative scenarios. Hence, the essential aspects of the GA parameters and GA metrics are discussed subsequently. The functionalities of the visual interface for viewing are demonstrated using not only the library setup, but various other possible interior space scenarios are used to demonstrate the various capabilities of the desktop virtual environments generated for this study. But, before looking at the visual interface, let us look at the GA-based MOO component.

4.2 Crucial GA operations for experiments

At the core of the GA based optimization are the important processes of crossover and mutation. Typically, the term "GA parameters" refers to the probabilities of crossover and mutation. However, before getting to the crossover probability and the mutation probability, the method of selecting the parent individuals for crossover needs to be discussed. The type of selection employed in this study for selecting the parent chromosomes for "mating" or crossover is called "tournament selection". This method involves selecting the parent chromosomes from a smaller subset of individuals from a generation. Typically, there is a stochastic element involved in this selection of the few chromosomes to serve as the pool for "parents-to-be". From this subset, the selection of the parent is classically based on its fitness value. Hence, the process of tournament selection is, in a way, characterized by stochastic as well as deterministic traits. One important feature in this process of tournament selection is the tournament size. When the tournament size is increased, the probability that an unfit chromosome (individual with lesser fitness value) is selected gets minimized. The tournament size employed in this study is 5. At random, five plans are selected from the parent generation. From this subset, two plans with the highest fitness values are selected as parent individuals for the subsequent step of crossover (mating). This is implemented as follows:

1. Random set containing five integer values are selected,
2. The five integer values are used to select the subset of individuals,

3. From this subset, individuals with highest fitness are selected as parents.

To prevent recurrence, a sub-routine (function) is executed in the program to ensure that the same chromosome is not selected again, as this would result in the children chromosomes being the same as the parents. As noted earlier, single-point crossover was employed for the reproduction process. Splitting refers to the case when a single artifact element occurring at non-continuous locations. The crossover point (after gene 50, G_{50}) represents the location where the portions are interchanged to generate the offspring.

The values (genes) in the first offspring before the crossover point are identical to the parent Individual no. 1 and the values after the crossover point are identical to the values in the parent individual (PI) no. 2 (51–100). The first part of the second offspring has values similar to PI2 and the other part replicates PI1's values. Subsequent to the reproduction operation, the chromosomes in the resultant population must be subjected to mutation. The mutation operation is carried out by comparing the result of the random value generated with the probability of mutation. In this study, a mutation probability of 0.05 is employed. In the natural evolution, mutations are considered to be a sporadic or rare occurrence and hence the mutation probability is assigned a low value. A random value (Ran_{Val}) in the range of 0–1 is produced corresponding to each gene in the two children chromosomes (offspring). The "crux" of the mutation process involves comparing this value hence generated with the "probability of mutation" (Mut_{prob}). If ($Mut_{prob} - Ran_{val} > 0$), then the integer value corresponding to the gene is changed to another random value between 1 and 12. Else, it remains non-mutated or integer value corresponding to the gene is not changed. It should not be the other way ($Ran_{val} - Mut_{prob} > 0$), as the probability of mutation is very low and then this may lead a very high frequency of mutation which contradicts the natural process of mutation. The process of mutation executed needs to be "constrained". This implies that the mutation process should not result in the "splitting of artifacts" mentioned earlier and it should not allot elements in the reserved cells (for safety features/emergency purposes).

4.3 GA validation

Kemenade et al. [33] examined the role of the parameters in the GA operational efficiency, their impact on the outcomes, and the various schemas for GA operations. Parameters of the GA such as the Generation size, tournament size, and the probabilities of crossover and mutation have a strong correlation with the overall performance of the GA. In this study, the GA was executed for varying values for the probability for mutation and tournament size and finally a medium tournament size of 5 and a low mutation probability (0.05) is employed.

The GA results are explained here from the perspective of the fitness of the generations and "Pareto-optimality". For repeated runs of the GA for at least 10 times, the average fitness of the generations showed consistent improvement explaining that the GA has significantly improved the fitness of the individuals. This implies that the final plans, whilst satisfying the constraints, have maximized the objectives incorporated within the GA. The result of the GA iterations is the Pareto-optimal set, which consists of the non-dominated solutions. However, practically only one solution can be implemented and may be more suitable for the specific problem on hand when taking in to consideration other factors that were not included in the GA. When there is an inventory list containing 12 elements, the random generation of the chromosomes cannot be expected to produce a gene allocation that matches the artifact attributes (or dimensions). Hence, a step is implemented before the actual execution of the genetic algorithms known as "selection of feasible solutions" is implemented to ensure that the genotypic representation corresponds to the realistic representation.

4.4 Visualization results

The GA results are input into a .pde file (Processing program) which converts these numerical results into the cells of a graphic interface (reference). Hence, the numerical array (Pareto plan) is provided as the input in the Processing file (.pde), which produces a output in the form of a graphic layout, which can now be used as a reference for constructing the 3D Virtual world scenarios depicting the Pareto-optimal plans. The floor is superimposed over this layout and the transparency value of the floor is adjusted so as to reveal the layout. In order to facilitate this process of further generating the virtual worlds from this 2D layout, the *x*, *y*, *z* axes are used initially as a reference and also a grid is superimposed over the scene to facilitate for properly situating the elements within the VR program code.

A virtual environment or a scene is composed of various scene objects, where the scene itself can be considered to be the root object. All the objects can be envisioned in the form of "parent-child" associations. Besides, as mentioned earlier, to be able to develop a framework that can facilitate visualizing the MOO results and further interacting with it generating customizable nodes, the PROTOS, is a crucial step. Table 2 illustrates selected artifact elements on the layout space and what virtual scene elements are corresponding to these index values. A more complex virtual artifact element leads to greater branching in terms of the scene hierarchy.

When evolutionary procedures are employed for multiobjective design optimization, it is imperative that the differences among the candidate solutions are meticulously evaluated to consider their suitability to the actual problem on hand. More importantly, it is not practically possible to include all possible or desired features in the form of objectives or constraints. A whole set of factors that have not been expressed as either objectives or constraints can be evaluated using the visual medium. Visualization empowers the decision makers to view and comprehend even hidden or ineffable information. The 3D virtual scenes besides enabling navigation also provide a sense of immersion whereby the planners can explore and evaluate the designs (Figure 10).

This study provides interactive capabilities within the virtual environment. In this research, Desktop VR is used for

Table 2. Index values of selected elements and scene hierarchies.

Artifact classification virtual scene	Level 1 of scene hierarchy	Level 2 of scene hierarchy
Index value – 1	IT lab	1. PC terminals 2. Furniture 1
Index value – 3	Refreshments/Rec. space	1. Vending machine 2. Snacks machine
Index value – 5	Group study room	1. Furniture 1 2. Table.Ch.1
Index value – 11	Ref/Info desk	1. Furniture 1 2. Table.Ch.1
Index value – 12	Book-shelf	1. Shelf

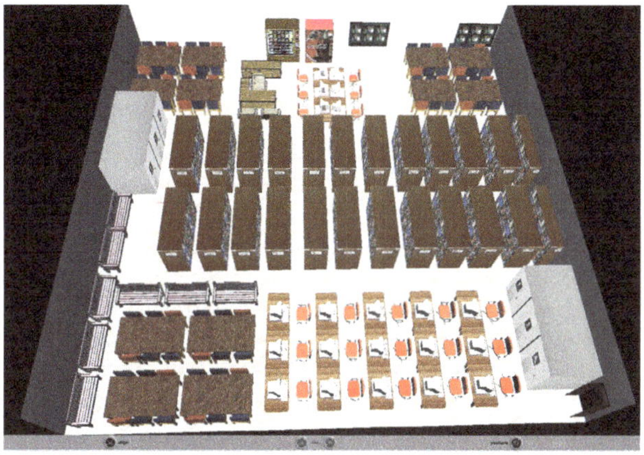

Figure 10. Desktop VR scene for a Pareto plan.

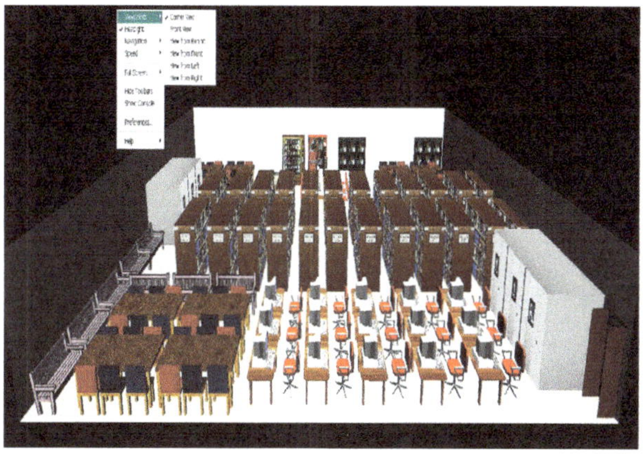

Figure 11. Multiple POVs (points of view).

the visualization component. By using 3D visual scene renderings, planners who are experts in the fields of design planning can identify desirable or undesirable patterns. It is very important to interact with the VR environment to perceive the unique aspects of the plan under study. The "point of view" is a feature that indicates the use of visualization to study the same scenario from different points of view (Figure 11).

The viewpoint is a preprogrammed spot within the VR environment that takes the viewer to the specific position and orientation wherefrom the scene can be viewed. This is very helpful when evaluating a scene with respect to a specific problem, so that the pros and cons of the design can be studied in detail. Also, another salient advantage while programing virtual scene representations is the capability to build the same by employing varying levels of detail. For instance, when viewing from a distance, the bigger or finer details need not be rendered for the smaller objects that are not quite as obvious. This notion can be used to efficiently model the scene so as to facilitate faster and smoother scene rendering. The virtual environment can be rendered in accordance with the viewer's position with the scene. Table 3 highlights few salient features provided by VR settings and explains how the VR visualization of Pareto plans enhances evaluating the various plans with respect to the objectives.

4.5 Interaction with the virtual scene

Virtual environments can be rendered interactive using different techniques. The EAI (External Authoring Interface) provides a window to access the nodes that constitute the various shapes within the virtual scene.

Typically, a VRML based virtual environment is composed of scene elements that can be broken up into nodes that constitute these shapes. These nodes (in these shapes) control how a scene object looks by programming their geometry and appearance. Interaction necessarily involves a change of state of the objects involved. The node that has to be selected and modified depends on the type of interaction desired. While the capabilities of the standalone virtual environment are limited, a plethora of possibilities can be opened up by employing the External Authoring Interface (EAI). Two important functionalities used in this study, from a design perspective, include "sensors" and "EAI". Sensors refer to the ability of a virtual environment to "sense" and respond to user actions. The primary means of interaction for a user with a desktop virtual environment involves the mouse and the keyboard. Examples of common mouse actions are as follows:

- Hover,
- Left-click,
- Right-click,
- Drag.

Theoretically, innumerable alternatives exist for a design problem, which is why an efficient tool such as GA is used to narrow down on a subset of solutions. It should however be noted that it is difficult to say that with 100% accuracy that the solution obtained is the "best" solution. This is so because

Table 3. Salient advantages of VR scene visualization of Pareto plans.

Feature	Advantage
Point of view (POV)	The scene can be evaluated from innumerable points of view.
Levels of detail (LOD)	The same scene can be created and viewed using varying levels of detail.
Navigation	Different browser/plug-in combinations offer various methods of navigation such as study, flying, and walk-through.
Controlling scene attributes	The scene attributes and the attributes of the VR scene objects such as Geometry Appearance can be controlled in multiple ways.

the final solution is the result of the interplay of a whole lot of factors such as GA formulation, constraints applied, various GA parameters such as rate of selection, probability of mutation, fitness function, and so on.

A different set of values for the aforementioned characteristics might lead to totally diverse set of solutions. Let *X* denotes the complete search space for a problem on hand. For design problems, depending on the number of various factors including objectives involved, the nature of objectives (e.g., conflicting), and constraints imposed etc. the search space may be enormous. The rationale behind using tools such as GA is to narrow down and identify the "optimal" solution.

Assuming that a robust GA has been employed with suitable fitness function and appropriate parameters, the user is still left with a pool of solutions (Pareto-optimal plans). From the above explanation it follows that for a different combination of factors, a different solution (or set of solutions) could have been attained. Also, most literature refer to terms such as "expert knowledge" or "domain-specific knowledge" or "higher level information" [29] that plays a crucial role in the selection of the final plan or design for implementation. This, quite frequently involves "intangible" or "ineffable" factors in the sense that some aspects may become obvious to the expert upon visualizing the design. Whilst, this can always be "expressed" once it has been "envisioned", it may be difficult to express prior to that.

The ultimate purpose of all these tools is to facilitate the process of "envisioning" the best or optimal design. The visual representation generated portrays a possible solution to the problem considered. However, when this visual environment lacks interactive capabilities, it is a major handicap as the user or the expert is not able to envision "What if" scenarios. It is extremely important that any transformation (translation or rotation) can occur only within the "constraints" already set in the GA. A transformation should not violate the GA constraints; else the entire process of MOO using the GA will be undermined. The following pseudo-code expresses this notion:

```
do{
// Object transformation within the Virtual Scene
} while (GAConstraintsMet == true);
```

In order to ensure that the conditions are always met, the virtual environment (and hence, the objects within the virtual scene) are programmed as follows:

- Not all objects are allowed to translate. Repositioning is enabled only for select objects within virtual scene.
- Areas where objects cannot be moved into are programmed in such a way that a "visual warning" (a transparent red object) is displayed when any infringement occurs.

The use of scripting here is to identify specific areas within the virtual environment that should not be transgressed or infringed upon (as the GA constraints are so formulated). In this process there are specific mouse activities that are performed such as hovering, clicking etc. When the mouse hovers upon the a specific scene area that has been categorized as a "restricted zone" (e.g., emergency exit as mentioned earlier), this area should undergo a change of state that is "visible". In other words, if the restricted area gets highlighted in a different color (e.g., red/transparent red), that indicates that the user should not move any object into this colored area.

In order to get this effect, a shape object should be embedded within this area and the actual work occurs in the MATERIAL node of the shape object. Let us assume that "Box" geometry is used to indicate the demarcated area. The MATERIAL node of this object should change the color to red when it "senses" mouse movement. This implies that, when the user hovers the mouse over this area, the object (upon sensing the mouse) should change the color of the Box object from "invisible" to "red". The state of invisibility is accomplished by turning the transparency to 100%. Inside the code snippet this is represented between values of 1 and 0, 1 being 100%. Any value in between is represented using corresponding decimals. In this study, JavaScript is used for scripting. A whole lot of interactions can be done with the virtual scene hence generated. These include transformations as well as modifying appearance related attributes.

On the whole, all the modifications can be classified either as geometry related modifications, appearance related modifications, and position related modifications. The last one also includes modifications related to the orientation and scale. Modifications to position, orientation, and scale are all grouped as transformational changes. In the virtual environment, the color that is seen or is visible finally is the result of various programmatic elements such as color, transparency, ambient intensity, etc. These can be controlled using a menu-driven environment, in which visible sensors can be activated to control the various values such as the transparency, color, shininess etc. This can be especially useful for designers focusing on the various design aspects involving light and those who study the impact of applying various -modulations to the light and color in the interior environment. Also, the ability to vary the intensity provides a valuable functionality to determine the most suitable form and intensity of lighting for the specific environment under study.

5 Conclusion

This research integrated the notion of GA-based MOO within the discipline of interior space planning. he results evinced the robustness of the GA and its efficiency in generating the Pareto plans. The study goes beyond presenting the results in the form of a Pareto set with a pool of candidate solutions, by providing a means to visualize potential solutions using a virtual reality imaging, and by experimenting with those plans through an active interaction capability. Experimenting indicates producing alternative representations corresponding to a different set of appearance and geometry variables and the subset of variables influencing the final VR Scene. The subset of variables mentioned above includes a wide range of values such as diffuseColor, transparency, specularColor, etc. Visualizing and manipulating the plans in this manner provides a means of more effectively exploring and navigating the potential solutions. This greatly facilitates the practice of informed decision-making and selecting the optimum plan, and allows for the inclusion of many more stakeholders in the final decision process.

The generation of the graphic designs is based on a web-friendly framework. Designs are generated using OpenSource software that are web-compatible so that the results can be hosted online (World Wide Web). For this reason the research has accomplished all programming procedures using web-compatible OpenSource software such as VRML, Java, and JavaScript. In order to more up-to-date with the web standards, the code snippets can also be easily converted to X3D if it is so required. This conversion is a seamless and straightforward process as X3D is considered the successor to VRML and is the XML-based 3D authoring language. Programming the EAI (External Authoring Interface) of the virtual worlds (.wrl or .x3d) with Java and JavaScript opens up the key graphics attributes include geometry and appearance.

References

1. Ballast DK. 2002. Interior construction & detailing for designers and architects. Professional Publications: Belmont, CA.
2. Pile JF. 2007. Interior design, 4th edn. Prentice Hall, Upper Saddle River, NJ.
3. Piotrowski CM. 2002. Professional practice for interior designers, 3rd edn. John Wiley & Sons: New York.
4. Vosinakis S, Azariadis P, Sapidis N, Kyratzi S. 2007. A virtual reality environment supporting the design & evaluation of interior spaces, 4th INTUITION Conf. on Virtual Reality & Virtual Environments, Athens.
5. Sims K. 1991. Artificial evolution for computer graphics. Computer Graphics, 24(4), 319–328.
6. Dawkins R. 1986. The blind watchmaker. Longman: Essex.
7. Graf J, Banzhaf W. 1995. Interactive evolution for simulated natural evolution, artificial evolution, in European Conf. (AE'95) Selected Papers, Alliot JM, Editor. Springer-Verlag: Berlin, Germany. p. 259–272.
8. Todd S, Latham W. 1992. Artificial life or surreal art? Proceedings of 1st European Conf. on Artificial Life. MIT Press, Cambridge, MA. p. 504–513.
9. Chen W, Shi YQ, Xuan G, Su W. 2008. Computer graphics identification using genetic algorithm. IEEE International Conference on Pattern Recognition (ICPR08). Tampa, Florida, USA.
10. Yu TT, Yang M, Chen CS. 2005. Automatic feature extraction & stereo image processing with genetic algorithms for lidar data. Proceeding of Computer Graphics, Imaging & Vision: New Trends, 307–309.
11. Nishino H, Sueyoshi T, Kagawa T, Utsumiya K. 2007. An interactive graphics rendering optimizer based on immune algorithm. Lecture Notes in Computer Science, Applications of Evolutionary Computing, Springer, 4448, 459–469.
12. Nishino H, Sueyoshi T, Kagawa T, Utsumiya K. 2008. An interactive 3D graphics modeler based on simulated human immune system. Journal of Multimedia, 3(3), 51–60.
13. Stewart TJ, Janssen R, van Herwijnen M. 2004. A genetic algorithm approach to multiobjective l & use planning. Computers & Operations Research, 31(14), 2293–2313.
14. Balling R, Powell B, Saito M. 2004. Generating future l&-use & transportation plans for high-growth cities using a genetic algorithm. Computer-Aided Civil & Infrastructure Engineering, 19(3), 213–222.
15. Chandramouli M, Huang B, Xue L. 2009. Spatial change optimization: integrating GA with visualization for 3D scenario generation. Journal of the American Society of Photogrammetry & Remote Sensing, 75(8), 1015–1023.
16. Karlen M. 1993. Space planning basics. Van Nostr & Reinhold Publishers: New York.
17. F. Ching. 2009. Architectural graphics, 5th edn. Wiley.
18. Bertoline GR, Wiebe EN, Hartman NW, Ross WA. 2009. Technical graphics communication, 6th edn. McGraw-Hill: Boston, MA.
19. Kopec D. 2006. Environmental psychology for design. Fairchild Publications, NY.
20. Kilmer R, Kilmer WO. 1992. Designing interiors. Thomson Learning, CT.
21. Panero J, Zelnik M. 1979. Human dimensions & interior space. Whitney Library of Design/Watson Guptill Publications: New York.
22. Rengel R. 2003. Shaping interior space. Fairchild Publications: New York.
23. Kubba S. 2003. Space planning for commercial & residential interiors. McGraw-Hill, New York.
24. Lok L. 2004. A critical survey of software packages for use by interior designers. MSc Thesis, Department of Computer Science, University of Wales.
25. Maher ML, Bilda Z, Gül LF. 2006. Impact of collaborative virtual environments on design behavior, in Design Computing & Cognition '06. Gero JS, Editor. Dordrecht, The Netherlands: Springer, p. 305–321.
26. Whyte J, Bouchlaghem N, Thorpe A, McCaffer R. 2000. From CAD to virtual reality: modeling approaches, data exchange, & interactive 3d building design tools. Automation in Construction, 10, 43–55.
27. Balling RJ, Taber JT, Brown MR, Day K. 1999. Multiobjective urban planning using genetic algorithm. Journal of Urban Planning & Development, 125, 86–99.
28. Chandramouli M, Bertoline R, Connolly P. 2009. Generating alternative engineering design by integrating desktop VR with

genetic algorithms. Engineering Design Graphics Journal, 73(3), 1–12.
29. Seixas J, Nunes JP, Louren P, Lobo F, Condado P. 2005. GeneticLand: modeling land use change using evolutionary algorithms. Proceedings of the 45th Congress of the European Regional Science Association, Land Use and Water Management in a Sustainable Network Society, Vrije Universiteit: Amsterdam, p. 23–27.
30. Connolly PE. 2005. Virtual reality and immersive technology in education. International Journal of Information & Communication Technology Education, 1(1), 12–18.
31. Reas C, Fry B. 2003. Processing: a learning environment for creating interactive web graphics, in Proceedings of the SIGGRAPH 2003 Conf. on Web Graphics, SESSION: Java applications. San Diego: ACM.
32. Ames AL, Nadeau DR, Moreland JL. 1997. VRML 2.0 sourcebook , 2nd edn. John Wiley & Sons: NY.
33. van Kemenade C, Kok J, Eiben A. 1995. Raising GA performance by simultaneous tuning of selective pressure & recombination disruptiveness, in Proceedings of the 2nd IEEE Conference on Evolutionary Computation. IEEE Press, p. 346–651.

12

A market model: uncertainty and reachable sets

Stanislaw Raczynski*

Universidad Panamericana, Facultad de Ingeniería, Augusto Rodin 498, 03920 Mexico DF, Mexico

Abstract – Uncertain parameters are always present in models that include human factor. In marketing the uncertain consumer behavior makes it difficult to predict the future events and elaborate good marketing strategies. Sometimes uncertainty is being modeled using stochastic variables. Our approach is quite different. The dynamic market with uncertain parameters is treated using differential inclusions, which permits to determine the corresponding reachable sets. This is not a statistical analysis. We are looking for solutions to the differential inclusions. The purpose of the research is to find the way to obtain and visualise the reachable sets, in order to know the limits for the important marketing variables. The modeling method consists in defining the differential inclusion and find its solution, using the differential inclusion solver developed by the author. As the result we obtain images of the reachable sets where the main control parameter is the share of investment, being a part of the revenue. As an additional result we also can define the optimal investment strategy. The conclusion is that the differential inclusion solver can be a useful tool in market model analysis.

Key words: Marketing, Market model, Simulation, Differential inclusions, Uncertainty.

Introduction

The market model itself is not the main topic of this article. Though the optimal strategy of the market with respect to the investment is calculated, this is not the main topic, either. The work is focused on the treatment of uncertainty by means of the reachable sets, rather than stochastic modeling. The model we are using (at least the model of the demand) is taken from the literature, with some dynamic properties added.

A lot of demand models that may be used for similar purpose can be found in Lilien and Kotler [1] (linear, non-linear, deterministic, stochastic, static, dynamic models, etc.). What we need is a model that may be adopted into the dynamic market simulation. The demand model we choose is taken from that book. The dynamics is added, to simulate the total accumulated revenue, and the market growth due to the investment. The growth is charged with certain inertia, which implies additional differential equations (see the next section for details).

The search for market models for simulation purpose dates from the early 1960s, and significant publications began to appear in the late 60s. King [2] gives a comparison of iconic, analog and symbolic models. Recall that an iconic model represents reality on a smaller scale, an analog model shows reality by means of maps and diagrams, and a symbolic model uses mathematical expressions to portray reality. Montgomery and Urban [3] discuss the *descriptive*, *predictive* and *normative* models, see Stanovich [4]. By descriptive we mean models that consist largely of diagrams and maps or charts designed to describe a real-world system. Predictive model is used in predictive analytics to create a statistical model of future behavior, and normative model evaluates alternative solutions to answer the question, "What is going on?" and suggests what ought to be done or how things should work according to an assumption or standard.

Optimal control in marketing is not a new topic. There are many publications in the field, like, for example Bertsimas and Lo [5]. In that article we can find the application of the Bellman's dynamic programming approach to the problem of the price impact on the dynamic trading on the stock exchange. A tutorial and survey on the relevant technical literature on models of economic growth can be found in Burmeister and Rodney [6]. Feichtinger et al. [7] consider a similar problem for a general market and optimal advertising policy. In that article a detailed formulation of an implementation of the Pontriagin's Maximum Principle is shown. Yuanguo [8] uses the Bellman's Principle of Optimality, and derive the principle of optimality for fuzzy optimal control. This is applied to a portfolio selection model.

The applications of differential inclusions (DIs) and reachable set determinations to marketing problems can hardly be found in the known literature. Perhaps the closer topics can

*e-mail: stanracz@stanr.com

be found in my own articles, see Raczynski [17] and Raczynski [9]. However, the problems considered in those articles are quite different from the problem statement of the present research (marketing game with competition and general remarks on DIs in modeling and simulation). Beard [10] presents an application od the DIs to a very specific problem of fertilizers carryover, based on the Liebig's law or the law of the minimum. That law has been developed in agricultural science by Justus von Liebig and states that growth is controlled not by the total amount of resources available, but by the scarcest resource (limiting factor). However, the Beard problem statement is not exactly in the field of marketing and no reachable sets of the model variables are considered.

Read et al. [11] approach the uncertainty in marketing by more heuristic way. Two groups of experts and managers have been asked to think aloud as they make marketing decisions in exactly the same unpredictable situation. The results show significant differences in heuristics used by the two groups. While those without entrepreneurial expertise rely primarily on predictive techniques, expert entrepreneurs tend to invert these. In particular, they use an effectual or non-predictive logic to tackle uncertain market elements and to construct novel markets with committed stakeholders.

Slovik [12] also addresses the problem of market uncertainty. The paper describes the market uncertainty theory comprising the market uncertainty theorem and the notion of heterogeneity of market uncertainty, as well as policy recommendations relevant for rating agencies, financial institutions, and public authorities.

Lien [13] also considers the problem od uncertainty from a cultural and an ethnographic point of view. However, as in the papers of Read and Slovik mentioned above, the problem is stated and described in a textual way only, with no models (logical or mathematical) proposed. This does not mean that their or our approach is better, but means rather that the research of such kinds cannot be compared to each other.

Simos et al. [14] propose a conceptual framework of antecedents and market performance consequences of emergent marketing strategies and test it with a sample of 214 UK enterprises. The results suggest that dimensions of market uncertainty and strategic feedback systems influence the formation of emergent marketing strategy.

Another approach to market uncertainty can be found in Courtney et al. [15]. In that article, four levels of uncertainty are considered: (1) a clear-enough future, (2) alternate futures, (3) a range of futures, and (4) true ambiguity, when multiple dimensions of uncertainty interact and create an environment that is virtually impossible to predict. Compared to a more mathematically rigorous DIs approach, the most similar is the level 3 of uncertainty, where a range of possible future outcomes for marketing data can be known. A heuristic strategies for the different levels of uncertainty are proposed. The general concepts of that work may be useful in the marketing practice, though no formal market model is used.

The article of Brace et al. [16] is an example of one of numerous works on stochastic market models using Heath-Jarrow-Morton (HJM) framework. This, and similar methods are used to analyse forward rates and pricing statistical behavior. It is difficult to compare the DI approach to those methods. As mentioned also in a comment to the Figure 2 of the present paper, our approach is very different. The main point is that, in marketing and in more general soft system models, not everything can be treated using statistical frameworks.

As stated earlier, the optimization is not our main subject. However, the presented method can also be used to obtain optimal investment strategy. Anyway, to determine reachable sets we use the methods of the optimal control theory.

The model

The demand model we use is a non-linear model of the demand and revenue which parameters are: price, advertising, seasonal index, market overall growth, consumer income, production or acquisition cost and the market elasticities with respect to the price, advertising and consumer income (Lilien and Kotler [1]). The demand model is as follows:

$$q(p,a,y,v,g,t) = q_0 v(t)s(t)\left(\frac{p(t)}{p(0)}\right)^{e_p}\left(\frac{a(t)}{a(0)v(t)s(t)}\right)^{e_a}\left(\frac{y(t)}{y(0)}\right)^{e_y} \quad (1)$$

where

q – demand, q_0 – initial or reference demand,
t – the time, $s(t)$ – overall market size,
$p(t)$ – price, $a(t)$ – advertising per time unit,
$y(t)$ – consumer income, $v(t)$ – seasonal index,
e_p, e_a, e_y – market elasticities with respect to price, advertising and consumer income, respectively.

In the model of Lilien and Kotler the market size $s(t)$ is supposed to be equal to $(1 + g)^t$, where g is the market growth rate. We use the variable s instead, to be able to link the market growth to the investment that the company may do in order to expand the market (new installations, infrastructure, etc.).

Normally, the price elasticity is negative (price increase means less demand), and the elasticity for advertising and consumer income are positive. Note that the term $v(t)s(t)$ multiplies the demand, and appears also in the denominator of the advertising impact term. This means that if the market and the seasonal index grow, then we must spend more on advertising to achieve the same effect. We do not consider any storage or warehouse mechanism, so it is supposed that the sales are equal to the demand. The revenue can be calculated as follows:

$$U = (1 - b)e - a, \quad \text{where } e = (p - c)q \quad (2)$$

where U is the *net revenue*, c is the unit cost of the product (production, acquisition cost), a is the advertising, b is the investment factor, p is the price to the public and q is given by the formula (1). The investment b determines what part of the *utility* e is being invested in the market growth, $0 \leq b \leq 1$. The market growth rate is proportional to the product be. It will be one of our control variables. In our equations the market size is relative with initial condition equal to 1. If the company does not invest in the market ($b = 0$), then the market size has no influence on the model trajectory (remains constant). The final *total revenue* is the

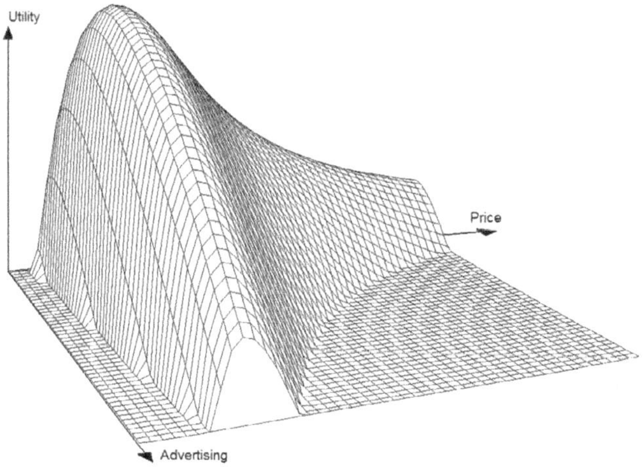

Figure 1. The revenue as function of the price and advertising. Static case, $b = 0$.

sum (integral) of U over the time interval under consideration.

Figure 1 shows the shape of the profit (utility) e as a function of the price and advertising, when the time and other variables are fixed.

It can be seen that the utility function has a maximum that determines the optimal price and advertising at the moment. This is a static case.

Now, consider a fixed time interval $[0, T]$. To calculate the total revenue we must integrate (2) over the time. The variable b controls the investment, $0 \leq b \leq 1$. The market size growth due to the investment. To make the model more realistic, we suppose that the market growth rate follows the investment with certain inertia. The basic equations are as follows:

$$\begin{cases} \frac{dr}{dt} = (1-b)\, v/v_0 \text{ accumulated revenue growth, relative} \\ \text{where } v = (p-c)\, q(z, w, y, v, s, t) - a,\ v_0 = \text{initial revenue} \\ \frac{ds}{dt} = (x-s)/T_c \text{ market size growth, relative} \\ \frac{dx}{dt} = g\, b\, v/v_0 \text{ accumulated investment, relative} \end{cases} \tag{3}$$

In the above equations r is the relative accumulated net revenue, s is the relative market size and x is an auxiliary variable. T_c is the time constant of the market growth inertia, and g is a constant that defines the impact of the investment in the market expansion. Namely, g tells what is the relative market growth per one currency unit invested. All other variables may change in time. Observe that the additional state variable x has been introduced to manage the inertia.

The final accumulated revenue and the market size strongly depend on the changes of the investment control variable b. So, it is important to be able to see the limits of the revenue and the market size when b changes due to the investment strategy. Note that if $b = 1$ all the time (maximal investment), then we have no revenue available, and if $b = 0$ then the market does not grow, which also reduces the available profit.

The aim of the present work is to calculate the reachable sets for all possible investment strategies, as well as to assess the impact of the uncertainty of important market variables.

This should be emphasized that our approach to uncertainty has nothing to do with randomness or stochastic models. If a value of a parameter is uncertain, this means that it can take values from some interval, but this does not mean that it is a random variable. Such a parameter does not have any probability distribution. The changes in its value can be caused by any internal or external agents. For example, in the stock market the information about the actual share price can be obtained by means of observations and predictions, but it also can be a false information introduced intentionally. To obtain the reachable sets, we use the differential inclusion solver, see Raczynski [17, 18].

Differential inclusion solver

The DI solver is a software that calculates and displays the solution to a given differential inclusion. The software was developed by S. Raczynski and published in several articles [18]. In Raczynski [9] an application to a marketing differential game is described. However, the topic of that article is different from that presented here. The present research can hardly be compared to other similar published research, because the DI solver had never be used before to marketing models and the shapes of the corresponding reachable sets have not been calculated and presented by anybody (as to the author's knowledge).

Differential inclusions

Let us consider a dynamic model given in the form of an ordinary differential equation (state equation):

$$\mathrm{d}\boldsymbol{x}/\mathrm{d}t = \boldsymbol{f}(\boldsymbol{x}, \boldsymbol{u}, t) \tag{4}$$

where $\boldsymbol{x} \in R^n$ is the state vector, t is the time, $\boldsymbol{f}$ is a vector-valued function and $\boldsymbol{u} \in R^m$ is an external variable, called *control* in the automatic control theory. Suppose that the value of the control $\boldsymbol{u}$ is restricted so that $\boldsymbol{u}(t) \in C(t)$, where $C(t) \subset R^m$ is a subset of the R^m space. For each fixed $\boldsymbol{x}$ and t, the function $\boldsymbol{f}$ maps the set C (all possible values of $\boldsymbol{u}$) into a set $F \subset R^n$. In this way we obtain the following condition.

$$\mathrm{d}\boldsymbol{x}/\mathrm{d}t \in F(\boldsymbol{x}, t) \tag{5}$$

What we obtained is a *differential inclusion* (DI). The right-hand side of the equation (4) determines the set F. However, this is merely one of various possible ways to represent the set. In this case it is parameterized by the control $\boldsymbol{u}$. In general, the DI (5) is not necessarily linked to an ODE model. Let us restrict to a time interval $t \in I = [0, T]$. Any function of time $x(t)$ that obeys the inclusion (5) for almost all $t \in I$ with a given initial condition, is a *trajectory* of the differential inclusion. Consider a set Z in $I \times R^n$ where the graphs of all possible trajectories of the DI that start from a given initial set belong. The set Z is a *solution* to the DI. The set Z is called *reachable set* (*RS*). Note that a trajectory of a DI is just one possible trajectory, and the solution to a DI with a fixed initial set is given by the reachable set. For more information on differential inclusions consult Aubin and Cellina [19] and Zaremba [20].

In this very natural way we see that the uncertainty in dynamic system modeling leads to differential inclusions as a corresponding mathematical tool. Namely, if any model parameter has uncertain value, it can be treated as a limited control that belongs to a given interval or set. This results in a differential inclusion.

A more extended survey on DI application in modeling and simulation can be found in Raczynski [17]. One of the best publications on DIs in finite-dimensional as well as abstract spaces is the book of Aubin and Cellina [19].

The solver

The DI solver algorithm is not just an extension of any numerical method for the ODEs. We cannot integrate an infinite number of trajectories to obtain the reachable set. One may suppose that RS can be obtained, up to a certain accuracy, integrating a sufficient number of trajectories, each one with different control, generated randomly. Unfortunately, this is not the case and may lead to completely wrong result. The comparison between such random shooting and the DI solver results will be shown in the next section.

One of the properties of the reachable sets is the fact that if a trajectory reaches a point on the boundary of the RS at the final time, then its entire graph must belong to the RS boundary. This fact is well known and used in the optimal control theory (see Pontryagin et al. [21]; Polak [22] and Lee and Markus [23]). Observe that any trajectory that reaches a point on the boundary of the RS is optimal in some sense. Such trajectories can be calculated using several methods, the main one being the Maximum Principle of Pontryagin. If we can calculate a sufficient number of trajectories that scan the RS boundary, then we can see its shape. This can be done by some kind of random shooting over the RS boundary. Such shooting has nothing to do with a *simple* or *primitive random shooting* mentioned earlier, when the trajectories are generated randomly inside the RS.

In few words, the DI solver works as follows. The user provides the DI in the form of an equivalent control system. To do it he/she must parameterize the right-hand size (the set F) using a m-dimensional auxiliary control variable $\boldsymbol{u}$. The DI solver automatically generates the equations of the so-called conjugate vector $\boldsymbol{p}$ and integrates a set of trajectories, each of them belonging to the boundary of the RS. To achieve this, over each trajectory the Hamiltonian $H(\boldsymbol{x}, \boldsymbol{p}, \boldsymbol{u}, t)$ is maximized. This procedure is similar to that used in dynamic optimization. In the optimal control problem the main difficulty consists in the boundary conditions for the state and conjugate vectors. For the state vector we have the initial condition given, and for the conjugate vector only the final conditions (at the end of the trajectory) are known, given by the *transversality conditions*. This means that the optimal control algorithm must resolve the corresponding two-point-boundary value problem. In the case of a DI we are in a better situation. There are no objective function and no transversality conditions. As the consequence, for the vector $\boldsymbol{p}$ we can define the final as well as the initial conditions. Anyway we obtain a trajectory which graph belongs to the RS boundary. With the initial conditions for $\boldsymbol{p}$ we integrate the trajectory only once, going forward. The only problem is how to generate these initial conditions in order to scan the RS boundary with nearly uniform density. The algorithm is quite simple: the initial conditions for $\boldsymbol{p}$ are generated randomly, due to a density function that is being automatically modified, covering with more density points that correspond to trajectories that fall into low density spots at the RS boundary. Trajectories that are very close to each other are not stored (storing only one from each eventual cluster). As the result we obtain a set of trajectories covering the RS boundary that can be observed in graphical form and processed.

Application to the market model

Note that our model (3) has the form of a control system, and defines the corresponding DI. In the investment strategy problem, the control variable is $b(t)$, the share of investment we apply out of the total revenue (Experiment 1, below). The uncertainty of the other model variables can be treated in a similar way, letting the variable change within a given interval. Below are presented the results of such analysis for the uncertainty in the values of e_p, – the market elasticity with respect to the price (Experiment 2).

Experiment 1

In this case our aim is to obtain the reachable set in the revenue/market size plane for all possible strategies of investment, with final time fixed. As an additional result, the DI solver provides the control function (investment as a function of time) for any point on the boundary of the reachable set. So, we can obtain the optimal strategy that maximizes the total revenue or the final market size.

The model is given by the equations (3), where b is the control variable, that can change within 0 and 1. The model parameters are as follows:

Initial sales = 70 000
Initial (reference) advertisement = 10 000 per time unit (a day)
Product unitary cost = 0.8
Initial relative market size = 1
Unit price = 1
Unit initial relative price = 1.2
Time constant for investment inertia T_c = 30 days
g = 0.006
Market elasticity for the product price = −2
Market elasticity for advertisement impact = 0.6
Final simulation time = 365 days

Note that the initial relative price is set equal to 1.2 and not 1. This is to make the demand change when the price elasticity changes. In this experiment both seasonal index and the consumer income are functions of time. The relative consumer income is equal to 1 except the interval (200, 214) where it is equal to 2.5. The seasonal index is equal to 3 within the interval (100, 128), equal to 0.5 in (160, 167) and equal to 1 elsewhere.

Figure 2 shows the image of the reachable set for the revenue and market size at the simulation final time, with the

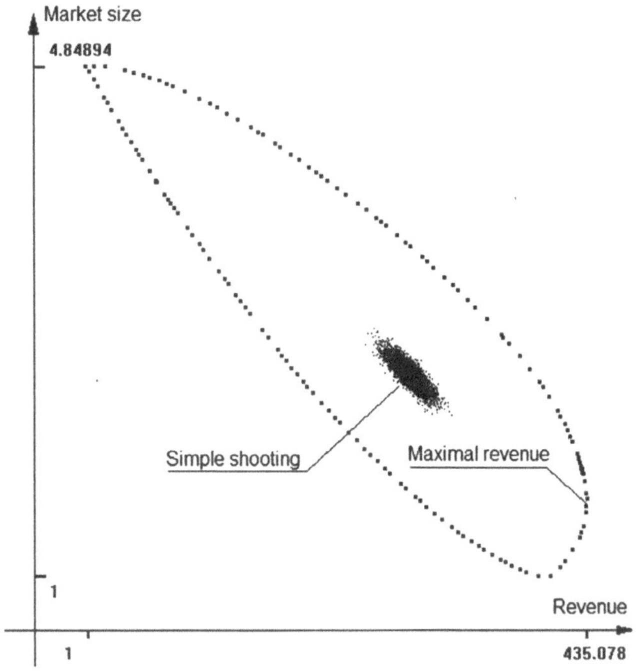

Figure 2. Reachable set for Experiment 1.

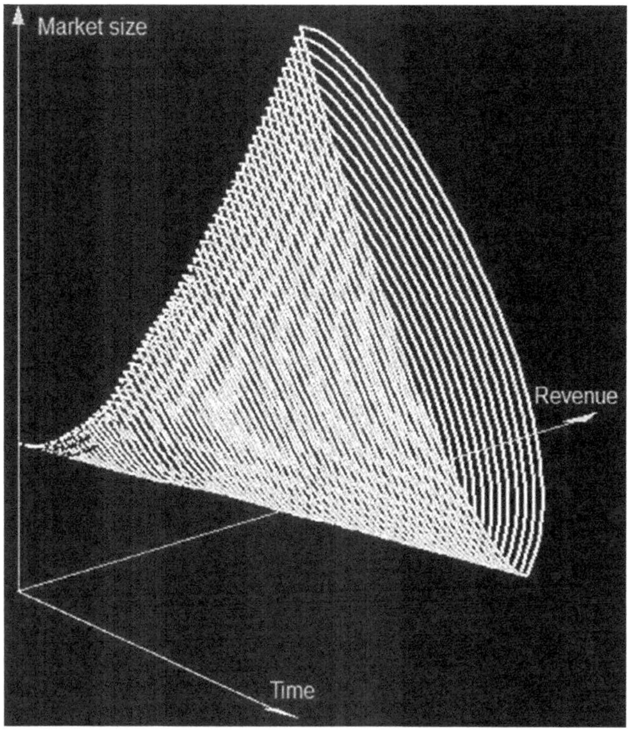

Figure 3. Experiment 1. 3D image of the reachable set.

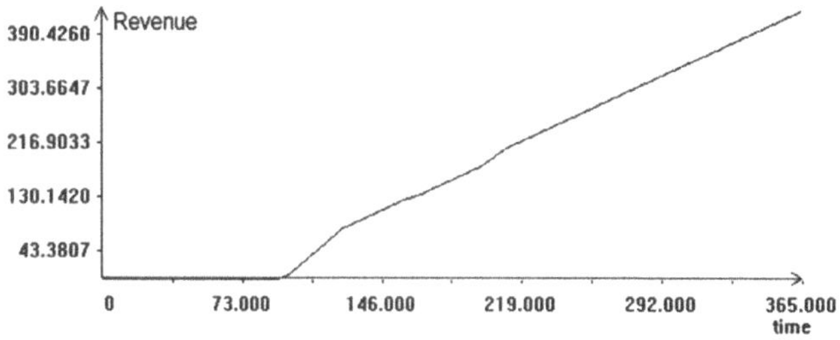

Figure 4. Net revenue accumulated with optimal investment strategy (relative).

investment control b included in [0, 1]. It should be noted that in our case the dimensionality of the state space is equal to 3. So, to show the complete reachable set we should display it in the 4-dimensional time-state space, which is quite difficult. If we limit the displayed state space components to market size-revenue, then the reachable set boundary is not always seen just one contour, because for each time instant the 2-dimensional image is a projection of a 3-dimensional reachable region. Fortunately, in this case the boundary is clearly seen, perhaps because the control vector dimensionality is equal to 1. On the same figure we also can see a set of reachable points obtained by simple random shooting (10 000 trajectories integrated). It is clear that the simple shooting provides a wrong assessment of the reachable set.

The result of Figure 2 shows the advantage of the DI approach compared to any other method which uses models with random variables to treat uncertainty (stochastic approach). Whatever the distribution of the uncertain parameter would be, the estimate of the reachable set will result to be wrong. Moreover, recall that uncertain parameters are not necessarily random ones. An uncertain parameter may have no probability distribution at all, and may have quite deterministic value, like, for example a false information generated intentionally (this occurs frequently in the stock market).

Figure 3 depicts the 3D image of the reachable set in the time-state space.

After the set has been displayed, the user can select any point of its boundary and see the corresponding control. Selecting the point with maximal revenue we obtain a control of "bang-bang" type, equal to 1 for time between 0 and 97 and equal to zero between 97 and 365 days. The changes in the corresponding accumulated net revenue are shown on Figure 4.

Experiment 2

Now, suppose that the value of the market elasticity with respect to the price e_p is uncertain and may change up to plus minus 20% of its original value. Using model parameters as in Experiment 1, with variable investment control in [0, 1] and uncertain (variable) e_p, we obtain a larger reachable set. In this case the set is more complicated (2D projection of a 3D region).

This results in a cloud of reachable points and not in a simple contour. A post-processing of this cloud provides an assessment of the reachable set, shown as the gray region on Figure 5. To compare with Experiment 1, the boundary points of the previous reachable set are also shown in the same figure. Observe both the net revenue and the market size may reach greater values than that obtained with fixed elasticity e_p.

In general, it can be seen that our market needs the optimized investment strategy to reach maximal revenue. Note that we have fixed final simulation time, as if the company sells

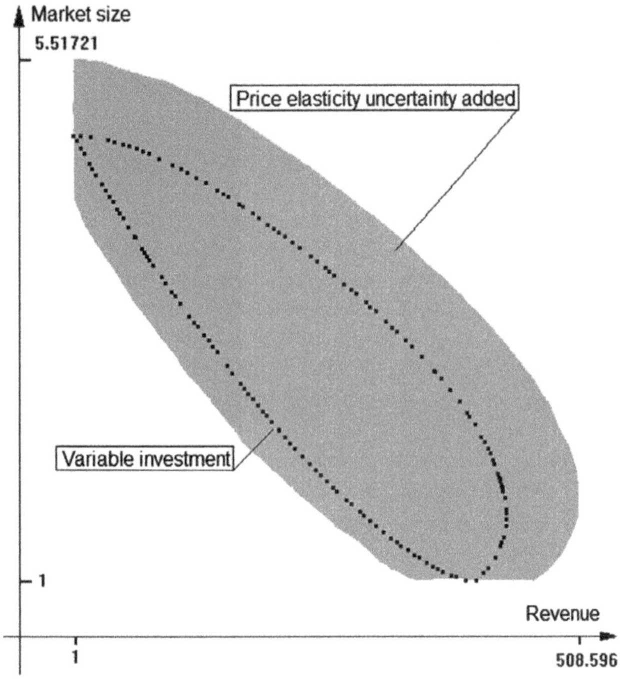

Figure 5. Reachable set with variable investment and uncertain price elasticity.

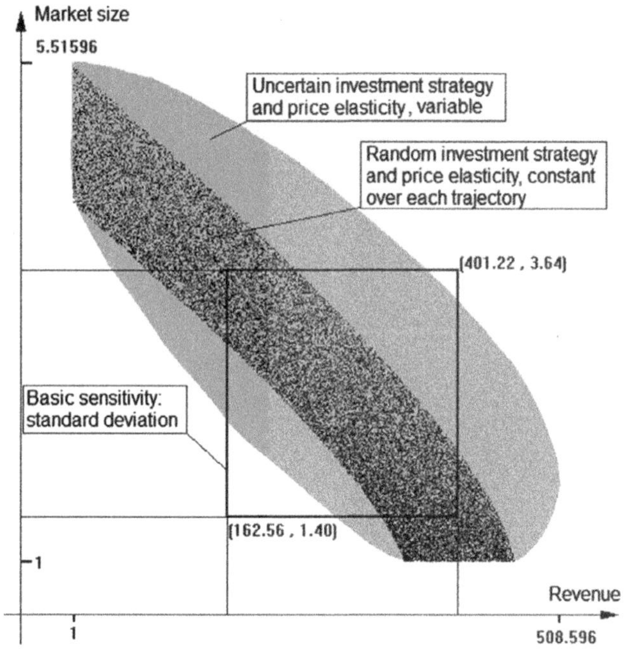

Figure 6. The reachable set for uncertain investment and price elasticity and sensitivity analysis results.

during the given time interval and then disappears. Experiments with moving time horizon can be done, but may be more complicated and time-consuming.

Experiment 3: Uncertainty and reachable set vs. sensitivity

Our approach to uncertainty is based on model reachable sets. As stated before, this is a deterministic approach. We do not deal with any probability distribution for the uncertain parameters, using rather the intervals where the parameters may belong. The influence of the parameters is being treated dynamically, i.e. we allow that the parameters are functions of time, changing over each model trajectory. The reachable sets obtained in this way show the influence of the uncertain parameters, somewhat similar to the sensitivity analysis. Let us recall that the basic sensitivity analysis shows the possible changes in the model state due to the changes of the initial conditions and the (constant) model parameters. Our problem statement and that of sensitivity methods are quite different to each other, so it is difficult to compare these methods. We do not pretend to prove if the DI applications are better or worse from the results of sensitivity methods. It is just interesting to compare them.

The sensitivity analysis is not the topic of this paper, so we will not provide any review of the (huge) literature in the field. For a comprehensive review see Hamby [24]. The common way to conduct sensitivity analysis is to assign probability density functions to each model parameter and assess the resulting outcome variance with respect to the corresponding parameters. Though the advanced methods is sensitivity research are quite sophisticated and perhaps more accurate, we will use in this comparison a basic sensitivity concepts. One of the commonly known methods is the general error propagation formula, i.e.

$$V(Y)=\sum_{i=1}^{n}\left(\frac{\partial Y}{\partial X_i}\right)^2 V(X_i) \qquad (6)$$

where Y is the model output (the outcome being analyzed), $V(Y)$ is the variance of Y, the model parameters are $X_1,\ldots,X_n$, and $V(X_i)$ is the variance of the parameter X_i. In our case Y may be the value of any of the model state variables at the end point of the simulated trajectory. We do not have any algebraic expression for Y, which value is the result of model simulation. However, as the calculation of a single trajectory take only a fraction of a second, the partial derivatives can be easily calculated by numerical differentiation. Let us do it for the values of the final revenue and market size, with respect to investment and the price elasticity parameter, with the model parameters as in the Experiment 2. The investment control variable is supposed to belong to [0, 1] and the price elasticity will change plus minus 20%, belonging to the interval $[-2.4, -1.6]$. Assuming the uniform probability distribution, we get the variance equal to 0.0833 and 0.0533 for the investment and price elasticity, respectively. The corresponding standard deviations are 0.288 and 0.23. The calculations are performed for the model trajectory obtained with the midpoint values for both parameters. The results are shown on the Figure 6. Bold rectangle shows the region of plus minus standard deviation for the outcome. It can be seen that this region is little informative, and obviously very different from the reachable set (gray area).

Formula (6) has been obtained using model linearization. As our model is non-linear, a better, direct way to see the influence of the parameters is simply integrating model equations, changing the parameters, each of them being constant over the trajectory. The result is also shown on Figure 6 as a dotted area.

Observe that the true reachable set is greater than the region obtained this way. For more complex models, in particular for oscillating systems, this difference is greater.

Conclusions

Calculating reachable sets is something more than obtaining the optimal control. Though we can get the optimal control strategy as an additional result, the possibility of viewing the shape of reachable sets provides much more information about the model behavior. The same simulation program can be used for many more experiments, looking for reachable sets with variable advertising and the uncertainty of any one of the model parameters. Adding more inertia to the model will result with higher state vector dimensionality and more complicated images of the reachable sets. For such sets the problem is not only the set determination, but also the way to display it. As the computer screen is (still) 2-dimensional, the good effects can be obtained while displaying the cloud of points in movement, rotating around one of the state space axes. Unfortunately, such moving displays can hardly be shown as figures of a printed article. The comparison of the DI approach with sensitivity analysis is interesting, but it should be remembered that the problem statement in both approaches is quite different. Most of the sensitivity methods are designed to find the variance propagation and statistical properties of the model output, while the DIs and the reachable sets are deterministic. The selection of an appropriate methodology always depend on which is the purpose of the research.

References

1. Lilien GL, Kotler P. 2010. Marketing decision making: a model-building approach, Edition 4. New York, Harper & Row, ISBN: 0060440767.
2. King WR. 1967. Quantitative analysis for marketing management. McGraw-Hill: New York.
3. Montgomery DB, Urban GL. 1969. Management science in marketing. Prentice Hall: Englewood Cliffs, NJ.
4. Stanovich KE. 1999. Discrepancies between normative and descriptive models of decision making and the understanding/ acceptance principle. Cognitive Psychology, 38, 349–385, Article ID cogp.1998.0700, available online at http:// www.idealibrary.com.
5. Bertsimas D, Lo AW. 1998. Optimal control of execution costs. Journal of Financial Markets, 1, 1–50.
6. Burmeister E, Rodney D. 1972. Guidance and optimal control of free-market economies: a new interpretation. IEEE Transactions on Systems, Man, and Cybernetics, SMC-2, 9–15.
7. Feichtinger G, Hartl RF, Suresh P. 1994. Dynamic optimal control models in advertising: recent developments. Management Science, 40(2), 195–226.
8. Yuanguo Z. 2009. A Fuzzy Optimal Control Model. Journal of Uncertain Systems, 3(4), 270–279, Online at: www.jus.org.uk.
9. Raczynski S. 2005. Reachable sets for a marketing game: looking for new modeling tools, Summer Computer Simulation Conference, The Society for Modeling and Simulation. Philadelphia PA, p. 47–54, ISBN: 1-56555-299-7.
10. Beard R. 2000. Optimal Fertilizer Carryover and von Liebigs Law of the Minimum. Conference (44th), January 23–25, 2000, Sydney, Australia 123600, Australia, Australian Agricultural and Resource Economics Society, p. 1–10.
11. Read S, Dew N, Sarasvathy SD, Song M, Wiltbank R. 2009. Marketing under uncertainty: the logic of an effectual approach. Journal of Marketing May 2009, 73(3), 1–18.
12. Slovik P. 2011. Market uncertainty and market instability. IFC Bulletin No 34, December 2011, p. 430–435.
13. Lien ME. 2004. The virtual consumer; construction of uncertainty in marketing discourse, in Market matters: exploring cultural processes in the global marketplace, Garsten C, de Montoya ML, Editors. New York, Palgrave Publishers, p. 1–23, ISBN: 1-4039-1757-4.
14. Simos C, Katsikeas CS, Balabanis G, Robson MJ. 2014. Emergent marketing strategies and performance: the effects of market uncertainty and strategic feedback systems. British Journal of Management, 25(2), 145–165.
15. Courtney H, Kirkland J, Vi P. 2000. Strategy under uncertainty. Harward Business Review, November 1997 reprint 97603, p. 2–15.
16. Brace A, Gaterek D, Musiela M. 1997. The market model of interest rate dynamics. Mathematical Finance, 7(2), 127–147.
17. Raczynski S. 1996. Differential inclusions in system simulation. Transactions of the Society for Computer Simulation, San Diego, California, 13(1), 47–54.
18. Raczynski S. 2002. Differential inclusion solver. International Conference on Grand Challenges for Modeling and Simulation, conference of The Society for Computer Simulation Int., San Antonio, TX, USA.
19. Aubin JP, Cellina A. 1984. Differential inclusions. Berlin-Heidelberg-New York-Tokyo, Springer Verlag.
20. Zaremba SK. 1936. Sur les équations au paratingent. Bulletin des Sciences Mathematiques, 60, 139–160.
21. Pontryagin LS, Boltyanskii VG, Gamkrelidze RV, Mishchenko EF. 1962. The mathematical theory of optimal processes (Russian), English translation, Interscience 1962. ISBN: 2-88124-077-1 and ISBN: 978-2-88124-077-5.
22. Polak E. 1971. Computational methods in optimization. Academic Press: New York, ISBN: 0125593503.
23. Lee EB, Markus L. 1967. Foundations of optimal control theory. Wiley: New York.
24. Hamby DM. 1994. A review of techniques for parameter sensitivity analysis of environmental models. Environmental Monitoring and Assessment, 32(2), 135–154.

13

The periodic resin configuration design for ceramic-resin composite structure using topology optimization

Ji-Hong Zhu*, Qing Li, and Wei-Hong Zhang

Engineering Simulation and Aerospace Computing (ESAC), The Key Laboratory of Contemporary Design and Integrated Manufacturing Technology, Northwestern Polytechnical University, Xi'an, Shaanxi 710072, P.R. China

Abstract – The ceramic-resin composite structure has broad prospects in many fields like rapid casting and thermal protection. Due to the coefficient of thermal expansion (CTE) of resin is about 10 times higher than that of ceramic, the excessive thermal expansion of the resin pattern will lead to the undesirable crack of the ceramic shell during the heating procedure. The proposed approach is to find a reasonable resin configuration minimizing the thermal stress in the ceramic shell. Simultaneously, the mechanical stiffness of the whole structure should be maintained at a certain level to resist the pressure applied on the ceramic shell. In the actual production, the periodic resin configurations are more operable than the disordered configuration, so finding reasonable periodic resin configurations for ceramic-resin composite structure has lots of significance. The purpose of this paper is to introduce the topology optimization method into the periodic resin configuration design for ceramic-resin composite structure. A structural topology optimization procedure in combination with thermo-mechanical finite element analysis has been developed. A single-layer periodic model and a double-layer periodic model are optimized and respective CAD models are rebuilt according to the optimal results. By comparing with the existing hexagonal honeycomb configuration, two optimal designs have shown better performances both in reducing the thermal stress in the ceramic shell and maintaining the stiffness of the whole structure.

Key words: Ceramic-resin composite structure, Thermal stress, Periodic resin configuration, Topology optimization.

1 Introduction

The ceramic-resin composite structure has broad prospects in many fields such as rapid casting and thermal protection. In rapid casting field, compared with the traditional investment casting processes, rapid casting based on the stereolithography (SL) prototype resin patterns provides a better precision and avoids the consumption of the metal. At the same time, the ceramic-resin composite structure can be used as the thermal protection tile in high speed aircrafts for its excellent heat resistance, corrosion resistance, good stiffness and relatively lightweight.

However, there are several problems stand in the way of the achievement of widespread use of the ceramic-resin composite structure. One of the key difficulties is to avoid the crack of the ceramic shell during the heating of the resin pattern. Under normal circumstances, as a thermoset material, the epoxy resin doesn't melt but continues expanding until to burns out at around 1000–1200 °C with sufficient oxygen in the procedure of heating. Since the coefficient of thermal expansion (CTE) of the resin is about 10 times higher than that of the ceramic shell, during the heating process, a significant thermal expansion of the resin pattern will generate thermal stresses in the ceramic shell. If the maximum thermal stress reaches a certain value before resin pattern burns out, the unfortunate crack of the ceramic shell will take place and lead to the fail of the rapid casting immediately. So the satisfactory resin structure should collapse under the influence of its own expansion before the ceramic shell crack, at the same time, the mechanical stiffness of the whole structure should be maintained at a certain level to resist the pressure applied on the ceramic shell. To solve this problem, reasonable resin configurations should be found to meet above requirements. Among various resin configurations, the periodic resin configurations have special advantages in the convenience of manufacturing and this is one of the emphases of this paper. The ceramic-resin composite structure with single-layer periodic resin configuration is shown in Figure 1.

During the past decades, some prior studies have been made intensively. Hauge and Dickens studied the reasons and solutions to the shell crack problem [1, 2]. Among others, Jacobs hypothesized that the crack of the ceramic shell could be prevented when the thickness of resin pattern shell is thinner than a so-called critical section thickness [3, 4]. Hague and Dickens conducted a series of researches on the heating experiments

*e-mail: jh.zhu@nwpu.edu.cn

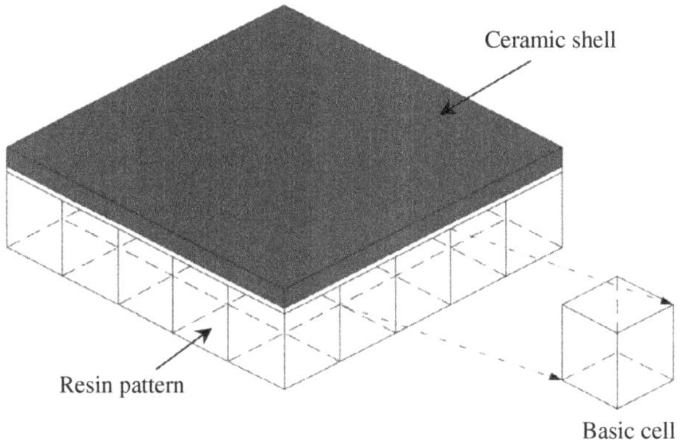

Figure 1. Ceramic-resin composite structure with single-layer periodic resin configuration.

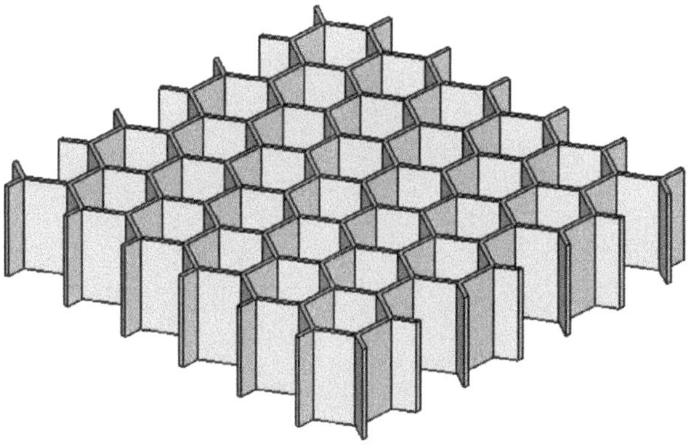

Figure 2. Hexagonal honeycomb resin configuration.

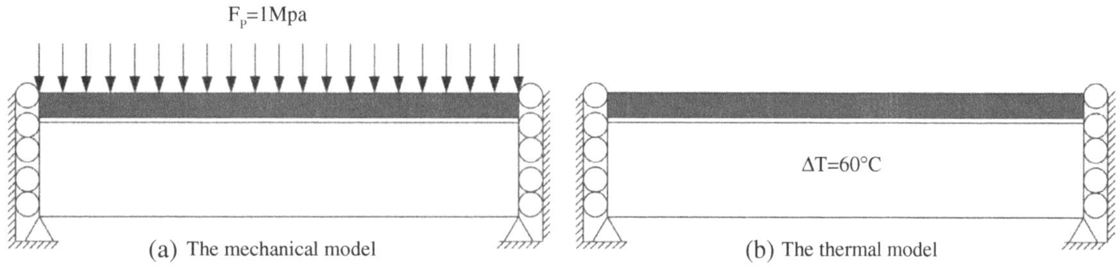

Figure 3. Two load cases of model.

with solid and hollowed SL patterns and observed that the shell cracking always took place at the temperatures below the glass transition temperature of epoxy resin [1, 2]. Later, 3D systems developed QuickCast platform and proposed several lattice styles [5]. Norouzi and Rehmati introduced a new octagonal internal lattice structure, which leads a great reduction in maximum stresses exerted on the ceramic shell [6]. Gu et al. [7] introduced the topology optimization method into the lattice structure configuration design for rapid casting patterns for the first time and proposed several 2D and 3D configurations. When used as thermal protection tile, the hexagonal honeycomb resin configuration as shown in Figure 2 is widely used for ceramic-resin composite structure in practical production.

In this paper, we propose a structure topology optimization procedure in combination with thermo-mechanical finite element analysis to find better periodic resin configurations for the ceramic-resin composite structure. Numerical modeling about the mathematical formulation of the objective function and design constraints is established and optimal periodic resin configurations for single-layer model and double-layer model are sought out. By comparing with the existing hexagonal honeycomb resin configuration, two optimal designs have shown better performances both in reducing the stress in the ceramic shell and maintaining the stiffness of the whole structure.

2 Numerical modeling

Optimal designs intend to avoid the crack of the ceramic shell by decreasing the maximum thermal stress inside. Besides, the whole structure must maintain the mechanical stiffness at a high enough level to support the external mechanical loads.

Table 1. Material properties used in the topology optimization.

Properties	Resin	Ceramic
Elastic modulus (Mpa)	700	3500
CTE (10^{-6}/°C)	80	7
Poisson's ratio	0.1	0.4

Thus the resin configuration design comes down to a topology optimization under the thermal-mechanical loads condition. Here, we can summarize the following goals that the topology optimization should attain:

1. Maintain the mechanical stiffness of the whole structure at a certain level.
2. Minimize the thermal stress in the ceramic shell.
3. Save the consumption of the resin material.

To carry out the topology optimization, two load cases, i.e. the thermal load and the pressure load are applied. In the mechanical model as shown in Figure 3a, a uniform pressure load of 1 Mpa is used to simulate the external load applied on the upper surface of the ceramic shell. In the thermal model as shown in Figure 3b, an isothermal temperature of 60 °C is applied on both the resin pattern and the ceramic shell to generate the critical thermal expansion. The two models share the same boundary conditions, suppose the composite structure is taken out from a large area, four bottom sides are fixed and four side faces are restrained to move within the corresponding faces, as shown in Figure 3.

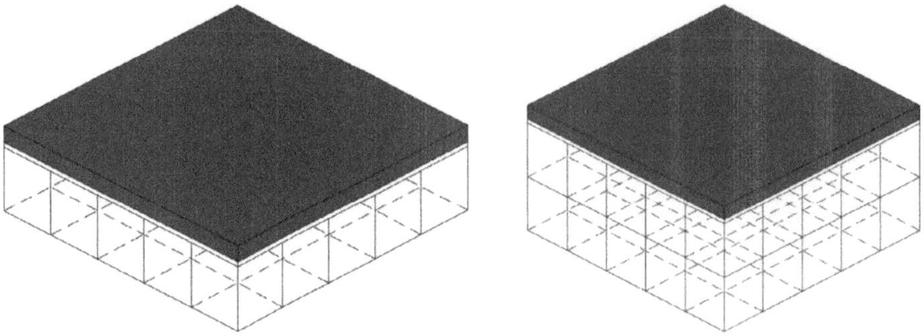

(a) Single-layer model (b) Double-layer model

Figure 4. Two simple models used to evaluate the stress level in the ceramic shell.

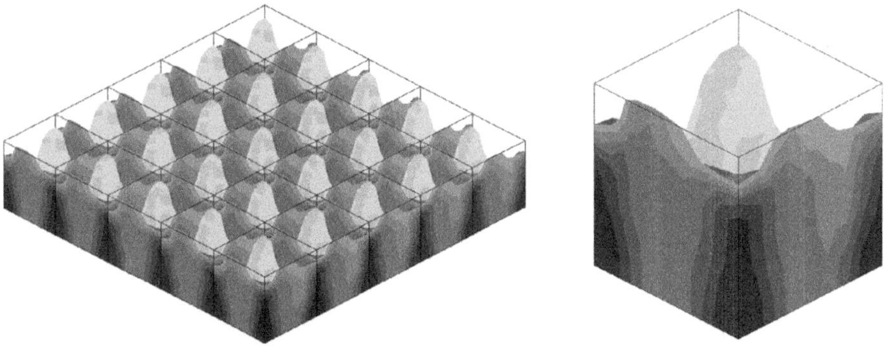

Figure 5. The optimal result obtained with topology optimization.

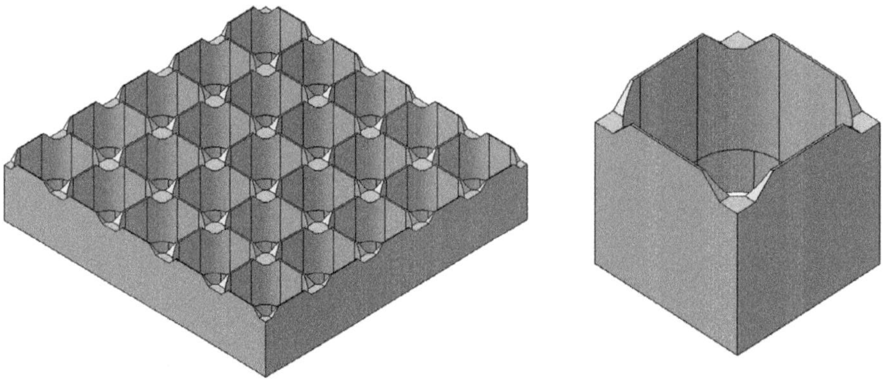

Figure 6. The rebuilt CAD model of periodic resin configuration.

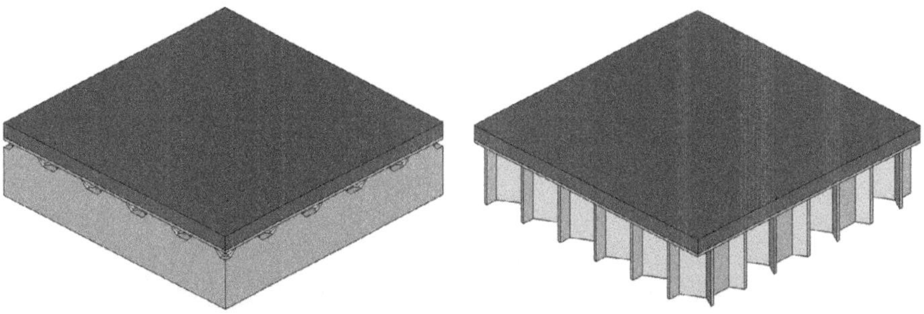

(a) The optimal model (b) The hexagonal honeycomb model

Figure 7. The optimal model and the hexagonal honeycomb model to be analyzed.

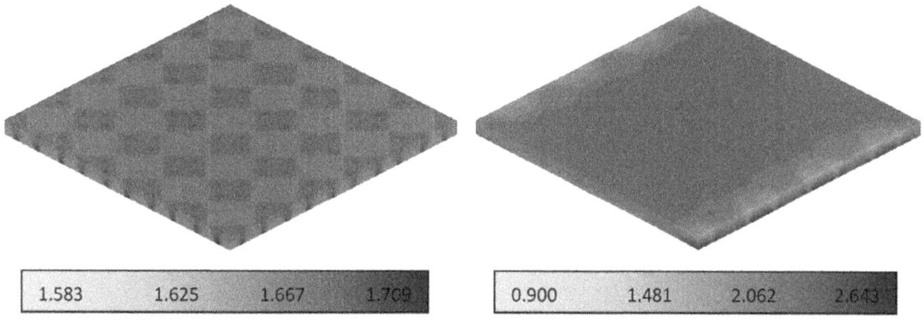

(a) The optimal model (b) The hexagonal honeycomb model

Figure 8. The thermal stress distribution in the ceramic shell under temperature load (Mpa).

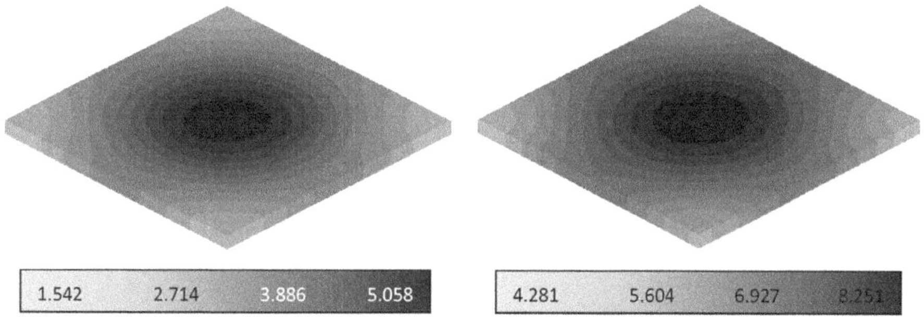

(a) The optimal model (b) The hexagonal honeycomb model

Figure 9. The deformation of the ceramic shell under pressure load (mm).

Table 2. Comparison of the optimal model and the hexagonal honeycomb model.

	Optimal model	Hexagonal honeycomb model
Maximum stress in the ceramic shell (Mpa)	1.709	2.643
Maximum deformation of the ceramic shell (mm)	5.508	8.251
Strain energy of the whole structure (J)	62.716	128.324
The volume of the periodic resin part (mm^3)	186511.572	187812.309

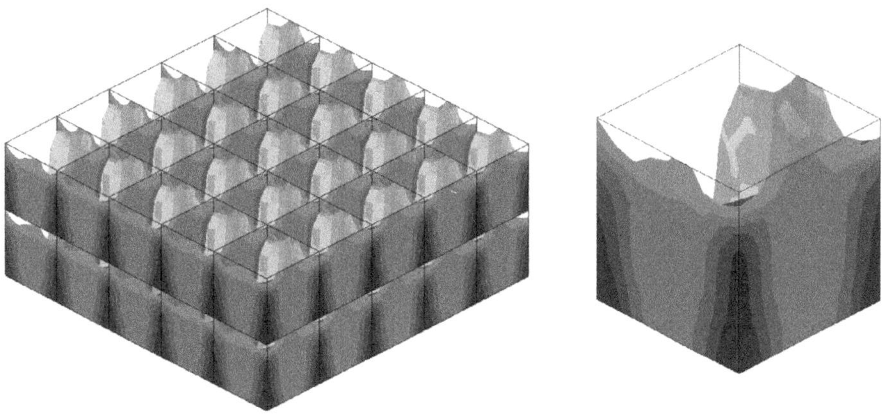

Figure 10. The optimal result obtained with topology optimization.

Table 3. Comparison of the optimal model and the hexagonal honeycomb model.

	Optimal model	Hexagonal honeycomb model
Maximum stress in the ceramic shell (Mpa)	1.709	2.845
Maximum deformation of the ceramic shell (mm)	5.545	7.499
Strain energy of the whole structure (J)	91.716	134.344
The volume of the periodic resin part (mm^3)	356172.138	375693.330

Therefore, the mathematical formulation of the topology optimization problem can be expressed as:

$$\begin{cases} \text{Find} : \mathbf{x} = (x_1, x_2, \ldots, x_n) \\ \text{Min } f(\mathbf{x}), = \max \ \sigma_C \\ \text{s.t. } C \leq C_{(U)} \\ V \leq V_{(U)} \\ 0 < x_{(L)} \leq x_i \leq 1, \ \ i = 1, 2, \ldots, n \end{cases} \tag{1}$$

where $\mathbf{x}$ is the vector of design variables, i.e. the pseudo-densities which vary between 0 and 1 to describe the material distribution over the design domain. σ_C is the Von-Mises stress in the ceramic shell due to the temperature effect. V is volume of the design domain. C is the strain energy of the whole structure related to the mechanical model and it is inversely proportional to the structure stiffness. $C_{(U)}$ and $V_{(U)}$ are the upper bounds of the strain energy and the material volume over the design domain, respectively. $x_{(L)}$ is the lower bound of the pseudo-densities to avoid the singularity of the element stiffness matrix when the element material is removed.

Since the material properties of the cured resin will change with the variation of temperature during the heating procedure, the material parameters and experiment conditions must be defined first. The results of preliminary experiment show that the thermal stress in the ceramic shell is maximized when temperature is about 60 °C. That's why we use the temperature at 60 °C as the thermal environment in thermal model for topology design optimization. Thus we use the temperature at 60 °C as the severe case to ascertain the material properties. The material properties are shown in Table 1, and the reference temperature is set to be 20 °C.

3 Numerical study

As shown in Figure 4, two simple models, i.e. the single-layer model and the double-layer model are optimized, respective CAD models are rebuilt according to the optimal results and then compared with the existing hexagonal honeycomb resin configuration.

A single-layer model

As shown in Figure 4a, the size of the periodic resin part is 200 mm × 200 mm × 40 mm, a 2 mm thick resin layer and a 10 mm thick ceramic shell cover on the upper surface successively. The periodic part is subdivided into 25 basic cells which share the same material distribution and every basic cell is a cube with 40 mm in side length.

By allowing the periodic part's volume fraction of 40% and constraining the strain energy of the whole structure to 40 J, adopt the material whose pseudo-density higher than 0.7, the optimal design of the resin configuration and a single basic cell are shown in Figure 5. Then a CAD model shown in Figure 6 is rebuilt accordingly.

To compare the optimal design with the existing hexagonal honeycomb resin configuration, the optimal model and the hexagonal honeycomb model shown in Figure 7 are analyzed respectively by finite element simulation. Two models share the same boundary conditions and load cases, and the optimal model uses less resin material than the hexagonal honeycomb model. The thermal stress distribution in the ceramic shell under the temperature load and the deformation of the ceramic shell under pressure load are shown in Figures 8 and 9 respectively. The overall comparison of two models is shown in Table 2.

As we can learn from Table 2, compared with the hexagonal honeycomb model, the optimal model reduce the maximum thermal stress in the ceramic shell from 2.643 Mpa to 1.709 Mpa, meanwhile, the strain energy of the whole structure decreases from 128.324 J to 62.716 J, and all these are in the precondition of less consumption of resin materials.

A double-layer model

As shown in Figure 4b, the size of the periodic resin part is 200 mm × 200 mm × 80 mm, the thicknesses of the resin layer and the ceramic shell are still 2 mm and 10 mm, respectively. The periodic part is subdivided into 5 × 5 × 2 basic cells with 40 mm in side length. By allowing the periodic part's volume fraction of 40% and constraining the strain energy of the whole structure to 55 J, adopt the material whose pseudo-density higher than 0.75, the optimal design of the periodic resin part and a single basic cell are shown in Figure 10, then the rebuilt optimal CAD model shown in Figure 11 is established. Be similar with the single-layer case, the double-layer optimal model and the hexagonal honeycomb model shown in Figure 12 are analyzed respectively. The boundary conditions and load cases remain the same. The thermal stress distribution in the ceramic shell under temperature load is shown in Figure 13, the deformation of the ceramic shell under pressure load is shown in Figure 14, and the overall comparison of two models is shown in Table 3.

As shown in Table 3, the optimal model uses less resin materials than the hexagonal honeycomb model, whereas the maximum thermal stress in the ceramic shell under temperature load decreases from 2.845 Mpa to 1.709 Mpa. In the

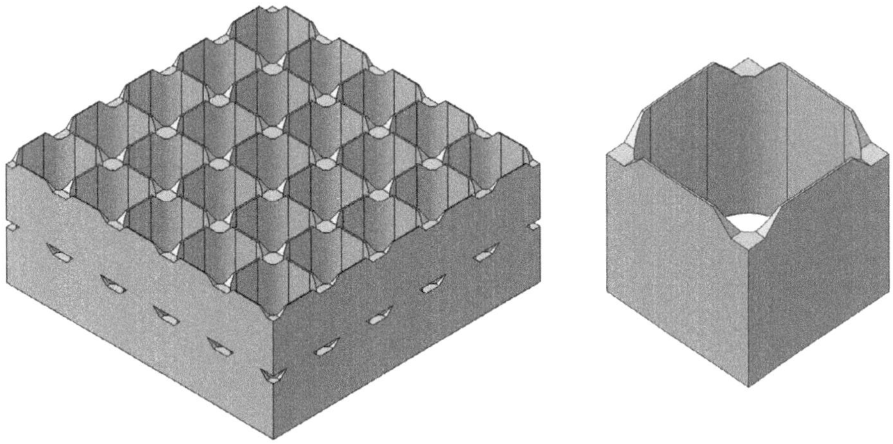

Figure 11. The rebuilt CAD model of periodic resin configuration.

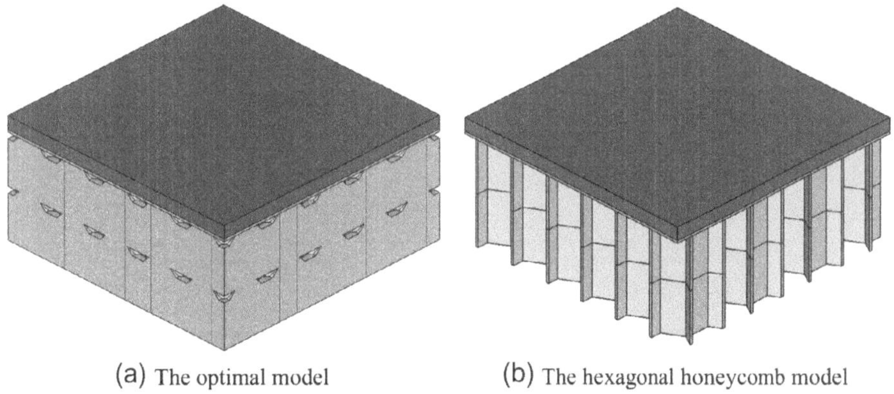

Figure 12. The optimal model and the hexagonal honeycomb model to be analyzed.

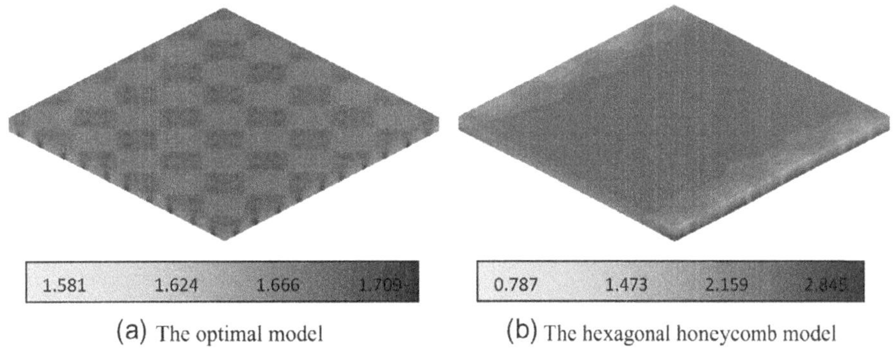

Figure 13. The thermal stress distribution in the ceramic shell under temperature load (Mpa).

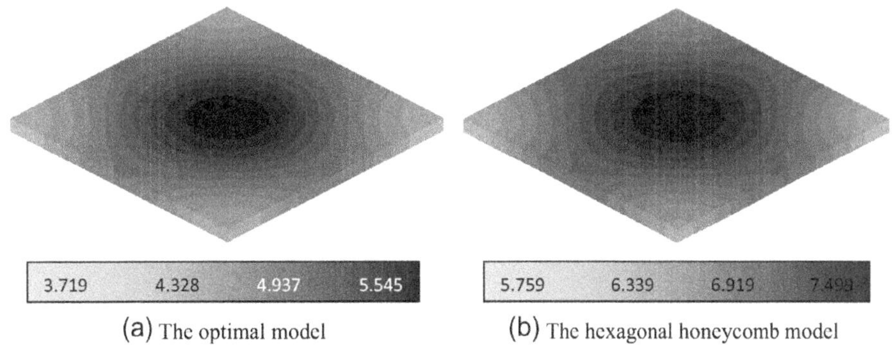

Figure 14. The deformation of the ceramic shell under pressure load (mm).

meantime, the strain energy of the whole structure under pressure load decreases from 14.84 J to 10.58 J.

4 Conclusion

In this paper, techniques of topology optimization are introduced to find better periodic resin configurations for the ceramic-resin composite structure. The optimization approach is carried out to solve the crack problem of the ceramic shell by reducing the maximum thermal stress in the shell. At the same time, the mechanical stiffness of the whole structure and the consumption of the resin material are taken into consideration. In the tested examples, reasonable optimal designs are obtained by topology optimization. Compared with the existing hexagonal honeycomb configuration, the optimal designs not only use less resin materials, but also are both stiffer and better in reducing the maximum thermal stress in the ceramic shell.

References

1. Hague R, Dickens PM. 1995. Stresses created in ceramic shells using QuickCast models, in Proceedings of the Solid Freeform Fabrication Symposium, Marcus H, Editors. University of Texas at Austin. p. 242–252.
2. Hague R, Dickens PM. 1996. Finite element analysis and strain gauging of the stereolithography/investment casting system, in Proceedings of the Solid Freeform Fabrication Symposium, Bourell D, Editors. University of Texas at Austin. p. 523–538.
3. Jacobs PF. 1996. Stereolithography and other RP&M technologies: from rapid prototyping to rapid tooling. ASME Press: New York, NY.
4. Jacobs PF. 1995. Rapid tooling. World Class Design to Manufacture, 2(6), 42–50.
5. Hague R. 2001. Structural design and resin drainage characteristics of QuickCast 2.0. Rapid Prototyping Journal, 7(2), 66–72.
6. Norouzi Y, Rehmati S. 2009. A novel lattice structure for SL investment casting patterns. Rapid Prototyping Journal, 4(14), 255–263.
7. Gu XJ, Zhu JH, Zhang WH. 2012. The lattice structure configuration design for stereolithography investment casting pattern using topology optimization. Rapid Prototyping Journal, 18(5), 353–361.

14

A particle swarm optimization approach for permutation flow shop scheduling problem

T. Radha Ramanan[1,*], Muhammed Iqbal[2] and K. Umarali[3]

[1] Assistant Professor, Department of Mechanical Engineering, National Institute of Technology Calicut, Kozhikode 673601, Kerala, India
[2] Senior Lecturer, Department of Mechanical Engineering, MEA Engineering College, Perinthalmanna 679325, Kerala, India
[3] Research Scholar, Department of Mechanical Engineering, National Institute of Technology Calicut, Kozhikode 673601, Kerala, India

Abstract – Flow shop scheduling problem (FSSP) is a combinatorial optimization problem. This work, with the objective of optimizing the makespan of an FSSP uses a particle swarm optimization (PSO) approach. The problems are tested on Taillard's benchmark problems. The results of Nawaz, Encore and Ham (NEH) heuristic are initialized to the PSO to direct the search into a quality space. Variable neighbourhood search (VNS) is employed to overcome the early convergence of the PSO and helps in global search. The results are compared with standalone PSO, traditional heuristics and the Taillard's upper bounds. Five problem set are taken from Taillard's benchmark problems and are solved for various problem sizes. Thus a total of 35 problems are solved. The experimental results show that the solution quality of FSSP can be improved if the search is directed in a quality space based on the proposed PSO approach (PSO-NEH-VNS).

Key words: Makespan, Flow shop scheduling, Particle swarm optimization, Variable neighbourhood search.

1 Introduction

Scheduling is the allocation of resources (e.g., machines) to tasks (e.g., jobs) in order to ensure the completion these tasks in a reasonable amount of time. In a flow shop, the technological constraints demand that the jobs pass between the machines in the same order; i.e., If J_1 must be processed on machine M_k before machine M_i then the same is true for all jobs. A permutation schedule is one on which each machine processes the jobs in the same order; i.e., if on machine M_1 job J_i is processed before J_k, then the same is true for all machines [1]. The objective is to find a sequence of jobs that minimizes the makespan. i.e., time required to complete all the jobs. This problem is found to be NP-hard [2].

The complexity of the flow shop scheduling problem renders exact solution methods impractical for instances of more than a few jobs and/or machines. This is the main reason for the various heuristic methods proposed in the literature.

The heuristics can be divided into either constructive heuristics or improvement heuristics, the former are heuristics that build a feasible schedule from scratch and the latter are heuristics that try to improve a previously generated schedule by normally applying some form of specific problem knowledge.

The objective of this work is to find a permutation of jobs that minimizes the makespan. The present study explores the further improvement in the solution quality, if the population in the improvement phase is provided with better initial solutions. The work describes the percentage improvement that the proposed method is able to obtain when compared with some of existing heuristics due to the improvisation in the search space of PSO. This approach can be considered as the significant contribution of this study. This work aims at directing the search space of PSO into certain quality. Then, VNS is employed to make the solution jump out of the local optima. This approach is termed as PSO-NEH-VNS heuristic. The solution thus obtained is compared with the results obtained using the PSO and the upper bounds of Taillard's [3] benchmark problems. The approach proves that results approach the optimum faster when the search space is nearer to the optimum.

The rest of the paper is organized as follows. Section 2 discusses the literature survey. The methodology is presented in Section 3. The results and discussions are made in Section 4. Conclusions and future scope is given in the Section 5.

2 Literature survey

2.1 Heuristics

Many constructive heuristic approaches have been proposed in the literature. Johnson's algorithm is the earliest known heuristic for the permutation flow shop problems.

*e-mail: radha_ramanan@nitc.ac.in

The two machine flow shop problem with the objective of minimizing makespan is also known as Johnson's [4] problems. An optimal sequence is found by following a heuristics of finding the minimum machining time and allotting the job to the machine in a preferential order is adopted.

Literature is replete with traditional heuristics [5–7] and nontraditional heuristics (Tabu search [8–10]); Simulated annealing [11–14]. Genetic algorithm [15–17], Ant colony optimization [18] to name a few.

Contrary to constructive heuristics, improvement heuristics start from an already built schedule and try to improve it by some given procedure. Many improvement heuristic approaches also have been proposed in the research [19–21].

2.2 Particle swarm optimization

Duda [22] evaluates a particle swarm optimization with variable neighbourhood search against a standalone VNS algorithm. The results of the conducted experiments indicate that hybridization of a meta-heuristic with a local search algorithms may not always bring additional performance benefit. Ali and Fawaz [23] address the flow shop scheduling problem with respect a due date-based performance measure, i.e., maximum lateness. Lian et al. [24] propose a novel particle swarm optimization algorithm to minimize the makespan of a flow shop.

Kuo et al. [25] propose a flow shop scheduling algorithm based on hybrid particle swarm optimization model. Liu et al. [26] propose a hybrid particle swarm optimization for flow shop scheduling with stochastic processing time. Chen [27] proposes an improved particle swarm optimization technique for flow shop scheduling. Sha and Lin [28] propose a PSO model for multi-objective flow shop scheduling. Zhang et al. [29] provide a hybrid alternate two phases particle swarm algorithm for flow shop scheduling problems. Deb [30] discusses about the evolutionary approaches used in solving the multi-objective optimization problems.

3 Methodology

The objective of the work is to minimize the makespan of the flow shop scheduling problems. This is attempted by developing a PSO approach. Five problem sets are taken from "Taillard" [2] benchmark problem and are solved for various sizes. A total of 35 problems are solved.

The results of PSO-NEH-VNS is compared with the upper bound (UB) of Taillard's benchmark and also compared with the results of CDS, NEH, PSO, PSO-NEH, and PSO-VNS.

The Taillards benchmark problem is first solved using NEH heuristics and a sequence is obtained. The PSO requires a population of sequences to be initialized. In this problem, the PSO is initialized with fifteen randomly generated populations. The sequence that is obtained by NEH heuristic is initialized as one of the initial population. This PSO is referred to as PSO initialized with NEH (PSO-NEH). This initialization is made with the hope that the search space is nearer to optimal solution. The results obtained by the PSO-NEH are further improved with the VNS. The VNS prevents the early convergence of PSO and jump out of local optima. This approach is termed as PSO-NEH-VNS. Hansen et al. [31] state that Variable neighbourhood search (VNS) is a meta-heuristic, or a framework for building heuristics, based upon systematic changes of neighbourhoods both in descent phase, to find a local minimum, and in perturbation phase to emerge from the corresponding valley.

3.1 PSO algorithm for FSSP

Particle swarm optimization (PSO), inspired by the motion of a flock of birds searching for food, was developed by Kennedy and Eberhart [32] for optimization of continuous nonlinear functions. To find the optimal solution each bird called a particle adjusts its searching direction according to two factors, its own best previous experience (pbest) and the best experience of all other members (gbest). The system is initialized with a population of random solutions and searches for optima by updating generations. In PSO potential solutions called particles, fly through the problem space by following the current optimum particles. PSO simulates the behaviors of bird flocking. In PSO each single solution is a "bird" in the search space. We call it "particles". All the particles have fitness value and velocities which direct the flying of the particles. The particles fly through the problem space by following the current optimum particles. Each particle is updated by following two "best" values. The first value is the best solution it has achieved so far known as "pbest". Second value is the best value obtained so far by any particle is the population called "gbest". After finding the two best values the particle updates its velocity and position with the following equations

$$V[\,] = w \times v[\,] - c1 \times \text{rand}(\,) \times (\text{pbest}[\,] - \text{present}[\,]) - c2 \times \text{rand}(\,) \times (\text{gbest}[\,] - \text{present}[\,]) \quad (1)$$

where, V[] = particle velocity, present [] = current particle solution, pbest [] = best solution among each particle, gbest [] = best among defined as before, w = inertia weights 0.8, $c1$ and $c2$ are learning factors. Usually $c1 = c2 = 2$.

This equation specifies that the velocity of a particle is determined by the previous velocity of the particle. Each particle moves to a new position according to the following equation.

$$\text{present}[\,] = \text{present}[\,] - v[\,]. \quad (2)$$

The out line of a PSO algorithm is as follows:

Step 1: (Initialization): randomly generate initial particles,
Step 2: (Fitness): evaluate the fitness of each particle in the population,
Step 3: (Update): calculate the velocity of each particle by the equation,
Step 4: (Construction): for each particle, moves to the next position according to the equation (2),
Step 5: (Termination): stop the algorithm if a specified stopping criterion is satisfied; return to Step 2 otherwise.

Table 1. Makespan of 20 jobs, 5, 10 and 20 machines.

		Makespan						
SL No.	Problem size	CDS	NEH	PSO	PSO NEH	PSO VNS	PSONEHVNS	UB*
1	20 × 5	1390	1286	1297	1286	1320	1278	1278
2	20 × 5	1424	1365	1373	1365	1368	1365	1359
3	20 × 5	1249	1159	1125	1132	1164	1100	1081
4	20 × 5	1418	1325	1364	1325	1349	1309	1293
5	20 × 5	1323	1305	1250	1263	1277	1250	1236
6	20 × 10	1757	1680	1660	1635	1673	1586	1582
7	20 × 10	1854	1729	1727	1722	1747	1684	1659
8	20 × 10	1645	1557	1517	1556	1588	1521	1496
9	20 × 10	1547	1439	1434	1419	1514	1399	1378
10	20 × 10	1558	1502	1492	1502	1538	1450	1419
11	20 × 20	2559	2410	2365	2386	2446	2330	2297
12	20 × 20	2285	2150	2177	2148	2195	2111	2100
13	20 × 20	2565	2411	2387	2399	2509	2342	2326
14	20 × 20	2434	2262	2304	2251	2410	2248	2223
15	20 × 20	2506	2397	2358	2388	2471	2302	2291

* Refers to upper bound of Taillard's benchmark problems.

Table 2. Makespan of 50 jobs, 5, 10 and 20 machines.

		Makespan					
SL No.	Problem size	CDS	NEH	PSO	PSO-NEH	PSO-NEH-VNS	UB*
1	50 × 5	2816	2733	2729	2729	2724	2724
2	50 × 5	3032	2843	2906	2843	2843	2834
3	50 × 5	2703	2640	2676	2621	2631	2621
4	50 × 5	2884	2782	2824	2782	2762	2751
5	50 × 5	3038	2868	2873	2864	2864	2863
6	50 × 10	3421	3135	3240	3134	3059	3025
7	50 × 10	3246	3032	3093	3025	2934	2892
8	50 × 10	3280	2986	3139	2965	2932	2864
9	50 × 10	3393	3198	3236	3172	3115	3064
10	50 × 10	3375	3160	3186	3115	3052	2986
11	50 × 20	4328	4082	4192	4061	4010	3875
12	50 × 20	4216	3921	4067	3918	3864	3715
13	50 × 20	4189	3927	3981	3907	3808	3668
14	50 × 20	4280	3969	4067	3957	3844	3752
15	50 × 20	4122	3835	3999	3832	3815	3635

* Refers to upper bound of Taillard's benchmark problems.

4 Results and discussions

The objective of this work is to build a PSO model to solve the flow shop scheduling problems for minimizing the makespan; Five problem set are taken from Taillard benchmark problem and are solved. The problems are 20 jobs, 5, 10 and 20 machines; 50 jobs, 5, 10 and and 20 machines and 100 jobs, 5 machines.

Table 1 shows the results of the five instances of 20 × 5, 20 × 10 and 20 × 20 problems that are solved using the various methods.

The proposed PSO-NEH-VNS clearly gives better results than other methods such as CDS, NEH, PSO, PSO-NEH, and PSO-VNS. This is possible because the search space of PSO-NEH-VNS is in the area of superior solution. The search space of PSO is initialized with the results of NEH heuristic; this ensures that the solution space is of certain quality. The VNS helps in the solution of PSO jumping out of the local optima. Thus, the interchange of VNS has helped the PSO to come out of the pre convergence.

In smaller size problems the average difference in makespan between Upper bound and PSO-NEH-VNS is less and this difference increases with increasing number of machines. Average difference in makespan between PSO-NEH-VNS and UB for 20 Job 5 machines is 12.6, 20 Job 10 machines is 21 and 20 Job 20 machines is 32.

Table 2 shows the makespan of 50 jobs and 5, 10 and 20 machines problem size. The proposed PSO-NEH-VNS clearly give better results than other methods. The makespan obtained by PSO-NEH-VNS is nearly equal (within 5%) to the upper bound of the Taillard's benchmark problems and in some cases it is equal to the upper bound. (For 50 × 5, SL No. 1). Similar

Table 3. Makespan of 100 jobs, 5 machine.

SL No.	Problem size	Makespan					
		CDS	NEH	PSO	PSO NEH	PSO NEH VNS	UB*
1	100 × 5	5592	5519	5527	5519	5493	5493
2	100 × 5	5563	5348	5327	5300	5290	5268
3	100 × 5	5493	5219	5253	5213	5213	5175
4	100 × 5	5273	5023	5078	5023	5023	5014
5	100 × 5	5461	5266	5323	5266	5253	5250

* Refers to Upper bound of Taillard's benchmark problems.

Table 4. Computational time.

SL No.	Problem size	CPU time in seconds					
		CDS	NEH	PSO	PSO NEH	PSO VNS	PSO NEH VNS
1	20 × 5	0.0156	0.0156	1.29	1.32	7.96	7.96
2	20 × 10	0.0156	0.0156	1.59	1.50	35.82	36.68
3	20 × 20	0.0156	0.0313	1.95	1.92	206.92	197.84
4	50 × 5	0.0156	0.0156	1.96	1.96	13.07	13.06
5	50 × 10	0.0156	0.0156	2.26	2.26	60.62	60.45
6	50 × 20	0.0156	0.0781	2.95	2.95	332.14	330.84
7	100 × 5	0.0313	0.0156	3.09	3.07	22.50	22.10

to the previous problem set, the average difference between upper bound and PSO-NEH-VNS increases with increasing number of machines. (For 50 Job 5 machines is 6.2; 50 job 10 machines is 52.2; 50 Job 20 machines is 139.2.)

Makespan values of 100 jobs and 5, 10 and 20 machines are shown in Table 3. For the 100 jobs problem as well the proposed PSO-NEH-VNS method gives better results when compared with other methods. In certain instances (for 100 × 5, SL No. 1), the PSO-NEH-VNS is able to give the makespan values equivalent to the upper bound of the Taillard's benchmark problems.

Consistently, the PSO-NEH-VNS is giving better results than the other methods taken for comparison. Hence, it can be stated that with a better search space provided for the PSO, the PSO with the help of VNS is able to avoid early convergence. It can also be inferred that the search is able to converge faster because of the initial quality of the search space.

In order to compare computational time required, various approaches are coded in matlab and run on a 3 G B RAM, 2.19 GHz, Dual Core system with Windows XP operating system. The CPU time for various size problems are shown in Table 4.

It can be observed that computational time for PSO NEH VNS approach is slightly lesser than PSO VNS for larger problem sizes.

Table 5 shows the percentage improvement of results of PSO VNS NEH for a 20 job problem size, when compared with other methods used in this work.

It can be observed that PSO do better than PSO-VNS. PSO when initialized with the results of NEH does better than both PSO and PSO-VNS algorithms. The VNS takes the PSO to early convergence. But when the same solution is initialized with quality solution space, obtained from the constructive heuristic (NEH), it is able to perform better.

Table 5. Improvement of makespan using PSO NEH VNS.

SL No.	Problem size	% Improvement of makespan over				
		CDS	NEH	PSO	PSO NEH	PSO VNS
1	20 × 5	8.08	0.00	0.85	0.00	2.64
2	20 × 5	4.32	0.00	0.58	0.00	0.21
3	20 × 5	13.54	5.36	2.27	2.90	5.81
4	20 × 5	8.32	1.22	4.20	1.22	3.05
5	20 × 5	5.84	4.40	0.00	1.04	2.16
6	20 × 10	10.78	5.92	4.66	3.08	5.48
7	20 × 10	10.09	2.67	2.55	2.25	3.74
8	20 × 10	8.15	2.36	0.26	2.30	4.40
9	20 × 10	10.57	2.85	2.50	1.42	8.22
10	20 × 10	7.44	3.58	2.89	3.58	6.06
11	20 × 20	9.26	2.90	0.98	1.87	4.44
12	20 × 20	87.73	1.36	2.64	1.27	3.48
13	20 × 20	8.82	2.29	1.27	1.78	6.44
14	20 × 20	8.27	0.62	2.49	0.13	7.20
15	20 × 20	7.73	3.05	1.37	2.66	6.23

Table 6. Comparison of results.

Problem size	Makespan		
	UB*	Kuo et al. HPSO	NEH-PSO-VNS
20 × 5	1278	1278	1278
20 × 10	1582	1587.3	1586
20 × 20	2297	2307	2310
50 × 5	2724	2724	2724

* Refers to Upper bound of Taillard's benchmark problems.

Table 6 shows the comparison of experimental results from literature The results show the PSO-NEH-VNS is providing results that are comparable to the Kuo et al. [25] HPSO and in one instance (20 × 10) it is able to provide better result. This indicates the PSO-NEH-VNS model is worth exploring.

Though the results show the proposed heuristic performs better than certain PSO models, the percentage deviation from the optimum still persists. VNS has an inherent characteristic that is not taking the results to the optimum. But, the objective of the work is to prove that if the initial solution space is provided with quality solutions, the PSO can be used for obtaining faster results.

5 Conclusions

A particle swarm optimization algorithm named PSO-NEH-VNS is used to solve the FSSP with the objective of minimizing makespan. The research shows that the PSO initialized with a better solution space obtained from the constructive heuristic (NEH), has the ability to provide solutions that are better than some of the existing heuristics and takes the solution to early convergence of results. The results show that the percentage deviation of the makespan increases with the increasing machine, proving that the complexity of the flow shop problems depends on the increasing machine size. The results also show that the percentage deviation of the makespan with respect to the increasing job is not significant.

Acknowledgements. The authors wish to thank the anonymous reviewers for giving their valuable suggestions in improving the quality of the paper.

References

1. French S. 1982. Sequencing and scheduling: an introduction to the mathematics of the job-shop. Ellis Horwood Limited: Chichester, West Sussex, England, UK.
2. Garey MR, Johnson DS, Sethi R. 1976. The complexity of flowshop and jobshop scheduling. Mathematics of Operations Research, 1(2), 117–129.
3. Taillard E. 1993. Benchmarks for basic scheduling problems. European Journal of Operation Research, 64, 278–285.
4. Johnson SM. 1954. Optimal two-and three-stage production schedules with set up times include. Naval Research Logistics Quarterly, 1, 61–68.
5. Palmer DS. 1965. Sequencing jobs through a multistage process in the minimum total time – a quick method of obtaining near optimum. Operational Research Quarterly, 16, 101–107.
6. Campbell HG, Dudek RA, Smith ML. 1970. A heuristic algorithm for the n job m machine sequencing problem. Management Science, 16(10), B630–B637.
7. Navaz M, Enscore EE, Ham I. 1983. A heuristic algorithm for the m-machine n-job flow shop sequencing problem. Omega, 11, 91–95.
8. Widmer M, Hertz A. 1989. A new heuristic for the flow shop sequencing problem. European Journal of Operational Research, 41, 186–193.
9. Nowicki E, Smutnicki C. 1996. A fast Tabu search algorithm for the permutation flow shop problem. European Journal of Operations Research, 91, 160–175.
10. Moccellin JV, Santos MO. 2000. An adaptive hybrid metaheuristic for permutation flow shop scheduling. Control and Cybernetics, 29(3), 761–771.
11. Osman I, Potts C. 1989. Simulated annealing for permutation flow shop scheduling. OMEGA, 17(6), 551–557.
12. Chen CL, Vempati VS, Aljaber N. 1995. An application of genetic algorithms for flow shop problems. European Journal of Operational Research, 80, 389–396.
13. Low C. 2005. Simulated annealing heuristics for flow shop scheduling problems with unrelated parallel machines. Computers & Operations Research, 32(8), 2013–2025.
14. Ishibuchi H, Misaki S, Tanaka H. 1995. Modified simulated annealing algorithms for the flow shop sequencing problem. European Journal of Operational Research, 81(2), 388–398.
15. Reeves C. 1995. A genetic algorithm for flow shop sequencing. Computers and Operations Research, 22(1), 5–13.
16. Chen CL, Vempati VS, Aljaber N. 1995. An application of genetic algorithms for flow shop problems. European Journal of Operational Research, 80, 389–396.
17. Zobolas GI, Tarantilis CD, Ioannou G. 2009. Minimizing makespan in permutation flow shop scheduling problems using a hybrid metaheuristic algorithm. Computers & Operations Research, 36(4), 1249–1267.
18. Yagmahan B, Yenisey MM. 2008. Ant colony optimization for multi-objective flow shop scheduling problem. Computers & Industrial Engineering, 54, 411–420.
19. Rajendran C. 1995. Theory and methodology heuristics for scheduling in flow shop with multiple objectives. European Journal of Operational Research, 82, 540–555.
20. Koulamas C. 1998. A new constructive heuristic for the flow shop scheduling problem. European Journal of Operational Research Society, 105, 66–71.
21. Suliman S. 2000. A two-phase heuristic approach to the permutation flow-shop scheduling problem. International Journal of Production Economics, 64, 143–152.
22. Duda J. 2006. Local search and nature based metaheuristics: a case of flow shop scheduling problem, in Proceedings of the International multiconference on Computer Science and Information Technology, PIPS. p. 17–24, ISSN 1896-7094.
23. Allahverdi A, Fawaz S, Anzi A. 2006. A PSO and a Tabu search heuristics for the assembly scheduling problem of the two-stage distributed database application. Computers & Operations Research, 33(4), 1056–1080.
24. Lian Z, Gu X, Jiao B. 2008. A novel particle swarm optimization algorithm for permutation flow-shop scheduling to minimize makespan. Chaos, Solitons & Fractals, 35(5), 851–861.
25. Kuo IH, Horng SJ, Kao TW, Lin TL, Fan P. 2007. An efficient flow-shop scheduling algorithm based on a hybrid particle swarm optimization model, in IEA/AIE 2007, LNAI 4570, Okuno HG, Ali M, Editors. Springer-Verlag: Berlin, Heidelberg. p. 03–312.
26. Liu B, Wang L, Jin Y. 2005. Hybrid particle swarm optimization for flow shop scheduling with stochastic processing time, in CIS 2005, Part I, LNAI 3801, Hao Y, et al., Editor. Springer-Verlag: Berlin, Heidelberg. p. 630–637.
27. Chen Q. 2010. Flow shop scheduling using an improved PSO, vol. 2, in International Conference on Measuring Technology and Mechatronics Automation. p. 296–299.
28. Sha DY, Lin HH. 2009. A particle swarm optimization for multi-objective flow shop scheduling. International Journal of Advanced Manufacturing Technology, 45, 749–758.
29. Zhang C, Ning J, Ouyang D. 2010. A hybrid alternative two phases particle swarm optimization algorithm for flow shop scheduling problem. Computers & Industrial Engineering, 58(1), 1–11.

30. Deb K. 2007. Current trends in evolutionary multi-objective optimization. International Journal for Simulation and Multi-disciplinary Design Optimization, 1, 1–8.
31. Hansen P, Mladenovíc N. 1997. Variable neighborhood search for the p-median. Location Science, 5, 207–226.
32. Kennedy J, Eberhart RC. 1995. Particle swam optimization, in Roc. IEEE International Conference on Neural Networks (Path, Australia). IEEE Service Center: Piscataway, NJ. p. 1942–1948,

15

Thermal modeling of hydrogen storage by absorption in a magnesium hydrides tank

Karim Lahmer[1,*], Rachid Bessaïh[1], Angel Scipioni[2], and Mohammed El Ganaoui[2]

[1] Département de Génie Mécanique, (LEAP), Université Constantine 1, Route d'Ain El Bey, Constantine 25000, Algeria
[2] Université de Lorraine, Institut Universitaire de Technologie Henri Poincaré de Longwy (LERMAB & GREEN), 186 Route de Lorraine, 54400 Cosnes-et-Romain, France

Abstract – This paper summarizes numerical results of hydrogen absorption simulated in an axisymmetric tank geometry containing magnesium hydride heated to 300 °C and at moderate storage pressure 1 MPa. The governing equations are solved with a fully implicit finite volume numerical scheme used by a commercial software FLUENT. The effect of the different kinetic reaction equations modeling hydrogen absorption was studied by the introduction of a specific subroutine at each time step in order to consider which one will provide results close to available experimental results. Spatial and temporal profiles of temperature and concentration in hydride bed are plotted. Results show that suitable method for our two-dimensional study is a CV-2D technique because it generates the smallest error especially during the beginning of the reaction. Also, its computational time is the shortest one compared to the other methods.

Key words: Hydrogen storage, Magnesium hydrides, CFD simulation, Absorption kinetic equations.

Nomenclature

C_p	specific heat, J/kg K^{-1}
C_a	kinetic coefficient, s^{-1}
ΔH	molar enthalpy of reaction at standard conditions, J/mol
K	permeability, m^2
P	hydrogen pressure, Pa
P_{eq} (T)	equilibrium pressure, Pa
R	universal gas constant = 8.314 J/mol K
S	source term of reaction, mol/m s
ΔS	molar entropy of reaction at standard conditions, J/mol K
wt	maximum weight percentage of hydrogen in the material, %

Greek symbols

α	absorption rate
λ	thermal conductivity, W/m K
ε	porosity
ρ	density, kg/m^3
$d\alpha/dt$	hydriding velocity, s^{-1}

Subscripts and superscripts

e	energy
eff	effective
eq	balance
g	gas
H	hydrogen
M	metal
m	mass
MH	metal hydride
i	initial

1 Introduction

Because of the huge amounts of dangerous gases emitted by industry, the predicted shortage of fossil fuels and the announced general climate change, SOFC fuel cells running with hydrogen could be the solution to the recurring problem of intermittent renewable energy. These kinds of systems are particularly interesting for electricity production in isolated areas, for the daily stabilization of electricity demand or as security high power kits. However, the major obstacle to the rapid growth of this technology is the hydrogen storage in the most compact volume. One of the most promising methods is that of storing hydrogen in metal hydrides. Indeed, magnesium hydride is a very good substance for reversible hydrogen

*e-mail: lahmer.karim@sonelgaz.dz

storage because of its highest capacity of storage (7.6%m H_2) compared to the others such as $TiVCrH_6$ (2%m H_2), $FeTiH_{1.95}$ (1.6%m H_2), $LaNi_5H_{6.7}$ (1.5%m H_2). Its volume capacity (111 kg/m^3) is greater than that of liquid hydrogen (71 kg/m^3), that's why it would be used in a large scale in the few next years. The hydrogen storage in a solid form offers significant advantages, like reducing the tank dimensions, resulting chemical reactions running at moderate temperatures 300 °C and at pressures about 1–10 bars; also, hydrogen density in the solid form is more important than the one in molecular form. Finally, this type of storage allows better safety compared to conventional methods such as compression or liquefaction of hydrogen.

Many experimental and computational studies were realized in order to improve the hydrogen tank performances. Jemni and Nasrallah [1] realized one of the first numerical studies of 2D-hydrogen reactor. The results showed the importance of the geometry, inlet pressure and inlet temperature choice. Kaplan et al. [2] presented a mathematical model for hydrogen storage in a metal hydride bed; the team concluded that a rapid charge needs efficient cooling. Muthukumar et al. [3–9] made a parametric investigation of a metal hydride hydrogen storage device, they showed that overall increasing heat transfer coefficient is not beneficial. Phate et al. [4] carried out a computational analysis of a cylindrical metal hydride bed; their conclusion is that the concentration gradient in the bed is the major driving force of hydrogen flow in the bed. Marty et al. [5–14] added an experimental validation to the computational simulations of the hydrogen storage tank with metal hydrides; their goal was to obtain performances according to the objectives imposed by a stationary cogeneration system. Askri et al. [6] made a numerical investigation of heat and mass transfer of a 3D annular tank. Results showed that the use of fins enhances heat transfer and consequently 40% improvement of the time required for 90% storage can be achieved over the case without fins. A CFD model for simulating hydrogen storage in an activated carbon tank was described by Chahine et al. [7]; this model showed that the amount of adsorbed hydrogen is greater than that of the compressed gas hydrogen.

Because of these many coupled phenomena, numerical simulation allows us to anticipate and understand the evolution of the hydrogen storage reaction. In addition, the numerical tool will save considerable time for the optimization of hydrogen tank design. The industrial software FLUENT was used to perform numerical simulations. One of the advantages of this code consists of the important part devoted to the modeling of thermodynamics and kinetics reaction between a gas and a porous medium. A file called (UDF: User Defined Function) in C grouping models was realized and included in the calculations. The purpose of this work is to find the best reaction kinetics equation which is the most adapted for phenomena modeling of the hydrogen absorption.

2 Geometry and mathematical formulation

The tank geometry (Figure 1) consists of a domain with (L = 450 mm) × (H = 7.5 mm) dimensions. It is surrounded by an aluminum shell which equalizes the temperature. Also,

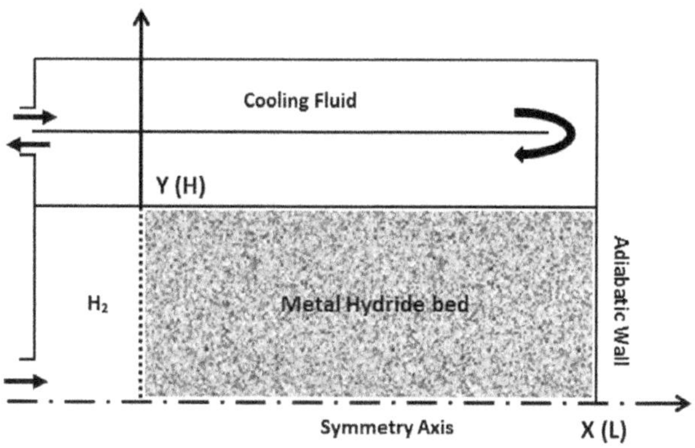

Figure 1. Schematic of a basic tank.

it is cooled by a process which ensures effective further reactions. Hydrogen enters from a lateral left side. The activated magnesium hydride porous powder (Metal alloy) is retained on the other side by a thin filter. The right side wall is considered adiabatic.

In order to simplify the model, some assumptions are adopted. First, hydrogen is considered as an ideal gas between the hydride pores and its generated flow before the absorption phase is negligible. Then, the hydrogen temperature is locally the same as that of the powder, the flow is laminar between the powder grains, the radiative transfer in the porous medium is neglected, the pressure and friction energies are negligible, the inlet temperature is constant. Finally, the resolution method used is the finite volume in a two-dimensional, axisymmetric, laminar regime and at unsteady flow.

The governing equations of the phenomenon are given in the following sections.

2.1 The equations of mass balance and Darcy

– Continuity equation in the gas phase

$$\varepsilon_{\mathrm{M}} \frac{\partial \rho_{\mathrm{g}}}{\partial t} + \mathrm{div}\left(\rho_{\mathrm{g}} \cdot \vec{U}\right) = - S_{\mathrm{m}} \cdot M_{\mathrm{H2}}.$$

– Continuity equation in the solid phase

Mass transfer is taken into account by Fluent with the continuity equation and the Darcy one:

$$\varepsilon_{\mathrm{MH}} \frac{\partial \rho_{\mathrm{MH}}}{\partial t} = S_{\mathrm{m}} \cdot M_{\mathrm{H2}}. \tag{1}$$

The term source of hydrogen mass depends on the reaction rate and can be expressed:

$$S_{\mathrm{m}} = \rho_{\mathrm{MH}} \cdot wt(1 - \varepsilon) \frac{\mathrm{d}\alpha}{\mathrm{d}t}. \tag{2}$$

The Darcy equation is: $\vec{\nabla} P_{\mathrm{H2}} = -\frac{\mu}{K} \cdot \vec{U}$.

This expression is the balance between viscous friction and pressure gradient where (K) is the permeability of the metal hydride. Considering a high porosity of the material and a low velocity of the flow, we can suppose that: $\vec{\nabla} P_{\mathrm{H2}} = 0$.

2.2 The energy equation

The energy conservation equation simplified and solved by fluent in both phases is:

$$\varepsilon\rho_g C_{p_g}\frac{\partial T}{\partial t}+(1-\varepsilon)\rho_s C_{p_s}\frac{\partial T}{\partial t}+\varepsilon\rho_g C_{p_g}\nabla(T\cdot\vec{U})$$
$$=\nabla(\lambda_{\text{eff}}\nabla T_{\text{MH}})+S_e. \tag{3}$$

- Since the solid medium (s) + gas (g) is treated as a continuous medium with:

$\rho_{\text{MH}}C_{p\text{MH}}=\varepsilon\rho_g C_{p_g}+(1-\varepsilon)\rho_m C_{p_m}=\sum_i \varepsilon_i\rho_i C_{p_i}$ and $\lambda_{\text{eff}}=\varepsilon\lambda_g+(1-\varepsilon)\lambda_m$.

So, to the absorption source term of the energy equation taken into account through a module in Fluent UDF is:

$$S_e=\frac{\Delta H}{M_{\text{H2}}}S_m. \tag{4}$$

The material thermal parameters are a function of the powder compactness, temperature and hydrogen inlet pressure and absorption rate. Also, a good knowledge of the thermal conductivity of the activated powder is particularly necessary to obtain perfect simulation accuracy. This value was already obtained experimentally and taken into account such as: $\lambda_{\text{eff}}=0.48$ W/m/K.

2.3 Initial and boundary conditions

2.3.1 Initial conditions

Initially, the powder temperature, the gas pressure and the hydride density of the reaction bed are considered uniform through the tank and the system was assumed under the P-c-T equilibrium

$$T_S=T_{\text{H2}}=T_i;\ \ P_{\text{H2}}=P_i;\ \ \rho_{\text{H2}}=\rho_i.$$

2.3.2 The boundaries conditions

The boundary conditions taken into consideration are:

- Hydrogen is supplied axially to the hydride bed through a porous filter.

$$\left.\frac{\partial T_{\text{H2}}}{\partial X}\right|_{x=0}=0;\quad \left.\frac{\partial P_g}{\partial X}\right|_{x=0}=0.$$

- Wall with heat transfer at the top of tank (forced convection by cold fluid)

$$-\lambda_{\text{eff}}\left.\frac{\partial T_{\text{MH}}}{\partial Y}\right|_{y=H}=h\,(T_{\text{MH}}-T_\infty);\quad \left.\frac{\partial P_g}{\partial Y}\right|_{y=H}=0.$$

- Adiabatic right tank wall:

$$\left.\frac{\partial T_{\text{MH}}}{\partial X}\right|_{x=L}=0;\quad \left.\frac{\partial P_{\text{H2}}}{\partial X}\right|_{x=L}=0.$$

Table 1. Different modeling mechanisms equations [8].

Models equation	Method
$kt=\alpha$	Chemisorption
$kt=[-\ln(1-\alpha)]^{1/2}$	2D/Johnson-Mehl-Avrami-Kolmogorov: two-dimensional growth of existing nuclei with constant interface velocity (JMAK-2D)
$kt=1-\ln(1-\alpha)^{1/3}$	3D/Contracting Volume with constant interface velocity (C V-3D)
$kt=1-\ln(1-\alpha)^{1/2}$	2D/Contraction Volume with constant interface velocity (C V-2D)

with: $k(T,P)=C_a\,e^{\frac{-E_a}{RT}}\frac{P-P_{eq}}{P_{eq}}$ (the Arrhenius law) and E_a, C_a, R are extracted values from the experiment study [11].

2.4 Modeling of thermodynamics and kinetic reaction

The modeling of thermodynamics and kinetics reaction permits the calculation of the spatial and temporal evolution of absorption rate and hydriding velocity into the material. These laws are included in the UDF that allows an iterative calculation at the end of each time step Δt).

There are different mechanisms that can model the reaction kinetics. In this study, we have compared the results of these models in order to know the best mechanism which closely approximates the results obtained from the experiences of the real phenomenon. The considered mechanisms are shown in Table 1.

The terms of the absorption rate ($0<\alpha<1$) and the hydriding velocity $\frac{d\alpha}{dt}$ are derived from the models equations of different reaction.

The law which defines the thermodynamic equilibrium pressure (P_{eq}) between the two domains of existence of Mg and MgH_2 is the Van't Hoff law. It is given by the following expression:

$$P_{eq}(T)=P^\circ\ e^{\left[\frac{\Delta H}{RT}-\frac{\Delta S}{R}\right]}, \tag{5}$$

where $P^\circ=P_{\text{atm}}$, ΔH: the absorption enthalpy, ΔS: the entropy of the reaction.

3 Validation

Our results were compared with the published experimental data. In fact, we performed FLUENT simulations and confronted them with those available in relevant studies such as [12]. Figure 2 shows the profile of the average temperature of the metal hydride bed as a function of elapsed hydrogen absorption time. Figures 3 and 4 present the comparison with the average temperature contours and the hydrogen capacity with the contribution of [13]. Our computational results showed good agreement with those available in literature.

This study aims to model the heat transfer and reaction kinetics during hydrogen loading magnesium hydride tank. FLUENT industrial software was chosen to model this problem. It uses the finite volume method to discretize the equations of

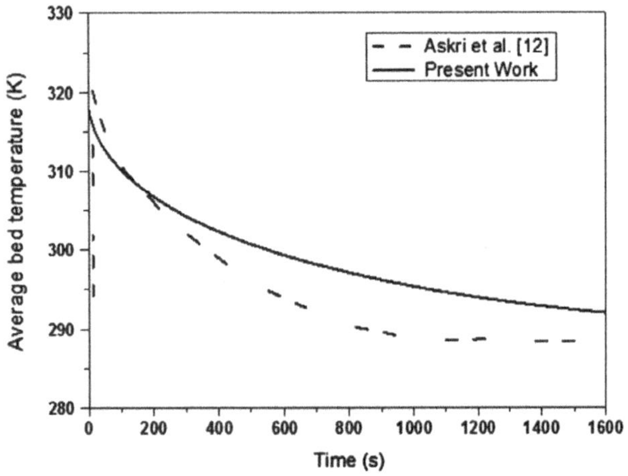

Figure 2. Absorption average temperature profiles with P_{H2} = 3 MPa and Δt = 0.1 s.

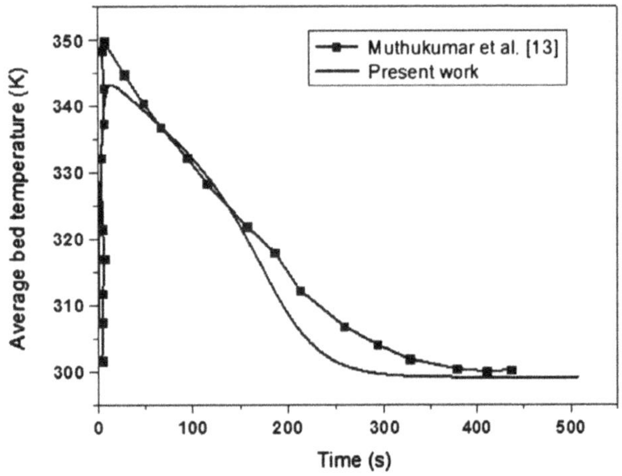

Figure 3. Absorption average temperature profiles with P_{H2} = 2 MPa and Δt = 0.1 s.

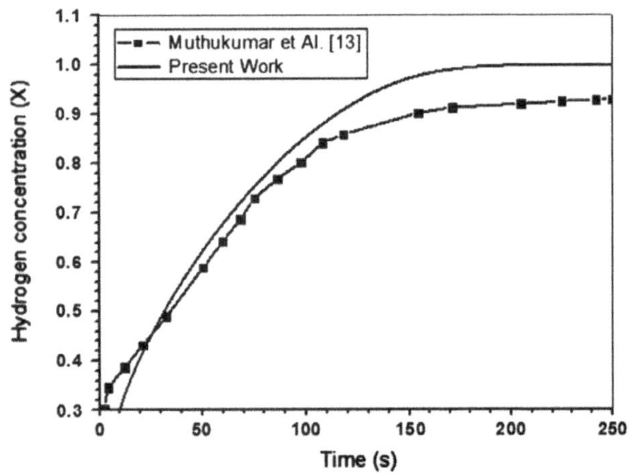

Figure 4. Variation of hydrogen concentration with time.

Table 2. Main parameter values used in computations [10].

Input boundary conditions	
T_i = 573 K	Initial bed temperature
T_f = 300 K	Cold fluid temperature
P_i = 1 MPa	Supply pressure
$h = 10^3$ W/m^2 K	Heat transfer coefficient
Properties of metal alloy	
ρ_m = 1945 kg/m^3	Metal density
Cp_{MH} = 1545 J/kg K	Specific heat of metal
λ_{eff} = 0.48 W/m K	Effective thermal conductivity
ε = 0.77	Porosity
E_a = 130 kJ/mol H2	Activation energy
ΔS = 135.6 J/mol H2 K	Entropy of formation
ΔH = 75 kJ/mol H2	Enthalpy of formation
Properties of hydrogen	
λ_g = 0.127 W/m K	Thermal conductivity of hydrogen
Cp_g = 14 283 J/kg K	Specific heat of hydrogen
ρ_g = 0.0838 kg/m^3	Hydrogen density
Constants	
R = 8.314 J/mol K	Universal gas constant
$k_c = 10^{11}$ s^{-1}	Reaction constant

the mathematical model. Then, it solves in an unsteady case the equations of mass, momentum and energy and includes the kinetic reaction differential equations with the appropriate physical parameters. These physical parameters used in the numerical simulations are presented in Table 2.

4 Results and discussion

Computations were carried out under windows with an Intel core i5 processor. The simulations were executed on several grids and time steps ranging from 0.01 s to 1 s. The mesh corresponding to 396 × 210 nodes and time step $\Delta t = 10^{-2}$ s was therefore adopted for all numerical simulations, in order to optimize the calculating time and the convergence criterion. At the end of the computations, we observed (Figure 5) that the hydrogen absorption phenomenon which starts with preheated discharged tank from 573 K at supply pressure of 10 MPa causes increase in the metal hydride temperature during the first ten seconds of loading and reaches different maximum temperatures from 625 K to 670 K depending on the used kinetic reaction model. This quick temperature increase during tank loading is due to absorption exothermic reactions and hydrogen compression process which produce heat. Then, it slows down gradually and becomes equal to the cooling temperature of 300 K.

Figure 6 shows stratified isothermal contours when x = 50 mm at 52 s using the kinetic reaction method of JMAK. In addition, significant temperature gradients were recorded between the tank surface and its center. These thermal gradients are axial and radial. The temperature reaches a maximum value at the center of the tank, and is lower in the area close to the wall. Finally, the results showed little differences between temperature contours when changing kinetic reaction models.

Figure 7 shows hydrogen capacity profiles with different kinetic reaction equations. Indeed, the hydrogen capacity

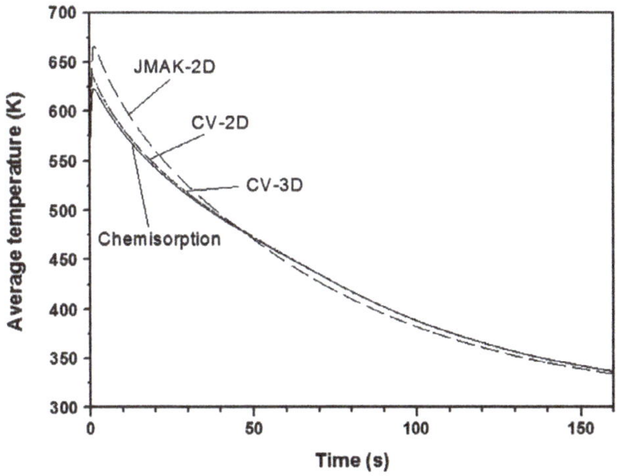

Figure 5. Absorption average temperature profiles with the different kinetic equations methods with $\Delta t = 0.01$ s and at supply pressure of 1 MPa.

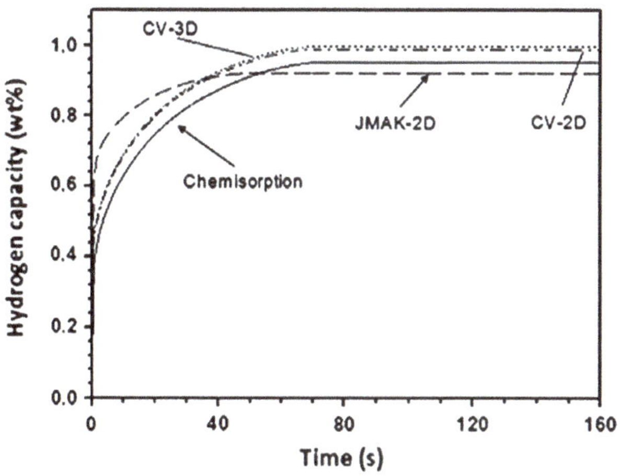

Figure 7. Hydrogen storage capacity profiles at supply pressure of 1 MPa with different methods.

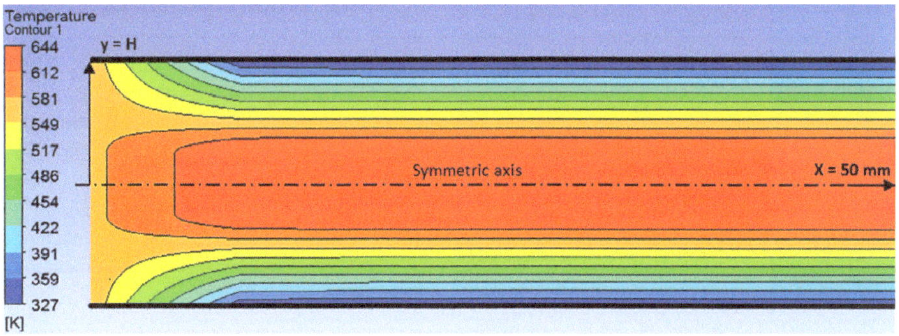

Figure 6. Average bed temperature contours at absorption time $t = 52$ s and $x = 50$ mm with JMAK method (horizontal line is the symmetric axis of geometry).

increases gradually during the first 73 s. Then, it becomes constant after reaching its maximum value.

Since the JMAK method is the mostly used in comparative studies with experimental data, the obtained results show that the CV-2D is the most suitable method in such a case as compared to the other models.

5 Conclusion

In this numerical study, thermal aspects of a two-dimensional metal hydride tank are explored by formulating transient heat transfer by conduction. Four different models for kinetic reaction in porous bed (metal hydride) were introduced by an UDF at Fluent simulations. Fluent uses the finite volume method to discretize the equations generated from the mathematical model. The validation was realized when comparing the obtained results with the experimental data and a good agreement has been observed.

Results have shown an important and quick generated heat at the beginning of the hydrogen absorption process into metal hydride. These temperature gradients between the center and the surface of the tank are axial and radial. Also, little differences between temperature contours were noticed when changing kinetic reaction models. Also, cooling tank becomes necessary to allow the continuation of hydrogen loading.

The use of different equations to model the kinetics of reaction has been found to produce almost similar results. However, the most suitable method for our two-dimensional study is a CV-2D technique because it generates the smallest error especially during the beginning of the reaction. Also, its computational time is the shortest one compared to the other methods (e.g., the JMAK)

Those results are the first step in our efforts to optimize the hydrogen storage tank. The Modeling continues in a 3D perspective and can be extended to predict the best tank cooling mode.

Acknowledgements. Special thanks to "Longwy IUT – France" staff for their welcoming, support and their important contribution to this work.

References

1. Jemni A, Ben Nasrallah S. 1995. Study of two dimensional Heat and mass transfer during absorption in a metal hydrogen reactor. International Journal of Hydrogen Energy, 20(1), 43–52.

2. Dogan A, Kaplan Y, Veziroglu TN. 2004. Numerical investigation of heat and mass transfer in a metal hydride bed, Applied Mathematics and Computation, 150, 169–180.
3. Muthukumar P, Madhavakrishna U, Dewan A. 2007. Parametric studies on a metal hydride based hydrogen storage device. International Journal of Hydrogen Energy, 32, 4988–4997.
4. Phate K, Maiya MP, Murthy SS. 2007. Simulation of transient heat and mass transfer during hydrogen sorption in cylindrical metal hydride beds. International Journal of Hydrogen Energy, 32, 1969–1981.
5. Botzung M, Chaudourne S, Gillia O, Perret C, Latroche M, Percheron-Guegan A, Marty P. 2008. Simulation and experimental validation of a hydrogen storage tank with metal hydrides. International Journal of Hydrogen Energy, 33, 98–104.
6. Askri F, Ben Salah M, Jemni A, Ben Nasrallah S. 2009. Heat and mass transfer studies on metal-hydrogen reactor filled with $MmNi_{4.6}Fe_{0.4}$. International Journal of Hydrogen Energy, 34, 6705–6711.
7. Ye F, Xiao J, Hu B, Bénard P, Chahine R. 2012. Implementation for model of adsoptive hydrogen storage using UDF in fluent. Physics Procedia, 24, 793–800.
8. Varin R, Czujkon T, Wronski ZS. 2009. Nanomaterials for Solid State Hydrogen Storage Fuel cells and hydrogen energy. Springer, ISBN 978-0-387-77711-5.
9. Muthukumar P, Prakash Maiya M, Murthy SS, Vijay R, Sundaresan R. 2008. Tests on mechanically alloyed Mg_2Ni for hydrogen storage. Journal of Alloys and Compounds, 452, 456–461.
10. Chaise Albin. 2008. Etude expérimentale et numérique des réservoirs à hydrures de magnésium. Ph.D. dissertation, Material Physics speciality, Joseph Fourier University.
11. Chaise A, Marty P, de Rango D, Fruchart D. 2009. A simple criterion for estimating the effect of pressure gradients during hydrogen absorption in a hydride reactor. International Jounal of Heat and Mass Transfer, 52, 4564–4572.
12. Askri F, Ben Salah M, Jemni A, Ben Nasrallah S. 2009. Heat and mass transfer studies on metal-hydrogen reactor filled with $MmNi_{4.6}Fe_{0.4}$. International Journal of Hydrogen Energy, 34, 6705–6711.
13. Muthukumar P, Venkata Ramana S. 2009. Numerical simulation of coupled heat and mass transfer in metal hydride-based hydrogen storage reactor. Journal of Alloys and Compounds, 472, 466–472.
14. Botzung M, Chaudourne S, Gillia O, Perret C, Latroche M, Percheron-Guegan A, Marty P. 2008. Simulation and experimental validation of a hydrogen storage tank with metal hydrides. International Journal of Hydrogen Energy, 33, 98–104.
15. Yang F, Zhang Z. Febuary2011. Simulation studies on the coupling process of heat/mass transfer in a metal hydride reactor. Mass Transfer in Multiphase Systems and its Applications, Pub. 11
16. Gambini M, Manno M, Vellini M. 2008. Numerical analysis and performance assessment of metal hydride-based hydrogen storage systems. International Journal of Hydrogen Energy, 33, 6178–6187.
17. Hermosilla-Lara G, Momen G, Marty PH, Le Neindre B, Hassouni K. 2007. Hydrogen storage by adsorption on activated carbon: investigation of the thermal effects during the charging process. Investigation of the International Journal of Hydrogen Energy, 32, 1542–1553.
18. Marty P, Fourmigue JF, de Rango P, Fruchart D, Charbonnier J. 2006. Numerical simulation of heat and mass transfer during the absorption of hydrogen in a magnesium hydride. Energy Conversion and Management, 47, 3632–3643.

16

Implementation of modified artificial bee colony (ABC) optimization technique for minimum cost design of welded structures

Vimal Savsani*

Department of Mechanical Engineering, Pandit Deendayal Petroleum University, Raysan, Gandhinagar 382 007, Gujarat, India

Abstract – Welded structures are widely used in many engineering load carrying structures such as columns, towers for wind turbine or water tanks, offshore and submarine structures, girders, stiffened doors, etc. welding is a costly fabrication process and proper sequence and welding process effect the cost of a huge structure. Keeping this in view two different problems from the literature for the cost optimization of welded structures are considered in this paper. The optimization procedure is carried out using artificial bee colony (ABC) optimization technique. Classical ABC is modified to increase the convergence rate of the original algorithm. Comparison of both the variants is experimented on many bench mark examples from the literature and also on two cost optimization problems of welded structures. The results of the considered techniques are compared with the previously published results. The considered techniques have given much better results in comparison to the previously tried approaches and also modified ABC has shown superiority over classical ABC.

Key words: Welded structures, Optimum design, Cost optimization, Artificial bee colony.

1 Introduction

Stiffeners are required to improve the load carrying capacity of structures. Stiffeners help to achieve lightweight and robust design of the structure. For this purpose they have wide use in structural engineering domain. Specially, stiffened plates are used in critical and sensitive structures such as in aircrafts, ship hulls and box girders in which safety and a perfect design is crucial. Buckling is the one of the most complex phenomenon that is inevitable for heavily axially loaded stiffened plate structures. The wide area of application of such structures is an aircraft industry where light, stiff and highly resistant structures are required. It is possible to manufacture aircraft elements as thin panels reinforced by stiffeners. The proper stiffeners arrangement in a domain of the structure decides about the effectiveness of the construction or about the effectiveness of reinforcement. Optimal properties of structures can be searched using optimization tools which are discussed in the later part of this paper.

For steel structures a general total cost function K_T can be defined in the following form [19]

$$K_T = K_m + K_f + K_t + K_e, \qquad (1)$$

where K_m is the material cost of structural members, K_f is the fabrication cost, K_t is the cost of transporting the fabricated pieces to the construction field, and K_e is the erection cost. Many studies are reported in literature for the cost optimization of different structures.

Anderson and Chong [4] presented the minimum cost design of homogeneous and hybrid stiffened steel plate girders according to the AISC code [2]. Douty [13] described the minimum cost design of three different types of bolted and welded connections used in steel frames based on the AISC specifications [1] The cost function is presented in terms of the connection variables, such as the diameter of the bolts, flange plate width and thickness, shear plate length and thickness, and the leg size of the fillet weld for connecting the shear plate to the column flange and for the flange plate moment connection. Lorenz [25] discussed the minimum cost design of composite beams based on the AISC load and resistance factor design (LRFD) code [3]. Farkas [15] presents closed-form solutions for optimum cost values of the cross-sectional variables for simply supported welded box girders subjected to a uniformly distributed load and constraints on bending stress, local flange and web buckling, shear fatigue for longitudinal fillet welds, and deflection. Farkas and Jarmai [16] presented the minimum cost design of laterally loaded welded cellular steel plates using

*e-mail: vimal.savsani@gmail.com

three different approaches: the backtracking method, the hill-climbing method, and feasible sequential quadratic programming. Bhatti [7] presented the minimum cost design of simply supported composite I-shaped steel beams with concrete slabs subjected to a uniformly distributed load, and strength, deflection, and vibration constraints using the Lagrange multiplier approach. Farkas [18] derived a differential equation for the radial shell deformation using the bending theory of cylindrical shells. The solution of this equation gives an approximate formula for the maximum radial deformation. Comparing this formula with the limiting deformation given by ECCS allows calculating the required shell thickness. Another thickness can be obtained using ECCS buckling formulae. Using an interpolation, a thickness can be calculated, which fulfills the constraints. Jarmai et al. [21] demonstrated the optimal design of a cylindrical orthogonally stiffened shell member of an offshore fixed platform truss, loaded by axial compression and external pressure. Ring stiffeners of welded box section and stringers of halved rolled I-section were used. The design variables considered in the optimization were the shell thickness as well as the dimensions and numbers of stiffeners. The design constraints relate to the shell, panel ring and panel stringer buckling, as well as manufacturing limitations. The cost function includes the cost of material, forming of plate elements into cylindrical shape, welding and painting. In the optimization a number of relatively new mathematical optimization methods such as leap-frog – LFOPC, Dynamic-Q, ETOPC, and Particle swarm optimization – PSO were used. Jarmai [20] presented the single and multiobjective optimization of a welded stringer-stiffened cylindrical steel shell. A column fixed at the bottom and free on the top was constructed of stringer-stiffened cylindrical shell and loaded by axial compression as well as by horizontal force acting on the top. Halved rolled I-section stringers were welded outside of the shell by longitudinal fillet welds. Design Variables taken were the shell thickness as well as dimension and number of stringers. Three different objective functions were considered viz. material cost, forming cost to form the shell elements into the cylindrical shape, assembly and welding, and painting cost. The original PSO algorithm was modified to handle multiobjective optimization techniques and to find discrete values of design variables. Luis et al. [26] presented the optimization of a uniaxially compressed stiffened plate subjected to static and fatigue loading. The design variables were the thickness of the base plate, the number and dimension of stiffeners of the orthogonally stiffened plate. Cost optimization was carried out considering material, assembly, welding and painting costs. Farkas and Jarmai [14] presented the cost minimization of a cantilever stub column of square box cross section with orthogonally stiffened side plates. Design variables include the thickness and width of the side plates as well as the dimensions and numbers of stiffeners. PSO was implemented for the cost optimization. Marcin et al. [27] presented the optimal configuration of stiffening ribs in the steel welded I-beams and columns. The configuration with diagonal ribs is proved better than traditional ones. Farkas et al. [17] used refined version of the original Snyman-Fatti (SF) global continuous optimization algorithm to the optimal design of welded square stiffened plates. Square plates of square symmetry subjected to uniformly distributed normal static loads, supported at four corners, and stiffened by a square symmetrical orthogonal grid of ribs was investigated. Halved rolled I-section stiffeners are used welded to the base plate by double fillet welds. The cost function includes material, welding as well as painting costs, and is formulated according to the fabrication sequence. Design variables include base plate thickness as well as the dimensions of the edge and internal stiffeners. Constraints on stress in the base plate and in stiffeners, as well as on deflection of edge stiffeners and of internal stiffeners were considered. PSO algorithm was applied to confirm the results given by the SF algorithm.

It has been observed from the review of research work on design optimization of welded structures that the recently used advanced optimization technique is PSO. PSO requires many controlling parameters such as inertia weight, maximum velocity, and learning factors which influence the performance of the algorithm. Therefore the efforts are continuing to use more different optimization algorithms, which are more powerful, robust and able to provide accurate solution. This paper is intended to apply one such optimization algorithms known as artificial bee colony (ABC) [22] optimization technique. In the present work, an effort is made to verify if any improvement in the solution is possible by employing ABC optimization technique to two different cost optimization problems for welded structures reported in the literature. Moreover ABC optimization technique is the most recently developed technique. It is not so far tried for the complex optimization problems for structural design. The next section gives a brief description of ABC optimization techniques.

2 Artificial bee colony (ABC) techniques

Artificial bee colony (ABC) Algorithm is an optimization algorithm based on the intelligent foraging behaviour of honey bee swarm. The colony of artificial bees consists of three groups of bees: employed bees, onlookers and scouts [5, 22].

The colony of the artificial bees is divided into two groups, first half of the colony consists of the employed artificial bees and the second half includes the onlooker bees. Scout bees are the employed bee whose food source has been abandoned. In ABC algorithm, the position of a food source represents a possible solution to the optimization problem (value of design variables) and the nectar amount of a food source corresponds to the quality of the associated solution (value of objective function). At the first step, the ABC generates a randomly distributed initial population P_{initial} of N solutions, where N denotes the size of population. Each solution x_i is an S-dimensional vector where S is the number of optimization parameters (design variables). After initialization, the population of the solutions is subjected to repeated cycles, $C = 1, 2, \ldots, G$, of the search processes of the employed bees, the onlooker bees and scout bees. An employed bee which is equal to $N/2$ produces a modification on the solution in her memory depending on the local information. If the objective function value (fitness) of the new solution is higher than that of the previous one, the bee memorizes the new position and forgets the old one. Otherwise she keeps the position of the previous one in her memory. After all employed bees complete the search process; they share the nectar information of the food sources and their position information with the onlooker bees on the dance area. An onlooker bee evaluates the fitness information taken from all employed bees and chooses a food source with a probability related to its fitness value. An onlooker bee also produces a new solution and it memorizes the new position if its fitness value is better than the previous position. An artificial onlooker bee chooses a food source depending on the probability value associated with that food source, p_i, calculated by the following expression

$$p_i = \frac{F_i}{\sum_{n=1}^{N_b} F_n}, \qquad (2)$$

where F_i is the fitness value of the solution i which is proportional to the nectar amount of the food source in the position i and N_b is the number of food sources which is equal to the number of employed bees. In order to produce a candidate food position from the old one in memory, the ABC uses the following expression (3):

$$v_{ij} = x_{ij} + R_{ij}(x_{ij} - x_{kj}), \qquad (3)$$

where $k \in \{1, 2, \ldots, N\}$ and $j \in \{1, 2, \ldots, D\}$ are randomly chosen indexes. Although k is determined randomly, it has to be different from i. R_{ij} is a random number between $(-1, 1)$. It controls the production of neighbour food sources around x_{ij} and represents the comparison of two solutions. As can be seen from (3), as the difference between the parameters of the x_{ij} and x_{kj} decreases, the perturbation on the position x_{ij} gets decrease, too. Thus, as the search approaches to the optimum solution in the search space, the step length is adaptively reduced.

If the position of the food source cannot be improved for some predetermined number of cycles than that food source is abandoned. The abandoned food source is replaced with a new food source by the scouts. In ABC, this is simulated by producing a position randomly and replacing it with the abandoned one. The value of predetermined number of cycles is an important control parameter of the ABC algorithm, which is called "*limit*" for abandonment. The value of limit is generally taken as *Number of employed bees**S* [23, 24].

Assume that the abandoned source is x_i and $j \in \{1, 2, \ldots, D\}$, then the scout discovers a new food source to be replaced with x_i. This operation can be defined as in (4)

$$x_i^j = x_{\min}^j + \text{rand}(0,1)(x_{\max}^j - x_{\min}^j). \qquad (4)$$

It is clear from the above explanation that there are three control parameters used in the ABC: The number of food sources which is equal to the number of employed or onlooker bees (N), the value of *limit*, the maximum cycle number (G).

As suggested by Karaboga [22] R_{ij} is uniformly distributed random number between –1 and 1. In this paper ABC is modified by using normally distributed random number with mean 0 and variance 1 instead of uniformly distributed random number between –1 and 1. The distributions of both the random numbers generated using MATLAB with 1e7 random points are shown in Figures 1 and 2 respectively.

Another modification is made in equation (3) by replacing $(x_{ij} - x_{ik})$ by $(x_{bestj} - x_{ij})$. Here x_{bestj} is the best solution from the population. The reason is that in the earlier formula the solution tries to move towards any random solution (x_{ik}). There is no guarantee for the x_{ik} to be better than x_{ij}. So solution can move towards worst solution also, which may require more computational time to reach the optimum solution. By replacing x_{ik} with x_{bestj}, the solution will always move towards the best solution in each iteration, which will lead to global optimum solution in less computational effort. Moreover effort is made to increase the convergence rate of the original ABC algorithm. Many experiments were conducted to change equation (3) so that algorithm converges to optimum solution in less number of generations. In this paper original ABC will be referred as ABC_1 and modified ABC as ABC_2. Experiments shows that by using ABC_2 solutions improves by approximately 87% and by using ABC_1 solutions improves by 39% in successive first five generations which indicates the fast convergence rate of modified ABC. Convergence rate for the benchmark Sphere function $\left(f(X) = \sum_{i=1}^{n} x_i^2, x \in [-100\,100],\ i = 20\right)$ for five different runs for both the variants of ABC is shown in Figures 3 and 4. From Figures 3 and 4 also it can be noted that modified ABC have better convergence than classical ABC.

The algorithm discussed above takes real number in the aspect of representation solution and updates themselves with

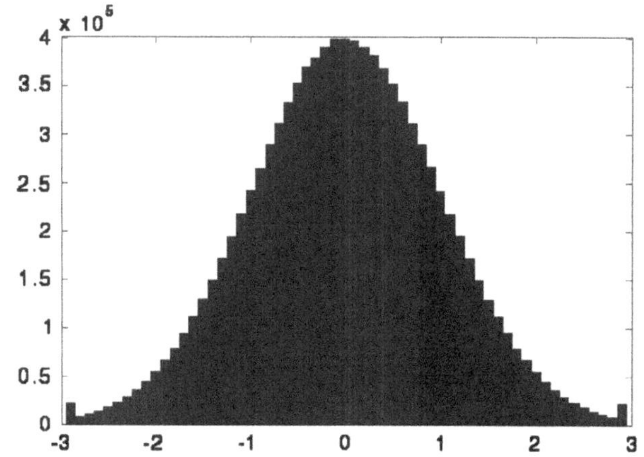

Figure 1. Normal distribution of 1e7 random numbers with mean 0 and variance 1.

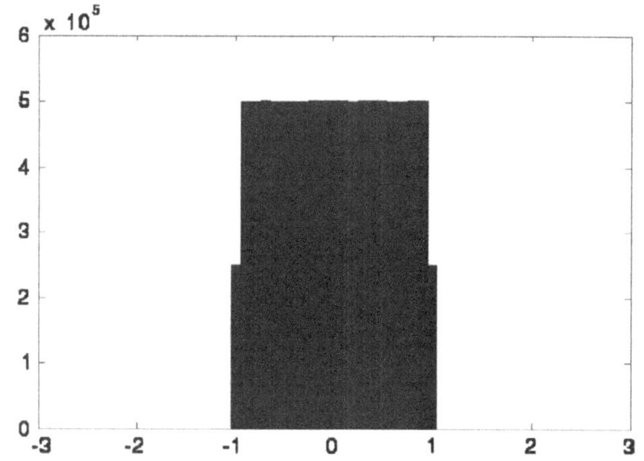

Figure 2. Uniformly distributed random number between –1 and 1 for 1e7 random points.

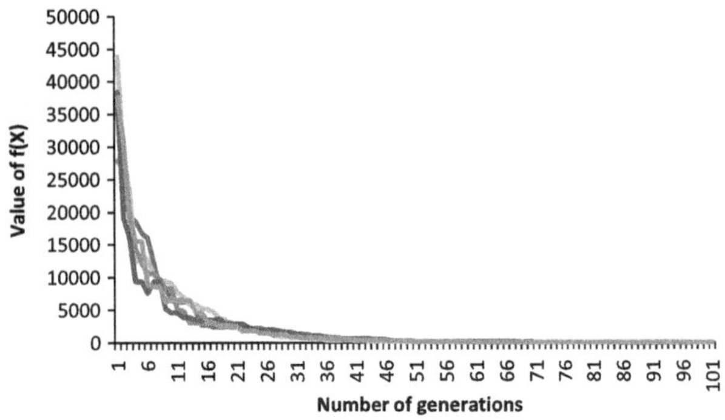

Figure 3. Variation of solution for the sphere function for classical ABC for five different runs.

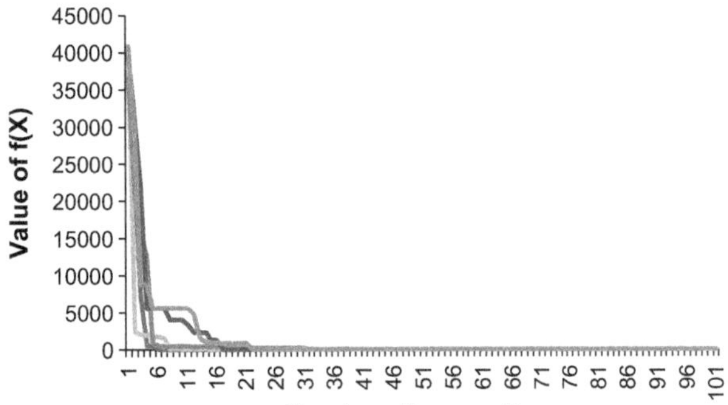

Figure 4. Variation of solution for the sphere function for modified ABC for five different runs.

certain rules based on the behavior of algorithm. In this algorithm, process looks only for the best solution and tends to converge to the best solution. In the implementation process, solutions are randomly generated at the beginning and usually violate the system constraints resulting in infeasible solutions. Therefore, the handling of system constraints, particularly nonlinear equation constraints, and the measurement and evaluation of infeasible solutions is very important. To cope with constrained problems with evolutionary computation, various approaches such as rejection of infeasible individuals, repair of infeasible individuals, replacement of individuals by their repaired versions, and penalty function methods can be adopted. Among them, the penalty function method is particularly promising as evidenced by recent developments [12] and the same is adapted in the present work. Penalty function changes the constrained optimization problem into unconstrained optimization problem. For example consider an optimization problem as

$$\text{Minimize}: f(X), \text{ Subjected to}: g_i(X) \geq 0. \quad (5)$$

This problem is converted in unconstrained form as

$$\text{Minimize}: f(X), \text{ if } g_i(X) \geq 0, \text{ else}, \quad (6)$$

$$\text{Minimize}: f(X) + R\sum_{i=1}^{n} g_i(X), \quad (7)$$

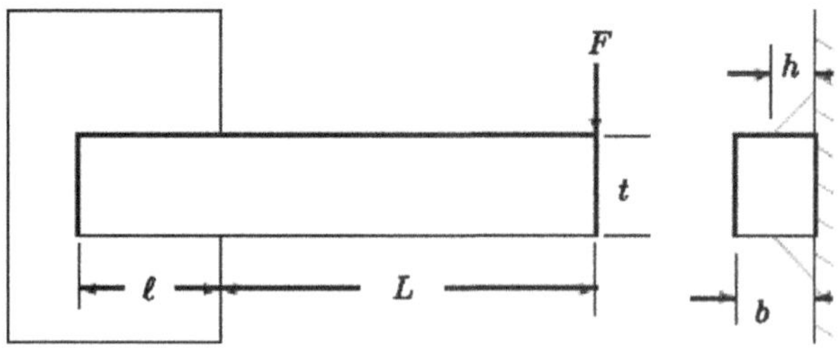

Figure 5. Welded beam design (Example 1).

where R is very large number and n is number of constraints.

The next section presents examples to demonstrate and validate the considered optimization techniques for the cost optimization of welded structures.

3. Examples

To check the performance of ABC some classical benchmark problems from Karaboga and Basturk [24] were experimented. In experiments, Schaffer function has two parameters ($-100 \leq X \leq 100$), while Sphere ($-100 \leq X \leq 100$), Griewank ($-600 \leq X \leq 600$), Rastrigin ($-5.12 \leq X \leq 5.12$) and Rosenbrock ($-50 \leq X \leq 50$) functions have 20 parameters each. Moreover some engineering benchmark problems were also taken for the investigation. They include pressure vessel problem [32] consisting of four design variable and four constraints, Spring weight minimization problem [6]) consisting of three design variable and four constraints, and speed reducer problem [28] consisting of 7 design variables and 11 constraints. As this paper focuses on the optimization of welded structures two special problems of welded structure optimization is taken from the literature and is given in detail as follows. More detailed information can be obtained from the mentioned reference. Experiments were conducted for different population size and number of generations. In this paper N50G200 indicates population size of 50 with number of generation of 200. Results were obtained for 50 different runs.

3.1 Example – 1 welded beam

The objective is to design a welded beam for minimum cost. There are four design variables as shown in Figure 5: h (x_1), l (x_2), t (x_3) and b (x_4). Design is subjected to the constraints on shear stress (τ), bending stress in the beam (σ), buckling load on the bar (P_c), end deflection of the beam (δ), and side constraints.

The problem can be stated as follows:

$$\text{Minimize}: f(x) = 1.10471x_1^2x_2 + 0.04811x_3x_4(14.0 + x_2). \quad (8)$$

Subject to:

$$g_1(x) = \tau(x) - \tau_{\max} \leq 0, \quad (9)$$

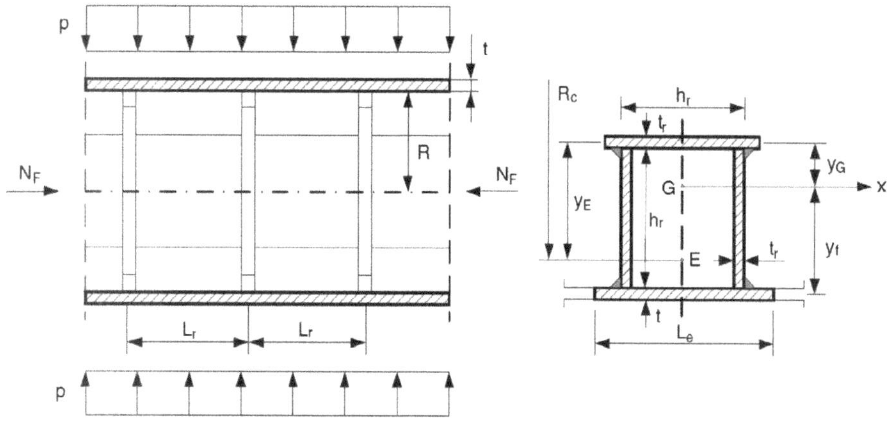

Figure 6. Orthogonally stiffened cylindrical shell with stringer and ring stiffener acted by compression and external pressure [21].

$$g_2(x) = \sigma(x) - \sigma_{\max} \le 0, \tag{10}$$

$$g_3(x) = x_1 - x_4 \le 0, \tag{11}$$

$$g_4(x) = 0.10471x_1^2 + 0.04811x_3x_4(14.0 + x_2) - 5.0 \le 0, \tag{12}$$

$$g_5(x) = 0.125 - x_1 \le 0, \tag{13}$$

$$g_6(x) = \delta(x) - \delta_{\max} \le 0, \tag{14}$$

$$g_7(x) = P - P_c(x) \le 0, \tag{15}$$

where

$$\tau(x) = \sqrt{(\tau')^2 + 2\tau'\tau''\frac{x_2}{2R} + (\tau'')^2}, \tag{16}$$

$$\tau' = \frac{P}{\sqrt{2}x_1x_2}, \tag{17}$$

$$\tau'' = \frac{MR}{J}, \tag{18}$$

$$M = P\left(L + \frac{x_2}{2}\right), \tag{19}$$

$$R = \sqrt{\frac{x_2^2}{4} + \left(\frac{x_1 + x_3}{2}\right)^2}, \tag{20}$$

$$J = 2\left[\sqrt{2}x_1x_2\left\{\frac{x_2^2}{12} + \left(\frac{x_1 + x_3}{2}\right)^2\right\}\right], \tag{21}$$

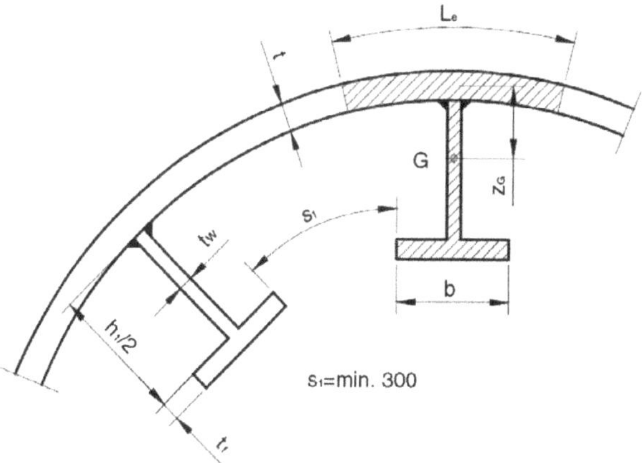

Figure 7. Halved rolled I – section as a stringer [21].

$$\sigma(x) = \frac{6PL}{x_4x_3^2}, \tag{22}$$

$$\delta(x) = \frac{4PL^3}{Ex_3^3x_4}, \tag{23}$$

$$P_c(x) = \frac{4.013E\sqrt{\frac{x_3^2x_4^6}{36}}}{L^2}\left(1 - \frac{x_3}{2L}\sqrt{\frac{E}{4G}}\right)$$
$$P = 6000lb, L = 14in, E = 30e6psi, G = 12e6psi,$$
$$\tau_{\max} = 13600psi, \sigma_{\max} = 30000psi, \delta_{\max} = 0.25in$$
$$0.1 \le x_1 \le 2.0, 0.1 \le x_2 \le 10.0, 0.1 \le x_3 \le 10.0, 0.1 \le x_4 \le 2.0 \tag{24}$$

3.2 Example-2 orthogonally stiffened cylindrical shell

Objective is to optimize cost of a cylindrical shell member that is orthogonally stiffened by using ring stiffeners of box cross section and stringers of halved I-section Figures 6 and 7.

The cost function includes material cost, manufacturing cost and painting cost.

Cost function is given by [21]

$$K = K_M + \sum_i K_{Fi} + K_P, \tag{25}$$

where K_M is the material cost, K_F is the fabrication cost and K_P is the painting cost.

The manufacturing cost is derived from the manufacturing sequence which is explained as follows:

(K_{F0}) Cylindrical shapes are formed from plate elements of 3 m length (L_s).

$$K_{F0} = 5k_F \theta e^{\mu}. \tag{26}$$

(K_{F1}) 3 m shell segments are welded from two curved plate elements with two butt welds using gas metal arc welding.

$$K_{F1} = 5k_F(\theta\sqrt{2\rho V_1} + (1.3)(0.1520X10^{-3}t^{1.9358})(2L_s). \tag{27}$$

(K_{F2}) 15 m un-stiffened shell is welded from five shell segments with four circumferential butt welds using gas metal arc welding.

$$K_{F2} = k_F(\theta\sqrt{25\rho V_1} + (1.3)(0.1520X10^{-3}t^{1.9358})(8R\pi). \tag{28}$$

(K_{F3}) nr ring stiffeners are welded from three plate elements with two fillet welds each using shielded metal arc welding.

$$K_{F3} = n_r k_F\left[3\sqrt{3\rho V_R} + 1.3X0.3394X10^{-3}a_{wr}^2 4\pi(R-h_r)\right]. \tag{29}$$

(K_{F4}) nr ring stiffeners are welded into the whole shell with 2 nr circulferential fillet welds using shielded metal arc welding.

$$K_{F4} = k_F\left[3\sqrt{(n_r+1)\rho(5V_1+n_rV_R)} + 1.3X0.3394X10^{-3}a_{wr}^2 n_r 4\pi R\right]. \tag{30}$$

(K_{F5}) ns stringers are welded into the shell with 2 ns fillet welds using shielded metal arc welding.

$$K_{F5} = k_F\left[3\sqrt{(n_r+n_s+1)\rho(5V_1+n_rV_R+n_sA_sL)} + 1.3X0.3394X10^{-3}a_{ws}^2 n_s 2L\right]. \tag{31}$$

The material cost is given by

$$K_M = k_{M1}5\rho V_1 + k_{M1}\rho n_r V_R + K_{M2}\rho n_s A_s L, \tag{32}$$

where V_1 is the volume of the shell segment and V_R is the volume of ring stiffener.

The painting cost is given by

$$K_P = k_P\Big[2R\pi L + 2R\pi(L-n_rh_r) + 2n_r\pi h_r(R-h_r) + 4\pi Rh_r\left(R-\frac{h_r}{2}\right) + n_sL(h_1+2b)\Big]. \tag{33}$$

Shell buckling constraint $g_1(X)$ associated with the design is given by [11]

$$\sigma e = \sqrt{\sigma_a^2 - \sigma_a\sigma_p + \sigma_p^2} \le f_{y1}/\sqrt{1+\lambda_s^4}, \tag{34}$$

where σ_e is the equivalent stress according to Det Norske Veritas rules. And $\sigma_a = N_F/2R\pi t_e$; $t_e = t + A_{ss}$; $s = 2R\pi/n_s$; $\sigma_p = p_F R/t(1+\alpha)$; $\alpha = A_R/L_{e0}t$; $L_{e0} = \min(L_r, L_{er}) = 1.56\sqrt{Rt}$; $L_r = \frac{L}{n_r} - 1$; $\lambda_s^2 = \frac{f_{y1}}{\sigma_e}\left(\frac{\sigma_a}{\sigma_{Eas}} + \frac{\sigma_p}{\sigma_{Eps}}\right)$; $\sigma_{Eas} = C_{as}\frac{\pi^2 E}{12(1-\gamma^2)}\left(\frac{t}{s}\right)^2$; $C_{as} = \Psi_{as}\sqrt{1+\left(\frac{\rho_{as}\xi_{as}}{\Psi_{as}}\right)^2}$; $\Psi_{as} = 4$; $Z_{as} = \frac{s^2}{Rt}\sqrt{1-\nu^2}$; $\xi_{as} = 0.702Z_{as}$; $\rho_{as} = 0.5\left(1+\frac{R}{150t}\right)^{-0.5}$; $\sigma_{Eps} = C_{ps}\frac{\pi^2 E}{10.92}\left(\frac{t}{s}\right)^2$; $C_{ps} = \Psi_{ps}\sqrt{1+\left(\frac{\rho_{ps}\xi_{ps}}{\Psi_{ps}}\right)^2}$; $\rho_{ps} = 0.$; $\xi_{ps} = 1.04\frac{s}{L_r}\sqrt{Z_{ps}}$; $Z_{ps} = Z_{as}$; $\Psi_{ps} = \left[1+\left(\frac{s}{L_r}\right)^2\right]^2$.

Stringer buckling constraint $g_2(X)$ is given by

$$\sigma_e \le \frac{f_{y1}}{\sqrt{1+\lambda_p^4}}, \tag{35}$$

where, $\lambda_p^2 = \frac{f_{y1}}{\sigma_e}\left(\frac{\sigma_a}{\sigma_{Eap}} + \frac{\sigma_p}{\sigma_{Epp}}\right)$; $\sigma_{Eap} = C_{ap}\frac{\pi^2 E}{10.92}\left(\frac{t}{L_r}\right)^2$; $C_{ap} = \Psi_{ap}\sqrt{1+\left(\frac{\rho_{ap}\xi_{ap}}{\Psi_{ap}}\right)^2}$; $\rho_{ap} = 0.5$; $\xi_{ap} = 0.702Z_{ap}$; $Z_{ap} = \frac{L_r^2}{Rt}0.9539$; $\Psi_{ap} = \frac{1+\gamma_s}{1+\frac{A_s}{s_e t}}$; $\gamma_s = 10.92\,\frac{I_{sef}}{st^3}$; $s_E = 1.9t\sqrt{\frac{E}{f_y}}$; $Z_G = \frac{\frac{h_1}{2}t_w\left(\frac{h_1}{4}+\frac{t}{2}\right)+bt_f\left(\frac{h_1+t+t_f}{2}\right)}{s_e t + bt_f + \frac{h_1 t_w}{2}}$; if $s_E \le s$, $s_e = s_E$, else $s_E \ge s$, $s_e = s$; $I_{sef} = s_e t z_G^2 + \left(\frac{h_1}{2}\right)^3\frac{t_w}{12} + \frac{h_1 t_w}{2}\left(\frac{h_1}{4}+\frac{t}{2}-z_G\right)^2 + bt_f\left(\frac{h_1+t+t_f}{2}-z_G\right)^2$; $A_s = bt_f + \frac{h_1 t_w}{2}$, $\sigma_{Epp} = C_{pp}\frac{\pi^2 E}{10.92}\left(\frac{t}{L_r}\right)^2$, $C_{pp} = \Psi_{pp}\sqrt{1+\left(\frac{\rho_{pp}\xi_{pp}}{\Psi_{pp}}\right)^2}$; $\xi_{pp} = 1.04\sqrt{Z_{pp}}$, $Z_{pp} = Z_{ap}$, $\rho_{pp} = 0.6, \Psi_{pp} = 2(1+\sqrt{1+\gamma_s})$.

Panel ring buckling constraints $g_3(X)$ and $g_4(X)$ are given by

$$A_{Rreq} \le A_R \text{ and } I_{Rreq} \le I_R$$

where, $\frac{1}{\delta_r} = 42\varepsilon$, $\varepsilon = \sqrt{\frac{235}{f_y}}$; $f_y = 355$; $\delta_r = 1/34$; $A_r = 3h_rt_r = 3\delta_r h_r^2$; $A_{Rreq} = \left(\frac{2}{Z^2}+0.06\right)L_r t$; $Z = \frac{L_r^2}{Rt}0.9539$; $L_e = \min\left(L_r, 2\times 1.56\sqrt{Rt}\right)$; $y_E = \frac{L_e t\left(h_r+\frac{t}{2}\right)+\delta_r h_r^3}{3\delta_r h_r^2 + L_e t}$; $I_R = \frac{\delta_r h_r^4}{6} + 2\delta_r h_r^2\left(\frac{h_r}{2}-y_E\right)^2 + \delta_r h_r^2 y_E^2 + L_e t\left(h_r+\frac{t}{2}-y_E\right)^2$; $I_{Rreq} = I_a + I_p$; $I_a = \frac{\sigma_a t\left(1+\frac{A_s}{st}\right)R_0^4}{500EL_R}$; $R_0 = R-(h_r-y_E)$; $I_p = \frac{p_F R R_0^2 L_r}{3E}\left[2+\frac{3Ey_E\delta_0}{R_0^2\left(\frac{f_y}{2}-\sigma_p\right)}\right]$; $\delta_0 = 0.005R$.

Manufacturing constraint $g_5(X)$ is given by

$$\frac{2\left(R-\frac{h_r}{2}\right)\pi}{n_s} - b \ge 300, \text{mm}. \tag{36}$$

Table 1. Comparison of two different variants of ABC for different benchmark problems.

		ABC_1			ABC_2		
Benchmark problems	Function evaluations	Best	Mean	*SD*	Best	Mean	*SD*
Sphere	10000	0.000088	0.00037112	1.95E-04	**0**	**0.0000636**	**1.13E-04**
Rastrigin	10000	12.144862	26.75682439	9.65E+00	**0**	**1.21746087**	**4.25E+00**
Rosenbrock	200000	17.093176	51.52220345	9.34E+01	**0.201002**	**12.22409523**	**7.05E+00**
Griewank	20000	0.001143	**0.010804619**	**9.20E-03**	**0.000373**	0.03364881	5.19E-02
Schaffer	10000	**0**	0.0085749	3.03E-03	**0**	**0.00609025**	**4.69E-03**
Pressure vessel	10000	**6059.714335**	**6059.71437**	**1.53E-04**	**6059.714335**	6061.566119	4.53E+00
Spring	5000	0.012676	0.012681389	**2.17E-06**	**0.012667**	**0.0126805**	4.94E-06
Gear box	5000	2996.351945	2996.360606	6.56E-03	**2996.348657**	**2996.349589**	**7.86E-04**

Table 2. Comparison of different variants of ABC algorithm for Example 1 (welded beam) with different function evaluations for the best solutions obtained in 100 runs.

	ABC_1			ABC_2		
FE	Best	Mean	*SD*	Best	Mean	*SD*
1000	2.118958	2.666595	0.314317	1.749212	2.222916	0.455776
2000	1.814226	2.392055	0.259713	1.725834	2.114724	0.480265
4000	1.790327	2.099398	0.194092	1.72486	2.029395	0.368859
2500	1.993107	2.453174	0.231802	1.728703	1.94462	0.249578
5000	1.892304	2.188238	0.162927	1.724973	1.881657	0.191329
10000	1.802669	1.999135	0.105969	1.724855	1.77547	0.104087

Table 3. Comparison of different variants of ABC algorithm for Example 2 (Orthogonally stiffened cylindrical shell) with different function evaluations for the best solutions obtained in 100 runs.

	ABC_1			ABC_2		
	Best	Mean	*SD*	Best	Mean	*SD*
1000	55146.36	56941.54	907.9698	54530.05	56518.38	963.8605
2000	54668.13	55832.04	702.165	54468.28	55730.39	732.8504
4000	54395.24	55025.25	488.9624	54370.12	55076.65	745.5146
2500	54821.67	56497.6	695.8313	54445.53	55616.16	845.8954
5000	54657.63	55523.57	476.3063	54372.41	54980.24	690.8988
10000	54429.48	54738.18	237.6145	54369.73	54686.88	593.0451

Table 4. Comparison of different variants of ABC algorithm for Example 1 (welded beam) with different function evaluations for the success rate obtained in 100 runs.

	ABC_1	ABC_2
1000	0	0
2000	0	0.19
4000	0	0.2
2500	0	0.08
5000	0	0.27
10000	0	0.61

Table 5. Comparison of different variants of ABC algorithm for Example 2 (Orthogonally stiffened cylindrical shell)) with different function evaluations for the success rate obtained in 100 runs.

	ABC_1	ABC_2
1000	0	0.07
2000	0.06	0.35
4000	0.59	0.78
2500	0.01	0.3
5000	0.09	0.73
10000	0.76	0.87

Example 1 is solved by many researchers using different optimization techniques, including Ragsdell and Phillips [30] using geometric programming, Deb [10] using Genetic Algorithm, Coello [9] using GA based coevolution model, Coello and Montes [8] using flexibility based tournament selection scheme, Ray and Liew [31] using society and civilization algorithm and Qie and Ling [29] using effective co-evolutionary particle swarm optimization.

Jarmai et al. [21] solved Example 2 using PSO technique for the cost optimization considering the above design variables, objective functions, and the constraints. Value of cognitive learning coefficient and social learning coefficient was taken as 2 and 1.4 respectively. Population size was varied from 1 to 500 and maximum allowable function evaluations were 100000 with no improvement termination criterion as 10 iterations. The same problem was also solved using three different

Table 6. Comparison of results obtained by different researchers using different optimization techniques for Example 1 (welded beam).

	Ragsdell & Phillips [30]	Deb [10]	Coello [9]	Coello & Montes [8]	Ray & Liew [31]	Qie & Ling [29]	Our approach 2009
x_1	0.2455	0.2489	0.2088	0.205986	0.244438	0.202369	**0.20573**
x_2	6.196	6.173	3.4205	3.471328	6.237967	3.544214	**3.470488**
x_3	8.273	8.1789	8.9975	9.020224	8.288576	9.04821	**9.03662**
x_4	0.2455	0.2533	0.21	0.20648	0.244566	0.205723	**0.20573**
$g_1(x)$	−5743.826	−5758.603	−0.337812	−0.074092	−5760.1104	−12.83979	**−0.0134**
$g_2(x)$	−4.715097	−255.5769	−353.9026	−0.266227	−3.245428	−1.247467	**−0.073**
$g_3(x)$	0	−0.0044	−0.0012	−0.000495	−0.000128	−0.001498	**0**
$g_4(x)$	−3.020289	−2.982866	−3.411865	−3.430043	−3.020055	−3.429347	**−3.433**
$g_5(x)$	−0.1205	−0.1239	−0.0838	−0.080986	−0.119438	−0.079381	**−0.0807**
$g_6(x)$	−0.234208	−0.23416	−0.235649	−0.235514	−0.234237	−0.235536	**−0.2355**
$g_7(x)$	−3604.275	−4465.27	−363.2323	−58.66644	−13.079305	−11.68135	**−0.0329**
$f(x)$	2.385937	2.433116	1.748309	1.728226	2.38119	1.728024	**1.724855**

Table 7. Top ten optimum solution obtained for example 1 using ABC.

	x_1	x_2	x_3	x_4	f(X)	Max. constraint violation
1	0.20573	3.470488	9.03662	0.20573	1.72486	–
2	0.20573	3.470484	9.036627	0.20573	1.72486	–
3	0.205725	3.470603	9.036626	0.20573	1.72486	–
4	0.205732	3.470446	9.036702	0.205731	1.72487	–
5	0.205723	3.470624	9.036645	0.205733	1.72489	–
6	0.205728	3.470573	9.036595	0.205734	1.72489	–
7	0.20572	3.470761	9.036576	0.205733	1.72489	–
8	0.205714	3.470808	9.036791	0.205734	1.72493	–
9	0.205712	3.471132	9.03687	0.20573	1.72496	–
10	0.205694	3.471154	9.037005	0.205733	1.72498	

Table 8. Top ten optimum solutions obtained for example 2 using ABC.

	x_1	x_2	x_3	x_4	x_5	$f(X)$	Max. constraint violation
1	13.8275	26.8855	8.3799	260.3097	223.9656	54444.50	–
2	13.767484	27.09977	8.75685	256.76358	214.443837	54377.00	(1)0.007
3	13.76841	26.8238	8.20406	262.09579	229.150404	54390.00	(1)0.003
4	13.755645	26.74911	8.05732	263.61189	233.347565	54393.00	(1)0.003
5	13.807123	27.08457	8.75527	256.77486	214.325228	54395.00	(1)0.003
6	13.807085	27.12217	8.80445	256.35323	213.007088	54397.00	(1)0.003
7	13.808004	26.97285	8.53223	258.85308	220.023952	54403.00	(1)0.002
8	13.812803	27.00257	8.59691	258.23552	218.318487	54405.00	(1)0.002
9	13.783138	27.238	8.81498	255.98918	212.613805	54429.00	(5)0.005
10	13.825026	26.92995	8.46453	259.48998	221.711473	54430.00	(1)0.0007

local optimization techniques viz. LFOPC, Dynamic – Q and ETOPC. The drawback of local optimization techniques is that it does not guarantee global optimum and also result changes with the change in starting point and thus they are time consuming with respect to global optimization techniques.

ABC algorithm used for the analysis was coded in MATLAB and all experiments were done on 1.4 GHz Intel Core 2 Duo Laptop machine. The performance of all the algorithms is compared for both the examples. To obtain the most suitable parameters a large number of experiments were conducted. For ABC the value of limit is set to $N \times S$ where N is the number of employed or onlooker bees and S is the dimension of design variable. One hundred trial runs were performed for each variant of ABC algorithm for each example. The performances of the different variants were compared based on different criteria. The first criterion for the comparison is the best solution obtained in one run. The algorithms are compared for the best value, mean value and the standard deviation for one hundred runs for the best solutions. In this comparison best value gives the global optima, and mean value serves the purpose for the average performance to search optimum result. Standard deviation gives the deviation of best result from the mean result for all the runs. The second criterion for the comparison is the success rate, which indicates the percentage success the technique has achieved in finding the global solution. Results of all the benchmark problems are shown in the Table 1.

It can be seen from the Table 1 that modified ABC is better than classical ABC in finding the best solution for both constraint and unconstraint benchmark problems. For Griewank and pressure vessel problem classical ABC has shown better result than modified ABC for the mean of the beat solutions. These results do not indicate that classical ABC is inferior to modified ABC, but main intension of comparison is that modified ABC can find better solution with less population size and number of generations which indicates better convergence rate of the modified ABC. If population size and number of generation is increased than both the variants converge to the same solution.

All the variants were experimented by taking different population size and different number of generations to study its effect on best solution, average solution, success rate and processing time for the considered two examples of welded structures.

Tables 2 and 3 show the best solutions for both the examples It is observed from the result that ABC_2 has shown better results than other two variants for best, mean solutions and standard deviations. All the variants have given improvement in the result with the increase in population size and number of generations. Moreover ABC_2 require less function evaluation to reach optimum solution than ABC_1. For example 2 also similar kinds of results are observed. Moreover ABC_2 has given the optimum solution using less than 10 000 function evaluations, which is better than the results of PSO implemented by Jarmai (2006), which requires between 26 184 and 100 000 function evaluations considering a population size of 500. Results show the better performance of ABC over PSO in terms of solution as well as in terms of function evaluations

Tables 4 and 5 show the success rate for both the examples. Here success is calculated if the results are within the tolerance of 1% of its global optima. It is observed that here also ABC_2 has better success rate with less population size and number of generations. With the increase in population size and number of generations success rate increases for all the variants. There is tremendous difference in the success rates of ABC_2 and ABC_1. It is clear from the above discussion than ABC_2 is a promising variant of ABC which serves all the requirement of a design engineer such as success rate and convergence of the solutions.

Example 1 has already been solved by many researchers and results from all the researchers are given in Table 6. It can be seen from the Table 5 that the best feasible solution found by ABC is better with respect to value of objective function than the best solutions found by other researchers using different optimization techniques. Results obtained by ABC are feasible because all the constraints are having negative values and thus satisfied.

More top ten solutions for both the examples are given in Tables 7 and 8 obtained by using ABC to help other researchers to check other optimization algorithms.

4 Conclusions

In the present work, cost optimization of welded structures has been presented. For orthogonally stiffened cylindrical shells cost function consist of material cost, manufacturing cost and painting cost. Shell buckling, stringer buckling, panel ring buckling and manufacturing limitations are considered as constraints. Classical ABC is modifies and it is applied on many constrained and unconstrained optimization benchmark problems and also to the optimization of welded structures. It is shown that ABC is capable of solving complicated cost optimization problems for welded structures. In terms of quality of solution, the results have shown that modified ABC has produced better results than classical ABC. It is also shown that ABC has given better results than other optimization techniques.

References

1. AISC. 1970. Specification for the Design, Fabrication and Erection of Structural Steel for Buildings, Manual of Steel Construction, 7th edn. Chicago, Illinois: American Institute of Steel Construction.
2. AISC. 1978. Manual of Steel Construction, 8th edn. Chicago, Illinois: American Institute of Steel Construction.
3. AISC. 1986. Manual of Steel Construction – Load and Resistance Factor Design, 8th edn. Chicago, Illinois: American Institute of Steel Construction.
4. Anderson KE, Chong KP. 1986. Least cost computer-aided design of steel girders. Engineering Journal. AISC, Fourth Quarter, 151–156.
5. Basturk B, Karaboga D. 2006. An Artificial Bee Colony (ABC) Algorithm for Numeric Function Optimization, IEEE Swarm Intelligence Symposium, May 12–14, Indianapolis, Indiana, USA.
6. Belegundu AD. 1982. A study of mathematical programming methods for structural optimization, Doctoral dissertation, Department of Civil and Environmental Engineering, University of Iowa, Iowa City, Iowa.
7. Bhatti MA. 1996. Optimum cost design of partially composite steel beams using LRFD. Engineering Journal. AISC, First Quarter. p. 18–29.
8. Coello CAC, Montes EM. 2002. Constraint-handling in genetic algorithms through the use of dominance-based tournament selection. Advanced Engineering Informatics, 16, 193–203.
9. Coello CAC. 2000. Use of a self-adaptive penalty approach for engineering optimization problems. Computers in Industry, 41, 113–127.
10. Deb K. 1991. Optimal design of a welded beam via genetic algorithms. AIAA Journal, 29(11), 2013–2015.
11. Det Norske Veritas (DNV). 1995. Buckling strength analysis, Classification Notes No. 30.1. Ed. Det Norske Veritas, Hovik, Norway.
12. Dong Y, Tang J, Xu B, Wang D. 2005. An application of swarm optimization to nonlinear programming. Computer and Mathematics with Applications, 49, 1655–1668.
13. Douty R. 1980. Design of steel connections by math programming. Journal of the Structural Division, ASCE, 106(ST5), 1135–1154.
14. Farkas J, Jarmai K. 2008. Minimum cost design of a square box column composed from orthogonally stiffened welded steel plates, International conference proceedings on design fabrication and economy of welded structures, Hungary. p. 70–71.
15. Farkas J. 1991. Fabrication aspects in the optimum design of welded structures. Structural Optimization, 3(1), 51–58.

16. Farkas J, Jarmai K. 1994. Minimum cost design of laterally loaded welded rectangular cellular plates. Structural Optimization, 8(4), 262–267.
17. Farkas J, Jarmai K, Snyman J. 2009. Global minimum cost design of a welded square stiffened plate supported at four corners, Structural and Multidisciplinary Optimization, 40, 477–489.
18. Farkas J. 2002. Thickness design of axially compressed unstiffened cylindrical shells with circumferential welds. Welding in the World, 46, 26–29.
19. Hojjat A, Kamal C. 2006. Cost Optimization of Structures, Fuzzy Logic, Genetic Algorithms and Parallel Computing. John Wiley: UK.
20. Jarmai K. 2007. Single- and multiobjective optimization of a welded stringer-stiffened cylindrical shell. Periodica Polytechnica, 51(2), 11–18.
21. Jarmai K, Snyman JA, Farkas J. 2006. Minimum cost design of a welded orthogonally stiffened cylindrical shell. Computers and Structures, 84, 787–797.
22. Karaboga D. 2005. An idea based on honey bee swarm for numerical optimization, Technical Report-TR06, Erciyes University, Engineering Faculty, Computer Engineering Department.
23. Karaboga D, Basturk B. 2007. Artificial bee colony (abc) optimization algorithm for solving constrained optimization problems, in IFSA 2007, LNAI 4529, Melin Pet al., Editors. Springer-Verlag: Berlin, Heidelberg. p. 789–798.
24. Karaboga D, Basturk B. 2008. On the performance of artificial bee colony (ABC) algorithm. Applied Soft Computing, 8, 687–697.
25. Lorenz RF. 1988. Understanding composite beam design methods using LRFD. Engineering Journal. AISC, First Quarter. p. 35–38.
26. Luis MCS, Farkas J, Jarmai K. 2008. Optimization of an orthogonally stiffened plate considering fatigue constraints, International conference proceedings on design fabrication and economy of welded structures, Hungary. p. 27–34.
27. Marcin C, Farkas J, Andrej G, Jarmai K, Katarzyna R. 2008. Optimization of steel beams and columns for variable rib configuration, International conference proceedings on design fabrication and economy of welded structures, Hungary. p. 113–118.
28. Montes EM, Coello CAC. 2005. Useful Infeasible Solutions in Engineering Optimization with Evolutionary Algorithms. LNAI, 3789, 652–662.
29. Qie H, Ling W. 2007. An effective co-evolutionary particle swarm optimization for constrained engineering design problems. Engineering Applications of Artificial Intelligence, 20, 89–99.
30. Ragsdell KM, Phillips DT. 1976. Optimal design of a class of welded structures using geometric programming. ASME Journal of Engineering for Industries, 98(3), 1021–1025.
31. Ray T, Liew K. 2003. Society and civilization: an optimization algorithm based on the simulation of social behavior. IEEE Transactions on Evolutionary Computation, 7, 386–396.
32. Sandgren E. 1988. Nonlinear integer and discrete programming in mechanical design, Proceedings of the ASME Design Technology Conference, Kissimine, FL. p. 95–105.

17

First-principles study on electronic, optic, elastic, dynamic and thermodynamic properties of RbH compound

Sinem Erden Gulebaglan[1,*], Emel Kilit Dogan[2], Mehmet Nurullah Secuk[2], Murat Aycibin[2] Bahattin Erdinc[2], and Harun Akkus[2]

[1] Department of Electric Program, Vacational School of Van, Yuzuncu Yil University, 65080 Van, Turkey
[2] Department of Physics, Faculty of Science, Yuzuncu Yil University, 65080 Van, Turkey

Abstract – We performed first-principles calculations to obtain the electronic, optical, elastic, lattice-dynamical and thermodynamic properties of RbH compound with rock salt structure. The ground-state properties, i.e., the lattice constant and the band gap were investigated using a plane wave pseudopotential method within density functional theory. The calculated lattice constant, bulk modulus, energy band gap and elastic constants are reported and compared with previous theoretical and experimental results. Our calculated results and the previous results which are obtained from literature are in a good agreement. Moreover, real and imaginary parts of complex dielectric function, reflectivity spectrum, absorption, extinction coefficient and loss function as a function of photon energy and refractive index with respect to photon wavelength were calculated. In addition, temperature dependent thermodynamic properties such as Helmholtz free energy, internal energy, entropy and specific heat have been studied.

Key words: Density functional theory, Rock salt structure, Optical properties, Thermodynamic and dynamic properties.

1 Introduction

Alkali hydrides (LiH, NaH, KH, RbH and CsH) have possible applications in nuclear and chemical industries [1]. Thus, this materials have been investigated both experimentally and theoretically for many years. Like alkali metal halides, alkali metal hydrides crystallize with the rock-salt structure. The X-ray experimental study [2] showed that RbH crystallize with the rock-salt (B1) structure at room temperature. Gaydon and Pearse [3] studied the band structure of RbH experimentally and theoretically. A structural phase transition from NaCl to CsCl phase was observed in RbH at high pressure [4]. Jain [5] investigated ground state of RbH using the Rydberg-Klein-Ress method. Stwalley et al. [6] reported structure of the alkali hydride diatomic molecules. Hooper and Zurek [7] investigated the structures of compressed rubidium polyhydrides, RbH_n with $n > 1$. Zrafi et al. [8] calculated the evaluation of the first and second derivative of nonadiabatic coupling between the several $^1\Sigma^+$ adiabatic states of LiH, RbH and CsH molecules. Ahuja [9] investigated the structure of RbH under high compressions by means of first principles self consistent total-energy calculations within local-density approximation. Fatmi et al. [10] showed structural, electronic, optical and thermodynamic properties of $Na_xRb_{1-x}H$ and $Na_xK_{1-x}H$ alloys. Hooper et al. [11] discussed three new structures of high-pressure alkali metal hydrides ($P6_3/mmc$, $I4_1/amd$, and Pnma) and found structural phase transitions via evolutionary methodologies. Sudha Priyonga et al. [12] calculated structural, electronic and elastic properties of alkali hydrides (MH: M = Li, Na, K, Rb, Cs).

*e-mail: sinemerden@gmail.com

2 Computational methods

The first-principles calculations are performed with the ABINIT [13] and the WIEN2K [14] code programs. The structural, electronic and optical properties of RbH crystal was investigated using WIEN2K package within the Generalized Gradient Approximation (GGA) based on the density functional theory (DFT). 286 k points were applied in the Brillouin zone for the structural, electronic and optical properties of RbH crystal.

In addition, the dynamic and thermodynamic properties of RbH crystal was investigated using ABINIT code. The calculated results were obtained using GGA included the exchange-correlation effects [15] of DFT. The self-consistent norm-conserving pseudopotentials of FHI-type with a

Troullier-Martins scheme [16] for all atoms of RbH crystal were used. The Kohn-Sham equations [17] were solved the conjugate gradient minimization method [18]. Plane waves were used for the electronic wave functions. For the true valence electron of Rubidium atom and for Hydrogen atom were deliberated the 5*s* electron and the 1*s* electron, respectively. Firstly, in all calculations total energy of this material was accommodated with respect to cutoff energy and Monkhorst-Pack mesh grid. The calculated total energy calculations were done to a good convergence at 30 hartree of cutoff energy. Besides, 60k points using $6 \times 6 \times 6$ Monkhorst-Pack mesh grid [19] in RbH crystal for structural optimization and the calculations of dynamic and thermodynamic properties.

3 Structural and electronic properties

RbH crystal contains two atoms per face centered cubic unit cell and crystallizes in the Rock salt-type structure. Rb atom is centered at (0, 0, 0) and H atom is located at (0.5, 0.5, 0.5) in the reduced coordinates. The volume of unit cell is optimized using WIEN2K code. The total energy of RbAu compound has been investigated as a function of the unit cell volume and fitted to the Murnaghan equation. Volume parameter was calculated as 376.39 (a.u^3) of cubic atomic. In this value, the total force is equal to zero at minimum total energy. The calculated values of lattice parameters for RbH compound is 11.461 bohr (6.064 Å), while experimental value is 6.037 Å [2]. As seen, there is a good agreement between calculated and experimental result for lattice parameters. Then, calculated result was used for calculation of all studied physical properties of this compound (Table 1).

The pseudopotential method based on density functional theory within general gradient approximation was used for electronic structure calculations of RbH. According to the band structure calculations, RbH crystal has a direct band gap at high symmetry point *L* with the value of 2.943 eV and experimental values are 2.96 eV. We found that RbH is a semiconductor. Electronic band structure was plotted as in Figure 1.

4 Optical properties

RbH crystal is a cubic crystal and space group number is 216. This structure is requested to investigate one dielectric component. Our calculation were performed 0–40 eV photon energy range. The calculated real (ε_1) and imaginary (ε_2) parts of dielectric function of this compound are illustrated in Figure 2. As seen from this figure, the real part of dielectric function, ε_1, goes to its maximum value at 4.8 eV of photon energy for RbH. There is no absorption through 2.93 eV. After this value, optic transitions from valance band to conduction band can be detected. Furthermore, real part goes to zero at 20.3133, 20.69429, 20.80311, 21.94599 and 21.9732 eV. The incident electromagnetic waves are completely reflected at the negative values of real part. The reflectivity (*R*) and absorption (α) active regions are from 12 eV to 24 eV and from 2 eV to 18 eV, respectively. The extinction coefficient (*k*) varies similar to the absorption coefficient (α). The region of the active damping is between 2 eV and 18 eV. Refractive index (*n*) is 1.75 for low and very low energy electromagnetic waves.

Table 1. Calculated lattice parameter a_0 (Å) and energy gap E_g (eV).

a_0 (Å)	E_g (eV)
6.064 [Present Work]	2.943 [Present Work]
6.037 [2] [exp]	2.960[20] [exp]
5.93 [21]	
6.04 [21]	
6.075 [20, 22]	
5.9918 [23]	3.0255 [23]
6.0611 [23]	2.201 [23]

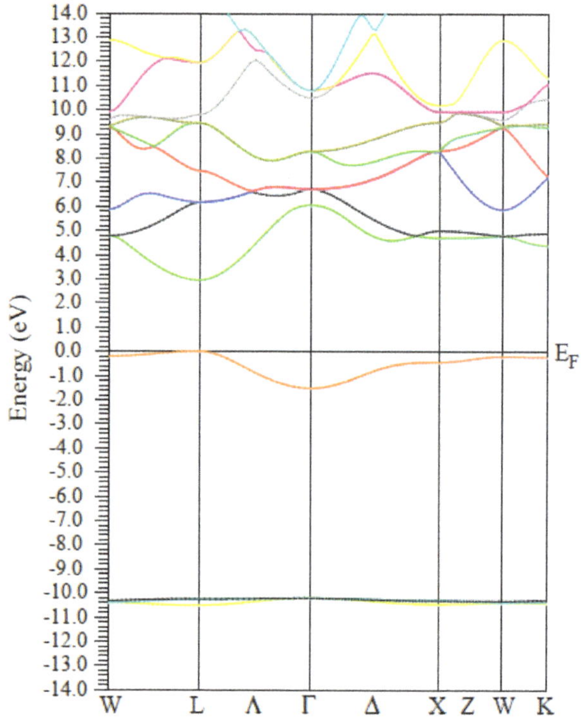

Figure 1. Plotted electronic band structure for RbH.

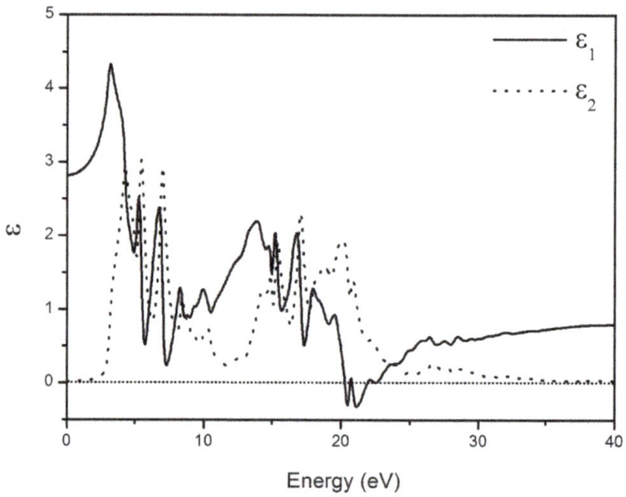

Figure 2. Real and imaginary parts of dielectric function for RbH.

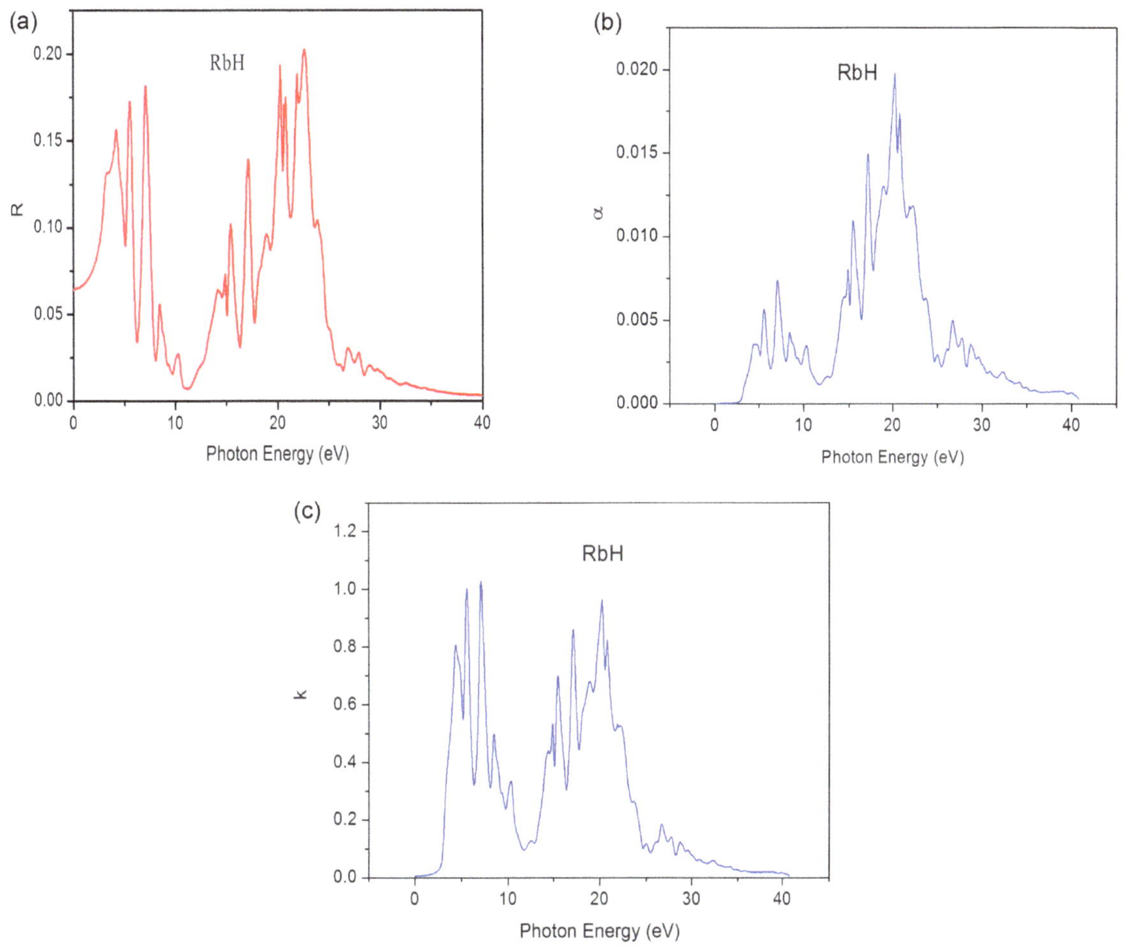

Figure 3. Calculated (a) reflectivity spectrum (R), (b) absorption (α) coefficient and (c) extinction coefficient (k) for RbH.

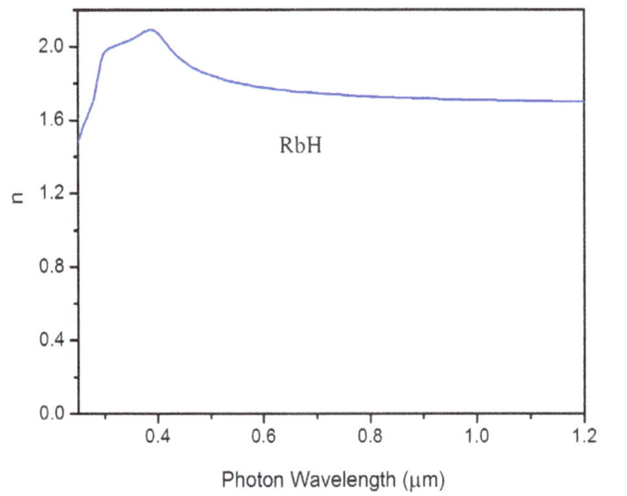

Figure 4. The calculated spectral dependence of refractive index (n) on wavelength of RbH.

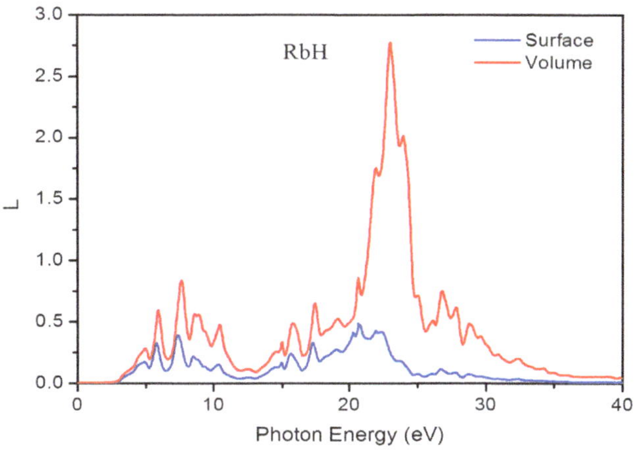

Figure 5. Volume and surface loss functions (L) for RbH as a function of photon energy for RbAu.

The range of the optical loss in surface and volume is maximum in the ultraviolet region of the electromagnetic spectrum. The calculated extinction coefficient, absorption coefficient, reflectivity and refractive index (n) for RbH are shown in Figures 3 and 4, respectively. The spectral dependence of calculated refractive index is in the wavelength range of 0–0.6 μm, and the calculated value of refractive indices is 1.75 for RbH. The energy loss functions (L), for both surface and volume, are illustrated in Figure 5. As seen in Figure 5, sharp peaks are incorporated with plasma oscillations concerned to the photon energy values.

Table 2. Calculated elastic constants for RbH with NaCl structure.

C_{11} (GPa)	C_{12} (GPa)	C_{44} (GPa)	Ref.
32.835	3.815	9.657	Present work
35.1	4.5	7.6	[21]
16.5	9.6	6.9	[22]
28.2	7.11	15.5	[25]
24.6	6.5	10.6	[26]

Table 3. Calculated single crystal bulk modulus (B_0), Young modulus (E), shear modulus (G), anisotropy factor (A), Poisson's ratio (v) obtained for RbH.

B_0 (GPa)	E (GPa)	G (GPa)	A	v	B/G	V_0 (Å^3)	Ref.
13.49	27.04	11.59	0.66	0.17	1.16	53.72	Present work
10.0	–	–	–	–	–	55.01	[27]
13.97	25.00	10.20	1.18	0.23	1.37	53.78	[23]

5 Elastic properties

To obtain elastic constants of this compound, we used the method developed by Morteza Jamal integrated in WIEN2 K code [24]. Here, to analyze the elastic properties, we calculated the elastic constants (C_{ij}), Young module, shear modulus, bulk modulus, and Poisson's coefficients with three different approximations that are namely Voight, Reuss and Hill using the mentioned method in reference [23]. The elastic constants are calculated to be C_{11} = 32.835 GPa, C_{12} = 3.815 GPa, and C_{44} = 9.657 GPa for RbH. In this case, RbH has a stable structure. The calculated elastic constants of RbH are listed in Table 2. The Born mechanical stability criteria for cubic structure is given below:

$$C_{11} > 0, C_{12} > 0, C_{44} > 0, C_{12} > C_{44}, C_{11} + 2\,C_{12} > 0, C_{11} - C_{12} > 0$$

The Young's modulus E, Zener anisotropy factor A and Poisson's ratio v, which have the plenty of exciting elastic properties for applications, are also computed in terms of the reckoned data using the following relations:

$$A = 2C_{44}/(C_{11} - C_{12})$$

$$v = \frac{1}{2}\left(\frac{(B - \frac{2}{3})G}{(B + \frac{1}{3}G)}\right)$$

and

$$E = \frac{9GB}{G + 3B}$$

where $G = (G_V + G_R)/2$ is the isotropic shear modulus. The G_V ($G_V = (C_{11} - C_{12} + 3\ C_{44})/5$) is Voigt's shear modulus concerned to the upstairs bound of G values, and G_R ($5/G_R = 4/(C_{11}\ C_{12}) + 3/C_{44}$) the Reuss's shear modulus concerned to the lower bound of G values. The calculated Zener anisotropy factor, Poisson's ratio, Young's modulus, and shear modulus for RbH are given in Table 3. The calculated and other results are in a good agreement.

6 Dynamic properties

Phonon dispersion curves and phonon density of states of RbH along the high-symmetry directions are calculated by using the ABINIT. In RbH crystal, unit cell involves two atoms at zero pressure. The Rb atom and H atom are positioned at the

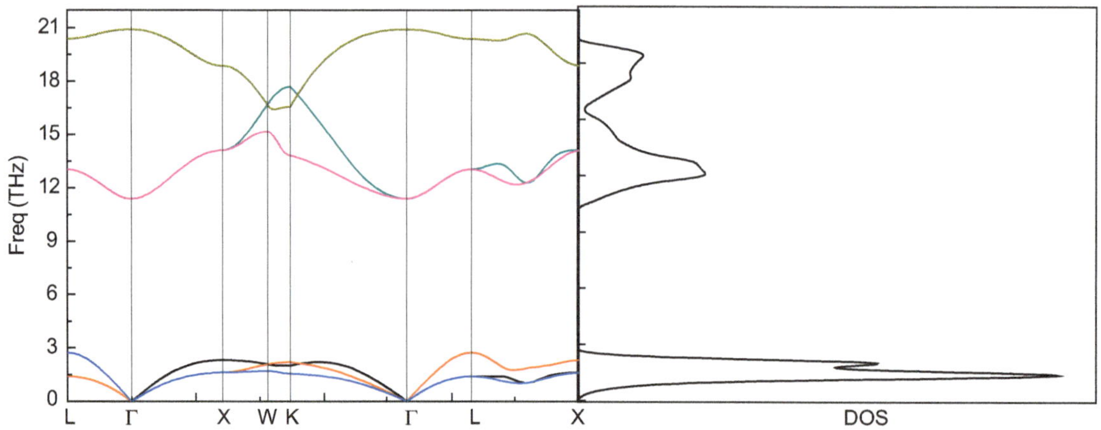

Figure 6. Phonon dispersion curves and phonon density of states for RbH.

Table 4. The calculated phonon frequencies at Γ, X and L points for RbH.

High symmetry point	LA (THz)	TA (THz)	LO (THz)	TO (THz)
Γ	0.00	0.00	11.38	20.92
X	1.62	2.31	13.03	18.88
L	1.39	2.68	13.03	20.53

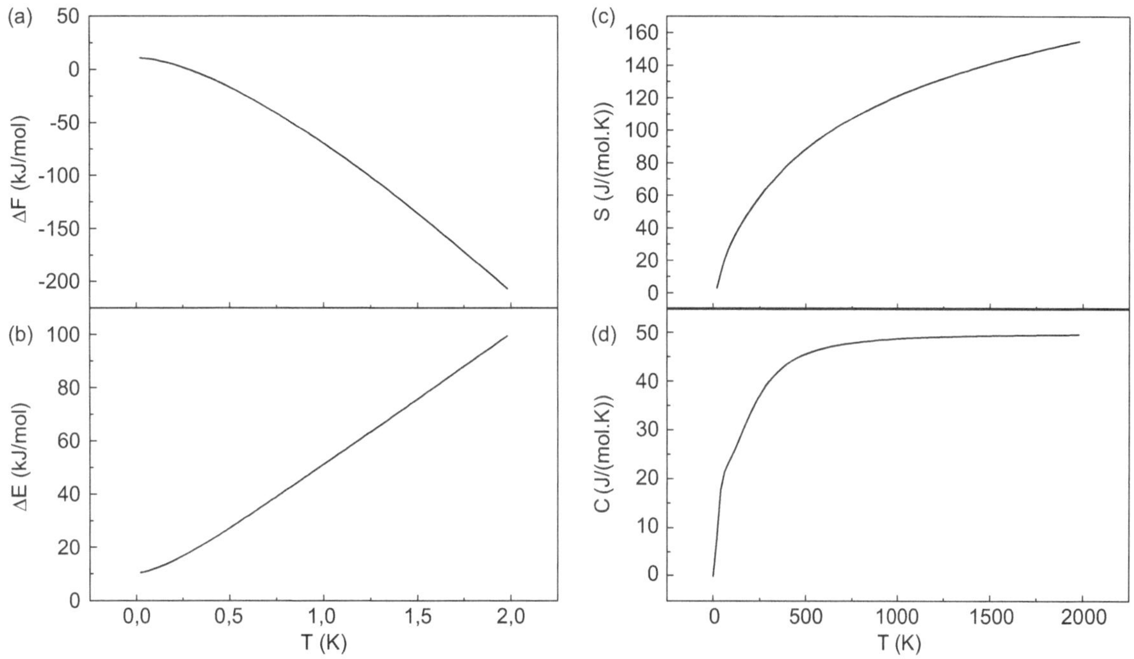

Figure 7. (a) Phonon contribution to the free energy as a function of temperature, (b) phonon contribution to the internal energy, (c) the entropy as a function of temperature and (d) constant volume specific heat as a function of temperature, as a function of temperature for RbAu.

point of (0.0, 0.0, 0.0) and (0.5, 0.5, 0.5) at zero pressure, respectively. The computed phonon dispersion curves of RbH, along the main symmetry direction of the Brillouin zone are illustrated in Figure 6. RbH has two atoms per unit cell so that six phonon branches occur. Three of them are acoustic and three of them are optical branches. Some of the branches are degenerate along more symmetrical directions (Γ-X and Γ-L) inside Brillouin zone of RbH crystal. The calculated phonon frequencies resulting from the high symmetry points (Γ, X and L) of the Brillouin zone are listed in Table 4.

There is a gap between the acoustic and the optic phonon branches which shows that RbH is a phononic crystal. Optical branches are almost horizontal around Γ symmetry point. Optical and acoustic branches degenerate at left and right side of Γ point. All branches are non-degenerate between X-W, W-K, K-Γ points. Slopes of acoustic branches around Γ point are steep, which shows large elastic constant for RbH. High phonon frequencies are attributed to H atoms, because its mass is small and can vibrate very rapidly.

7 Thermodynamic properties

After calculating the phonon band structure, we have inquired the thermodynamic properties of RbH using these phonon band structure calculations. In terms of thermodynamic properties, we have examined the entropy, constant-volume specific heat and the phonon contributions to the Helmholtz free energy and internal energy. We have plotted all these thermodynamic quantities as functions of temperature as can be seen in Figure 7. As seen from this figure, the values of free energy and internal energy are different from zero at zero temperature which shows zero point oscillations. The values of free energy and internal energy at zero temperature are very close to each other that is 9.92 kJ/mol for free energy and 10.47 kJ/mol for internal energy. The constant-volume specific heat of RbH approaches a limit value of 49.34 J/mol K at 1466.30 K.

8 Conclusion

We have investigated the dynamic and thermodynamic properties of Rubidium Hydride (RbH) by an ab-initio pseudopotential method applying density functional theory within the generalized gradient approximation. RbH show semiconducting property and the value of band gap is 2.943 eV. We have declared real and imaginary parts of dielectric function, reflectivity spectrum, absorption coefficient, extinction coefficient and loss function as a function of photon energy and refractive index as a function of photon wavelength. During the investigation of dynamical properties of RbH, we have calculated and plotted phonon density of states function graphs. While examining the thermodynamic properties of RbH, we have calculated the phonon contribution to the Helmholtz free energy and to the internal energy, the entropy and constant-volume specific heat as functions of temperature.

References

1. Latroche MJ. 2004. Structural and thermodynamic properties of metallic hydrides used for energy storage. Phys. Chem. Solids, 65, 517.
2. Zinti E, Harder A. 1931. Uber Alkalihydride. Z. Phys. Chem. Abt. B, 14, 265.
3. Gaydon AG, Pearse RWB. 1939. The spectrum of rubidium hydride, RbH. II. Potential curves, wave functions and intensity distribution. Proc. Roy. Soc. A, 173, 28–37, 37–47.
4. Hochheimer HD, Strossner K, Hale W, Baronowski B. 1987. High-pressure studies of NaH to 54 GPa. Phys. Rev. B, 36, 7664–7667.
5. Jain DC. 1963. Franck-Condon Factors and r-centroids for the Band System of the RbH Molecule. Proc. Phys. Soc., 81, 171–174.
6. Stwally WC, Zemke WT, Yang SC. 1991. Spectroscopy and structure of the alkali hydride diatomic molecules and their ions. J. Phys. Chem. Ref. Data, 20, 153.
7. Hooper J, Zurek E. 2012. Rubidium polyhydrides under pressure: emergence of the linear H_3^- species. Chem A Eur J, 18, 5013.
8. Zrafi W, Ovjia B, Berriche H, Gadea FX. 2006. Ab initio adiabatic and diabatic energies and dipole moments of the RbH molecule. J. Mol. Struct.: Theochem, 777, 87–97.
9. Ahuja R, Eriksson O, Johansson B. 1999. Theoretical search for the CrB-type high-pressure phase in LiH, NaH, KH and RbH. Physica B, 265, 87.
10. Fatmi M, Ghebouli B, Ghebouli MA, Bouhemadou A, Binanran S. 2012. Structural, electronic, optical and thermodynamic properties of $NaxRb_{1-x}H$ and $NaxK_{1-x}H$ alloys. J. Phys. Chem. Solids, 73, 1–7.
11. Hooper J, Baettig P, Zurek E. 2012. Pressure Induced Structural Transitions in KH, RbH and CsH. J. Appl. Phys., 111, 112611.
12. Sudba Priyanga G, Asuini Meenaatci AT, Rajeswara Palanichamy R, Iyakutti K. 2014. Structural, electronic and elastic properties of alkali hydrides (MH: M = Li, Na, K, Rb, Cs): Ab initio study. Comput. Mater. Sci., 84, 206–216.
13. Gonze X, Beuken JM, Caracas R, Detraux F, Fuchs M, Rignanese GM, Sindic L, Verstraete M, Zerah G, Jollet F, Torrent M, Roy A, Mikami M, Ghosez P, Raty JY, Allan DC. 2002. First-principles computation of material properties: the ABINIT software Project. Comput. Mater. Sci., 25, 478.
14. Blaha KSP, Madsen GKH, Kvasnicka D, Luitz J. 2001. An augmented plane wave + local orbitals program for calculating crystal properties, Universität Wien, Austria.
15. Perdew JP, Burke K, Ernzerhof M. 1996. Generalized gradient approximation made simple. Phys. Rev. Lett., 77, 3865.
16. Troullier N, Martins JL. 1991. Efficient pseudopotentials for plane-wave calculations. Phys. Rev. B, 43, 993.
17. Kohn W, Sham LJ. 1965. Self-consistent equations including exchange and correlation effects. Phys. Rev., 140, A1133.
18. Payne MC, Teter MP, Allan DC, Arias TA, Joannopoulos JD. 1992. Iterative minimization techniques for ab initio total-energy calculations: molecular dynamics and conjugate gradients. Rev. Mod. Phys., 64, 1045.
19. Monkhorst HJ, Pack JD. 1976. Special points for Brillouin-zone integrations. Phys. Rev. B, 13, 5188.
20. Zinenko VI, Fedorov AS. 1994. First principle calculations of alkali hydride electronic structures. Sov. Phys. Solid State, 36, 742.
21. Barrera GD, Colognesi D, Mitchell PCH, Ramirez Cuesta AJ. 2005. LDA or GGA? A combined experimental inelastic neutron scattering and ab initio lattice dynamics study of alkali metal hydrides. Chem. Phys., 317, 119.
22. Novakovic N, Radisavljevic I, Colognesi D, Ostojic S, Ivanovic N. 2007. First principle calculations of alkali hydride electronic structures. J. Phys. Condens. Matter, 19, 406211.
23. Sudha Priyanga G, Asvini Meenaatci AT, Rajeswara Palanichamy R, Iyakutti K. 2014. Structural, electronic and elastic properties of alkali hydrides (MH: M = Li, Na, K, Rb, Cs): Ab initio study. Comput. Mater. Sci., 84, 206.
24. Reshak H, Jamal M. 2012. DFT calculation for elastic constants of orthorhombic structure within WIEN2K code: A new package (ortho-elastic). J. Alloys Compd., 543, 147.
25. Engebretsen L. 1995. Electronic structure calculations of the elastic properties of alkali hydrides. Graduation Thesis, Royal institute of technology KTH, Stockholm.
26. Zhang J, Zhang L, Lui T, Ma Y. 2007. Phonon and elastic instabilities in rocksalt alkali hydrides under pressure: First-principles study. Phys. Rev. B., 75, 104115.
27. Hochheimer HD, Strossner K, Honle W, Baranowski B, Filipek F. 1985. High pressure X-ray investigation of the alkali hydrides NaH, KH, RbH and CsH. Less J. Comm. Met., 107, L13.

18

Stiffness optimization of geometrically nonlinear structures and the level set based solution

Qi Xia and Tielin Shi

The State Key Laboratory of Digital Manufacturing Equipment and Technology, Huazhong University of Science and Technology, Wuhan, P.R. China

Abstract – Load-normalized strain energy increments between consecutive load steps are aggregated through the Kreisselmeier-Steinhauser (KS) function, and the KS function is proposed as a stiffness criterion of geometrically nonlinear structures. A topology optimization problem is defined to minimize the KS function together with the perimeter of structure and a volume constraint. The finite element analysis is done by remeshing, and artificial weak material is not used. The topology optimization problem is solved by using the level set method. Several numerical examples in two dimensions are provided. Other criteria of stiffness, i.e., the end compliance and the complementary work, are compared.

Key words: Topology optimization, Geometrically nonlinear, Level set method, Stiffness, Kreisselmeier-Steinhauser function.

1 Introduction

Maximizing stiffness of structures is a major topic of structural optimization. However, most of the studies on stiffness maximization consider linear elastic structures, and the studies on geometrically nonlinear structures are limited.

As the magnitude of external load increases slowly from zero to the target load level (i.e., under quasi-static condition), the configuration of a geometrically nonlinear structure changes, as a result the stiffness of structure changes as well. For instance, the distance from the external load to the support of a structure may become shorter when the structure deforms, thus the stiffness of structure increases. One can see that the stiffness of a geometrically nonlinear structure depends on its configuration. Therefore, evaluation of the stiffness of a geometrically nonlinear structure is ambiguous. It may refer to an instantaneous stiffness of a specific configuration in the whole course of deformation, or it may have a more comprehensive meaning that involves with many configurations.

Stiffness of a structure is often represented by the compliance (denoted as W), the strain energy (denoted as U), or the complementary work (denoted as U_C) [1–4]. For a linear elastic structure, minimizing any one of the three functions leads to the same result because $W = 2U = 2U_C$. For a geometrically nonlinear structure, however, different results may be obtained if a different function is used in optimization because $W \neq 2U \neq 2U_C$ [2]. The compliance, the strain energy, the complementary work, along with a typical load-displacement curve of a geometrically nonlinear structure are illustrated in Figure 1. Area of the rectangle is equal to the compliance; area of the region under or above the curve is respectively equal to the strain energy or the complementary work.

Compliance or the so-called end compliance is used in most of the studies that aim to maximize the stiffness of geometrically nonlinear structures. It characterizes the stiffness of the final equilibrium configuration of a geometrically nonlinear structure, and the result of compliance minimization depends on load level [2]. In other words, the shape and topology of the optimal structure will change if the magnitude of design load is changed. More interesting, it was observed that the optimal structure obtained with a big design load is not stiff for smaller design load [2]. Therefore, it was suggested that the end compliance may not be the best criterion of stiffness for geometrically nonlinear structures [2, 5].

In order to overcome such a problem, Buhl et al. [2] proposed to minimize the complementary work and numerically approximated it through a weighted combination of displacements of many deformed configurations including the final equilibrium configuration and those went through before the final equilibrium. The strain energy was also used in the

*e-mail: qxia@mail.hust.edu.cn

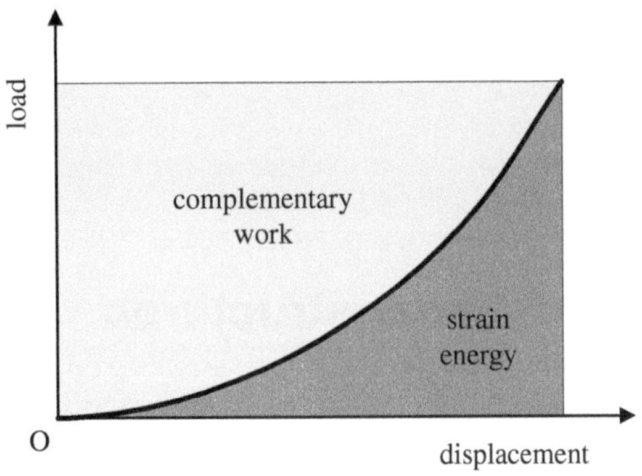

Figure 1. A load-displacement curve of geometrically nonlinear structure.

optimization of geometrically nonlinear structures [6]. Again, it was numerically approximated through a weighted combination of displacements of many deformed configurations [6]. Minimizing the complementary work or the strain energy can be regarded as a multi-objective optimization whose objectives are the displacements of many deformed configurations. To conclude, the end compliance describes only the instantaneous stiffness of the equilibrium configuration, but the complementary work and the strain energy give more comprehensive descriptions of stiffness since they involve with many configurations.

Although minimization of the complementary work or the strain energy offers a control over the displacements of many deformed configurations, the fixed weights of displacements in the weighted combinations are not perfect when they are viewed in the perspective of multi-objective optimization. Therefore, in the present study, we propose to extend the inspiring previous studies by considering load-normalized strain energy increments between consecutive load steps. The increments are aggregated through the Kreisselmeier-Steinhauser (KS) function [7], and the KS function is proposed as a stiffness criterion of geometrically nonlinear structures. Then, a topology optimization problem is defined. The different criteria of stiffness are investigated and compared in the present paper.

Another important issue in the topology optimization of geometrically nonlinear structures is concerned with the numerical instability that may arise in the finite element analysis. In most topology optimizations, a method employing a fixed mesh and ersatz material is used for the finite element analysis. Such a method uses an artificial weak material to represent void. However, in the topology optimization of geometrically nonlinear structures, if an element filled with the artificial weak material undergoes excessively large deformation, its tangent stiffness matrix may become negative, and this will cause numerical instability. In the literature, several methods have been proposed to deal with such numerical instability. Buhl et al. [2] observed that the numerical instability arised at the nodes surrounded by "void" elements and proposed to eliminate such nodes from the convergence criterion. Bruns and Tortorelli [8] presented a systematic method for removing and reintroducing low density elements. Yoon and Kim [9] proposed a method that uses connectivity between elements as the design variables of topology optimization. In addition, Lahuerta et al. [10] and Wang et al. [11] proposed to tailor the material model of the artificial weak material such that the numerical instability can be eliminated. Meshless methods were also proposed by Cho and Kwak [12] and He et al. [13] to overcome the numerical instability. In a level set based solution to the topology optimization of geometrically nonlinear structures, remesh is proposed by Ha and Cho [14] to obtain a mesh that is conformal to the geometry of the structure, thus eliminating the artificial weak material as well as the numerical instability coming with it. In the present study, being similar to the method proposed by Ha and Cho [14], we do the finite element analysis by modifying a fixed background mesh and do not use the artificial weak material, as proposed in our previous study [15].

The paper is organized as follows. In Section 2 some basics of geometrically nonlinear structure are outlined. In Section 3 the end compliance, the strain energy, the complementary work, and the proposed maximum of load-normalized strain energy increments are discussed; formulation of the topology optimization problem is described. In Section 4 sensitivity analysis for the optimization problem is described. In Section 5 the level set method is briefly described. Section 6 describes the finite element analysis. Section 7 gives numerical examples and discussions. Section 8 concludes this paper.

2 Analysis of geometrically nonlinear structure

The open bounded set occupied by a structure is denoted as $\Omega \subset \mathbb{R}^d$ (d = 2 or 3). It is required that during optimization all admissible structures stay in a fixed reference domain $\mathscr{D} \subset \mathbb{R}^d$, i.e., $\Omega \subseteq \mathscr{D}$. The boundary of Ω consists of three disjoint segments, i.e.,

$$\partial\Omega = \Gamma_N \cup \Gamma_H \cup \Gamma_D, \tag{1}$$

where an external load is applied on the Neumann boundary Γ_N; a displacement constraint is enforced on the Dirichlet boundary Γ_D; Γ_H denotes the free boundary. In the present study, only Γ_H is subjected to optimization, while both Γ_N and Γ_D are fixed during the optimization.

In total Lagrangian formulation, weak form of the equilibrium equation of a geometrically nonlinear structure is given by

$$a(\mathbf{u}, \bar{\mathbf{u}}) = \ell(\bar{\mathbf{u}}), \quad \forall\, \bar{\mathbf{u}} \in \mathscr{U}, \tag{2}$$

where

$$a(\mathbf{u}, \bar{\mathbf{u}}) = \int_{{}^0\Omega} S_{ij}(\mathbf{u})\hat{\varepsilon}_{ij}(\mathbf{u}; \bar{\mathbf{u}})\mathrm{d}\Omega, \tag{3}$$

$$\ell(\bar{\mathbf{u}}) = \int_{{}^0\Gamma_N} t_i\bar{u}_i\mathrm{d}\Gamma, \tag{4}$$

where ${}^0\Omega$ and ${}^0\Gamma_N$ are the structure domain and the Neumann boundary at initial undeformed configuration (a left superscript on Ω and Γ_N denotes the configuration of structure); $\mathbf{u}$ is displacement; $\bar{\mathbf{u}}$ is virtual displacement; $\mathscr{U}$ is the space of kinematically admissible displacement fields, i.e.,

$\mathscr{U} = \{\bar{\mathbf{u}} \in H^1(\Omega)^d : \bar{\mathbf{u}} = 0 \quad \text{on} \quad \Gamma_D\}$; $S_{ij}(\mathbf{u})$ is the second Piola-Kirchhoff stress tensor; $\hat{\varepsilon}_{ij}(\mathbf{u}; \bar{\mathbf{u}})$ is the virtual Green-Lagrange strain tensor; t_i is surface traction. In the present study, surface traction are assumed to be independent of deformation (e.g., a point load that does not change its direction during deformation of the structure), and body force is not considered. Details of the analysis are referred to [16, 17].

In the present study, although a structure undergoes large displacement, the strain is assumed to be small. In such case, the St.Venant-Kirchhoff material model is used, and the constitutive relation is given by

$$S_{ij}(\mathbf{u}) = C_{ijkl}\varepsilon_{kl}(\mathbf{u}), \tag{5}$$

where C_{ijkl} is the elasticity tensor, and in the present study the material is assumed to be isotropic. The Green-Lagrange strain $\varepsilon_{kl}(\mathbf{u})$ is given by

$$\varepsilon_{ij}(\mathbf{u}) = \frac{1}{2}\left(\frac{\partial u_i}{\partial^0 x_j} + \frac{\partial u_j}{\partial^0 x_i} + \frac{\partial u_k}{\partial^0 x_i}\frac{\partial u_k}{\partial^0 x_j}\right), \tag{6}$$

the virtual Green-Lagrange strain is given by

$$\hat{\varepsilon}_{ij}(\mathbf{u}; \bar{\mathbf{u}}) = \frac{1}{2}\left(\frac{\partial \bar{u}_i}{\partial^0 x_j} + \frac{\partial \bar{u}_j}{\partial^0 x_i} + \frac{\partial \bar{u}_k}{\partial^0 x_i}\frac{\partial u_k}{\partial^0 x_j} + \frac{\partial u_k}{\partial^0 x_i}\frac{\partial \bar{u}_k}{\partial^0 x_j}\right). \tag{7}$$

Because $a(\mathbf{u}, \bar{\mathbf{u}})$ is nonlinear in the argument $\mathbf{u}$, the Newton-Raphson method is used to solve the nonlinear equilibrium equation, i.e., equation (2), in an iterative manner. The displacement ${}^{m+1}\mathbf{u}$ of the $(m+1)$-th iteration is computed as the sum of ${}^{m}\mathbf{u}$ and an increment of displacement $\Delta\mathbf{u}$

$${}^{m+1}\mathbf{u} = {}^{m}\mathbf{u} + \Delta\mathbf{u}. \tag{8}$$

The increment of displacement $\Delta\mathbf{u}$ is obtained by solving a linearized equilibrium equation given by

$$a^*({}^{m}\mathbf{u}, \bar{\mathbf{u}}; \Delta\mathbf{u}) = R({}^{m}\mathbf{u}, \bar{\mathbf{u}}), \quad \forall\, \bar{\mathbf{u}} \in \mathscr{U}, \tag{9}$$

where

$$a^*({}^{m}\mathbf{u}, \bar{\mathbf{u}}; \Delta\mathbf{u}) = \int_{{}^0\Omega} C_{ijkl}\hat{\varepsilon}_{kl}({}^{m}\mathbf{u}; \Delta\mathbf{u})\hat{\varepsilon}_{ij}({}^{m}\mathbf{u}; \bar{\mathbf{u}})\mathrm{d}\Omega + \int_{{}^0\Omega} C_{ijkl}\varepsilon_{kl}({}^{m}\mathbf{u})\hat{\eta}_{ij}(\Delta\mathbf{u}; \bar{\mathbf{u}})\mathrm{d}\Omega, \tag{10}$$

$$R({}^{m}\mathbf{u}, \bar{\mathbf{u}}) = \int_{{}^0\Gamma_N} t_i\bar{u}_i\mathrm{d}\Gamma - \int_{{}^0\Omega} C_{ijkl}\varepsilon_{kl}({}^{m}\mathbf{u})\hat{\varepsilon}_{ij}({}^{m}\mathbf{u}; \bar{\mathbf{u}})\mathrm{d}\Omega, \tag{11}$$

and

$$\hat{\eta}_{ij}(\Delta\mathbf{u}; \bar{\mathbf{u}}) = \frac{1}{2}\left(\frac{\partial \Delta u_k}{\partial^0 x_i}\frac{\partial \bar{u}_k}{\partial^0 x_j} + \frac{\partial \Delta u_k}{\partial^0 x_j}\frac{\partial \bar{u}_k}{\partial^0 x_i}\right). \tag{12}$$

The residual $R({}^{m}\mathbf{u}, \bar{\mathbf{u}})$ in equation (11) defines the difference between $a({}^{m}\mathbf{u}, \bar{\mathbf{u}})$ and $\ell(\bar{\mathbf{u}})$, i.e.,

$$R({}^{m}\mathbf{u}, \bar{\mathbf{u}}) = a({}^{m}\mathbf{u}, \bar{\mathbf{u}}) - \ell(\bar{\mathbf{u}}). \tag{13}$$

The equilibrium of a structure is found when R becomes a zero vector.

3 Criteria of stiffness and optimization problem

In order to quantify the stiffness of a geometrically nonlinear structure, several criteria were used and discussed in the literature. They include the end compliance, the strain energy, and the complementary work [1–4]. The relation between the three criteria is given by

$$W(\mathbf{u}) = U(\mathbf{u}) + U_C(\mathbf{u}), \tag{14}$$

where W is the end compliance; U is the strain energy; U_C is the complementary work.

The end compliance is used in most of the studies on stiffness optimization of geometrically nonlinear structures. It is defined by

$$W(\mathbf{u}) = \int_{{}^0\Gamma_N} t_i\, u_i\mathrm{d}\Gamma = \int_{{}^0\Omega} C_{ijkl}\varepsilon_{kl}(\mathbf{u})\hat{\varepsilon}_{ij}(\mathbf{u}; \mathbf{u})\mathrm{d}\Omega. \tag{15}$$

The last equality is obtained by substituting $\mathbf{u}$ for $\bar{\mathbf{u}}$ in equation (2). The form of end compliance given by equation (15) is the same as that of the virtual work of external loads (i.e., Eq. (4)), however, the end compliance is not the true work of external loads (under quasi-static condition).

The true work of external loads under quasi-static condition is equal to the strain energy stored in a structure, and it is given by

$$U(\mathbf{u}) = \frac{1}{2}\int_{{}^0\Omega} C_{ijkl}\varepsilon_{kl}(\mathbf{u})\varepsilon_{ij}(\mathbf{u})\mathrm{d}\Omega. \tag{16}$$

It can be approximated by using a trapezoidal scheme as [1, 6]

$$U(\mathbf{u}) = \sum_{s=1}^{n}\int_{{}^0\Gamma_N}\frac{1}{2}\left(t_i^s + t_i^{s-1}\right)\left(u_i^s - u_i^{s-1}\right)\mathrm{d}\Gamma = \sum_{s=1}^{n}\Delta U^s, \tag{17}$$

where t^s and u^s denote respectively the surface traction and displacement of the s-th load step (here we use a right superscript to denote the index of load step in the numerical solution of equilibrium equation); the displacement u_i^s or $\mathbf{u}^s$ is the converged solution obtained through the Newton-Raphson method for the s-th load step t_i^s; ΔU^s denotes the increment of strain energy from the load step $s-1$ to s.

Assuming that the increments of external load between consecutive load steps are the same, we use Δt_i to denote the increment. In the present study, because only dead load is considered, such an assumption is valid, but it will not be valid for follower load. Based on this assumption, the strain energy can be rewritten as

$$U(\mathbf{u}) = -\frac{1}{2}\int_{{}^0\Gamma_N} u_i^0\Delta t_i\mathrm{d}\Gamma - \sum_{s=1}^{n-1}\int_{{}^0\Gamma_N} u_i^s\Delta t_i\mathrm{d}\Gamma + \left(n - \frac{1}{2}\right)\int_{{}^0\Gamma_N} u_i^n\Delta t_i\mathrm{d}\Gamma. \tag{18}$$

The complementary work U^C of external load is obtained by substituting equations (15) and (16) into equation (14) as [3]

$$U_C(\mathbf{u}) = U(\mathbf{u}) + \frac{1}{2}\int_{{}^0\Omega} C_{ijkl}\varepsilon_{kl}(\mathbf{u})\hat{\eta}_{ij}(\mathbf{u}; \mathbf{u})\mathrm{d}\Omega. \tag{19}$$

It can also be approximated by using a trapezoidal scheme as [2]

$$U_C(\mathbf{u}) = \sum_{s=1}^{n} \int_{{}^0\Gamma_N} \frac{1}{2}(u_i^s + u_i^{s-1})(t_i^s - t_i^{s-1})\mathrm{d}\Gamma = \sum_{s=1}^{n} U_C^s$$
$$= \frac{1}{2}\int_{{}^0\Gamma_N} u_i^0 \Delta t_i \mathrm{d}\Gamma + \sum_{s=1}^{n-1} \int_{{}^0\Gamma_N} u_i^s \Delta t_i \mathrm{d}\Gamma + \frac{1}{2}\int_{{}^0\Gamma_N} u_i^n \Delta t_i \mathrm{d}\Gamma, \tag{20}$$

where ΔU_C^s denotes the increment of complementary work from the load step $s-1$ to s. From equation (19), one can see that the complementary work U_C of a geometrically nonlinear structure is not equal to the strain energy U.

As can be seen in equations (18) and (20), the strain energy and the complementary work can be regarded as a weighted combination of displacements. For instance, $\Delta t_i/2$ or Δt_i are the weights in the complementary work. Minimization of U or U_C offers a control over the displacements of intermediate load steps or the stiffness of intermediate configurations. Therefore, as a stiffness criterion of geometrically nonlinear structures, the strain energy and the complementary work give a more comprehensive description of stiffness than the end compliance. However, the fixed weights of displacements in the weighted combinations are not perfect when they are viewed in the perspective of multi-objective optimization.

In the present study, load-normalized strain energy increments between consecutive load steps are aggregated through the Kreisselmeier-Steinhauser (KS) function [7], and the KS function is proposed as a stiffness criterion of geometrically nonlinear structures. Here, the KS function is defined as

$$\mathrm{KS}(\Delta u) = \frac{1}{\rho} \ln \left[\sum_{s=1}^{n} \exp(\rho \Delta u^s) \right], \tag{21}$$

where $\rho > 0$ is a parameter, and Δu^s is defined by

$$\Delta u^s = \int_{{}^0\Gamma_N} l_i (u_i^s - u_i^{s-1}) \mathrm{d}\Gamma, \tag{22}$$

where l_i, $i = 1, \ldots d$ are the weighting parameters, and they are regarded as the components of an unit fictitious load whose direction is the same as that of the external load t. Because only dead load (magnitude and direction of the load are independent of displacement) is considered in the present study, the weighting parameters l_i are fixed during the optimization.

Comparing equation (22) with equation (17), one can see that Δu^s can be regarded as a load-normalized strain energy increment, i.e., the increment of strain energy ΔU^s is normalized by the average load $(t_i^s + t_i^{s-1})/2$. In addition, suppose that a point load along the negative direction of y-axis is applied to a geometrically nonlinear structure. Then, in equation (22), the components of fictitious load are set to $l_1 = 0$ and $l_2 = -1$. In this case, Δu^s can also be regarded as the increment of displacement at the traction boundary.

The KS function has been used in many optimization problems to aggregate a family of functions into a single function [18–21]. It was proved that the KS function is always bigger than the maximum of the aggregated functions and that the limit of KS function will be equal to the maximum when the parameter ρ approaches positive infinity [22], i.e.,

$$\mathrm{KS}(\Delta u) \geq \max\{\Delta u^s\}, \quad \text{for all } \rho > 0, \tag{23}$$

and

$$\lim_{\rho \to +\infty} \mathrm{KS}(\Delta u) = \max\{\Delta u^s\}. \tag{24}$$

Therefore, when the parameter ρ is very big, the KS function can be regarded as the maximum of load-normalized strain energy increments.

Now, the optimization problem is defined as

$$\begin{aligned} \min \quad & J = \mathrm{KS}(\Delta u) + \alpha P(\Omega) \\ \text{s.t.} \quad & a(\mathbf{u}, \mathbf{v}) = \ell(\mathbf{v}), \quad \forall \bar{\mathbf{u}} \in \mathscr{U} \\ & V(\Omega) - \overline{V} \leq 0 \end{aligned}, \tag{25}$$

where $V(\Omega) = \int_\Omega \mathrm{d}x$ is the volume of a structure; α in the objective function is a fixed positive parameter; $P(\Omega)$ is the perimeter of a structure that is used to regularize the optimization problem [23–26]

$$P(\Omega) = \int_{\partial\Omega} \mathrm{d}\Gamma. \tag{26}$$

The detailed discussions about the perimeter-based regularization in a level set based structure optimization are referred to our previous studies [25, 26] and the references therein.

4 Sensitivity analysis

4.1 Sensitivity of KS (Δu)

We consider first the sensitivity of a generic function of displacement given by

$$\psi(\mathbf{u}) = \int_{{}^0\Gamma_N} w_i u_i \mathrm{d}\Gamma, \tag{27}$$

where w_i is the weight of displacement; u_i is the displacement of an equilibrium configuration (equilibrium configuration of an intermediate load step or the final equilibrium configuration). For the end compliance, the weight is the external load t_i. For the strain energy and the complementary work, the weights can be inferred from equation (18) to equation (20). For the Δu^s in equation (22) the weights are the fictitious load l_i.

The Lagrangian concerned with the generic function $\psi(\mathbf{u})$ is defined as

$$\mathscr{L}({}^0\Omega, \mathbf{u}, \mathbf{v}) = \psi(\mathbf{u}) + \ell(\mathbf{v}) - a(\mathbf{u}, \mathbf{v}), \tag{28}$$

where $\mathbf{v}$ is a Lagrange multiplier for equation (2), and it is regarded as the adjoint displacement.

The partial derivative of $\mathscr{L}$ with respect to $\mathbf{u}$ in the direction $\bar{\mathbf{u}}$ is

$$\left\langle \frac{\partial \mathscr{L}}{\partial \mathbf{u}}, \bar{\mathbf{u}} \right\rangle = \left\langle \frac{\partial \psi}{\partial \mathbf{u}}, \bar{\mathbf{u}} \right\rangle - \left\langle \frac{\partial a}{\partial \mathbf{u}}, \bar{\mathbf{u}} \right\rangle, \tag{29}$$

where

$$\left\langle \frac{\partial \psi}{\partial \mathbf{u}}, \bar{\mathbf{u}} \right\rangle = \int_{{}^0\Gamma_N} w_i \bar{u}_i \mathrm{d}\Gamma = \psi(\bar{\mathbf{u}}), \tag{30}$$

$$\begin{aligned} \left\langle \frac{\partial a}{\partial \mathbf{u}}, \bar{\mathbf{u}} \right\rangle &= \int_{{}^0\Omega} C_{ijkl} \hat{\varepsilon}_{kl}(\mathbf{u}; \mathbf{v}) \hat{\varepsilon}_{ij}(\mathbf{u}; \bar{\mathbf{u}}) \mathrm{d}\Omega \\ &+ \int_{{}^0\Omega} C_{ijkl} \varepsilon_{kl}(\mathbf{u}) \hat{\eta}_{ij}(\mathbf{v}; \bar{\mathbf{u}}) \mathrm{d}\Omega \\ &= a^*(\mathbf{u}, \mathbf{v}; \bar{\mathbf{u}}). \end{aligned} \tag{31}$$

The stationary condition of equation (29) leads to the adjoint equation

$$a^*(\mathbf{u}, \mathbf{v}; \bar{\mathbf{u}}) = \psi(\bar{\mathbf{u}}), \quad \forall \bar{\mathbf{u}} \in \mathscr{U}. \tag{32}$$

Equation (32) is solved for the adjoint displacement $\mathbf{v}$, and the discretized version of $a^*(\mathbf{u}, \mathbf{v}; \bar{\mathbf{u}})$ is the tangent stiffness matrix constructed when the Newton-Raphson algorithm arrives at a converged solution $\mathbf{u}$.

Ignore the shape derivatives defined on boundary ${}^0\Gamma_N$ since ${}^0\Gamma_N$ is not subject to optimization, and note that only ${}^0\Gamma_H$ is optimized, we have the partial derivative of $\mathscr{L}$ with respect to ${}^0\Omega$ in the direction θ

$$\begin{aligned} \frac{\partial \mathscr{L}}{\partial {}^0\Omega}({}^0\Omega, \mathbf{u}, \mathbf{v})(\theta) = - \int_{{}^0\Gamma_H} & \left(C_{ijkl} \varepsilon_{kl}(\mathbf{u}) \hat{\varepsilon}_{ij}(\mathbf{u}; \mathbf{v}) \right) \\ & \times \theta \cdot n \, \mathrm{d}\Gamma, \end{aligned} \tag{33}$$

and shape derivative of the function $\psi(\mathbf{u})$ is

$$\psi' = \frac{\partial \mathscr{L}}{\partial {}^0\Omega}({}^0\Omega, \mathbf{u}, \mathbf{v})(\theta). \tag{34}$$

Through simple derivation, the shape derivative of strain energy can be obtained as

$$U'(\mathbf{u}^n) = \int_{\Gamma_H} \left(\sum_{s=0}^{n} G^s \right) \theta \cdot n \, \mathrm{d}\Gamma, \tag{35}$$

where

$$G^s = -C_{ijkl} \varepsilon_{kl}(\mathbf{u}^s) \hat{\varepsilon}_{ij}(\mathbf{u}^s; \mathbf{v}^s), \tag{36}$$

and $\mathbf{u}^s$ and $\mathbf{v}^s$ are the primary and adjoint displacements of the s-th load step. The shape derivative of complementary work has the same form as equation (35), except that the adjoint displacements $\mathbf{v}^s$ are the solutions to different adjoint equations.

The shape derivative of the KS function equation (21) is

$$\mathrm{KS}' = \sum_{s=1}^{n} \frac{\partial \mathrm{KS}(\Delta u)}{\partial \Delta u^s} (\Delta u^s)' = \sum_{s=1}^{n} \beta^s (\Delta u^s)', \tag{37}$$

where

$$\beta^s = \frac{\exp(\rho \Delta u^s)}{\sum\limits_{s=1}^{n} \exp(\rho \Delta u^s)}, \tag{38}$$

$$\Delta(u^s)' = \int_{{}^0\Gamma_H} (G^s - G^{s-1}) \theta \cdot n \, \mathrm{d}\Gamma, \tag{39}$$

where the adjoint displacements $\mathbf{v}^s$ in G^s and $\mathbf{v}^{s-1}$ in G^{s-1} are the solutions of the following two adjoint equations

$$a^*(\mathbf{u}^s, \mathbf{v}^s; \bar{\mathbf{u}}) = \int_{{}^0\Gamma_N} l_i \bar{u}_i \mathrm{d}\Gamma, \quad \forall \bar{\mathbf{u}} \in \mathscr{U}, \tag{40}$$

$$a^*\left(\mathbf{u}^{s-1}, \mathbf{v}^{s-1}; \bar{\mathbf{u}}\right) = \int_{{}^0\Gamma_N} l_i \bar{u}_i \mathrm{d}\Gamma, \quad \forall \bar{\mathbf{u}} \in \mathscr{U}. \tag{41}$$

The final result of shape derivative is rewritten as

$$\mathrm{KS}' = \int_{{}^0\Gamma_H} \left\{ \sum_{s=1}^{n} \beta^s \left(G^s - G^{s-1} \right) \right\} \theta \cdot n \, \mathrm{d}\Gamma. \tag{42}$$

4.2 Sensitivity of the perimeter term and volume constraint

Now, we consider the objective function J augmented with the volume constraint

$$\bar{J} = \mathrm{KS} + \alpha P + \lambda \left(V - \overline{V} \right), \tag{43}$$

where λ is the Lagrange multiplier for the volume constraint, and it is updated during the optimization according to the the augmented Lagrange multiplier method [27].

Note that only ${}^0\Gamma_H$ is optimized, the shape derivative of $\bar{J}$ is

$$\bar{J}' = \int_{{}^0\Gamma_H} G \theta \cdot n \mathrm{d}s, \tag{44}$$

where

$$G = \sum_{s=1}^{n} \beta^s \left(G^s - G^{s-1} \right) + \alpha \kappa + \lambda, \tag{45}$$

where κ is the curvature of boundary.

According to equation (45), we can readily obtain the steepest descent direction θ by setting

$$\theta(x) \cdot n(x) = -G, \quad \forall x \in {}^0\Gamma_H. \tag{46}$$

It can be seen that such θ will yield $\bar{J} \leq 0$ which implies the descent of $\bar{J}$.

The properties of the optimization problem and the optimization algorithm in the present work do not guarantee global optima. The optimization problem in the present study is not convex, and there are many solutions such as one global and many local minima. The optimization algorithm, i.e., the steepest descent, is based on the gradient at the current point in the design space, thus it sees no global information about the objective function and constraints, and all improvements of the structure is made based on local information at the current point in the design space. Therefore, it cannot perform a global search [28].

5 The level set method

In the level set method [29, 30], the boundary of a shape Ω is represented implicitly through a level set function $\Phi(x)$ as its zero level set, i.e., $\{x \in \mathbb{R}^d \mid \Phi(x) = 0\}$, and $\Phi(x)$ is used to define the inside and outside regions with respect to the boundary as follows

$$\begin{aligned} \Phi(x) = 0 &\Leftarrow \forall x \in \partial\Omega \cap \mathscr{D} \\ \Phi(x) < 0 &\Leftarrow \forall x \in \Omega \quad . \\ \Phi(x) > 0 &\Leftarrow \forall x \in (\mathscr{D} \backslash \overline{\Omega}) \end{aligned} \tag{47}$$

In the present study, Φ is constructed as the signed distance function to the free boundary ${}^0\Gamma_H$ of a structure.

Propagation of ${}^0\Gamma_H$ is modeled by the Hamilton-Jacobi equation

$$\Phi_t + a|\nabla\Phi| = b\kappa|\nabla\Phi|, \tag{48}$$

where

$$a = -\sum_{s=1}^{n} \beta^s \left(G^s - G^{s-1}\right) - \lambda, \quad b = \alpha. \tag{49}$$

A structured grid and the finite difference method (the first order upwind spatial differencing and forward Euler time differencing) are used to solve the equation, and a reinitialization procedure is periodically performed. A PDE based method is used for velocity extension [31]. In the velocity extension, we first compute a and b on Γ_H, then we extend these quantities to the entire reference domain $\mathscr{D}$. More details of the numerical computations of level set can be found in [29, 30].

A move limit strategy [32] is applied to the extended velocity to improve the stability and convergence rate of the level set based optimization. This is achieved by applying the following equation to the extended velocity

$$\tilde{v}^e(x) = v^e(x)\, e^{-(\Phi(x)/h)^2/\gamma}, \tag{50}$$

where v_e is the original extended velocity; $\tilde{v}^e$ is the extended velocity after applying the move limit; h is the grid size of level set computation, $\gamma > 0$ is a parameter that controls the size of move limit. The term $\Phi(x)/h$ stands for the normalization of $\Phi(x)$. The normalization is introduced here so that the value of γ is independent of the grid size h. When the value of γ decreases, the size of move limit decreases. In our present work, γ is set to 0.1.

6 Finite element analysis

Conventionally, an Eulerian method employing a fixed mesh and ersatz material is used for the finite element analysis in the level set based method of topology optimization. Instead of solving the equilibrium equation on Ω, it solves the equation on the entire reference domain $\mathscr{D}$ with the void $\mathscr{D}\backslash\overline{\Omega}$ being mimicked by an artificial weak material. However, in the present study of topology optimization of geometrically nonlinear structures, if the elements filled by the artificial weak material undergo excessively large deformation, their tangent stiffness matrix may become negative, and this will cause numerical instability.

In the present work, being similar to the method proposed by Ha and Cho [14], we do the finite element analysis by modifying a fixed background mesh and do not use the artificial weak material, as proposed in our previous study [15] and used in [25, 33]. A similar method was also proposed by Allaire et al. [34]. Before the optimization, we set up a background triangle mesh for the reference domain $\mathscr{D}$. In each iteration of the optimization, modification of the fixed background mesh is performed to obtain a mesh that is conformal to the geometry of the structure. The detail of the modification is referred to [15].

7 Numerical examples

In this section, several examples in two dimensions are given, although there exists no conceptual difficulty to do the numerical examples in three dimensions. In these examples, it is assumed that the solid material has a Young's modulus E = 3 GPa and Poisson's ratio $\nu = 0.4$. For all the examples, the plane stress state is assumed; thickness is 0.1 m; 3-node triangle finite element is used.

7.1 Example 1

The optimal design problem of the first example is shown in Figure 1. The reference domain is a rectangle of size 1 m × 0.25 m. The left side is fixed. The upper bound of volume is 50% of the reference domain. There are 12,318 triangle elements in the background mesh for the reference domain $\mathscr{D}$. A 500 × 125 rectilinear grid is used for level set computations. The initial design is shown in Figure 1.

We set load to 100, 200, 300, and 500 kN respectively in five optimizations, and in each optimization the load is evenly divided into five load steps. The parameter in the KS function is set to ρ = 20, and the parameter for the perimeter term in the objective function is set to $\alpha = 1 \times 10^{-4}$ because the order of magnitude of KS function is 10^{-1}. The optimal structures are shown in Figure 3. The history of convergence with the load in optimization being 100 kN is provided as an example in Figure 4.

In order to compare the proposed KS function with the end compliance and the complementary work, optimizations are also done by replacing the KS function in the objective function with W or U_C. In these optimizations, the parameter for the perimeter term in the objective function is set to $\alpha = 1$ because the order of magnitude of W and U_C is

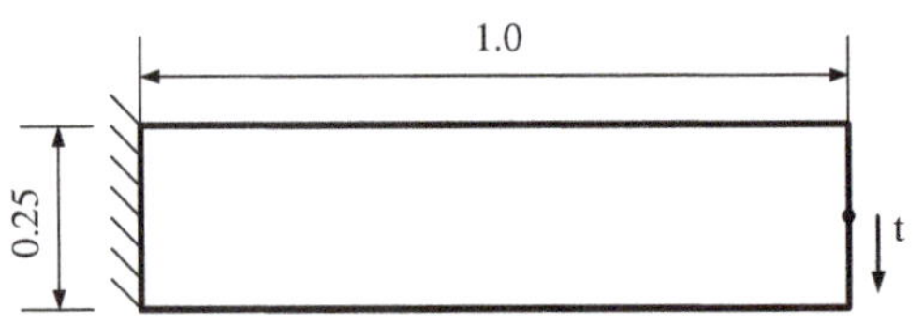

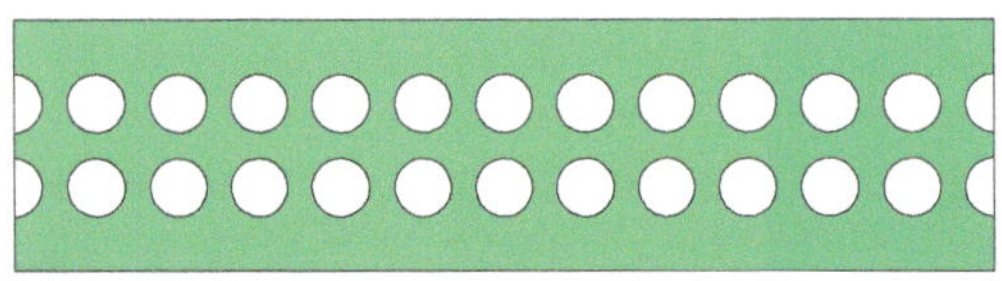

Figure 2. (a) Design problem and (b) initial design.

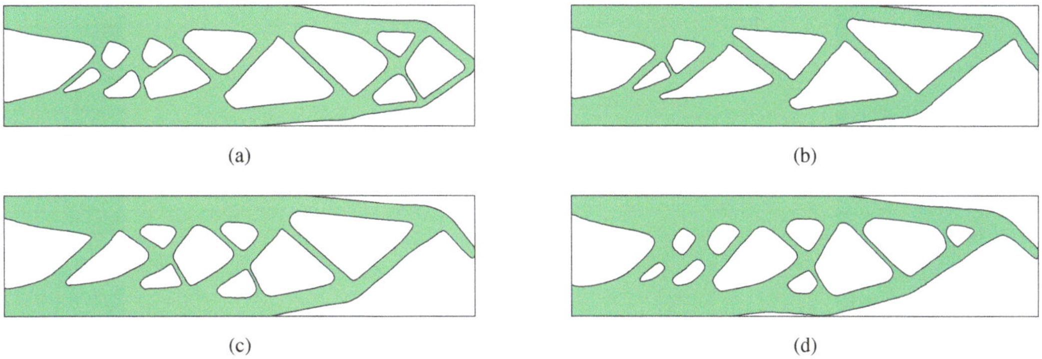

Figure 3. Optimal structures with the KS function being used as the stiffness criterion. (a) t = 100 kN, (b) t = 200 kN, (c) t = 300 kN, (d) t = 500 kN.

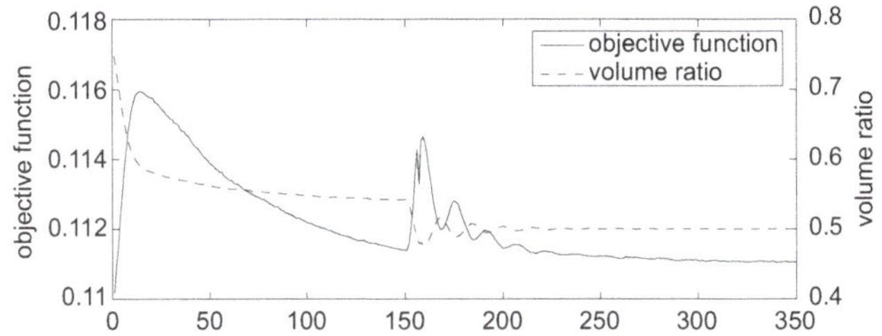

Figure 4. History of convergence.

10^4–10^5. The optimal structures obtained when the end compliance is used as the stiffness criterion are shown in Figure 5, and those of the complementary work are shown in Figure 6. The deformed shape of the optimal structures are shown in Figure 7. Figures 7a, d, g, j are the deformed optimal structures obtained when the KS function is chosen as the stiffness criterion; Figure 7b, e, h, k are of the end compliance; Figure 7c, f, i, l are of the complementary work.

From these results one can see that the shape and topology of all the optimal structures depend on the load level. In contrast, for a linear elastic structure, the results will be independent of load level due to the principle of superposition [5].

When the load is 500 kN, all the optimal structures have a slender beam at the right end except that their lengths are different. Such a slender beam is not stiff, but it is able to effectively reduce the length of load arm in the final equilibrium configuration, as shown in Figure 7, thus effectively increasing the stiffness of the final equilibrium configuration. Such a phenomenon is most prominent in the optimal structures obtained when the end compliance is chosen as the stiffness criterion, since the end compliance concerns only with the stiffness of the final equilibrium configuration. However, the slender beam at the right end, for instance the one shown in Figure 5d, is not efficient for smaller loads, for instance an 100 kN load. One can imagine that even if the slender beam in Figure 5d is changed to connect with the remaining structure *through a hinge*, the end compliance will not increase significantly.

Detailed quantitative comparisons are given in Table 1. As one can see that when the end compliance is chosen as the stiffness criterion in optimization, the corresponding optimal structures have the smallest end compliance and the smallest strain energy (except when load is 200 kN), and that such a trend is more prominent when the load level is high. But as explained above, this phenomenon only indicates that the end compliance is a good stiffness criterion for the final equilibrium configuration. Actually, it was suggested that the end compliance may not be the best criterion of stiffness for geometrically nonlinear structures [2, 5].

In order to overcome such a problem, Buhl et al. [2] proposed to minimize the complementary work. Minimizing the complementary work can be regarded as a multi-objective optimization whose objectives are the displacements of many

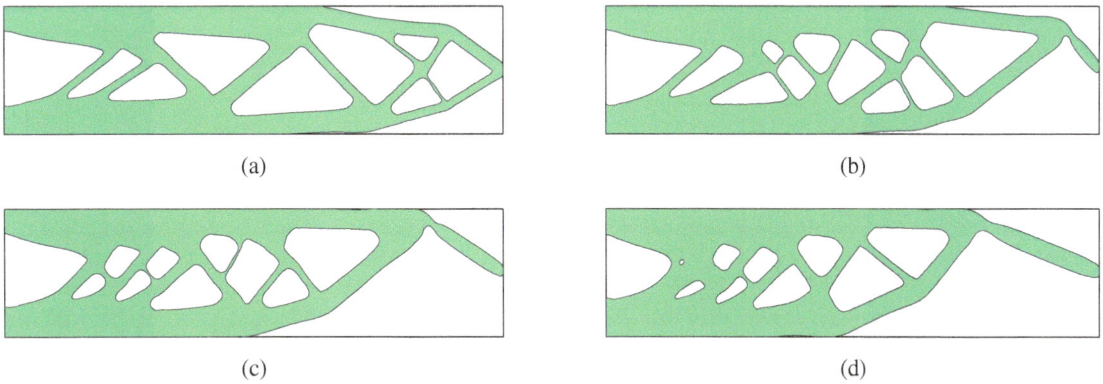

Figure 5. Optimal structures with W being used as the stiffness criterion. (a) t = 100 kN, (b) t = 200 kN, (c) t = 300 kN, (d) t = 500 kN.

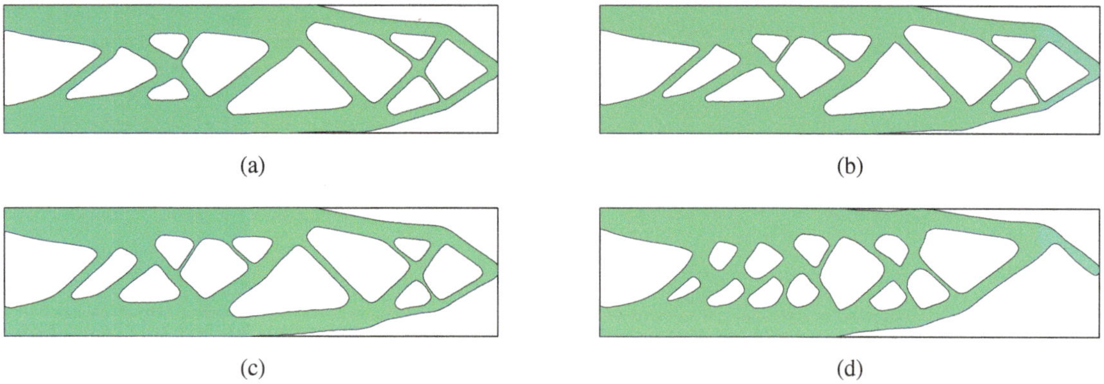

Figure 6. Optimal structures with UC being used as the stiffness criterion. (a) t = 100 kN, (b) t = 200 kN, (c) t = 300 kN, (d) t = 500 kN.

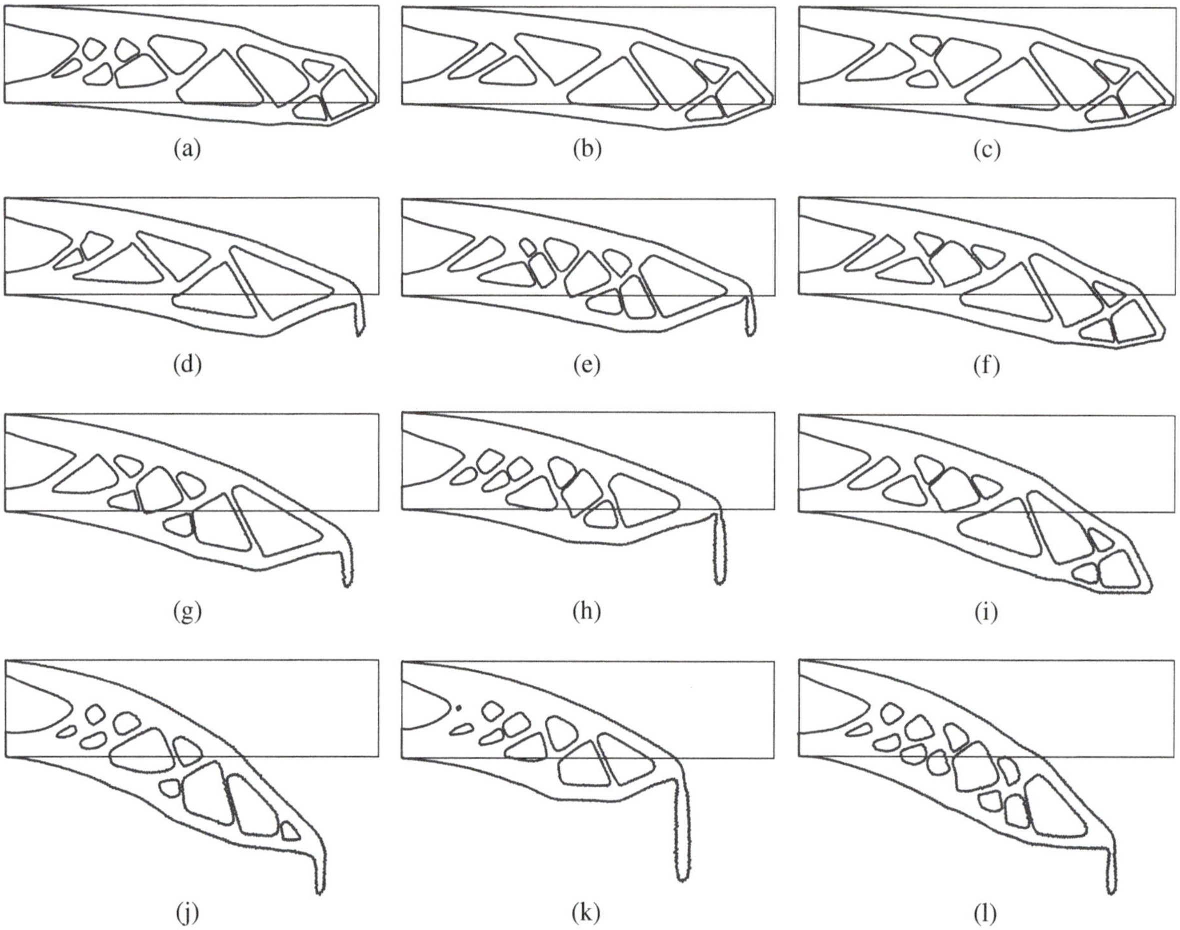

Figure 7. Deformed optimal structures of different stiffness criteria. (a) KS (t = 100 kN), (b) U (t = 100 kN), (c) U_C (t = 100 kN), (d) KS (t = 200 kN), (e) U (t = 200 kN), (f) U_C (t = 200 kN), (g) KS (t = 300 kN), (h) U (t = 300 kN), (i) U_C (t = 300 kN), (j) KS (t = 500 kN), (k) U (t = 500 kN), (l) U_C (t = 500 kN).

deformed configurations. As one can see in the results shown in Figure 6, when the load is 200 kN or 300 kN there is no slender beam at the right end, and when the load is 500 kN the length of the slender beam is much shorter than that shown in Figure 5d. In addition, as can be seen in Table 1, when U_C is chosen as the stiffness criterion, the optimal structures have the smallest U_C (except when load is 200 kN).

The inspiring previous studies are extended in our present work by considering the aggregated load-normalized strain energy increments between consecutive load steps. From the optimal structure shown in Figure 3d, one can see that when the load is 500 kN the length of slender beam is much shorter than that shown in Figure 5d and is comparable to that shown in Figure 6d. More importantly, the quantitative results given in

Table 1. Quantitative results of optimization with different criteria of stiffness. (Although there is no parameter ρ in the optimization of end compliance and complementary work, ρ is set to 20 for the comparison. The volume of all the optimal structures are 50% of the reference domain. The unit of W, U, U_C is J.)

Stiffness criterion in optimization	100 kN				200 kN			
	KS	W	U	U_C	KS	W	U	U_C
KS	0.11	12,083	5997	6086	0.13	45,210	21,279	23,931
W	0.11	11,845	5875	5969	0.13	45,518	20,740	24,777
U_C	0.11	11,846	5882	5964	0.13	45,786	22,345	23,441
Stiffness criterion in optimization	300 kN				500 kN			
KS	0.14	96,090	44,111	51,978	0.18	239,634	104,774	134,860
W	0.15	95,395	38,009	57,385	0.22	220,999	77,043	143,956
U_C	0.15	98,435	46,911	51,523	0.18	240,152	104,942	135,209

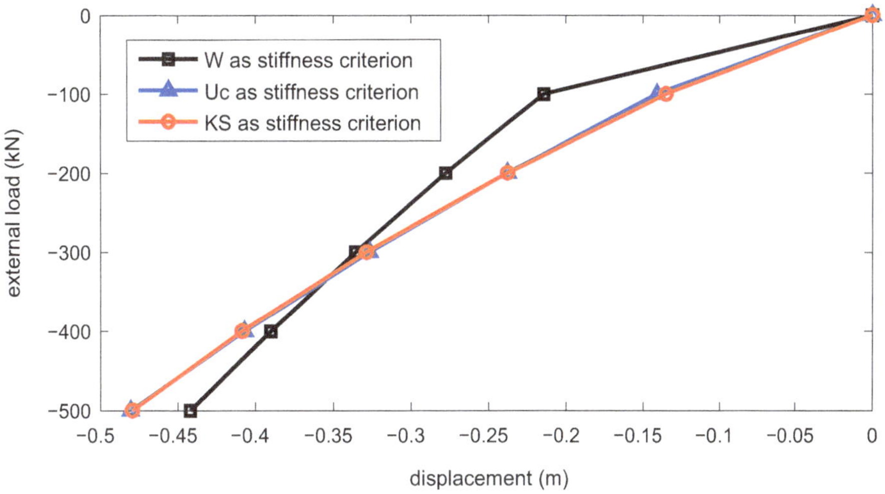

Figure 8. Load-displacement curves of the optimal structures (load is 500 kN).

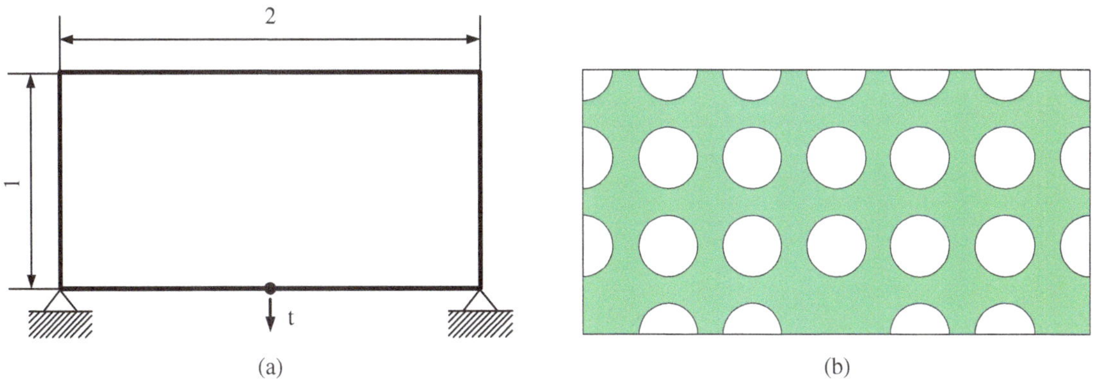

Figure 9. (a) Design problem and (b) initial design of the second example

Table 1 show that when the KS function is chosen as the stiffness criterion in optimization, the optimal structures have smaller end compliance and smaller strain energy as compared to the optimal structures obtained when the complimentary work is chosen as the stiffness criterion (except when load is 100 kN).

Comparison of load-displacement curves of the optimal structures obtained when load is 500 kN in the optimization are shown in Figure 8. As can be seen in this figure, when the load level is below 300 kN, the displacement of the optimal structure obtained when the end compliance is chosen as the stiffness criterion is the largest, which indicates that the slender beam at the right end shown in Figure 5d is not stiff. The red curve is almost the same as the blue curve, nevertheless Δu^1 of the red curve is smaller than that of the blue curve.

As mentioned in Section 3, when the parameter ρ in the KS function approaches positive infinity, the limit of the KS function will be equal to the maximum of Δu^s. In this example,

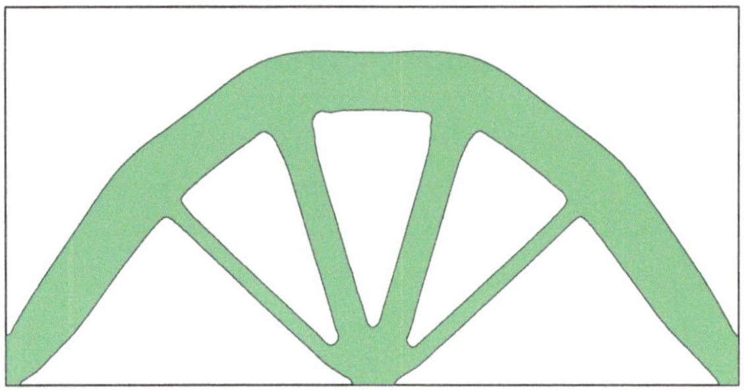

Figure 10. The optimal structure of the second example.

we observed that Δu^1 is always the maximal one of Δu^s because when the structure deforms the length of load arm decreases and hence stiffness increases. Therefore, a bigger value of ρ in this example will let Δu^1 get a more significant role in the optimization. It seems that a higher value of ρ can eliminate the slender beam at the right end. However, we also observed that when the optimal structure obtained with an 100 kN load is subjected to a 500 kN load, the structure buckles and is not stable. Therefore, in the optimization when load is 500 kN, Δu^1 should not dominate the optimization (recall that Δu^1 is the load normalized strain energy increment when load is 100 kN), and the value of ρ should not be too big. In the present example, we see that $\rho = 20$ is a good choice. For the structure shown in Figure 3d, the Δu^s are: $\Delta u^1 = 0.1345$, $\Delta u^2 = 0.1033$, $\Delta u^3 = 0.0907$, $\Delta u^4 = 0.0800$, $\Delta u^5 = 0.0709$, and the KS function is 0.1829. In fact, the parameter ρ in the KS function offers to designer a way to adjust the weights of each Δu^s, but the weights in the complementary work are fixed.

7.2 Example 2

The optimal design problem of the second example is shown in Figure 8. The reference domain is a rectangle of size 2 m × 1 m. The upper bound of volume is 30% of the reference domain. There are 10,934 triangle elements in the background mesh for the reference domain $\mathscr{D}$. A 600 × 300 rectilinear grid is used for level set computations. The initial design is shown in Figure 8. The load is 1200 kN, and it is evenly divided into five load steps. The parameter in the KS function is set to $\rho = 400$. The value of ρ is much bigger than

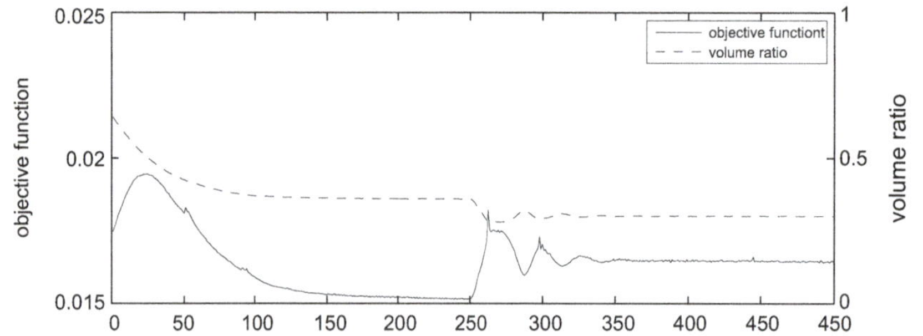

Figure 11. History of convergence of the second example.

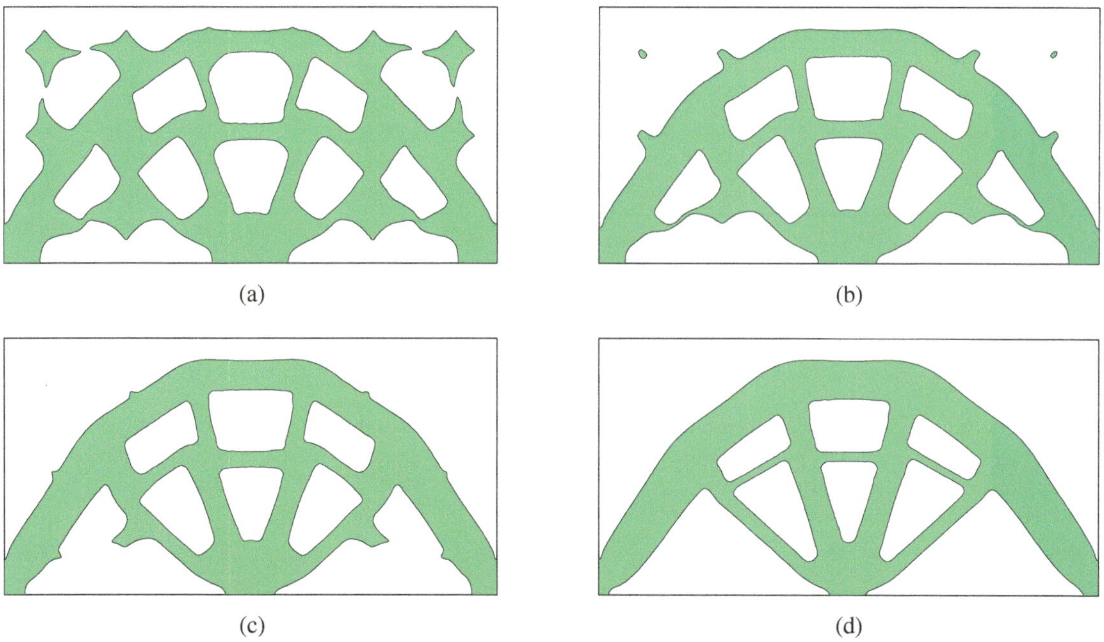

Figure 12. Intermediate designs of the optimization of the second example. (a) Step 55, (b) step 90, (c) step 110, (d) step 200.

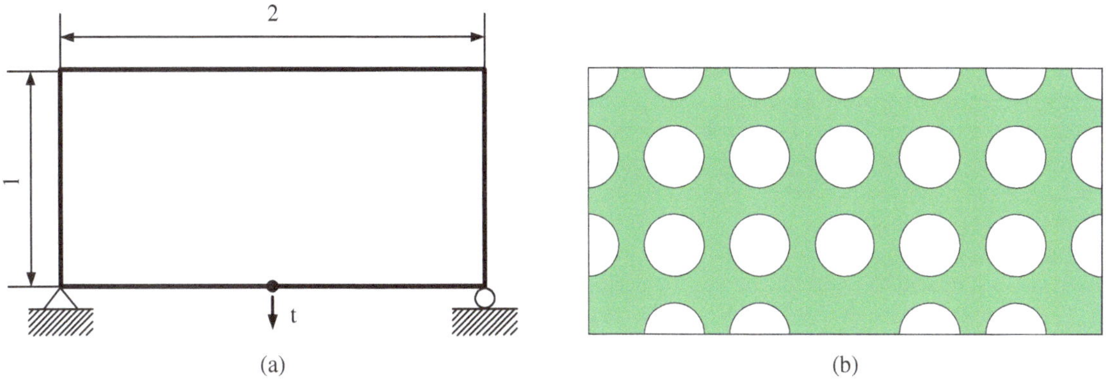

Figure 13. (a) Design problem and (b) initial design of the third example.

Figure 14. The optimal structure of the third example.

that of the first example because the Δu^s in this example are closer to zero, and we observed that when the values of the aggregated functions are closer to zero, ρ should be bigger to ensure the KS function reasonably approximate the maximum. The parameter for the perimeter term in the objective function is set to $\alpha = 1 \times 10^{-4}$.

The optimal structure is shown in Figure 10. The history of convergence is shown in Figure 11, from which one can see that the convergence is smooth. Four intermediate designs of the optimization are shown in Figure 12, from which one can see that the shape and topology were changed during the optimization. The end compliance is $W = 67{,}823$ J; the strain energy is $U = 34{,}673$ J; the complementary work is $U_C = 33{,}150$ J; the KS function is KS = 0.015. The Δu^s are: $\Delta u^1 = 0.0108$, $\Delta u^2 = 0.0109$, $\Delta u^3 = 0.0111$, $\Delta u^4 = 0.0115$, $\Delta u^5 = 0.0121$; the volume ratio is 30%. From the values of Δu^s, one can see that the optimal structure is softening as the load increases, which is different from the hardening behavior in the first example.

If the structure is linear elastic, a well-known Michell's solution to the current design problem is available in the literature [35–37]. The shape and topology of the optimal geometrically nonlinear structure is almost the same as the well known solution.

7.3 Example 3

The optimal design problem of the third example is shown in Figure 12. The upper bound of volume is 40% of the reference domain. There are 10,934 triangle elements in the background mesh for the reference domain $\mathscr{D}$. A 600 × 300 rectilinear grid is used for level set computations. The initial design is shown in Figure 12. The load is 1500 kN, and it is evenly divided into five load steps. The parameter in the KS function is set to $\rho = 400$. The parameter for the perimeter term in the objective function is set to $\alpha = 1 \times 10^{-4}$.

The optimal structure is shown in Figure 14. The history of convergence is shown in Figure 15, from which one can see that the convergence is smooth. The end compliance is $W = 124{,}948$ J; the strain energy is $U = 64{,}229$ J; the complementary work is $U_C = 60{,}719$ J; the KS function is KS = 0.021. The Δu^s are: $\Delta u^1 = 0.0157$, $\Delta u^2 = 0.0160$, $\Delta u^3 = 0.0164$, $\Delta u^4 = 0.0171$, $\Delta u^5 = 0.0181$; the volume ratio is 40%. From the values of Δu^s, one can see that the optimal structure is softening as the load increases.

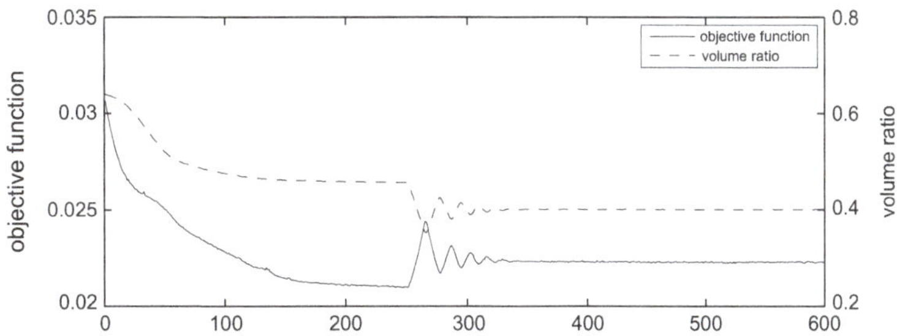

Figure 15. History of convergence of the third example.

8 Conclusion

In the present paper, the aggregated load-normalized strain energy increments between consecutive load steps is proposed as a stiffness criterion of geometrically nonlinear structures. A topology optimization problem is then defined to minimize the KS function together with the perimeter of structure and a volume constraint. The finite element analysis is done by remeshing, and artificial weak material is not used. Therefore, the numerical instability related to the artificial weak material is eliminated. The shape derivative of free boundary is derived by using a Lagrangian function and the adjoint method. The topology optimization problem is solved by using the level set method. Several numerical examples in two dimensions are provided. Other criteria of stiffness, i.e., the end compliance and the complementary work, are compared. The results shown that the proposed KS function gives a comprehensive description of the stiffness of geometrically nonlinear structures.

Acknowledgements. This research work is supported by the National Natural Science Foundation of China (Grant No. 51575203).

References

1. Yuge K, Iwai N, Kikuchi N. 1999. Optimization of 2-D structures subjected to nonlinear deformation using the homogenization method. Struct. Optim., 17, 286–299.
2. Buhl T, Pedersen CBW, Sigmund O. 2000. Stiffness design of geometrically nonlinear structures using topology optimization. Struct. Multidiscip. Optim., 19, 93–104.
3. Gea HC, Luo J. 2001. Topology optimization of structures with geometrical nonlinearities. Comput. Struct., 79, 1977–1985.
4. Kemmler R, Lipka A, Ramm E. 2005. Large deformations and stability in topology optimization. Struct. Multidiscip. Optim., 30, 459–476.
5. Bruns TE, Tortorelli DA. 2001. Topology optimization of nonlinear elastic structures and compliant mechanisms. Comput. Methods Appl. Mech. Eng., 190(26–27), 3443–3459.
6. Huang X, Xie XM. 2008. Topology optimization of nonlinear structures under displacement loading. Eng. Struct., 30, 2057–2068.
7. Kreisselmeier G, Steinhauser R. 1979. Systematic control design by optimizing a vector performance index, International Federation of Active Controls Symposium on Computer-Aided Design of Control Systems, Zurich, Switzerland.
8. Bruns TE, Tortorelli DA. 2003. An element removal and reintroduction strategy for the topology optimization of structures and compliant mechanisms. Int. J. Numer. Meth. Eng., 57, 1413–1430.
9. Yoon GH, Kim YY. 2005. Element connectivity parameterization for topology optimization of geometrically nonlinear structures. Int. J. Solids Struct., 42, 1983–2009.
10. Lahuerta RD, Oes ETS, Campello EMB, Pimenta PM, Silva ECN. 2013. Towards the stabilization of the low density elements in topology optimization with large deformation. Comput. Mech., 52, 779–797.
11. Wang F, Lazarov BS, Sigmund O, Jensen JS. 2014. Interpolation scheme for fictitious domain techniques and topology optimization of finite strain elastic problems. Comput. Methods Appl. Mech. Eng., 276, 453–472.
12. Cho S, Kwak J. 2006. Topology design optimization of geometrically non-linear structures using meshfree method. Comput. Methods Appl. Mech. Eng., 195, 5909–5925.
13. He Q, Kang Z, Wang YQ. 2014. A topology optimization method for geometrically nonlinear structures with meshless analysis and independent density field interpolation. Comput. Mech., 54, 629–644.
14. Ha SH, Cho S. 2008. Level set based topological shape optimization of geometrically nonlinear structures using unstructured mesh. Comput. Struct., 86, 1447–1455.
15. Xia Q, Shi T, Liu S, Wang MY. 2012. A level set solution to the stress-based structural shape and topology optimization. Comput. Struct., 90–91, 55–64.
16. Bathe KJ. 2004. Finite element procedures. Prentice Hall: Upper Saddle River, New Jersey.
17. Reddy JN. 2004. An introduction to nonlinear finite element analysis. Oxford University Press: Oxford.
18. Yang RJ, Chahande AI. 1995. Automotive applications of topology optimization. Struct. Optim., 9, 245–249.
19. Akgun MA, Haftka RT, Wu KC, Walsh JL, Garcelon JH. 2001. Efficient structural optimization for multiple load cases using adjoint sensitivities, AIAA J., 39, 511–516.
20. Poon NMK, Martins JRRA. 2007. An adaptive approach to constraint aggregation using adjoint sensitivity analysis. Struct. Multidiscip. Optim., 34, 61–73.
21. Luo YJ, Wang MY, Kang Z. 2013. An enhanced aggregation method for topology optimization with local stress constraints. Comput. Methods Appl. Mech. Eng., 254, 31–41.
22. Raspanti CG, Bandoni JA, Biegler LT. 2000. New strategies for flexibility analysis and design under uncertainty. Comput. Chem. Eng., 24, 2193–2209.
23. Allaire G, Jouve F, Toader AM. 2004. Structural optimization using sensitivity analysis and a level-set method. J. Comput. Phys., 194, 363–393.
24. Wang MY, Wang XM. 2004. PDE-driven level sets, shape sensitivity and curvature flow for structural topology optimization. CMES-Comput. Model. Eng. Sci., 6, 373–395.
25. Xia Q, Wang MY, Shi TL. 2014. A level set method for shape and topology optimization of both structure and support of continuum structures. Comput. Methods Appl. Mech. Eng., 272, 340–353.
26. Xia Q, Wang MY, Shi TL. 2015. Topology optimization with pressure load through a level set method. Comput. Methods Appl. Mech. Eng., 283, 177–195.
27. Nocedal J, Wright SJ. 1999. Numerical optimization. Springer: New York.
28. Rozvany GIN. 2001. Aims, scope, methods, history and unified terminology of computer-aided topology optimization in structural mechanics. Struct. Multidiscip. Optim., 21, 90–108.
29. Sethian JA. 1999. Level set methods and fast marching methods: evolving interfaces in computational geometry, fluid mechanics, computer vision, and materials science, 2nd edn. Cambridge Monographs on Applied and Computational Mathematics, Cambridge University Press: Cambridge, UK.
30. Osher S, Fedkiw R. 2002. Level set methods and dynamic implicit surfaces. Springer-Verlag: New York.
31. Peng DP, Merriman B, Osher S, Zhao HK, Kang M. 1999. A PDE-based fast local level set method. J. Comput. Phys., 155, 410–438.
32. Xia Q, Wang MY, Shi TL. 2013. A move limit strategy for level set based structural optimization. Eng. Optim., 45, 1061–1072.

33. Xia Q, Shi T, Liu S, Wang MY. 2013. Shape and topology optimization for tailoring stress in a local region to enhance performance of piezoresistive sensors. Comput. Struct., 114–115, 98–105.
34. Allaire G, Dapogny C, Frey P. 2014. Shape optimization with a level set based mesh evolution method. Comput. Methods Appl. Mech. Eng., 282, 22–53.
35. Michell AGM. 1904. The limits of economy of material in frame-structures. Phil. Mag., 8, 589–597.
36. Hemp WS. 1973. Michell's structural continua, in Optimum Structures. Clarendon Press: Oxford, Ch. 4.
37. Xie YM, Steven GP. 1993. A simple evolutionary procedure for structural optimization. Comput. Struct., 49, 885–896.

19

Current source converter based D-STATCOM for voltage sag mitigation

Moirangthem Deben Singh*, Ram Krishna Mehta, and Arvind Kumar Singh

Department of Electrical Engineering, North Eastern Regional Institute of Science & Technology, 791 109 Nirjuli, Arunachal Pradesh, India

Abstract – This paper presents a novel method of realizing one of the custom power controllers, the distribution static synchronous compensator (D-STATCOM) using current source converter (CSC) topology. Almost all the custom power controllers such as dynamic voltage restorer (DVR), unified power quality conditioner (UPQC) including D-STATCOM are generally designed and implemented by using voltage source converters (VSC) and not much research publications with CSC based approach has been reported over the last one decade. Since the D-STATCOM is a current injection device, its performance can be improved when realized by a current-source converter which can generate a controllable current directly at its output terminals and offers many advantageous features. In this paper, an attempt has been made to study the performance of a CSC based D-STATCOM suitable for use in the power distribution system in order to mitigate voltage sag and improve power quality. The proposed model uses a three leg CSC whose switching strategy is based on sinusoidal pulse width modulation (SPWM). The model has been simulated in the Matlab/Simulink environment. The results of the simulation runs under steady state and dynamic load perturbation provide excellent voltage and current waveforms that support the justification of the proposed model.

Key words: Current source converter, Custom power controller, Distribution static synchronous compensator, Proportional integral, Sinusoidal pulse width modulation.

1 Introduction

The emergence deregulated power system operation coupled with the increasing applications of power electronics technology in power transmission and distribution system has led to the research and development of new engineering systems capable of solving various types of power quality problems. The power electronics plays a vital role in FACTS and custom power (CP) devices [1, 2]. The quality of electrical power which is supplied to the customers can be judged in terms of constant voltage magnitude i.e., no voltage sags or swell, constant frequency, constant power factor, balanced phases, sinusoidal waveform i.e., no harmonic content, lack of interruptions and capability to withstand faults and to recover quickly [3]. The VSC topology has been using as the basic building block of the new generation of power electronics controllers emerging from FACTS and custom power research due to the following reasons:

(a) The CSC topology is more complex than a VSC topology in both power and control circuits. Filter capacitors are used at the ac terminals of a CSC to improve the quality of the output ac current waveforms. This adds to the overall cost of the converter. Furthermore, filter capacitors resonate with the ac-side inductances. As a result, some of the harmonic components present in the output current might be amplified, causing high harmonic distortion in the ac-side current. Besides, conventional bi-level switching scheme cannot be used in CSC.

(b) Unless a switch of sufficient reverse voltage withstanding capability such as Gate-Turn-Off (GTO) thyristor is used, a diode has to be placed in series with each of the switches in CSC. This almost doubles the conduction losses compared with the case of VSC.

*e-mail: mdsingh2007@gmail.com

(c) The dc-side energy-storage element in CSC topology is an inductor, whereas that in VSC topology is a capacitor. The power loss of an inductor is expected to be larger than that of a capacitor because of the need to store energy by circulating current in inductors which are more lossy than capacitive energy storage [4]. Thus, the efficiency of a CSC is expected to be lower than that of a VSC [5].

However, recent advancement in power semiconductor switching devices and control strategies of CSC topology may overcome the drawbacks mentioned above in the near future because of the following reasons:

(i) The CSC is usually more reliable and fault tolerant than a VSC due to the presence of large series inductor which limits the rate of rise of current in the event of a fault [4].
(ii) Due to the presence of the ac-side capacitors both voltage and current waveforms at the output terminals of a CSC are good sinusoids. The capacitors are the inherent filter for the CSC. The problem of the resonance between the capacitances and inductances on the ac-side can be overcome by careful design of the capacitor based filter circuit and introduction of sufficient damping using proper control methods [5]. Furthermore, all the switching problems faced in the early stages of CSC development can be overcome by employing tri-level switching scheme which has become a standard technique in the control of CSC [6].
(iii) Integrated Gate Commutated Thyristor (IGCT) having high ratings, high reverse voltage blocking capability, low snubber requirements, lower gate-drive power requirements than GTO, and also has higher switching speed than GTO, is the optimum combination of the characteristics demanded in high-power applications [7]. Using the state-of-the-art technology of the semiconductor switches, there will be no need for the series diode in the CSC topology anymore.
(iv) The dc-side losses are expected to be minimized by using superconductive materials in the construction of the dc-side reactor [5].

In comparison with the VSC topology, the application of CSC topology in D-STATCOM is expected to achieve many advantages. The direct output of a CSC is a controllable ac current, whereas that of a VSC is a controllable ac voltage. When operated under SPWM technique [8], the magnitudes of the harmonic components in both converters are directly proportional to the magnitudes of the fundamental components of their direct output quantities. Under the normal operating conditions, the current injected by D-STATCOM is a small percentage of the line current. Hence, when CSC topology is used, the current harmonics are also small. However, when VSC topology is used, for a small injected current, the output voltage of VSC is large and very close to the system voltage. This results in large voltage harmonics, leading to current harmonics that are larger than those generated by CSC, and thus more costly to filter. The other aspect of comparison is the dc-side energy storage requirement. When the D-STATCOM is realized by a CSC, the dc-side current is just larger than the peak value of the required injected current which is a small percentage of the line current. However, when a VSC is used to inject reactive power to the system, the dc-side voltage must be larger than the peak value of the system line-to-line voltage so that the reactive power can be exchanged between the D-STATCOM and the ac distribution system. Hence, the dc energy storage requirement of the CSC topology is expected to be lower than that of the VSC topology when it is used to implement a D-STATCOM system for mitigation of voltage sag.

Although, D-STATCOM can be used for the improvement of quality and reliability of power supplied to consumers in many ways, the aim of this paper is restricted to investigate the performance of a CSC based D-STATCOM for mitigating voltage sag phenomenon which is considered to be one of the well known unwanted power quality problem resulting from sudden change in load connected to a power distribution system as a first step. The simulation results of the proposed model are also presented to validate the model for its suitability in mitigating the voltage sag and provide improved power quality to the end-users.

2 Power quality problems

The term power quality is concerned with deviations of the voltage from its ideal waveform (voltage quality) and deviations of the current from its ideal waveform (current quality). Such deviation is known as "power quality phenomenon" or "power quality disturbance" [9]. Some examples of the power quality problem include- impulsive and oscillatory transients, short duration voltage variations (sag or dip, swell, interruption), long duration voltage variation (undervoltage, overvoltage, sustained oscillation), voltage imbalance, waveform distortion (harmonics, notching, dc offset), voltage flicker, etc. These problems are generally caused by the nature, faults on transmission or distribution system and also by the power consumers. The power transmission lines are exposed to the forces of nature and its loadability limit is usually determined by either stability considerations or by thermal limits. Although the power quality problem is a distribution side problem, transmission lines frequently have an impact on the quality of power supplied. It is however to be noted that while most of the problems associated with transmission systems arise due to the forces of nature or due to the interconnection of power systems, individual customers are responsible for a more substantial fraction of the power quality problems in the distribution side [4]. The flexible ac transmission system (FACTS) and the custom power devices are the two major power electronics based initiatives to counter the power quality problems. Although FACTS and custom power initiations share the same technological base they have different technical and economic objectives. The FACTS controllers are aimed at the transmission level whereas custom power controllers are aimed at the distribution level particularly at the point of connection of the electricity distribution on company with clients having sensitive loads and independent generators. Custom power focuses primarily on the reliability and power quality. However, voltage regulation, voltage balancing and harmonic cancellation may also get benefited from this technology [3].

2.1 Voltage sag and interruption

Although electric power consumers are very much concerned about the issues of power quality, still there is no consensus amongst the various agencies in defining the term power quality. However, it has become a widely used term and it is the best term available at the moment. The term power quality has gained some official status in IEEE through the name of SSC 22 (Standards Coordinating Committee). But the international standards setting organization in electrical engineering, the IEC does not use this term in any of its standard documents instead it uses the term electromagnetic compatibility which is not the same as power quality. However, these two terms has a strong overlap relationship. Power quality is the combination of voltage quality and current quality. It is concerned with the deviations of voltage and/or current from the ideal state. It should be noted that power quality has nothing to do with the deviations of the product of voltage and current (i.e. power) from any ideal shape [9].

The voltage sag is a power quality problem phenomenon which falls under the category of short duration voltage variation. Any variation in the supply voltage (rms) for duration not exceeding 1 min is called a short duration voltage variation. Usually such variations are caused by system faults, energization of heavy loads that require large inrush currents and intermittent loose connection in the power wiring. Voltage sag is a fundamental frequency decrease in the supply voltage for a short duration. The duration of voltage sag varies between five cycles to a minute [4]. The interest in voltage sag is mainly due to the problems caused by several types of equipments viz. adjustable speed drives (ASD), process control equipment, and computers which are highly sensitive. Some of the loads may trip when the rms voltage drops below 90% for longer than 1 or 2 cycles. If this is the process control equipment of a paper mill, one can imagine that the damage due to voltage sags can be enormous. Of course, voltage sag is not as damaging to industry as a (long or short) interruption. But as there are far more voltage sags than interruptions, the total damage due to voltage sags is still larger. Short interruptions and most long interruptions are originated from the local distribution network. However, voltage sags at equipment terminals can also be due to short circuit faults 100 s of kilometers away in the transmission system. Hence, voltage sag is much more of a "global" problem than an interruption. Reducing the number of interruptions typically requires improvements on one feeder and reducing the number of voltage sags requires improvements on several feeders and often even at transmission lines far away [9].

The voltage interruption according to IEEE standard 1159 or just "interruption" according to IEEE standard 1250 is a condition in which the voltage at the supply terminals is close to zero. Close to zero is by the IEC defined as lower than 1% of the declared voltage and by the IEEE as lower than 10%. The voltage interruptions are generally caused by faults which subsequently trigger protection measures. Some other causes include power system protection operation when there is no fault present, broken conductors not triggering protective measures and operator interventions. The voltage interruption can be subdivided based on their durations. The IEC uses the term long interruptions for interruptions longer than 3 min and the term short interruptions for interruptions lasting up to 3 min. However, in IEEE the terms momentary, temporary and sustained interruptions are used [9]. The problem of voltage sag at distribution network can be mitigated by using VSC based custom power devices such as D-STATCOM or dynamic voltage restorer (DVR). In this paper, a CSC based D-STATCOM is proposed for mitigating the voltage sag.

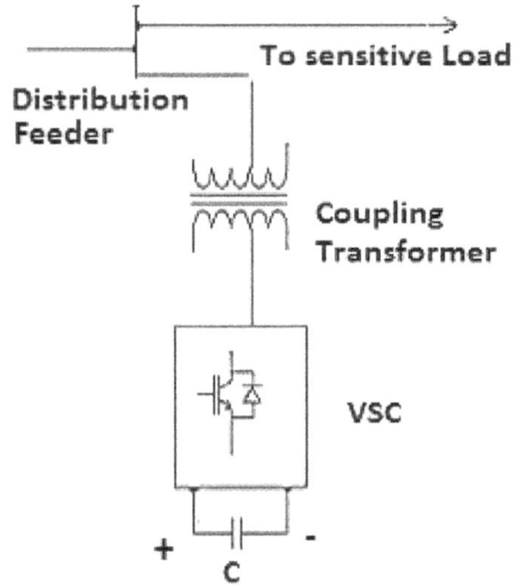

Figure 1. Schematic diagram of a VSC based D-STATCOM.

2.2 D-STATCOM configuration

The D-STATCOM is a shunt connected custom power device. If the FACTS device static synchronous compensator (STATCOM) is used in the low voltage distribution system, it is identified as D-STATCOM. However, there is a substantial difference in the operating characteristics of a STATCOM and a D-STATCOM. The STATCOM is required to inject a set of three balanced quasi-sinusoidal voltages which are phase displaced by 120° but the D-STATCOM must be able to inject an unbalanced and harmonically distorted current to eliminate the unbalance or distortions in the load current or the supply voltage. Hence, its control is significantly different from that of the STATCOM. A D-STATCOM is usually configured by using a voltage source inverter circuit interfaced with a coupling transformer and energy storage element viz. capacitor in case of VSC based D-STATCOM as shown in Figure 1. The D-STATCOM used in custom power applications uses PWM switching control as opposed to the fundamental frequency switching strategies which is preferably used in FACTS applications. PWM switching is practically used in custom power applications as it is at relatively low power level [3, 4, 10].

3 CSC based D-STATCOM

The CSC based D-STATCOM can be realized by modifying the basic configuration of the VSC based one. It also consists of a current source inverter circuit which is interfaced

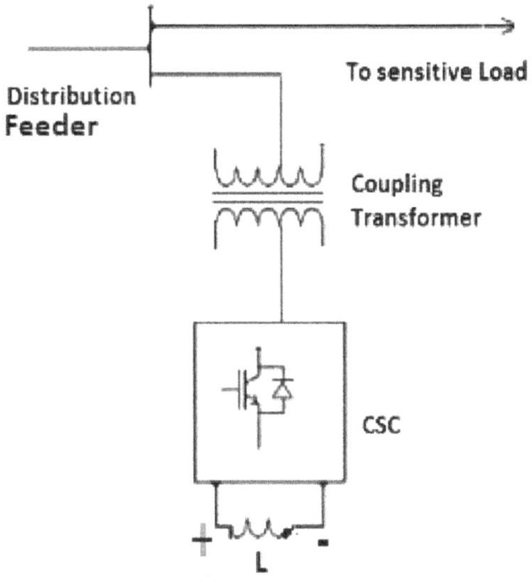

Figure 2. Schematic diagram of a CSC based D-STATCOM.

with a coupling transformer and a dc link reactor/inductor as shown in Figure 2.

On the ac side of the CSC based D-STATCOM capacitors are connected in shunt acting as low pass filter which ensures to provide good sinusoidal output voltage and current waveforms.

3.1 Design considerations of CSC for D-STATCOM

The working principle and control of CSC for use in D-STATCOM is similar to those of the CSC employed in ac motor drives but its design strategy is different to certain extent as listed below:

(a) The CSC circuit will be subjected to voltage regulation problem on the ac side when the reactive power is varied from full inductive to full capacitive, This will affect the voltage rating of the switching devices as well as the design of the input low pass filter (LPF) circuit.

(b) Since the dc link circuit consists only inductor and its internal resistance, the electrical time constant will be high and thus affects the design process of this circuit.

(c) The design objectives of the CSC for use in D-STATCOM is not only to filter harmonics but also to achieve optimal sizing of the CSC so as to meet the control range requirements of the D-STATCOM in both the capacitive and inductive load operating ranges [11]. The selection of switching device and modulation scheme depends on the application voltage and power rating [12]. For applications at transmission level voltage such as FACTS devices, GTO switches are used. At distribution level applications such as D-STATCOM, IGBT switches and PWM scheme can be chosen if VSC topology is used. The converter topology for the proposed model being CSC, switches having sufficient reverse voltage blocking capability should be selected. Unless a switch of sufficient reverse voltage withstanding capability such as Gate-Turn-Off (GTO) thyristor is used, a diode has to be placed in series with each of the switches in CSC. This almost doubles the conduction losses compared with the case of VSC. Hence, GTO switches are used in the proposed model. The schematic circuit diagram of the model consists six number of fully controllable GTO switches represented by S_1, S_2, S_3, S_4, S_5, and S_6 as depicted at Figure 3. These switches are configured in bridge fashion and fed from a dc link reactor acting as a energy storage element. The output of the CSC is filtered by a three-phase LPF comprising of three capacitors connected in shunt manner. The dc link reactor approximates the dc link current I_{dc} to a level current waveform in the steady state. The nearly level dc-link current is converted to bidirectional current pulses alternating at supply frequency in the ac lines of CSC by switching power semiconductor switches in accordance with a pre-specified pattern [13, 14].

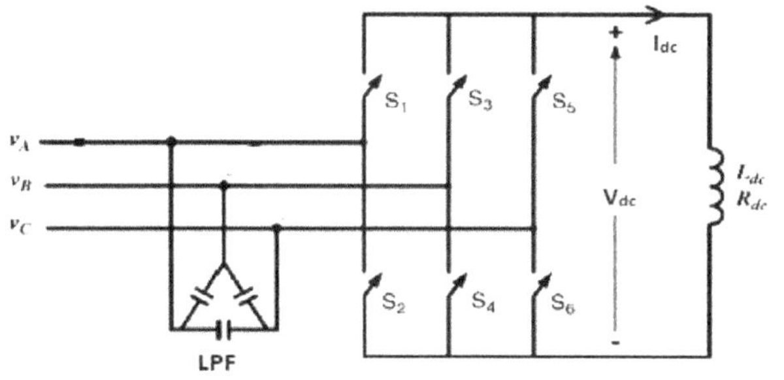

Figure 3. Schematic circuit diagram of CSC.

3.2 Design of dc link reactor and input LPF

In the CSC circuit the energy storage element is an inductor with its internal resistance. The amount of reactive power to be generated by the CSC can be computed from the relation:

$$Q = \sqrt{(3/2)}VMI_{dc}\cos\theta \qquad (1)$$

where V = rms value of the fundamental component of converter input line-to-line voltage, M = modulation index, I_{dc} = mean dc-link current and θ = phase shift [11]. Equation (1) indicates that Q is independent of the dc-link inductance L_{dc}. The value of L_{dc} affects the response time of D-STATCOM against the variations of reactive power demand of the load. The time constant of the dc-link circuit τ_{dc} is L_{dc}/R_{dc}. Hence, the value of L_{dc} should be selected as small as possible for allowing rapid rise or decay of mean dc-link current against the rapid changes in Q if the phase-shift angle control at fixed modulation index is used for controlling the reactive power.

The capacitor input low pass filter (LPF) is connected at the output of the CSC for feeding sinusoidal ac voltage after separating higher order harmonic components to the coupling transformer. The leakage impedance of the coupling

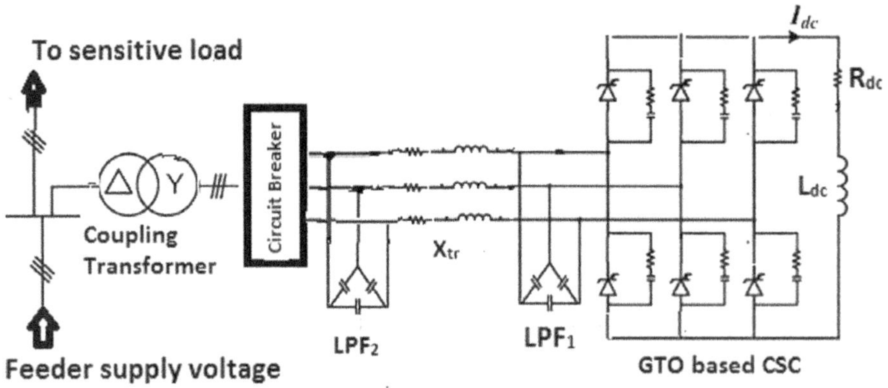

Figure 4. Simplified structure of the proposed D-STATCOM implementation.

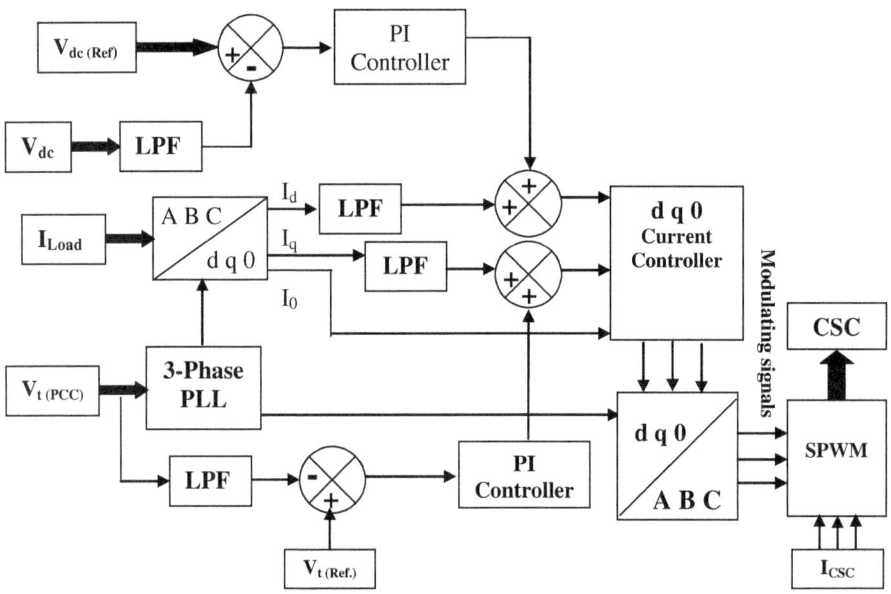

Figure 5. Control scheme block diagram for the proposed D-STATCOM.

transformer also behaves as a part of the LPF. External series reactors (i.e. X_{tr} in Figure 4) have been used on the low voltage side of the coupling transformer for adjusting the corner frequency of the input filter to an optimum value for the fixed shunt connected capacitor in implementing the proposed D-STATCOM. The corner frequency should be chosen as small as possible for better performance of the filter circuit. The detailed design strategies of the LPF are available at [14].

3.3 Control scheme

For generating proper gating signals of the GTO switches used in the CSC, SPWM control scheme is chosen for the proposed model. The SPWM switching strategy has constant switching frequency capability. This constant switching frequency reduces stress levels on the converter switches [16]. The control scheme will able to maintain constant voltage magnitude at the point where a sensitive load is connected under system disturbances. The control system only measures the rms voltage at the point of common coupling (PCC) and no reactive power measurements are carried out. The CSC switching strategy is based on a sinusoidal PWM technique which offers simplicity and good response. As the distribution network operates at a relatively low-power application, such method offers a more flexible option than the fundamental frequency switching method used in FACTS applications. The block diagram of the control scheme designed for the proposed model is shown in Figure 5. The commonly used control schemes for the generation of reference source currents in most of the VSC based D-STATCOM include instantaneous reactive power theory (IRPT), synchronous reference frame theory (SRFT), unity power factor (UPF) based, instantaneous symmetrical components based, etc. [16]. The control scheme chosen for the proposed D-STATCOM model is based on SRFT. The load currents, I_{Load} (i_a, i_b, & i_c), the PCC voltage $V_{t(PCC)}$ (v_a, v_b, & v_c), and the reactor dc voltage (v_{dc}) of D-STATCOM are sensed and used as feedback signals. The load currents from the *a-b-c* frame are first converted to the α-β-0 frame

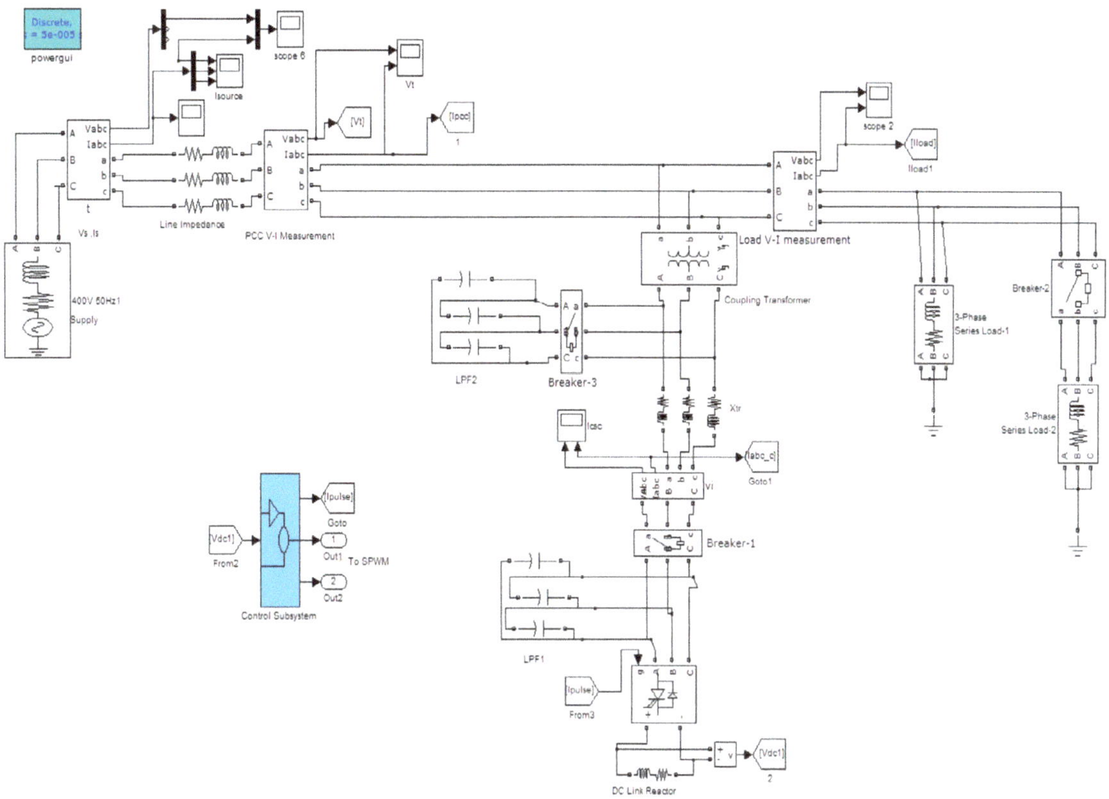

Figure 6. Test set up of the proposed CSC based D-STATCOM.

Table 1. Various parameter settings of the model under simulation.

Sl. no.	Name of the component	Parameter values
1.	Feeder system voltage	$V_{t(PCC)} = 400$ V (rms), 50 Hz
2.	Feeder system reference voltage	$V_{t(Ref.)} = 328$ V(rms), 50 Hz
3.	Feeder impedance	$R_f = 0.01\ \Omega$, $L_f = 2$ mH
4.	DC reference voltage	$V_{dc(Ref.)} = 1$ kV
5.	DC link reactor of CSC	$L_{dc} = 8000$ mH, $R_{dc} = 0.01\ \Omega$
6.	AC side LPF_1 shunt filter capacitor of CSC	$C = 2.85$ mF
7.	AC side LPF_2 shunt filter capacitor of CSC	$C = 8.5$ mF
8.	AC side external reactor, X_{tr}	$R = 0.01\ \Omega$, $L = 1$ mH
9.	Inductive $Load_1$	$R = 0.06\ \Omega$, $L = 8000$ mH
10.	Inductive $Load_2$	$R = 0.06\ \Omega$, $L = 98\,000$ mH
11.	DC link voltage PI controller	$K_{P(dc)} = 0.25$, $K_{i(dc)} = 0.14$
12.	AC system voltage (at PCC) PI controller	$K_{P(q)} = 0.4$, $K_{i(q)} = 0.5$
13.	SPWM switching frequency	$f_s = 20$ kHz

and then to *d*-*q*-0 frame by using the Park's transformation relation as:

$$i_d = \frac{2}{3}\left[i_a \sin\theta + i_b \sin(\theta - 2\pi/3) + i_c \sin(\theta + 2\pi/3)\right] \quad (2)$$

$$i_q = \frac{2}{3}\left[i_a \cos\theta + i_b \cos(\theta - 2\pi/3) + i_c \cos(\theta + 2\pi/3)\right] \quad (3)$$

$$i_0 = \frac{1}{3}\left[i_a + i_b + i_c\right] \quad (4)$$

where cos θ and sin θ are obtained by using the three-phase phase locked loop (PLL). The PLL receives signal from PCC terminal voltage $V_{t(PCC)}$ for generation of fundamental unit vectors for conversion of sensed currents to the *d*-*q*-0 reference frame. The SRF controller extracts dc quantities by a LPF and removes the harmonics from the reference signal. The distribution feeder terminal voltage $V_{t(PCC)}$ is regulated by a PI controller after comparing $V_{t(PCC)}$ with a reference terminal voltage $V_{t(Ref.)}$ which produces a i_q signal and adds with i_q generated at equation (3) and acts as reference component for the current controller. The error signal output after comparing with the dc link reactor voltage V_{dc} with a

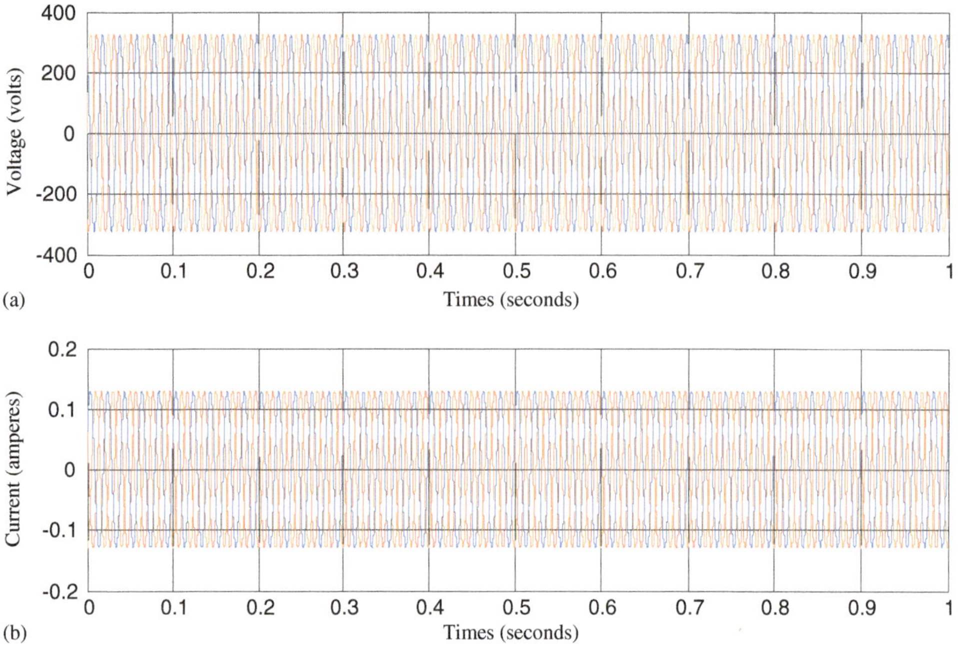

Figure 7. Load voltage and current waveforms under steady state condition.

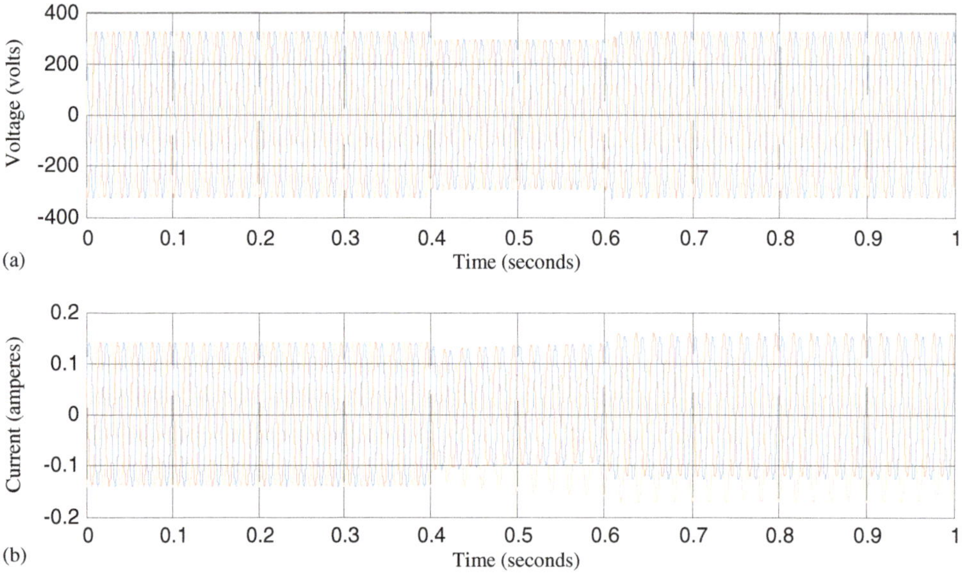

Figure 8. Load voltage and current waveforms under abnormal load condition without D-STATCOM.

$V_{dc(Ref.)}$ is processed by another PI controller to regulate the dc link voltage and produces i_d current component. This current is added with the i_d available at equation (2). The reference source current must be in phase with the voltage at the PCC but with no zero-sequence component and it can be obtained by using reverse Park's transformation process as discussed in reference [15]. The sensed current and the reference source current are compared and a proportional controller is used for amplifying current error in each phase in the current controller. The SPWM generates gate drive pulses for the six number of GTO switches used in the CSC based D-STATCOM.

4 Model simulation and results

The Matlab/Simulink model of the proposed CSC based D-STATCOM shown in Figure 6 has been simulated under steady state (with normal load) and dynamic (sudden change in load) conditions. The various parameter settings of the model are listed at Table 1.

The output voltage and current waveforms under different conditions are presented below:

(a) *Simulation results under steady state* (*with normal load*): The D-STATCOM and the second heavily inductive

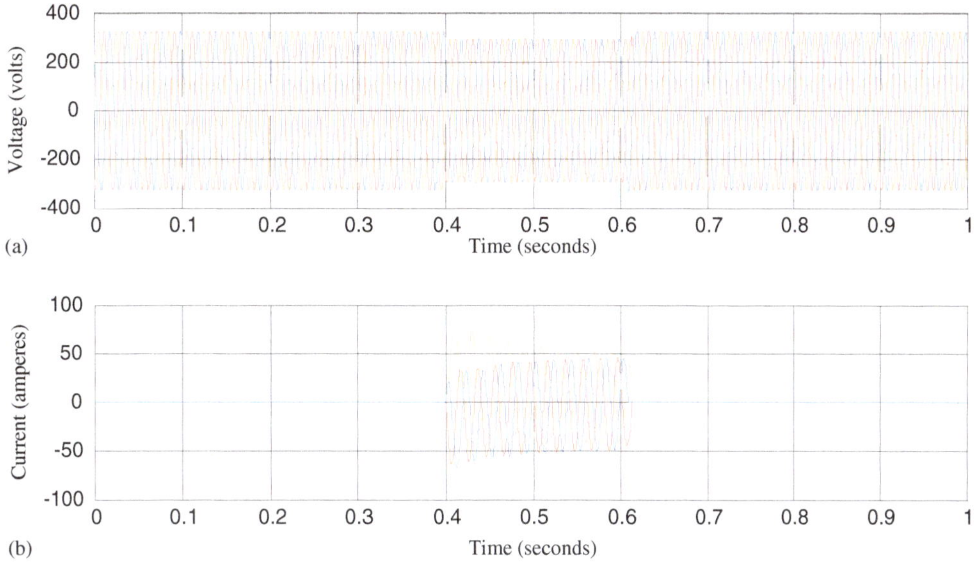

Figure 9. Voltage and current waveforms under heavily inductive load condition at PCC without D-STACOM.

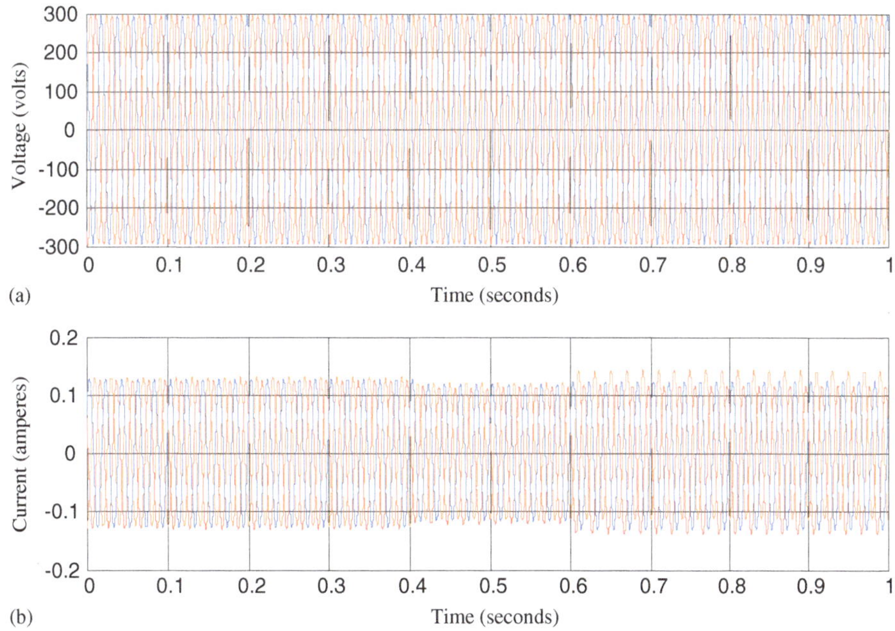

Figure 10. Voltage and current waveforms under sudden change in load condition with D-STATCOM.

three-phase load are disconnected from the power distribution system by opening the three-phase circuit breaker 1 and 2 respectively and the model is simulated. Figure 7 shows the voltage and current waveforms under steady state condition. Figure 7a depicts the voltage waveform across the load and Figure 7b shows the current waveform in the load. From this simulation result, it is observed that there is no voltage sag taking place across the load. The magnitude of voltage and current are found to be 360 V and 0.14 A respectively.

(b) *Simulation results under load perturbation* (*with sudden change in load*): For observing the profile of the voltage and the current waveforms under sudden change in load condition, the D-STATCOM is disconnected by opening the circuit breaker 1 and the second heavily inductive three-phase load is connected by closing the three-phase circuit breaker 2. The circuit breaker 2 closing transition time is set from 0.4 to 0.6 s. Figures 8a and 8b show the load voltage and current waveforms respectively. Under this condition, a voltage sag with magnitude of 340 V which is 20 V less than the non-sag voltage magnitude for a duration from 0.4 s to 0.6 s appears across the load as shown in Figure 8a. Figure 9 shows the voltage and current waveforms at the point of common coupling (PCC) without operating the D-STATCOM.

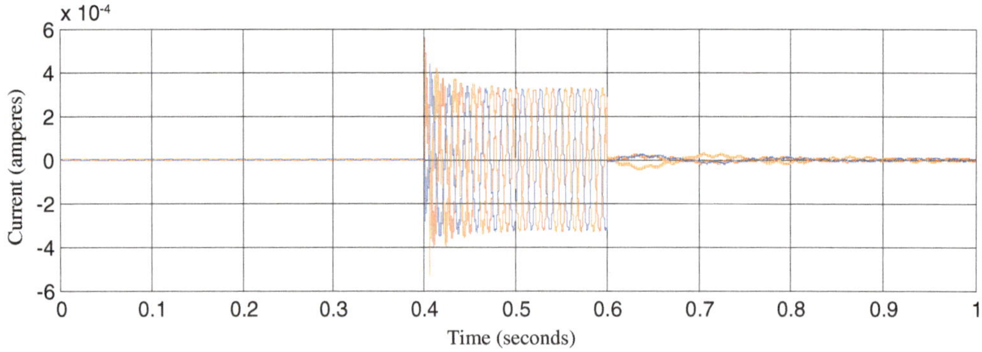

Figure 11. Injected current waveform from the D-STATCOM during voltage sag mitigation process.

(c) *Simulation results under load perturbation with the D-STATCOM*: In order to mitigate the voltage sag taking place due to the introduction of sudden change in load, the D-STATCOM is brought to operation by closing the circuit breaker 1 with its transition time setting from 0.4 s to 0.6 s and the model is simulated. Figures 10a and 10b depict the load voltage and current waveforms. It is observed that the load voltage sag during the period from 0.4 s to 0.6 s have been effectively mitigated by the proposed D-STATCOM under this condition. The voltage profile of the load can be improved by connecting the LPF_2 though the operation of circuit breaker 3 of the circuit diagram. For reducing the overall cost of the ac side filter, LPF_2 may be eliminated or kept as optional. Figure 11 shows the injected current waveform from the D-STATCOM during the process of mitigating the load voltage sag.

5 Conclusions

In this paper, a current source converter based D-STATCOM has been modeled and simulated with the objective of mitigating voltage sag occurring in the event of sudden change in load condition at power distribution system. The use of CSC topology in the custom power applications which has not been the focus of many researchers for a long time due various reasons is explored through this paper. From the simulation results of the proposed model, it is learnt that the voltage sag at the distribution level voltage under load perturbation has been successfully mitigated by introducing a CSC based D-STATCOM system instead of using a VSC based one. This paper will pave the way for encouraging the application of the CSC topology in other types of custom power devices for solving power quality problems. Some examples of the possible power quality problems where CSC topology can replace the VSC include voltage swell, unbalance, harmonics, power factor correction, etc.

References

1. Hingorani NG. 1995. Introducing custom power. IEEE Spectrum, 32, 41–48.
2. Hingorani NG, Gyugyi L. 2013. Understanding FACTS: concepts & technology of flexible AC transmission systems. John Wiley & Sons Inc: UK.
3. Deben Singh M, Khumanleima Chanu L. 2015. Power electronics technology for power quality improvement. International Journal of Advanced Research in Electrical, Electronics and Instrumentation Engineering, 4, 2073–2080. DOI: 10.15662/ijareeie.2015.0404031.
4. Ghosh A, Ledwich G. 2009. Power quality enhancement using custom power devices. Springer International Edition: New Delhi, India.
5. Ye Y, Kazerani M, Quintana VH. 2005. Current-source converter based STATCOM: modeling and control. Proceedings of IEEE Transactions on Power Delivery, 20, 795–800.
6. Wang X, Ph.D. thesis, Advances in Pulse Width Modulation Techniques, Dept. of Electrical Engineering, McGill University, Canada, 1993.
7. Steimer PK, Gruning HE, Werninger J, Carroll E, Klaka S, Linder S. 1999. IGCT – a new emerging technology for high power, low cost inverters. IEEE Industrial Application Magazine, 5, 12–18.
8. Mohan N, Undeland TM, Robbins WP. 1989. Power electronics: converters, applications, and design. John Wiley & Sons Inc., UK: New York, NY, USA.
9. Bollen MHJ. 2013. Understanding power quality problems. John Wiley & Sons Inc: UK.
10. Acha E, Agelidis VG, Anaya-Lara O, Miller TJE. 2006. Power electronic control in electrical systems. Elsevier-Newnes Power Engineering Series: UK.
11. Bilgin HF, Ermis M. 2010. Design and implementation of a current source converter for use in industry applications of D-STATCOM. Proceedings of IEEE Transactions on Power Electronics, 25, 1943–1957.
12. Han B, Moon S, Karady G. 2000. Static synchronous compensator using thyristor PWM current source inverter. Proceeding of IEEE Transactions on Power Delivery, 15, 1285–1290.

13. Bilgin HF, Ermis M, Kose KN, Cetin A, Cadirci I, Acik A, Demirci T, Terciyanli A, Kocak C, Yorukoglu M. 2007. Reactive power compensation of coal mining excavators by using a new generation STATCOM. Proceedings of IEEE Transactions on Industrial Applications, 43, 97–110.
14. Bilgin HF, Ermis M. 2011. Current source converter based STATCOM: operating principles, design and field performance. Elsevier Journal on Electric Power Systems Research, 81, 478–487.
15. Geddada N, Karanki SB, Mishra MK. 2012. Synchronous reference frame based current controller with SPWM switching strategy for DSTATCOM Applications, in Proceedings of IEEE International Conference on Power Electronics, Drives and Energy Systems (PEDES) 2012, December16–19, Bengaluru, India.
16. Singh B, Jayaprakash P, Kothari DP. 2008. A T-connected transformer and three-leg VSC based DSTATCOM for power quality improvement. Proceedings of IEEE Transactions on Power Electronics, 23, 2710–2718.

Challenges of additive manufacturing technologies from an optimisation perspective

Sofiane Guessasma[1,*], Weihong Zhang[2], Jihong Zhu[2], Sofiane Belhabib[3], and Hedi Nouri[4]

[1] INRA, UR1268 Biopolymères Interactions Assemblages, 44300 Nantes, France
[2] Laboratory of Engineering Simulation & Aerospace Computing-ESAC, Northwestern Polytechnical University, 710072 Xian, Shaanxi, P.R. China
[3] LUNAM University of Nantes Angers Le Mans, CNRS, GEPEA, UMR 6144, IUT de Nantes, 2 avenue du Professeur Jean Rouxel, 44475 Carquefou Cedex, France
[4] Department of Polymers and Composites Technology & Mechanical Engineering, Mines Douai, 941 rue Charles Bourseul, CS 10838, 59508 Douai, France

Abstract – Three-dimensional printing offers varied possibilities of design that can be bridged to optimisation tools. In this review paper, a critical opinion on optimal design is delivered to show limits, benefits and ways of improvement in additive manufacturing. This review emphasises on design constrains related to additive manufacturing and differences that may appear between virtual and real design. These differences are explored based on 3D imaging techniques that are intended to show defect related processing. Guidelines of safe use of the term "optimal design" are derived based on 3D structural information.

Key words: Additive Manufacturing, Topological optimisation, Process-induced defects, X-ray micro-tomography.

1 Introduction

Additive Manufacturing (AM) is a collection of versatile techniques of rapid prototyping that allow material design from 3D digital models [1–3]. The term AM comes also under different other nicknames such as direct digital manufacturing or solid freeform fabrication [4, 5]. AM is rated as one of the most promising technology for design [6], presented as a new industrial revolution [7], and a vector for creativity [8], impact [9] and interrogations [10]. The laying down of the material in different states including liquid, powder and fused material defines roughly categories of AM [2, 7, 11–13]. More accurate classifications do exist such as the ASTM F42 reported in reference [14]. Wide varieties of materials can be processed using additive manufacturing including metals [15–18], alloys [19–22], ceramics [23–26], polymers [27–30], composites [31–34], airy structures [35, 36] and multi-phase materials [37–39].

The main characteristic of AM is the reduced number of manufacturing steps that stands between the virtual design and the ready-to-use part [40]. In addition, one major advantage of AM reported in the literature is the ability to process complex shapes that are not easy to design using traditional ways such as extrusion and moulding [3, 41, 42]. AM potential, as a leading design technique, is enormous and the related applications are huge [7, 43–47]. Different printing techniques are used in the biomedical sector [41, 48] more particularly for tissue engineering [5, 35, 49–51]. Preform design of composites is evidently an inspiring topic for AM [31] because of the wide possibilities in structuring yarns and reinforcing composite structures [34]. Aerospace applications of AM are the most challenging because of the extreme performance that need to be achieved under fine scale monitoring and in-service validation [52, 53]. Recent contributions by NPU demonstrates the central role of topology optimisation in additive manufacturing for aerospace applications [54, 55]. Micro-fabrication technologies emerge also as an efficient way to produce functional micro-components for microelectronics systems [42]. The scaling down of AM is now possible thanks to cutting edge processes that allow material design at a very fine scale like with different types of lithography [55, 56].

The idea behind AM is the direct import of CAD (Computer-Aided-Design) object as machine instructions (Figure 1). The preferred way to achieve this import is the transformation of surface tessellation representing the geometry of the virtual part into a set of toolpaths. One starts from an

*e-mail: sofiane.guessasma@nantes.inra.fr

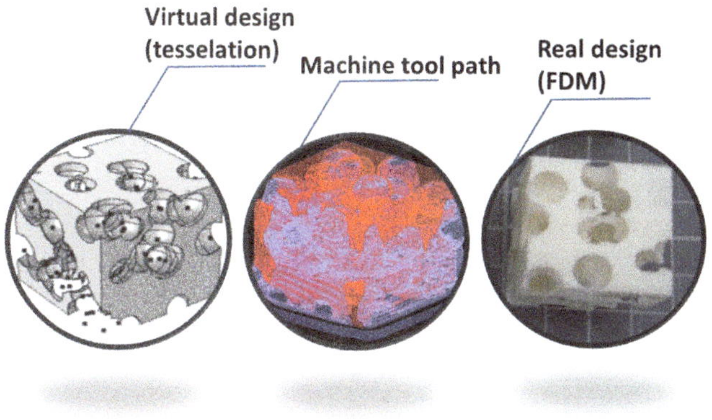

Figure 1. The process chain in typical AM.

STL or Standard Tessellation Language or STereoLithography file. All external boundaries and internal surfaces appear smooth and continuous using STL format. Generation of the toolpaths represents the first challenge and actually a limitation for AM [57]. The reason is that the building process in most 3D printing technologies relies on a successive layer-by-layer building process. So, starting from 3D space tessellation and ending with 2D building strategy is a first drawback. It is even worse when droplet based printing is considered because the fused matter is no more connected in any direction. Discontinuities may appear in all building directions (Figure 2) as a result of the layer-by-layer laying down process (Figure 3) [27]. The consequence of this appears to be the development of dimensional inaccuracy, inacceptable finishing state, structural and mechanical anisotropies, which are continuously addressed in many research contributions [12, 28, 58–64]. Anisotropy can be also inferred to the development of particular grain texture as revealed by laser melting deposition or arc welding alloying of metals [61, 65–68]. Reduction of anisotropy can be achieved by selecting the appropriate orientation of the virtual design [69–71]. Numerous papers mention strategies of common sense to build parts with acceptable performance, which are nearly equivalent or even superior to other design techniques. Scudino and others [67, 72] report that better ductility of metallic materials can be achieved using selective laser melting compared to casting as a result of the fine grain structure driven by AM. One particular feature highlighted in these contributions refers to avoid building the part along its largest surface. In other words, if the successive layers of the part exhibit a lack of cohesion, the large contact area between layers drives worse performance under tension. Some other strategies rely on reducing the lack of cohesion between layers or filaments by further processing of the real design. An example of radiation treatment is proposed in the work of Shaffer et al. [28].

In addition to the role of anisotropy, differences between the virtual and real design can be striking knowing that AM resolution is finite due to available tooling [64]. If we consider the example of fused deposition modelling, which is a popular AM technique [12], the toolpath generation is referred as collection of filament paths of finite dimensions (Figure 3). This has three main consequences on the real design: internal structural features may not be well captured depending on their size; discontinuities may appear depending on local curvature; and the surface finishing state may be limited due to rough profiles [62]. These limitations are illustrated in Figure 2, which highlights simple and complex geometries and the corresponding toolpath generation using two software, one is Repetier from Hot-World GmbH & Co, Germany and the other is CatalystEx from Stratasys Inc. Eden Prairie MN, USA.

All limitations mentioned earlier assume implicitly the role of defects induced by AM. These need to be faced in order to deliver a design representing, with much accuracy, the result of an optimisation procedure. These defects are related to the porosities that develop as a consequence of the discontinuous process of printing and other issues related to process errors [73]. A large number of contributions is dedicated to the evaluation of the effect of porosities in printed parts. One particular consequence of the role of porosity is that a large amount of them reduces the mechanical performance. Such reduction is represented by a theoretical linear decrease of stiffness with the increase porosity level (if limited stress transfer between layers is neglected). Under tension, porosities act as stress concentrators and may induce lower mechanical strength by enhancing the development of damage in the form of micro-cracks. Under compression, different considerations can be pointed out. Even if the porosities are closed during compressive loading, lateral expansion due to Poisson's effect may cause failure driven by opening mode or shearing effects that are dramatically enhanced by the anisotropy [27, 58].

Porosity should not be considered systematically as a negative issue in AM since it can be a positive driving factor for permeability [74].

Another type of defects is the presence of support material trapped between internal surfaces. The material is needed to withstand the fragile printed structure during the printing process. While this material is studied to provide limited adhesion to the deposited materials, its residual amount contributes in increasing the weight of the structure and modifies the load bearing distributions. These two drawbacks alter the expected performance of the optimal design. In addition, none-optimised support deposition affects finishing state, material consumption, fabrication time, etc. [75]. Strategies exist to reduce the dependence of AM to the presence of a support material by operating smart or slimming support generation [75, 76]. For some strategies, the part orientation is continuously adapted during the processing [57]. Curved regions can be processed smoothly under continuous deposition mode and the presence of support material is no more needed. For other strategies, building complex shape without the support material relies on the intrinsic properties of the deposited material itself. These materials exhibit generally rapid cooling kinetics, which allow them to support their own weight and prevent the structure collapsing. This kind of strategies obviously limits the spectrum of materials that can be printed.

2 Optimisation in additive manufacturing

In this paper, our focus goes towards optimisation difficulties that are inferred to AM. With regards to the large number

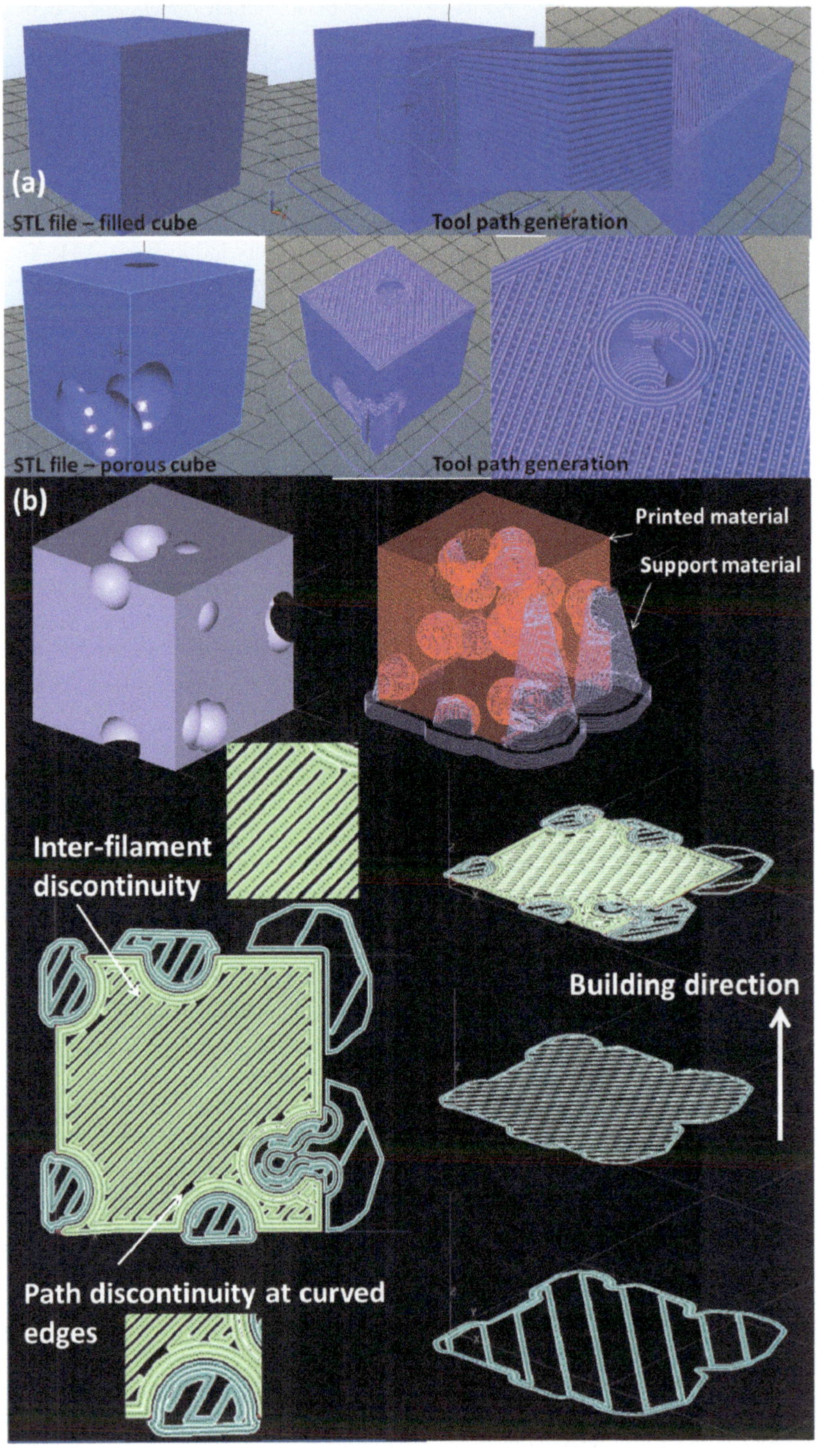

Figure 2. Typical examples of CAD objects transformed into collection of toolpaths using (a) Repetier and (b) CatalystEX software.

of disseminated works on the subject, common characteristics of optimisation in AM are highlighted in this section and described through selective literature work. In particular, several aspects of additive manufacturing can be optimized. Some of these aspects are related to design optimisation, more particularly topology optimisation. Some others like geometry accuracy, finishing state are tackled through process planning optimisation.

For most contributions, optimisation in AM is classically considered as a process parameter optimisation as it is the case for many design techniques [77, 78]. Raster angle, building direction, layer dimensions are some of the main parameters that find some interest in the literature. For instance, Garg et al. [79] present genetic programming approach as an intelligence tool to relate the AM process parameters to physical and structural outputs. While this is an important issue from the

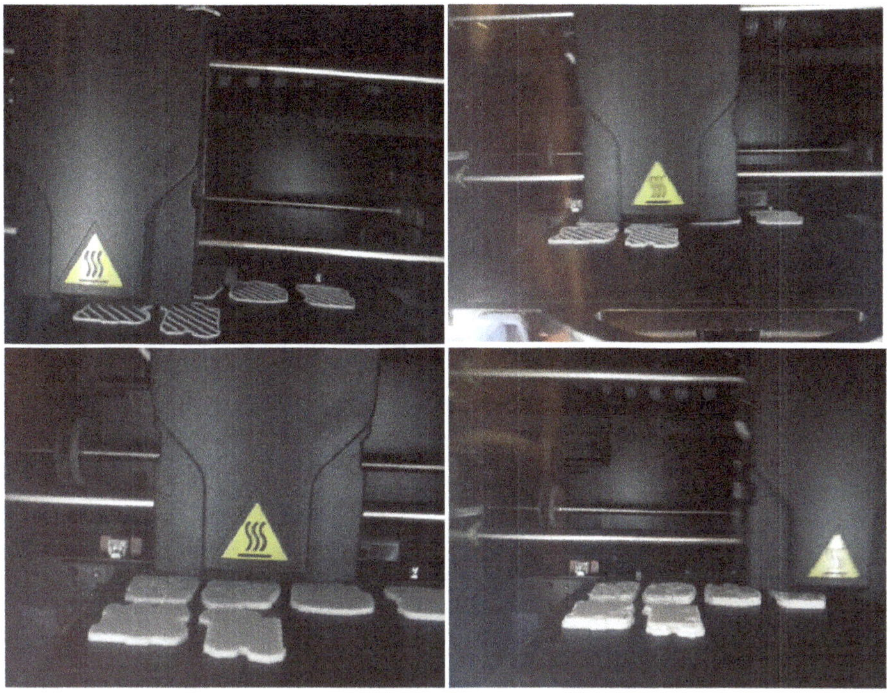

Figure 3. Laying down process of fused ABS polymer in typical FDM equipment.

processing viewpoint, it is less attractive from a numerical analysis perspective, where strong and robust optimisation tools need to address more significant challenges. This does not diminish the purpose of the earlier approach. Accurate dimension, acceptable roughness and processing time are some of the important outcomes that justify the continuity of the research effort in this particular field. A typical example showing the importance of the process optimisation is provided in recent works [54, 55, 78]. The paper by Zhou et al. [78] introduces the concept of pixel blending to define the effect of neighbour pixels light intensity in solid freeform fabrication using photopolymerisation medium. The optimisation in such kind of studies is meaningful as the achievement of shape accuracy relies on the precise control of light intensity over a pixel-based image.

The paper by Yang and Zhao [3] report one of the most recent review on AM-enabled design, which is the closest subject to topology optimisation. In their review, design guidelines are exposed and the focus on structure optimisation methodology is explored through different contributions. This kind of methodologies needs to take care about the specificities of the design in terms of material combination, shape complexity and the targeted in-service performance. In the same review [3], the authors bring to our attention the possibilities of process combination involving more conventional or AM-based techniques [80–82]. We consider that this research direction associating various processes is out of the scope of this paper from an angle view of topology optimisation. However, the other considerations discussed there are central to the topology optimisation such as those related to design simplicity, material choice efficiency, improved multi-functionality, integrated technical solutions [83], etc. The concepts of multiphase material [37] and functionally graded materials [84, 85] emerge as a direct consequence to point by point material control in AM [86]. Also, designers are no more bonded by the tooling which needs to be a factor in the design with traditional processing [87]. This opens new chances for simplifying the design but also for increasing the creativity [7]. This particular point helps significantly the designers who need generally to reshape the result of the topology procedure to fit processing constrains like the development of particular tooling. A simple example would be the conversion of a material deposition probability into a graded material [4] instead of thresholding to solid or air phase [88]. In addition, if the conditions of blending at the microscale or nanoscale are met, the achievement of multi-material design is not a threat for the design but mostly an advantage because of the possibility to further improve performance at a lower material consumption rate [4].

Considerations related to topology optimisation are numerous [37, 88]. Some of them depend on the level of access to the AM processing technology itself which is not systematically provided by the commercial solutions. Here are some of these considerations that are central to the development of robust AM solutions:

- Geometry of the CAD model: this is direct target of the topology optimisation, which needs to predict what would be the exact geometry that satisfies all model constrains. Successful examples of topology optimisation can be found in the literature for cellular materials [88, 89], multiphase materials [37], implants [90].
- Path generation: this is an important parameter that affects geometry accuracy, cohesion in the part, residual stresses, and the finishing state [20, 63, 91]. Path generation needs to handle as much the change in process speed and the transition time at the borders [92]. A specific

Figure 4. Two examples of airy randomly structured polymers designed using (a) fused disposition modelling (uPrint equipment in GEPEA lab., university of Nantes, France) and (b) photolithography (SPS350B equipment in ESAC lab, NPU, Xian, China).

branch of research is dedicated to the optimisation of tool generation path including studies related to the improvement of the scanning mode [93, 94], geometry slicing strategies [95, 96], optimal material deposition [97, 98], multi-directional AM [99], tool path anticipation procedures [100].

- Process selection: different technologies are developed to enhance the capabilities of AM like jet printing [26, 101], friction stir AM [102], welding based AM [94], ultrasonic AM [103], electrochemical AM [104], micro-plasma powder deposition [105], Solid freeform fabrication [106] and related variants such as selective laser sintering [107, 108], or directed light fabrication [20], selective infiltration manufacturing [109]). The outcomes of these technologies diverge. This reinforces specific aspects of AM like the material type, printing size, accuracy, speed, cost, etc. [110]. Figure 4 shows two examples of technologies applied for the design of airy structures, one based on FDM (uPrint® SE from Stratasys) and the other is a photolithography equipment (SPS350B from XJRP company). The earlier one is restricted to printing ABS polymer where the last one can process only photosensitive resins. The ability to control the process parameters is crucial to decide on the relevance of a particular AM process [111]. And perhaps, a badly performing design reflects only the lack of knowledge of the process. Thus, the mastering of the process parameters is a criterion for process selection.
- Process resolution: this is also an important aspect that guarantees the accuracy of the AM process [14, 112] and the development of appropriate scaling solutions for nano-, micro- and macro-features [38, 64, 113–116]. Some studies focus on slicing techniques to increase accuracy such as adaptive slicing proposed by Siraskar et al. [117] using volume decomposition by octants.
- Feed material properties: rheological and phase change properties of the feed materials are essential to the success of AM building capabilities especially for achieving standards in material selection [118]. This is a continuous research direction aiming at increasing the spectrum of printable materials and optimising the intrinsic material properties for a better performance during fabrication step [14, 19, 119–121].
- Support material: optimisation of the support material is now fully integrated in numerous commercial solutions, which require supporting of the part during processing. This optimisation relies on different options such as the

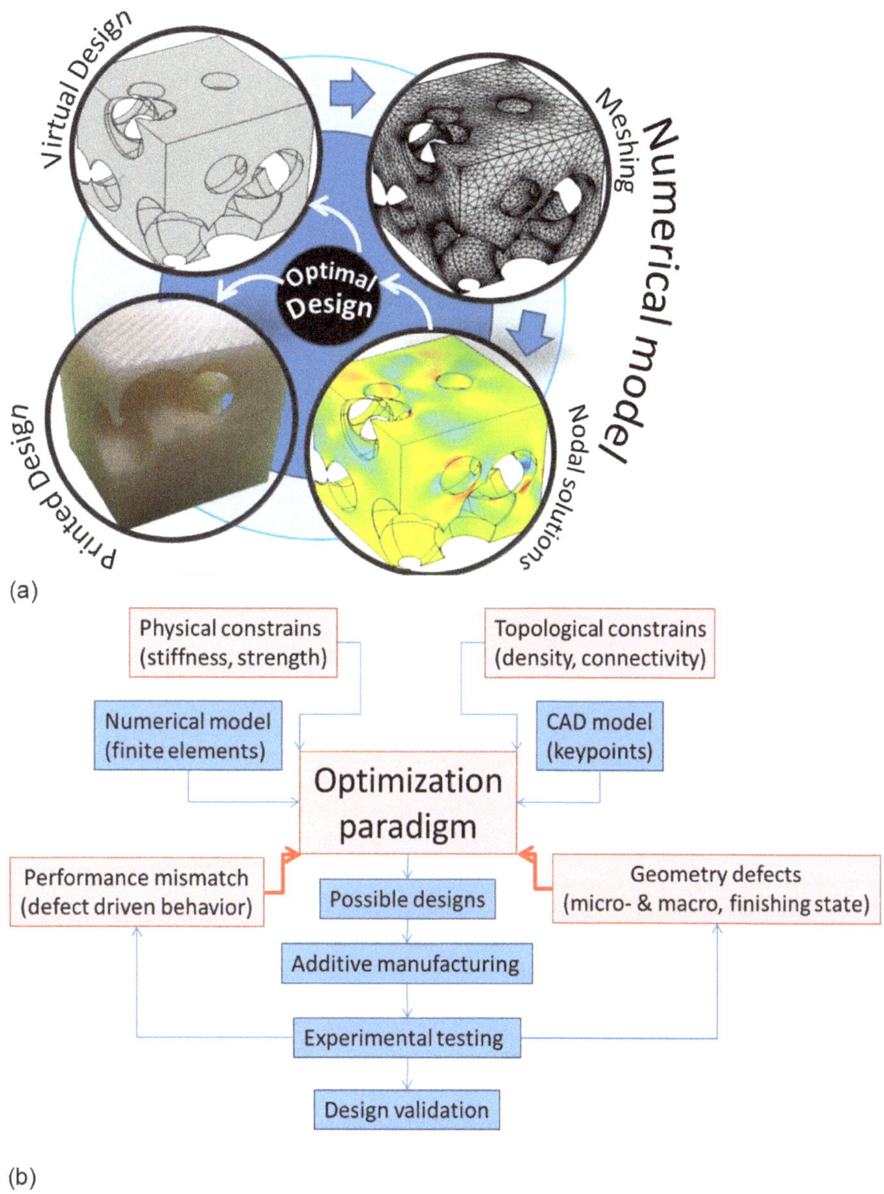

Figure 5. (a) Classical view of topological optimisation, (b) Modified scheme with defect sensitive topology optimisation of AM.

smart deposition for which only reduced amount of support material is needed [76]. Some research works still contribute in this particular area to improve the spatial distribution of airy support material [75, 122].

The logic behind topology optimisation is illustrated in Figure 5. The geometry of the CAD model can be something to discover through the optimisation procedure or assisted by imaging tools such as micro-CT scanning [123]. Since the optimisation combines physical and geometrical constrains, numerical solutions need to be available to predict what would be the result of the part performance [14, 112]. This is generally addressed using finite element computation [53]. The numerical model needs to converge in all design situations within a short time because the process is meant to be repeated several times. Depending on the nature of the physical constrain, the numerical model can be more or less sophisticated. For instance, residual stress analysis requires most of the time solving a multi-physics problems [124, 125]. If numerical models are able to handle technological, physical, and geometrical constrains, design guidelines can be adopted by coupling the optimisation tool to decision making paradigms. Some studies show that AM process simulation is possible [125, 126] but the ultimate goal would be to bridge such realistic tools with the optimisation paradigm. Recent works prove unfortunately that we are far from such ideal situation [127].

Topology optimisation needs to cope with the specificities of AM. As this process generates a complex network of 3D defects, numerical models need to integrate the result of defect in the analysis as an implicit performance perturbation or explicit defect influence. This is the main difference between the two schemes presented in Figure 5. The classical scheme (Figure 5a) does not handle the defects induced by processing,

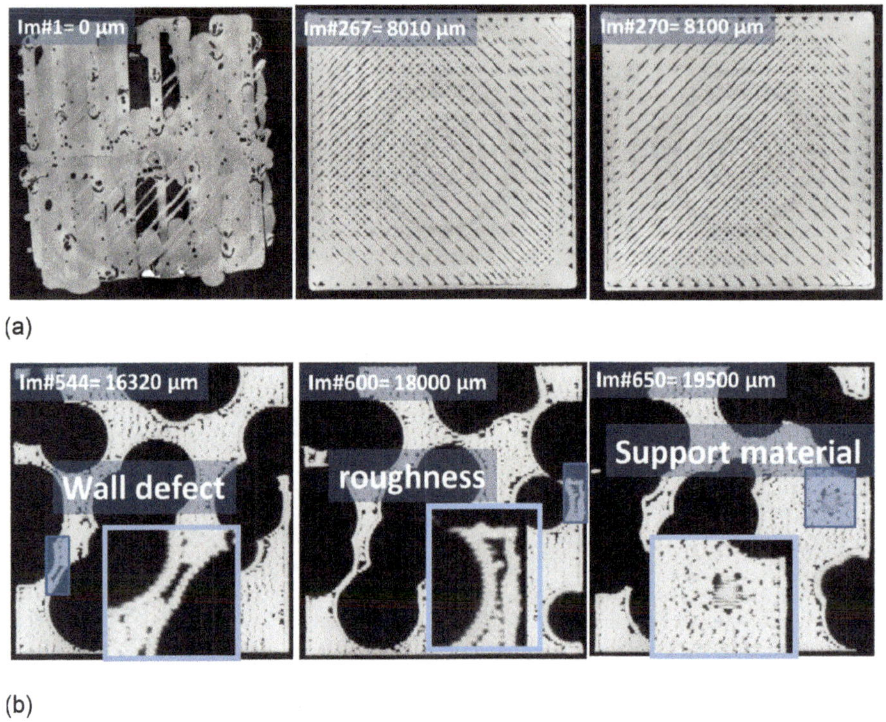

Figure 6. Microstructural defects in ABS polymer printed using FDM. Analysis is performed using X-ray micro-tomography. Two examples are shown one is (a) a dense ABS block and the other is (b) airy ABS materials exhibiting more than 60% of porosity.

which makes any deviation from the optimal virtual design a cause of failure. Numerical sizing in aerospace applications [128] is a typical example where such defects can be an issue to validate the final design in airframe development. In the second scheme (Figure 5b), the corrections introduced by the monitoring of the defects helps in guiding the optimisation tool towards the best realistic solution. This resolution is directly related to the AM tooling constrains, for instance the choice of the tip size. If the second scheme is used to consider appropriate selection for tooling options (like nozzle diameter), then such process parameter can be considered as a discrete variable. Optimal design can be searched in a larger space depending on the possibilities offered to select a certain number of available nozzles. Real-time control of AM like the optical tomography [129, 130], thermographic analysis [131] or ultrasonic monitoring [132, 133] helps in gaining valuable information about the structural defects that develop during AM processing and their direct consequence on failure of the designed part [73]. This is still a challenging issue as it appears that adequate non-destructive techniques are not yet fully available to evaluate properly AM part performance [52]. This situation can be improved through the development of standards which is still an ongoing process for the validation of testing techniques applicable to AM [134]. One of the most promising techniques to analyse microstructural defects in AM parts is X-ray micro-tomography [135]. This technique is able to provide precise information about the porous network induced by processing, surface roughness, part volume, amount of support material and any other microstructural defect [27]. As the technique relies on transformation of 2D projections into 3D image [136, 137], structural anisotropy effects can be quantified. Figure 6 shows two examples of defects revealed in ABS polymer printed using fused deposition modelling. Cross-section views refer to a dense block of ABS ($30 \times 30 \times 30$ mm^3) analysed using X-ray micro-tomography. This block is oriented at 0° in the printing plateau, but we notice clearly the crossing of filaments at an angle of 45° and the presence of bounding layer. In Figure 6a, the resolution of the image is $1077 \times 1062 \times 1059$ voxels, where a voxel is a graphical unit in 3D. The physical size of the voxel determines the accuracy of the structural defect evaluation. In Figure 5a, the voxel size is 30 µm. More information about the operating conditions can be found in reference [27]. Figure 5a shows the lack of cohesion between successive filaments and tendency to flattening because the filament diameter is tripled during the laying down of the fused matter. The subsequent porosity forms as a regular network of micro-sized defects, and appears to be highly connected. Also, residual support material can be found at the borders, which reveals difficulties of support material removal. The automatic cleaning process is generally followed by manual removal step to ensure that no residual support material is left behind. The second example highlighted in Figure 6b shows other imperfections that infiltrate the design of a two phase material. These imperfections are related to discontinuities of matter in the cell walls, the change of geometry due to design mismatch effects and the presence of support material trapped in closed pores of small size.

In several contributions, the red links in Figure 5b are ignored as if the approximate result of the optimisation is unavoidably accepted. More recent contributions tackle such links by suggesting corrections of the design based on manufacturability considerations [64]. This is performed, however,

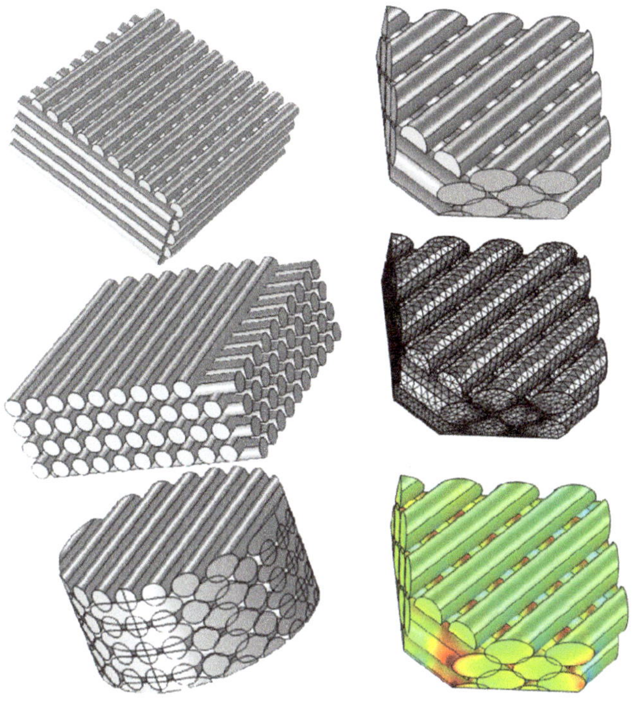

Figure 7. Towards modelling of AM process.

independently from the optimisation tool itself and does not involve defects induced by processing.

3 Challenges for AM topology optimisation

One of the important issues that topology optimisation needs to address is the pertinence of the constitutive laws representing the behaviour of the printed materials. Unfortunately, material law implementation is not yet fully revisited leaving an open area for research in this direction [138]. A typical research direction would be to explore interfacial effect in terms of limited load transfer and damage kinetics at the light of results achieved for composite materials [139, 140]. This direction is fully justified by the fact that lack of performance is more associated to the weak adhesion between filaments. Thus, failure mechanisms are likely to be affected by the arrangement of such weak regions [27].

Another concern is the embedding of the microstructural details in the topology optimisation. During the past decades, this opportunity to tune microstructurally the design was out of reach because of computation resource limitations. Now, this is accessible at the cost of using efficient paradigms that avoid unrealistic configurations and constrain the search volume to design-effective solutions. Recent experimental achievements show the potential of AM to tune locally the performance of multi-material parts [39]. The next realistic step would be to promote this kind of experimental attempts to fully automated procedures. The role of microstructural details can be even determinant in hierarchical structures. Indeed, previous studies show large possibilities of airy arrangement using hybrid optimisation strategies [141]. One can imagine the large possibilities of pore connectivity tailoring driven by AM if micro-porosity is considered.

Topology optimisation is not yet ready to provide systematic process error detection for AM. We know that AM processes are exposed to inaccuracy in terms of geometry imperfections, volume mismatch, and undesirable surface texture. All these drawbacks can be properly addressed by a tool that apprehends the limits of the AM processing. Realistic designs with acceptable defects are better than ideal designs with unmeasurable bias. Figure 7 shows some clues about how topology optimisation can achieve a higher sensitivity to defects in AM. A better understanding of the AM defects is a matter of scaling down the numerical model to the size of heterogeneities that are the birth sites of the process-induced defects. Explicit implementation of discontinuities can be handled as well as lack of bonding between layers or filaments. Also more elaborated constitutive laws can be considered in order to take into account anisotropies that are subsequent to the rapid cooling and stretching of the matter.

A straight-A learning paradigm is not also accessible for topology optimisation. The near future developments will meet substantial use of in-situ or in-line monitoring procedures for AM [16, 142]. Topology optimisation can integrate some of these procedures to learn from the design. This requires the combination with another class of algorithms that are derived from artificial intelligence [143]. As an end, the optimal design can be performed using optimal process parameters, which saves a considerable amount of time.

A bigger tool for higher perspectives is what process managers expect from optimisation tools. Topology optimisation can be part of it if other considerations are handled carefully like cost effectiveness with a large material catalogue, material saving logics, automated process selection, scenarios of durability, recyclability and projections of life time. Fully automated decision making processes can be then launched starting from the idea of design to the post-mortem step of the AM part.

Acknowledgements. The authors would like to thank Omar Ben Hassana, Pierre Malige and Julien Grison, from IUT of Nantes, and Kaike Yang from ESAC lab, NPU for their technical assistance on the 3D printing process and mechanical testing. The authors gratefully acknowledge the International Campus on Safety and Intermodality in Transportation (CISIT), the Nord-Pas-de-Calais Region and the European Community (FEDER funds) for partly funding the X-ray tomography equipment. The authors are thankful to IBSM federation for the financial support and to the Nord-Pas-de-Calais Region (France) for funding Hedi Nouri's post-doctoral grant (contract No 14002212). The second author, ZHANG Weihong, thanks the support by National Natural Science Foundation of China (11432011).

References

1. Yan X, Gu P. 1996. A review of rapid prototyping technologies and systems. Computer-Aided Design, 28(4), 307–318.
2. Pham DT, Gault RS. 1998. A comparison of rapid prototyping technologies. International Journal of Machine Tools & Manufacture, 38(10–11), 1257–1287.
3. Yang S, Zhao YF. 2015. Additive manufacturing-enabled design theory and methodology: a critical review. International Journal of Advanced Manufacturing Technology, 80(1–4), 327–342.

4. Chiu WK, Yu KM. 2008. Direct digital manufacturing of three-dimensional functionally graded material objects. Computer-Aided Design, 40(12), 1080–1093.
5. Shim JH, Kim AJ, Park JY, Yi N, Kang I, Park J, Rhie JW, Cho DW. 2013. Effect of solid freeform fabrication-based polycaprolactone/poly(lactic-co-glycolic acid)/collagen scaffolds on cellular activities of human adipose-derived stem cells and rat primary hepatocytes. Journal of Materials Science-Materials in Medicine, 24(4), 1053–1065.
6. Huang SH, Liu P, Mokasdar A, Hou L. 2013. Additive manufacturing and its societal impact: a literature review. International Journal of Advanced Manufacturing Technology, 67(5–8), 1191–1203.
7. Zhai YW, Lados DA, Lagoy JL. 2014. Additive manufacturing: making imagination the major limitation. JOM, 66(5), 808–816.
8. Schelly C, Anzalone G, Wijnen B, Pearce JM. 2015. Open-source 3-D printing technologies for education: bringing additive manufacturing to the classroom. Journal of Visual Languages and Computing, 28, 226–237.
9. Gress DR, Kalafsky RV. 2015. Geographies of production in 3D: theoretical and research implications stemming from additive manufacturing. Geoforum, 60, 43–52.
10. Kietzmann J, Pitt L, Berthon P. 2015. Disruptions, decisions, and destinations: enter the age of 3-D printing and additive manufacturing. Business Horizons, 58(2), 209–215.
11. Shirazi SFS, Gharehkhani S, Mehrali M, Yarmand H, Metselaar HSC, Kadri NA, Abu Osman NA. 2015. A review on powder-based additive manufacturing for tissue engineering: selective laser sintering and inkjet 3D printing. Science and Technology of Advanced Materials, 16(3), 033502-1–033502-20.
12. Turner BN, Gold SA. 2015. A review of melt extrusion additive manufacturing processes: II. Materials, dimensional accuracy, and surface roughness. Rapid Prototyping Journal, 21(3), 250–261.
13. Gong XB, Anderson T, Chou K. 2012. Review on powder-based electron beam additive manufacturing technology. Proceedings of the ASME/ISCIE International Symposium on Flexible Automation, Isfa 2012, St. Louis, Missouri, USA, 507–515.
14. Huang Y, Leu MC, Mazumder J, Donmez A. 2015. Additive manufacturing: current state, future potential, gaps and needs, and recommendations. Journal of Manufacturing Science and Engineering-Transactions of the ASME, 137(1), 014001-1–014001-10.
15. Frazier WE. 2014. Metal additive manufacturing: a review. Journal of Materials Engineering and Performance, 23(6), 1917–1928.
16. Tapia G, Elwany A. 2014. A review on process monitoring and control in metal-based additive manufacturing. Journal of Manufacturing Science and Engineering-Transactions of the ASME, 136(6), 060801-1–060801-10.
17. Vayre B, Vignat F, Villeneuve F. 2012. Metallic additive manufacturing: state-of-the-art review and prospects. Mechanics & Industry, 13(2), 89–96.
18. Nie B, Yang LM, Huang H, Bai S, Wan P, Liu J. 2015. Femtosecond laser additive manufacturing of iron and tungsten parts. Applied Physics A-Materials Science & Processing, 119(3), 1075–1080.
19. Finke S, Feenstra FK. 2002. Solid freeform fabrication by extrusion and deposition of semi-solid alloys. Journal of Materials Science, 37(15), 3101–3106.
20. Milewski JO, Dickerson PG, Nemec RB, Lewis GK, Fonseca JC. 1999. Application of a manufacturing model for the optimization of additive processing of Inconel alloy 690. Journal of Materials Processing Technology, 91(1–3), 18–28.
21. Xu W, Brandt M, Sun S, Elambasseril J, Liu Q, Latham K, Xia K, Qian M. 2015. Additive manufacturing of strong and ductile Ti-6Al-4V by selective laser melting via in situ martensite decomposition. Acta Materialia, 85, 74–84.
22. Tang HP, Yang GY, Jia WP, He WW, Lu SL, Qian M. 2015. Additive manufacturing of a high niobium-containing titanium aluminide alloy by selective electron beam melting. Materials Science and Engineering A-Structural Materials Properties Microstructure and Processing, 636, 103–107.
23. Deckers J, Vleugels J, Kruthl JP. 2014. Additive manufacturing of ceramics: a review. Journal of Ceramic Science and Technology, 5(4), 245–260.
24. Chappell WJ, Reilly C, Halloran J, Katehi LPB. 2003. Ceramic synthetic substrates using solid freeform fabrication. IEEE Transactions on Microwave Theory and Techniques, 51(3), 752–760.
25. Tang HH, Yen HC. 2015. Slurry-based additive manufacturing of ceramic parts by selective laser burn-out. Journal of the European Ceramic Society, 35(3), 981–987.
26. Gaytan SM, Cadena MA, Karim H, Delfin D, Lin Y, Espalin D, MacDonald E, Wicker RB. 2015. Fabrication of barium titanate by binder jetting additive manufacturing technology. Ceramics International, 41(5), 6610–6619.
27. Guessasma S, Belhabib S, Nouri H. 2015. Significance of pore percolation to drive anisotropic effects of 3D printed polymers revealed with X-ray μ-tomography and finite element computation. Polymer, 81, 29–36.
28. Shaffer S, Yang KJ, Vargas J, Di Prima MA, Voit W. 2014. On reducing anisotropy in 3D printed polymers via ionizing radiation. Polymer, 55(23), 5969–5979.
29. Hatzenbichler M, Geppert M, Seemann R, Stampfl J. 2013. Additive manufacturing of photopolymers using the texas instruments DLP lightcrafter. Emerging Digital Micromirror Device Based Systems and Applications V, 8618, 86180A-1–86180A-8.
30. Puppi D, Mota C, Gazzarri M, Dinucci D, Gloria A, Myrzabekova M, Ambrosio L, Chiellini F. 2012. Additive manufacturing of wet-spun polymeric scaffolds for bone tissue engineering. Biomedical Microdevices, 14(6), 1115–1127.
31. Quan Z, Wu A, Keefe M, Qin X, Yu J, Suhr J, Byun J-H, Kim B-S, Chou T-W. 2015. Additive manufacturing of multidirectional preforms for composites: opportunities and challenges. Materials Today, 18, 503–512.
32. Ning FD, Cong WL, Qiu JJ, Wei JH, Wang SR. 2015. Additive manufacturing of carbon fiber reinforced thermoplastic composites using fused deposition modeling. Composites Part B-Engineering, 80, 369–378.
33. Gu DD, Chang F, Dai DH. 2015. Selective laser melting additive manufacturing of novel aluminum based composites with multiple reinforcing phases. Journal of Manufacturing Science and Engineering-Transactions of the ASME, 137(2).
34. Tekinalp HL, Kunc V, Velez-Garcia GM, Duty CE, Love LJ, Naskar AK, Blue CA, Ozcan S. 2014. Highly oriented carbon fiber-polymer composites via additive manufacturing. Composites Science and Technology, 105, 144–150.
35. Lee JS, Cha HD, Shim JH, Jung JW, Kim JY, Cho DW. 2012. Effect of pore architecture and stacking direction on mechanical properties of solid freeform fabrication-based scaffold for

bone tissue engineering. Journal of Biomedical Materials Research Part A, 100A(7), 1846–1853.

36. Sa AME, Mello VM, Echavarria KR, Covill D. 2015. Adaptive voids primal and dual adaptive cellular structures for additive manufacturing. Visual Computer, 31(6–8), 799–808.
37. Takezawa A, Kobashi M, Kitamura M. 2015. Porous composite with negative thermal expansion obtained by photopolymer additive manufacturing. APL Materials, 3(7), 076103-1–076103-6.
38. Watjen AM, Gingter P, Kramer M, Telle R. 2014. Novel prospects and possibilities in additive manufacturing of ceramics by means of direct inkjet printing. Advances in Mechanical Engineering, 2014, 141346.
39. Sugavaneswaran M, Arumaikkannu G. 2014. Modelling for randomly oriented multi material additive manufacturing component and its fabrication. Materials & Design, 54, 779–785.
40. Herderick ED. 2015. Progress in additive manufacturing. JOM, 67(3), 580–581.
41. Giannatsis J, Dedoussis V. 2009. Additive fabrication technologies applied to medicine and health care: a review. International Journal of Advanced Manufacturing Technology, 40(1–2), 116–127.
42. Vaezi M, Seitz H, Yang SF. 2013. A review on 3D micro-additive manufacturing technologies. International Journal of Advanced Manufacturing Technology, 67(5–8), 1721–1754.
43. Venekamp NJR, Le Fever HT. 2015. Application areas of additive manufacturing from curiosity to application. IEEE Technology and Society Magazine, 34(3), 81–87.
44. Lipton JI, Cutler M, Nigi F, Cohen D, Lipson H. 2015. Additive manufacturing for the food industry. Trends in Food Science & Technology, 43(1), 114–123.
45. Heinrich A, Maillard P, Suckow A, Grzesiak A, Sorg P, Berger U. 2015. Additive manufacturing – a new approach for individualized optical shape metrology. Optical Measurement Systems for Industrial Inspection Ix, 9525, 95251T-1–95251T-10.
46. Dickson MN. 2015. Soft strain sensors fabricated through additive manufacturing. MRS Bulletin, 40(6), 463.
47. Setaki F, Tenpierik M, Turrin M, van Timmeren A. 2014. Acoustic absorbers by additive manufacturing. Building and Environment, 72, 188–200.
48. Ho CMB, Ng SH, Yoon YJ. 2015. A review on 3D printed bioimplants. International Journal of Precision Engineering and Manufacturing, 16(5), 1035–1046.
49. Butscher A, Bohner M, Hofmann S, Gauckler L, Muller R. 2011. Structural and material approaches to bone tissue engineering in powder-based three-dimensional printing. Acta Biomaterialia, 7(3), 907–920.
50. Cox SC, Thornby JA, Gibbons GJ, Williams MA, Mallick KK. 2015. 3D printing of porous hydroxyapatite scaffolds intended for use in bone tissue engineering applications. Materials Science & Engineering C-Materials for Biological Applications, 47, 237–247.
51. Jiang CP, Chen YY, Hsieh MF, Lee HM. 2013. Solid freeform fabrication and in-vitro response of osteoblast cells of mPEG-PCL-mPEG bone scaffolds. Biomedical Microdevices, 15(2), 369–379.
52. Waller JM, Saulsberry RL, Parker BH, Hodges KL, Burke ER, Taminger KM. 2015. Summary of NDE of Additive Manufacturing Efforts in NASA. 41st Annual Review of Progress in Quantitative Nondestructive Evaluation, Vol. 34, 1650, 51–62.
53. Zhu WJ, Zhao XL, Zhang W, Ren K, Zhang ZY, Li DC. 2014. Design and evaluation of fully configured models built by additive manufacturing. AIAA Journal, 52(7), 1441–1451.
54. Zhu J-H, Zhang W-H, Xia L. 2015. Topology optimization in aircraft and aerospace structures design. Archives of Computational Methods in Engineering, 1–28.
55. Gu XJ, Zhu JH, Zhang WH. 2012. The lattice structure configuration design for stereolithography investment casting pattern using topology optimization. Rapid Prototyping Journal, 18(5), 353–361.
56. Engstrom DS, Porter B, Pacios M, Bhaskaran H. 2014. Additive nanomanufacturing – a review. Journal of Materials Research, 29(17), 1792–1816.
57. Chakraborty D, Reddy BA, Choudhury AR. 2008. Extruder path generation for curved layer fused deposition modeling. Computer-Aided Design, 40(2), 235–243.
58. Lee CS, Kim SG, Kim HJ, Ahn SH. 2007. Measurement of anisotropic compressive strength of rapid prototyping parts. Journal of Materials Processing Technology, 187, 627–630.
59. Ahn SH, Montero M, Odell D, Roundy S, Wright PK. 2002. Anisotropic material properties of fused deposition modeling ABS. Rapid Prototyping Journal, 8(4), 248–257.
60. Lee PH, Chung H, Lee SW, Yoo J, Ko J. 2014. Review: dimensional accuracy in additive manufacturing processes. Proceedings of the ASME 9th International Manufacturing Science and Engineering Conference, 2014, Detroit, Michigan, USA, vol. 1, V001T04A045.
61. Zhu YY, Tian XJ, Li J, Wang HM. 2015. The anisotropy of laser melting deposition additive manufacturing Ti-6.5Al-3.5Mo-1.5Zr-0.3Si titanium alloy. Materials & Design, 67, 538–542.
62. Rosa B, Mognol P, Hascoet JY. 2015. Laser polishing of additive laser manufacturing surfaces. Journal of Laser Applications, 27, S29102-1–S29102-7.
63. Pinto JM, Arrieta C, Andia ME, Uribe S, Ramos-Grez J, Vargas A, Irarrazaval P, Tejos C. 2015. Sensitivity analysis of geometric errors in additive manufacturing medical models. Medical Engineering & Physics, 37(3), 328–334.
64. Nelaturi S, Kim W, Kurtoglu T. 2015. Manufacturability feedback and model correction for additive manufacturing. Journal of Manufacturing Science and Engineering-Transactions of the ASME, 137(2), 021015-1–021015-9.
65. Wang T, Zhu YY, Zhang SQ, Tang HB, Wang HM. 2015. Grain morphology evolution behavior of titanium alloy components during laser melting deposition additive manufacturing. Journal of Alloys and Compounds, 632, 505–513.
66. Ma Y, Cuiuri D, Hoye N, Li HJ, Pan ZX. 2015. The effect of location on the microstructure and mechanical properties of titanium aluminides produced by additive layer manufacturing using in-situ alloying and gas tungsten arc welding. Materials Science and Engineering A-Structural Materials Properties Microstructure and Processing, 631, 230–240.
67. Hong C, Gu DD, Dai DH, Alkhayat M, Urban W, Yuan PP, Cao S, Gasser A, Weisheit A, Kelbassa I, Zhong ML, Poprawe R. 2015. Laser additive manufacturing of ultrafine TiC particle reinforced Inconel 625 based composite parts: tailored microstructures and enhanced performance. Materials Science and Engineering A-Structural Materials Properties Microstructure and Processing, 635, 118–128.
68. Carroll BE, Palmer TA, Beese AM. 2015. Anisotropic tensile behavior of Ti-6Al-4V components fabricated with directed

energy deposition additive manufacturing. Acta Materialia, 87, 309–320.

69. Thrimurthulu K, Pandey PM, Reddy NV. 2004. Optimum part deposition orientation in fused deposition modeling. International Journal of Machine Tools & Manufacture, 44(6), 585–594.
70. Rua Y, Muren R, Reckinger S. 2015. Limitations of additive manufacturing on microfluidic heat exchanger components. Journal of Manufacturing Science and Engineering-Transactions of the ASME, 137(3), 034504-1–034504-5.
71. Carneiro OS, Silva AF, Gomes R. 2015. Fused deposition modeling with polypropylene. Materials & Design, 83, 768–776.
72. Scudino S, Unterdorfer C, Prashanth KG, Attar H, Ellendt N, Uhlenwinkel V, Eckert J. 2015. Additive manufacturing of Cu-10Sn bronze. Materials Letters, 156, 202–204.
73. Foster BK, Reutzel EW, Nassar AR, Dickman CJ, Hall BT. 2015. A brief survey of sensing for metal-based powder bed fusion additive manufacturing. Dimensional Optical Metrology and Inspection for Practical Applications Iv, 9489, 94890B-1–94890B-9.
74. Furumoto T, Koizumi A, Alkahari MR, Anayama R, Hosokawa A, Tanaka R, Ueda T. 2015. Permeability and strength of a porous metal structure fabricated by additive manufacturing. Journal of Materials Processing Technology, 219, 10–16.
75. Hu KL, Jin S, Wang CCL. 2015. Support slimming for single material based additive manufacturing. Computer-Aided Design, 65, 1–10.
76. Jin YA, He Y, Fu JZ. 2015. Support generation for additive manufacturing based on sliced data. International Journal of Advanced Manufacturing Technology, 80(9–12), 2041–2052.
77. Mohamed OA, Masood SH, Bhowmik JL. 2015. Optimization of fused deposition modeling process parameters: a review of current research and future prospects. Advances in Manufacturing, 3(1), 42–53.
78. Zhou C, Chen Y, Waltz RA. 2009. Optimized mask image projection for solid freeform fabrication. Journal of Manufacturing Science and Engineering-Transactions of the ASME, 131(6), 061004-1–061004-12.
79. Garg A, Lam JSL, Savalani MM. 2015. A new computational intelligence approach in formulation of functional relationship of open porosity of the additive manufacturing process. International Journal of Advanced Manufacturing Technology, 80(1–4), 555–565.
80. Mun J, Yun BG, Ju J, Chang B. 2015. Indirect additive manufacturing based casting of a periodic 3D cellular metal – flow simulation of molten aluminum alloy. Journal of Manufacturing Processes, 17, 28–40.
81. Giannitelli SM, Mozetic P, Trombetta M, Rainer A. 2015. Combined additive manufacturing approaches in tissue engineering. Acta Biomaterialia, 24, 1–11.
82. Rogers CM, Morris GE, Gould TWA, Bail R, Toumpaniari S, Harrington H, Dixon JE, Shakesheff KM, Segal J, Rose FRAJ. 2014. A novel technique for the production of electrospun scaffolds with tailored three-dimensional micro-patterns employing additive manufacturing. Biofabrication, 6(3), 035003.
83. Newman ST, Zhu ZC, Dhokia V, Shokrani A. 2015. Process planning for additive and subtractive manufacturing technologies. CIRP Annals-Manufacturing Technology, 64(1), 467–470.
84. Jin GQ, Li WD. 2013. Adaptive rapid prototyping/manufacturing for functionally graded material-based biomedical models. International Journal of Advanced Manufacturing Technology, 65(1–4), 97–113.
85. Zhou MY, Xi JT, Yan JQ. 2004. Modeling and processing of functionally graded materials for rapid prototyping. Journal of Materials Processing Technology, 146(3), 396–402.
86. Roper DA, Good BL, McCauley R, Yarlagadda S, Smith J, Good A, Pa P, Mirotznik MS. 2014. Additive manufacturing of graded dielectrics. Smart Materials and Structures, 23(4), 045029.
87. Becker R, Grzesiak A, Henning A. 2005. Rethink assembly design. Assembly Automation, 25(4), 262–266.
88. Zhang P, Toman J, Yu Y, Biyikli E, Kirca M, Chmielus M, To AC. 2015. Efficient design-optimization of variable-density hexagonal cellular structure by additive manufacturing: theory and validation. Journal of Manufacturing Science and Engineering-Transactions of the ASME, 137(2), 021004-1–021004-8.
89. Yang L, Harrysson O, West H, Cormier D. 2015. Mechanical properties of 3D re-entrant honeycomb auxetic structures realized via additive manufacturing. International Journal of Solids and Structures, 69–70, 475–490.
90. Chen JN, Ahmad R, Suenaga H, Li W, Sasaki K, Swain M, Li Q. 2015. Shape optimization for additive manufacturing of removable partial dentures – a new paradigm for prosthetic CAD/CAM. PLos One, 10(7), 0132552.
91. Denlinger ER, Heigel JC, Michaleris P, Palmer TA. 2015. Effect of inter-layer dwell time on distortion and residual stress in additive manufacturing of titanium and nickel alloys. Journal of Materials Processing Technology, 215, 123–131.
92. Thompson B, Yoon HS. 2014. Efficient path planning algorithm for additive manufacturing systems. IEEE Transactions on Components Packaging and Manufacturing Technology, 4(9), 1555–1563.
93. Kim HC, Choi KH, Doh YH, Kim DS. 2009. Fabrication of parts and their evaluation using a dual laser in the solid freeform fabrication system. Journal of Materials Processing Technology, 209(10), 4857–4866.
94. Ding DH, Pan ZX, Cuiuri D, Li HJ. 2015. Wire-feed additive manufacturing of metal components: technologies, developments and future interests. International Journal of Advanced Manufacturing Technology, 81(1–4), 465–481.
95. Ren L, Sparks T, Ruan JZ, Liou F. 2008. Process planning strategies for solid freeform fabrication of metal parts. Journal of Manufacturing Systems, 27(4), 158–165.
96. Paul R, Anand S. 2015. A new Steiner patch based file format for additive manufacturing processes. Computer-Aided Design, 63, 86–100.
97. Xu AP, Shaw LL. 2005. Equal distance offset approach to representing and process planning for solid freeform fabrication of functionally graded materials. Computer-Aided Design, 37(12), 1308–1318.
98. Ding DH, Pan ZX, Cuiuri D, Li HJ. 2015. A multi-bead overlapping model for robotic wire and arc additive manufacturing (WAAM). Robotics and Computer-Integrated Manufacturing, 31, 101–110.
99. Song X, Pan YY, Chen Y. 2015. Development of a low-cost parallel kinematic machine for multidirectional additive manufacturing. Journal of Manufacturing Science and Engineering-Transactions of the ASME, 137(2), 021005-1–021005-13.
100. Arntz K, Wegener M, Liu Y. 2015. Computer aided manufacturing supported process planning of additive manufacturing by laser deposition welding. Journal of Laser Applications, 27, S14002-1–S14002-5.

101. Seifert T, Sowade E, Roscher F, Wiemer M, Gessner T, Baumann RR. 2015. Additive manufacturing technologies compared: morphology of deposits of silver ink using inkjet and aerosol jet printing. Industrial & Engineering Chemistry Research, 54(2), 769–779.
102. Palanivel S, Nelaturu P, Glass B, Mishra RS. 2015. Friction stir additive manufacturing for high structural performance through microstructural control in an Mg based WE43 alloy. Materials & Design, 65, 934–952.
103. Monaghan T, Capel AJ, Christie SD, Harris RA, Friel RJ. 2015. Solid-state additive manufacturing for metallized optical fiber integration. Composites Part A-Applied Science and Manufacturing, 76, 181–193.
104. Sundaram MM, Kamaraj AB, Kumar VS. 2015. Mask-less electrochemical additive manufacturing: a feasibility study. Journal of Manufacturing Science and Engineering-Transactions of the ASME, 137(2), 021006-1–021006-9.
105. Wang H, Jiang W, Valant M, Kovacevic R. 2003. Microplasma powder deposition as a new solid freeform fabrication process. Proceedings of the Institution of Mechanical Engineers Part B-Journal of Engineering Manufacture, 217(12), 1641–1650.
106. Leong KF, Cheah CM, Chua CK. 2003. Solid freeform fabrication of three-dimensional scaffolds for engineering replacement tissues and organs. Biomaterials, 24(13), 2363–2378.
107. Cai K, Guo D, Huang Y, Yang JL. 2003. Solid freeform fabrication of alumina ceramic parts through a lost mould method. Journal of the European Ceramic Society, 23(6), 921–925.
108. Guo D, Cai K, Nan C, Li LT, Gul ZL. 2002. Gelcasting based solid freeform fabrication of piezoelectric ceramic objects. Scripta Materialia, 47(6), 383–387.
109. Sohn H, Yang DY. 2002. Selective infiltration of superheated micro-sized droplets of a tin-lead alloy into a preheated powder layer for a novel solid freeform fabrication process. Journal of Materials Science Letters, 21(21), 1717–1719.
110. Mancanares CG, Zancul ED, da Silva JC, Miguel PAC. 2015. Additive manufacturing process selection based on parts' selection criteria. International Journal of Advanced Manufacturing Technology, 80(5–8), 1007–1014.
111. Price S, Cheng B, Lydon J, Cooper K, Chou K. 2014. On process temperature in powder-bed electron beam additive manufacturing: process parameter effects. Journal of Manufacturing Science and Engineering-Transactions of the ASME, 136(6), 061019-1–061019-10.
112. Wei C, Dong JY. 2014. Development and modeling of melt electrohydrodynamic-jet printing of phase-change inks for high-resolution additive manufacturing. Journal of Manufacturing Science and Engineering-Transactions of the ASME, 136(6), 061010-1–061010-7.
113. Kang HW, Seol YJ, Cho DW. 2009. Development of an indirect solid freeform fabrication process based on microstereolithography for 3D porous scaffolds. Journal of Micromechanics and Microengineering, 19(1), 015011.
114. Mondrinos MJ, Dembzynski R, Lu L, Byrapogu VKC, Wootton DM, Lelkes PI, Zhou J. 2006. Porogen-based solid freeform fabrication of polycaprolactone-calcium phosphate scaffolds for tissue engineering. Biomaterials, 27(25), 4399–4408.
115. Choi H, Yang Y, Li XC. 2005. Experimental study on integration of laser-based additive/subtractive processes for meso/micro solid freeform fabrication. International Journal of Advanced Manufacturing Technology, 26(4), 335–341.
116. Yoshida H, Igarashi T, Obuchi Y, Takami Y, Sato J, Araki M, Miki M, Nagata K, Sakai K, Igarashi S. 2015. Architecture-scale human-assisted additive manufacturing. ACM Transactions on Graphics, 34(4), 88.
117. Siraskar N, Paul R, Anand S. 2015. Adaptive slicing in additive manufacturing process using a modified boundary octree data structure. Journal of Manufacturing Science and Engineering-Transactions of the ASME, 137(1), 011007-1–011007-11.
118. Slotwinski JA, Garboczi EJ, Stutzman PE, Ferraris CF, Watson SS, Peltz MA. 2014. Characterization of metal powders used for additive manufacturing. Journal of Research of the National Institute of Standards and Technology, 119, 460–493.
119. Sakar-Deliormanli A, Celik E, Polat M. 2008. Rheological behavior of PMN gels for solid freeform fabrication. Colloids and Surfaces A-Physicochemical and Engineering Aspects, 324(1–3), 159–166.
120. Yen HC. 2015. Experimental studying on development of slurry-layer casting system for additive manufacturing of ceramics. International Journal of Advanced Manufacturing Technology, 77(5–8), 915–925.
121. Shiva S, Palani IA, Mishra SK, Paul CP, Kukreja LM. 2015. Investigations on the influence of composition in the development of Ni-Ti shape memory alloy using laser based additive manufacturing. Optics and Laser Technology, 69, 44–51.
122. Strano G, Hao L, Everson RM, Evans KE. 2013. A new approach to the design and optimisation of support structures in additive manufacturing. International Journal of Advanced Manufacturing Technology, 66(9–12), 1247–1254.
123. Cooper D, Thornby J, Blundell N, Henrys R, Williams MA, Gibbons G. 2015. Design and manufacture of high performance hollow engine valves by additive layer manufacturing. Materials & Design, 69, 44–55.
124. Zohdi TI. 2015. Modeling and simulation of cooling-induced residual stresses in heated particulate mixture depositions in additive manufacturing. Computational Mechanics, 56(4), 613–630.
125. Bai XW, Zhang HO, Wang GL. 2015. Modeling of the moving induction heating used as secondary heat source in weld-based additive manufacturing. International Journal of Advanced Manufacturing Technology, 77(1–4), 717–727.
126. Brant AM, Sundaram MM, Kamaraj AB. 2015. Finite element simulation of localized electrochemical deposition for maskless electrochemical additive manufacturing. Journal of Manufacturing Science and Engineering-Transactions of the ASME, 137(1), 011018-1–011018-9.
127. Kranz J, Herzog D, Emmelmann C. 2015. Design guidelines for laser additive manufacturing of lightweight structures in TiAl6V4. Journal of Laser Applications, 27, S14001-1–S14001-16.
128. Grihon S, Krog L, Bassir D. 2011. Numerical optimization applied to structure sizing at AIRBUS: a multi-step process. International Journal for Simulation and Multidisciplinary Design Optimization, 3(4), 432–442.
129. Zenzinger G, Bamberg J, Ladewig A, Hess T, Henkel B, Satzger W. 2015. Process monitoring of additive manufacturing by using optical tomography. 41st Annual Review of Progress in Quantitative Nondestructive Evaluation, Vol. 34, 1650, 164–170.
130. Carl V. 2015. Monitoring system for the quality assessment in additive manufacturing. 41st Annual Review of Progress in

Quantitative Nondestructive Evaluation, Vol. 34, 1650, 171–176.

131. Krauss H, Zeugner T, Zaeh MF. 2015. Thermographic process monitoring in powderbed based additive manufacturing. 41st Annual Review of Progress in Quantitative Nondestructive Evaluation, Vol. 34, 1650, 177–183.

132. Rieder H, Dillhofer A, Spies M, Bamberg J, Hess T. 2015. Ultrasonic online monitoring of additive manufacturing processes based on selective laser melting. 41st Annual Review of Progress in Quantitative Nondestructive Evaluation, Vol. 34, 1650, 184–191.

133. Everton S, Dickens P, Tuck C, Dutton B. 2015. Evaluation of laser ultrasonic testing for inspection of metal additive manufacturing. Laser 3d Manufacturing Ii, 9353, 935316-1–935316-8.

134. Monzon MD, Ortega Z, Martinez A, Ortega F. 2015. Standardization in additive manufacturing: activities carried out by international organizations and projects. International Journal of Advanced Manufacturing Technology, 76(5–8), 1111–1121.

135. Raguvarun K, Balasubramaniam K, Rajagopal P, Palanisamy S, Nagarajah R, Hoye N, Curiri D, Kapoor A. 2015. A study of internal structure in components made by additive manufacturing process using 3D X-ray tomography. 41st Annual Review of Progress in Quantitative Nondestructive Evaluation, Vol. 34, 1650, 146–155.

136. Guessasma S, Nouri H. 2015. Compression behaviour of bread crumb up to densification investigated using X-ray tomography and finite element computation. Food Research International, 72, 140–148.

137. Ayadi A, Nouri H, Guessasma S, Roger F. 2015. An original approach to assess elastic properties of a short glass fibre reinforced thermoplastic combining X-ray tomography and finite element computation. Composite Structures, 125, 277–286.

138. Adamczak S, Bochnia J, Kaczmarska B. 2015. An analysis of tensile test results to assess the innovation risk for an additive manufacturing technology. Metrology and Measurement Systems, 22(1), 127–138.

139. Guessasma S, Bassir DH. 2010. Identification of mechanical properties of biopolymer composites sensitive to interface effect using hybrid approach. Mechanics of Materials, 42(3), 344–353.

140. Hbib M, Guessasma S, Bassir D, Benseddiq N. 2011. Interfacial damage in biopolymer composites reinforced using hemp fibres: finite element simulation and experimental investigation. Composites Science and Technology, 71(11), 1419–1426.

141. Guessasma S, Bassir D. 2010. Optimization of the mechanical properties of virtual porous solids using a hybrid approach. Acta Materialia, 58(2), 716–725.

142. Xiong J, Zhang GJ. 2014. Adaptive control of deposited height in GMAW-based layer additive manufacturing. Journal of Materials Processing Technology, 214(4), 962–968.

143. Witherell P, Feng S, Simpson TW, Saint John DB, Michaleris P, Liu ZK, Chen LQ, Martukanitz R. 2014. Toward metamodels for composable and reusable additive manufacturing process models. Journal of Manufacturing Science and Engineering-Transactions of the ASME, 136(6), 061025-1–061025-9.

21

Failure mechanism of PMI foam core sandwich beam in bending

Bo Wang[1,*], Yunfeng Shi[1], Caihua Zhou[1], and Tong Li[2]

[1] Faculty of Vehicle Engineering and Mechanics, Dalian University of Technology, Dalian 116024, China
[2] School of Chemistry, Physics and Mechanical Engineering, Queensland University of Technology, 2 George St., Brisbane, QLD 4001, Australia

Abstract – Polymethacrylimide (PMI) foams have been widely applied in aerospace engineering as the core material of sandwich structures. This paper proposes a modified model to predict the constitutive relation of PMI foams and compares it to existing testing data. The study is then applied to the investigation of the failure mechanism of PMI foam core sandwich beams in bending. Corresponding bending tests were carried out where a complex failure process was observed through a high-speed camera. Numerical model of the foregoing sandwich beam is developed, in which the maximum principal stress criteria is used to predict damage propagation in PMI foam core. Both results from tests and numerical simulation validate the reliability of the theoretical prediction of the failure of PMI foam core sandwich beam using the proposed modified model of PMI foams. This study provides a theoretic tool for the design of sandwich structures with PMI foam core.

Key words: Foams, Damage mechanics, Finite element analysis, Process monitoring.

1 Introduction

The ROHACELL WF series polymethacrylimide (PMI) foam has been introduced as the material of structural sandwich core for aerospace structures since 1971 and today this material is listed in more than 170 aerospace specifications worldwide [1]. Besides its high specific strength and specific stiffness, the inotropic and closed-cell character of PMI foams provides a dramatic improvement over other foam materials like aluminum foams [1]. However, despite the outstanding properties of PMI foams, difficulties exist in the full understanding the failure mechanism of sandwich structures with PMI foam cores, especially under complex loading conditions. Considerable studies have been carried out on the failure analysis of PMI foams. Li and Mines have studied the compressive process of PMI foam under uniaxial compression [2] and the crush behavior under a series of tests [3]. Chen et al. have investigated the influence of foam structure on the macroscopic properties of PMI foam [4]. The fatigue behaviors of the PMI foam in tension, compression and shear were investigated by Burman and Zenkert [5–7] as well as the impact responses [8–11]. In addition, the basic mechanical properties of sandwich structures with PMI foam core were also characterized [5, 7, 12–22], where the sandwich beams or panels with PMI foam core have been tested in standard or specialized tests such as shearing [7, 14, 20], three point bending [16, 18, 20], four point bending [5, 12, 14], edgewise compression [20, 22] and indentation [17, 21]. According to these studies, the foam's density has significant influence on the failure modes of the sandwich structures [17]. The most common failure mode of the PMI foam core sandwich structures is core shearing [7, 12, 13, 16, 22], followed by local indentation collapse [16–18, 21] and face yielding [7, 22]. However, among these studies, there is a lack of systematical research on the PMI foam core sandwich structures' failure mechanism in terms of the mechanical properties of foam, such as the foam density. Limited study was conducted on the modeling of damage propagation in PMI foam cores. For the above-mentioned reasons, this paper aims at investigating the failure mechanism of sandwich beams with PMI foam core in three point bending test by applying the cellular solids theory [23]. Corresponding numerical simulations and tests were carried out to validate the theoretical prediction.

In this paper, the constitutive relation of the PMI foam is studied based on the microscopic mechanics of cellular materials introduced by Gibson and Ashby [23]. The conclusion of constitutive relationship is then applied to the theoretical investigation of failure mechanism of the sandwich beams with PMI foam core in three point bending. Moreover, three point bending tests of the sandwich beams with PMI foam core and carbon fiber skin sheets were carried out with a high speed camera to capture the beams' failure process.

*e-mail: wangbo@dlut.edu.cn

Nomenclature

A_1, A_2, B_1, B_2	Modified geometric coefficients of foam cell
C_1, C_1', C_5, C_5''	Geometric coefficients of foam cell
E^*	Young's modulus of foam material
E_f	Young's modulus of fiber's principal direction
E_{f3}	Young's modulus of fiber's out-plane direction
E_s	Young's modulus of foam's cell-wall material
$(EI)_{eq}$	Equivalent flexural rigidity of sandwich beam
M	Moment applied on the sandwich beam
p_0	Initial fluid pressure of foam cell
p_{at}	Atmospheric pressure
P	Concentrated force applied on the sandwich beam
Q	Shear applied on the sandwich beam
ε_D	Critical strain at which the densification of foam starts
v^*	Poisson's ratio of foam material
ρ^*	Density of foam material
ρ_s	Density of foam's cell-wall material
σ_c	Normal stresses in the core of sandwich beam
σ_f	Normal stresses in the faces of sandwich beam
σ_{pl}^*	Compressive crushing strength of foam material
σ_{wf}	Wrinkling stress in the face material of sandwich beam
σ_{ys}	Yield strength of foam's cell-wall material
σ_{yf}	Yield strength of the face material of sandwich beam
τ_c	Shear stress in the core of sandwich beam
φ	Geometric coefficients of foam cell

Corresponding numerical models are also developed based on the maximum principal stress criteria to predict the damage in the PMI foam core. The results obtained from different methods are compared and discussed and the conclusions are made at last.

2 Theoretical predictions of the failure of sandwich beam with PMI foam core in three point bending

2.1 The mechanical behavior of PMI foams under compression

There are two approaches to model the constitutive behavior of the PMI foam materials. The first approach is continuum modeling, namely critical state theory that has been used in the standard finite element analysis (FEA). The second approach is micromechanical modeling, in which the actual cellular structure is modeled. Gibson and Ashby [23], Chen and Lakes [24] have developed the theory of micro-mechanics of cellular materials, in which the crush behaviors of the foams are dependent on the properties of its parent material. Since the major part of PMI foams is in compression [23], this section discusses the constitutive relation of the ROHACELL WF series PMI foams under compressive load on the basis of the micro-mechanics of cellular materials.

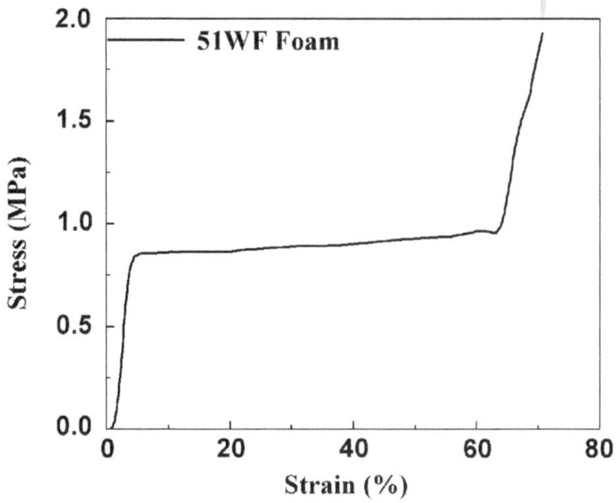

Figure 1. Stress-strain curve of ROHACELL-51WF foam tested under uniaxial compressive load.

Figure 1 shows the typical engineering compressive stress-strain curve of ROHACELL 51WF foam in compression test. The foam material shows elastic behavior under low stress condition followed by a long collapse plateau which is truncated by a high stress densification region. Theory developed by Gibson and Ashby predicts:

$$\frac{E^*}{E_s} = C_1\phi^2\left(\frac{\rho^*}{\rho_s}\right)^2 + C_1'(1-\phi)\frac{\rho^*}{\rho_s} + \frac{p_0(1-2v^*)}{E_s(1-\rho^*/\rho_s)} \tag{1.1a}$$

$$\frac{\sigma_{pl}^*}{\sigma_{ys}} = C_5\left(\phi\frac{\rho^*}{\rho_s}\right)^{3/2} + C_5''(1-\phi)\frac{\rho^*}{\rho_s} + \frac{p_0 - p_{at}}{\sigma_{ys}} \tag{1.1b}$$

$$\varepsilon_D = 1 - 1.4\frac{\rho^*}{\rho_s} \tag{1.1c}$$

where E^* and E_s are the Young's modulus of the foam and the cell-wall material, respectively, σ_{pl}^* and σ_{ys} are the compressive crushing strength of the foam and the yield strength of the cell-wall material, respectively, ε_D is the critical strain at which the curve's slope becomes E_s, ρ^* and ρ_s are the density of the foam and the cell-wall material, respectively, p_0 and p_{at} is the initial fluid pressure and the atmospheric pressure, respectively, C_1, C_1', φ, C_5 and C_5'' are the corresponding coefficients that contain all the geometric constants of the microscopic structure of foam cell [23]. The study of Simone and Gibson [25] shows that for most situations, these parameters can be identified as $C_1 \approx C_1' \approx 1$, $\varphi \approx 0.68$, $C_5 \approx 0.3$, $C_5'' \approx 0.44$. The first and second terms of equations (1.1a) and (1.1b) describe the combined effect of cell-edge bending and cell-face stretching while the third term is caused by the compression of cell fluid. In man-made foams, p_0 is usually close to p_{at} that is insignificant compared to E^* and σ_{pl}^*, so the third terms' contribution is negligible [23].

In Chen and Lakes's micro-mechanics model [24] for ROHACELL closed-cell foams based on the tetrakaidecahedral

Table 1. Comparison between micro-mechanics prediction and experimental results.

Type	Density (Kg/m^3)	E^* (MPa)				σ^*_{pl} (MPa)			
		Equation (1.1a)	Equation (1.2a)	Equation (1.3a)	Test result	Equation (1.1b)	Equation (1.2b)	Equation (1.3b)	Test result
51WF	52	76	117	76	75	0.69	0.82	0.79	0.8
71WF	75	113	171	113	105	1.03	1.22	1.68	1.7
110WF	110	173	257	172	180	1.58	1.91	3.48	3.6
200WF	205	354	503	352	350	3.23	4.10	10.42	9.0

unit cell shape, equations (1.1a) and (1.1b) can be written as follows:

$$\frac{E^*}{E_s} = \frac{0.343 + 0.823(\rho^*/\rho_s)}{0.684 + \rho^*/\rho_s} \times \frac{\rho^*}{\rho_s} \quad (1.2a)$$

$$\frac{\sigma^*_{pl}}{\sigma_{ys}} = 0.454\left(0.417 + \frac{\rho^*}{\rho_s}\right)\frac{\rho^*}{\rho_s} \quad (1.2b)$$

where E^*, E_s, σ^*_{pl}, σ_{ys}, ρ^* and ρ_s represent the same as in equations (1.1a)–(1.1c).

However, data derived from equations (1.1) and (1.2) show some disagreements compared to the test results of the ROHACELL WF series foams. One of the probable reasons is that the unit cell models used in these theories are different from the real situation, which means the geometric coefficients of the microscopic foam cell C_1, C_1', φ, C_5 and C_5'' shall be reconsidered. Assuming that for each type of ROHACELL WF series foams the foregoing geometric coefficients share the same value since nothing but the foams' porosities changed during the manufacturing processes of foams. Therefore, the Young's moduli of foams and compressive crushing strengths are only dependent on their relative densities. Equations (1.1a) and (1.1b) can be rewritten as:

$$\frac{E^*}{E_s} = A_1\left(\frac{\rho^*}{\rho_s}\right)^2 + B_1\frac{\rho^*}{\rho_s} \quad (1.3a)$$

$$\frac{\sigma^*_{pl}}{\sigma_{ys}} = A_2\left(\frac{\rho^*}{\rho_s}\right)^{32} + B_2\frac{\rho^*}{\rho_s} \quad (1.3b)$$

where A_1, B_1, A_2, B_2 can be determined by two sets of test data. Based on the test results of the ROHACELL 51WF and ROHACELL 71WF PMI foam [26], coefficients A_1, B_1, A_2 and B_2 are 0.44, 0.32, 2.32 and −0.28, respectively. Table 1 shows the Young's modulus (E^*) and the compressive crushing strength (σ^*_{pl}) of different types of ROHACELL-WF foams given by equation (1.1)–(1.3)and from tests results under uniaxial compressive load. It can be seen that the test results are in good agreement with the predictions from equation (1.3).

For closed-cell foams the post-collapse plateau rises because the enclosed gas is compressed when the cells collapse, creating a restoring pressure. The post-collapse curve can be described by [23]:

$$\sigma^* = \sigma^*_{pl} + \frac{p_0\varepsilon}{1 - \varepsilon - \rho^*/\rho_s} \quad (1.4)$$

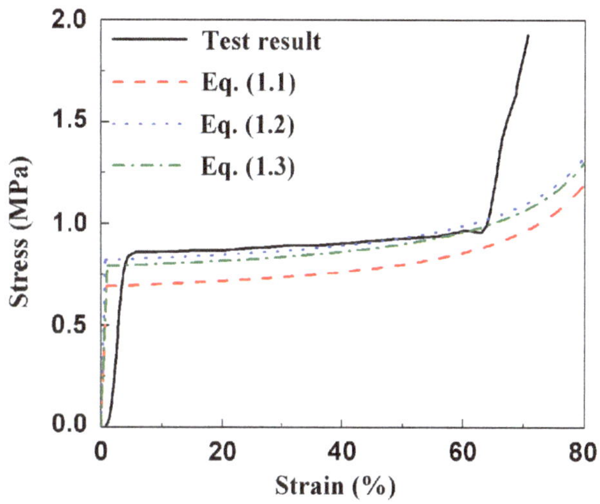

Figure 2. Compress stress-strain curve of ROHACELL-51WF PMI foam.

the second term in the right-hand side of the equation causes the plateau stress to rise with strain. Base on the data derived from equations (1.1)–(1.3) and equation (1.4), the compressive stress-strain curves of ROHACELL 51WF foams can be predicted as shown in Figure 2.

2.2 Failure mechanism of the sandwich beam with PMI foam core in three point bending

Basic mechanical properties of sandwich beams have been described in many researches [23, 27, 28]. A sandwich beam model in three-point bending is provided in Figure 3. To analyze the failure mechanism of beams, the normal stress and shear stress (results of specific moment and shear being applied) acting in the face and core needs to be characterized. Considering the shear stress as linear through the faces and constant through the core since the faces are much stiffer and thinner than the core, the normal and shear stresses can be expressed as [28]:

$$\sigma_f = \frac{ME_f \times y}{(EI)_{eq}} \quad (2.1a)$$

$$\sigma_c = \frac{ME^* \times y}{(EI)_{eq}} \quad (2.1b)$$

$$\tau_c = \frac{Q}{bc} \quad (2.1c)$$

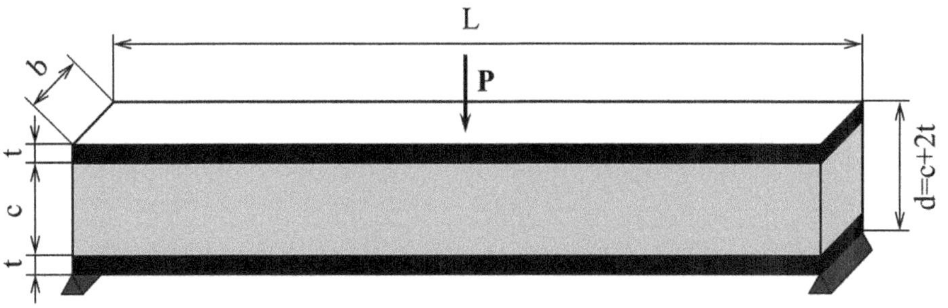

Figure 3. A sandwich beam in three-point bending load.

Table 2. Formulas of three failure modes' critical load.

Failure mode	Critical load	Equation number
Face yielding	$P = \frac{4btc}{l} \times \sigma_{yf}$	(2.6a)
Face wrinkling	$P = \frac{4btc}{l} E_f^{1/3} E_s^{2/3} \left[0.28 \left(\frac{\rho^*}{\rho_s} \right)^{4/3} + 0.20 \left(\frac{\rho^*}{\rho_s} \right)^{2/3} \right]$	(2.6b)
Core shearing	$P = 2bc \left[2.32 \left(\frac{\rho^*}{\rho_s} \right)^{3/2} - 0.28 \frac{\rho^*}{\rho_s} \right] \sigma_{ys}$	(2.6c)

where σ_f and σ_c are the normal stresses in faces and core respectively, τ_c is the shear stress in core; M and Q are the applied moment and shear, y is the distance from the neutral axis. $(EI)_{eq}$ is the equivalent flexural rigidity, which can be expressed as [28]:

$$(EI)_{eq} = \frac{E_f bt^3}{6} + \frac{E^* bc^3}{12} + \frac{E_f btd^2}{2} \approx \frac{E_f bc^2}{2} \tag{2.2}$$

The maximum stress occurs in the cross section which has the maximum moment. In the load case of three-point bending, the maximum moment and the shear can be easily acquired in terms of the concentrated load P, that

$$M = \frac{Pl}{4}, Q = \frac{P}{2} \tag{2.3}$$

Thus considering equations (2.2) and (2.3), the normal stresses in equation (2.1) can be replaced by the following:

$$\sigma_f = \frac{Pl}{4btc} \tag{2.4a}$$

$$\sigma_c = \frac{Pl}{4btc} \times \frac{E^*}{E_f} \tag{2.4b}$$

$$\tau_c = \frac{P}{2bc} \tag{2.4c}$$

Several failure modes have been identified for sandwich beams in three point bending [23, 27]: (a) plastic yielding of faces, (b) wrinkling of the compressive face, (c) shearing of the core, (d) face debonding, (e) indentation of the loading point. Since the strength of the adhesive (like epoxy resin) is usually higher than the foam core, the mode of face debonding rarely occurs as long as the adhesive is free of defects. While the last mode only occurs when the loading is extremely localized and can be avoided by increasing the loading area. Hence, the failure modes of (d) and (e) will not be discussed in this paper.

For the first failure mode (a), when the maximum normal stress in faces reaches the yield strength of the face material the faces start to yield. In this case:

$$\sigma_f = \sigma_{yf} \tag{2.5a}$$

where σ_{yf} is the yield strength of face material.

For the second failure mode (b), when the maximum normal stress in faces reaches the local elastic instability stress the faces start to wrinkle. In this case:

$$\sigma_f = \sigma_{wf} \tag{2.5b}$$

where σ_{wf} is wrinkling stress in the face material which can be expressed as follow [28]:

$$\sigma_{wf} = \frac{3E_f^{1/3} E^{*2/3}}{\left[12(3 - \upsilon^*)^2 (1 + \upsilon^*)^2\right]^{1/3}} \tag{2.5c}$$

For the last failure mode (c), if the shearing stress in the core is larger than the normal stress in faces during the loading process, the core shearing failure will happen when the shearing stress in the core reaches the yield strength of core material:

$$\tau_c = \sigma_{pl}^* \tag{2.5d}$$

where σ_{pl}^* is the yielding strength of core materials mentioned in equation (1.3b).

Based on equations (2.5), (2.4) and (1.3) and assuming that the Poisson's ratio ν^* of PMI foam is zero, the critical loads of different failure modes of the sandwich beams with PMI foam core in three point bending tests can be identified, as listed in Table 2. It can be seen that each equation contains two types of variables: one is related the properties

Table 3. Information on beam's geometry and face material's properties.

Item	Beam's geometry (mm)			Face material (MPa)			Core material (kg/m^3, MPa)			
	b	c	t	E_f	E_{f3}	σ_{yf}	ρ^*	ρ_s	E_s	σ_{ys}
51WF	60	20	1.2	97.9	42.1	792	0.052	1.2	5200	90
110WF							0.11			

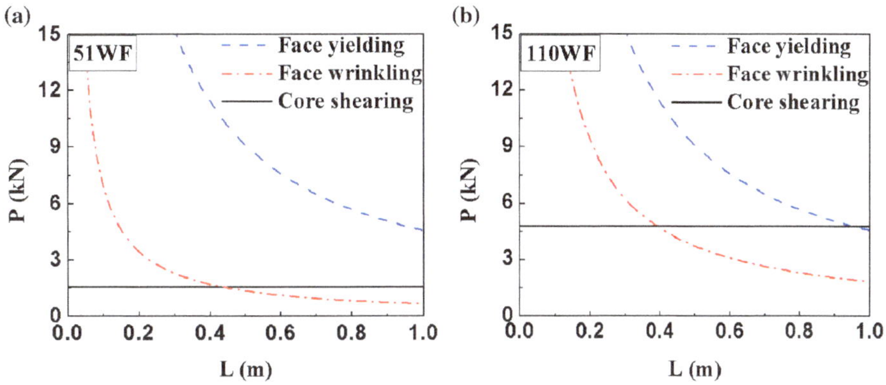

Figure 4. The PMI foam core sandwich beams' failure loads of different failure modes. (a) PMI foam of ROHACELL 51WF; (b) PMI foam ROHACELL 110WF.

of the beam's materials, such as σ_{ys}, σ_{yf}, E_s and E_f; the other is related the beam's geometrical sizes, including b, t, c and l. Besides, from Table 2 it also can be implied that the critical loads of the face yielding and face wrinkling decrease as the span length l increase, while the critical load of the core shearing failure remains saturated.

Now take the sandwich beam with PMI foam core and carbon fiber skin sheets as example. Utilizing the formulas from Table 2 the failure load and failure mode of the above sandwich beam can be predicted. Two types of PMI foam are selected as the core material: ROHACELL 51WF and ROHACELL 110WF. According to Table 2, for the face material, the Young's modulus and yielding strength along the beam's extension direction are needed to calculate the critical loads of face yielding and face wrinkling. For the PMI foam core, the foam density as well as the density of parent material, Young's modulus and yielding strength are needed to calculate the critical load of core shearing. Information on beam's geometry and material's properties are listed in Table 3. Figure 4 plots the above PMI foam core sandwich beams' failure loads of different failure modes in terms of the span length.

According to Figure 4, it can be implied that the sandwich beams with both types (51WF and 110WF) of PMI foam cores fail in the mode of core shearing (the black line) under small span situation and in the mode of face wrinkling (the red dash-point curve) under large span situation. The transition span length for the sandwich beam with ROHCELL 51WF foam is 443 mm and for the sandwich beam with ROHACELL 110WF is 391 mm. In addition, the failure mode of face yielding has less chance to happen in these sandwich beams as the corresponding failure limit is much higher than the other two modes.

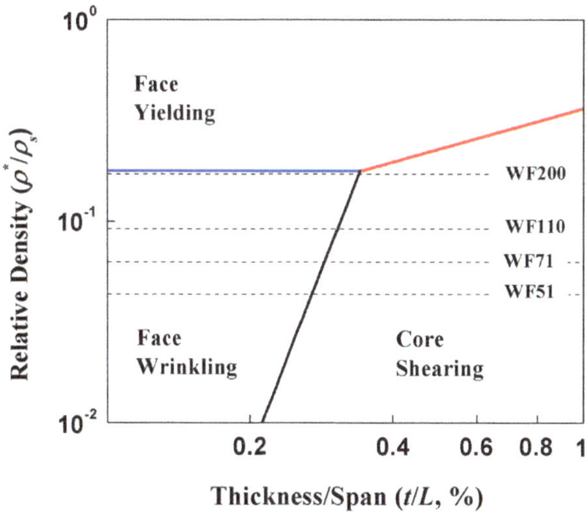

Figure 5. Failure mode map of sandwich beam with PMI foam core and carbon fiber skin sheets.

Given a specific value of the beam's core thickness and width, the failure mode map of the sandwich beam with PMI foam core and carbon fiber skin sheets can be obtained in terms of the relative density of foam and the ratio of beam thickness to span length, as shown in Figure 5. Here the relative density of foam is defined as the ratio of density of foam to density of parent material. From the figure it can be seen that all the ROHACELL WF series foams are out of the face yielding zone, therefore, it can be implied that for any type of PMI foam, the foregoing sandwich beam has only two failure modes, which are face wrinkling and core shearing.

Table 4. The failure loads of the tested sandwich beams.

Span (mm)	Specimen I	Specimen II	Specimen III	Specimen IV	Average
140	4.48	4.47	4.92	4.19	4.52
500	3.29	3.91	3.69	3.6	3.62

3 Tests of PMI foam core sandwich beam in three point bending

The testing sandwich beams consist of ROHACELL 110WF PMI foam core and carbon fiber skin sheets and the core was bonded to the skin layers using a filled epoxy resin. Geometry and materials properties of the specimens can be found in Table 4. The specimens are cut up from a whole sandwich panel and divided into two groups with span lengths of 140 mm and 500 mm. Each group contains four specimens. All tests were performed on an MTS test machine at a quasi-static loading rate of 2 mm/min according to the standard GB/T 1456-2005 of China [29], as shown in Figures 6a and 6b. The displacement and reacting force at the midpoint can be measured using the transducers set in the loading cell. An Olympus high-speed camera device was set up to capture the failure process in beams, as shown in Figure 6c. The camera's maximal frame rate is 33,000 frames per second. To guarantee the image quality, the shooting frame rate is finally set to 10,000 frames per second so that all details can been caught during the failure process.

The force-displacements curves of the testing sandwich beams with span lengths of 140 mm and 500 mm are shown in Figure 7, where a non-linear feature is observed at the beginning of the tests, which is caused by the seating effect on the loading cell and supports. However, the non-linear feature does not affect the measurement of the failure limit of beams and the critical failure load of each testing sandwich is determined by the maximum value of the force-displacements curve, as listed in Table 4.

Figure 8 shows the image of the 140 mm-span length beam tested in three-point bending load until failure. From the image it can be seen that there is a 45° crack progressed from the loading cell to the bottom face, which is followed by a horizontal crack that ends at the support point. It is clear that the beam fails in the core shearing mode, which is same as the theoretical prediction in the previous section. The failure of material occurs almost instantaneous and the high speed camera was unable to capture the failure process under the setting of 10,000 frames per second, which means that the failure duration is less than 0.1 ms. In addition, the failure only occurs on one side of the beam since the loading situation can hardly be exactly symmetrical.

Figure 9 shows a group of images of failure process in the testing sandwich beams with 500 mm span length, which are captured by the high speed camera as above-mentioned. The failure occurs on one side of the beam because of the asymmetrical loading situation. The images show that the upper face adjacent to the loading cell wrinkled at the beginning of the failure process and the wrinkle deformation caused the upper face debond from the core and a crack thereby initiated between the upper face and the core. The crack propagates along the axial direction towards the support end and large area of the upper face was separated from the core material. Figure 10 gives the failure of the 500 mm span length sandwich beam. It can be found from Figure 10 that a thin PMI foam layer with numerous surface irregularities adhere to the upper face, which indicates that the adhesive (epoxy resin) between the face and core remain intact and the face debonding failure mode does not occur. Hence, the principle failure mode of sandwich beam with 500 mm span length is face wrinkling according to the experimental findings.

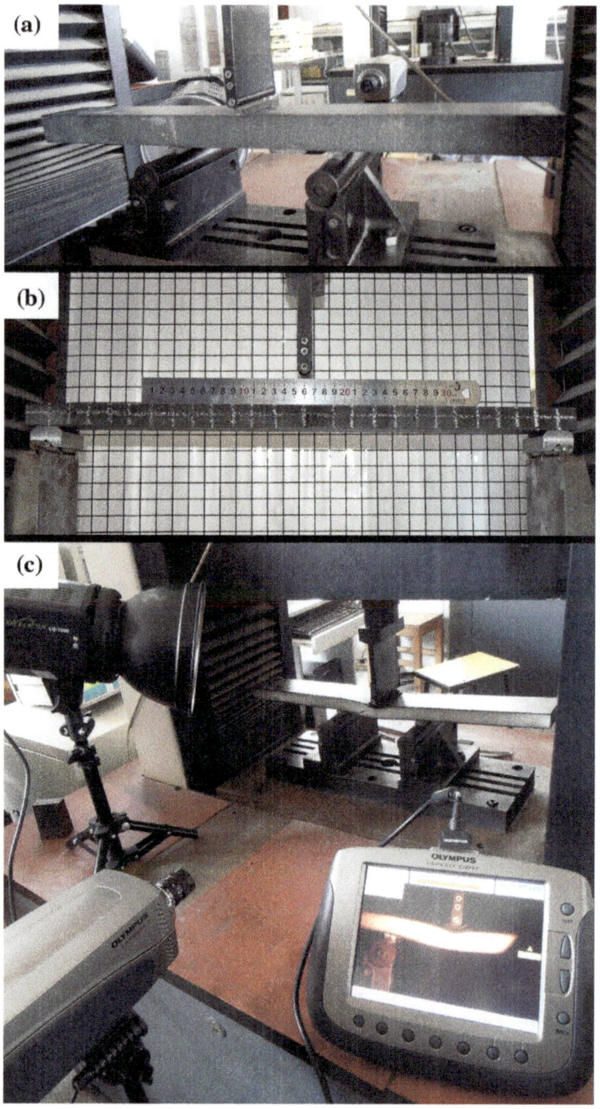

Figure 6. Three-point bending load tests of PMI foam core sandwich beams. (a) 500 mm span length, (b) 140 mm span length (c) the testing system and the high speed camera.

The sandwich beams consisted of ROHACELL 110WF PMI foam core and carbon fiber skin sheets in three point bending show different failure mode under different span length. The sandwich beam with 140 mm span length fails in the core shearing mode, while the sandwich beam with 500 mm span length shows a complex failure process: The face adjacent to the loading cell wrinkled at first and the excessive wrinkling deformation caused damage of the nearby foam

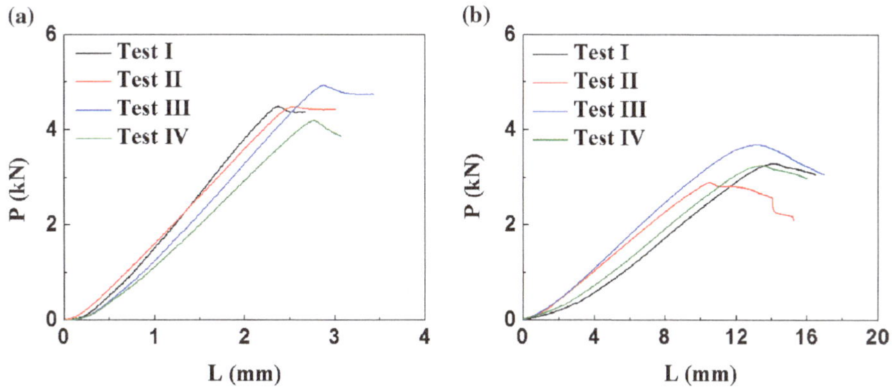

Figure 7. Force-displacement curve of the tested sandwich beams (a) span length of 140 mm (b) span length of 500 mm.

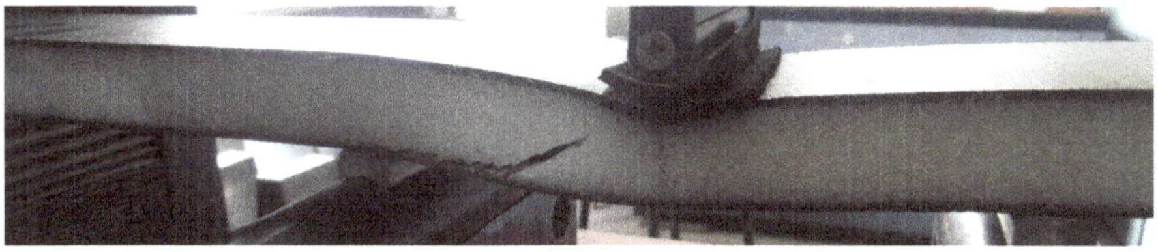

Figure 8. The failure of the sandwich beam of 140 mm span length.

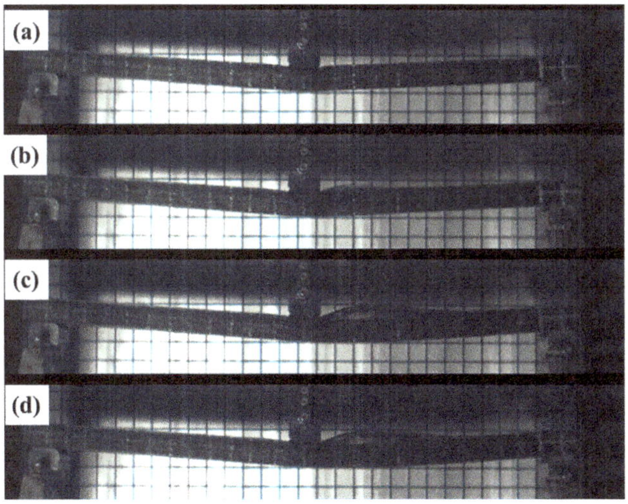

Figure 9. The failure process of the tested sandwich beam with 500 mm span length captured by high-speed camera.

material until most of the upper face was separated from the core. However, the principle failure mode of sandwich beams with 500 mm span length is face wrinkling. These above tests results are in agreement with the theoretical predictions that the beam fails in core shearing mode under small span situation, and in face wrinkling mode under large span situation as discussed in Section 1.

4 Numerical simulations of the sandwich beams with PMI foam core in three point bending

ABAQUS/Explicit is used here to simulate the failure of the sandwich beam consists of ROHACELL 110WF PMI foam core and carbon fiber skin sheets. The 2-D plain strain models of the sandwich beams with 140 mm and 500 mm span lengths are developed. The mechanical behavior of the PMI foam is described by the "low density foam" material model where the maximum principal stress criteria is used to model the failure in the core material while the carbon fiber skin sheets are considered as isotropic materials.

4.1 Numerical model

Since the face debond failure is not considered as discussed in Section 1, the adhesive in the beam structure is not included in the model. Thus the core and faces of sandwich beam are modeled as one part, which means the core and faces share

Figure 10. The failure of the sandwich beam of 500 mm span length.

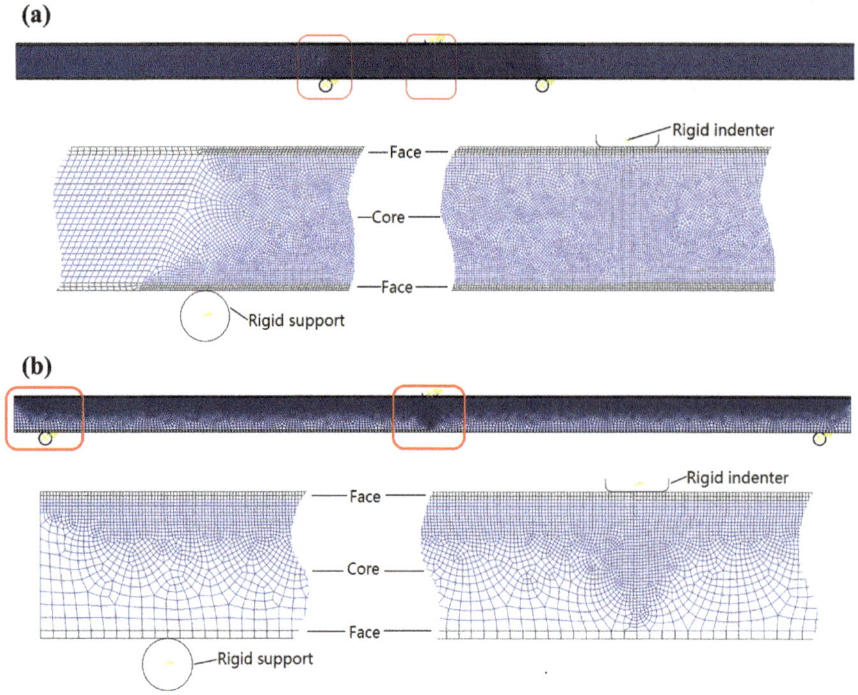

Figure 11. Numerical models of the PMI foam core sandwich beams (a) span length of140 mm (b) span length of 500 mm.

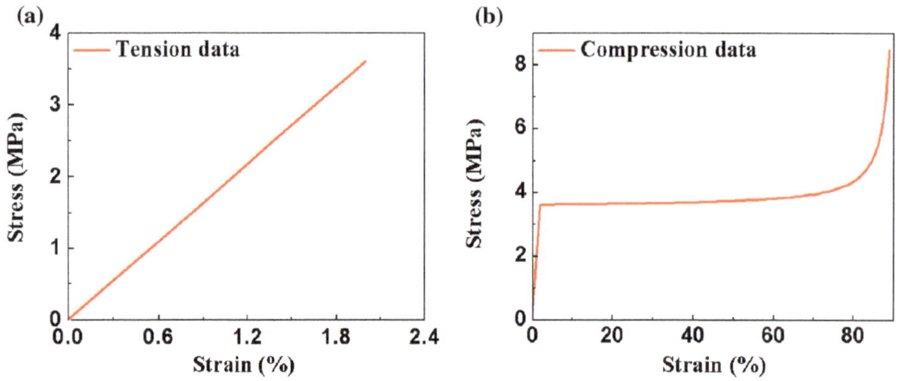

Figure 12. Stress-strain curves of ROHACELL 110WF under uniaxial tension and compression load.

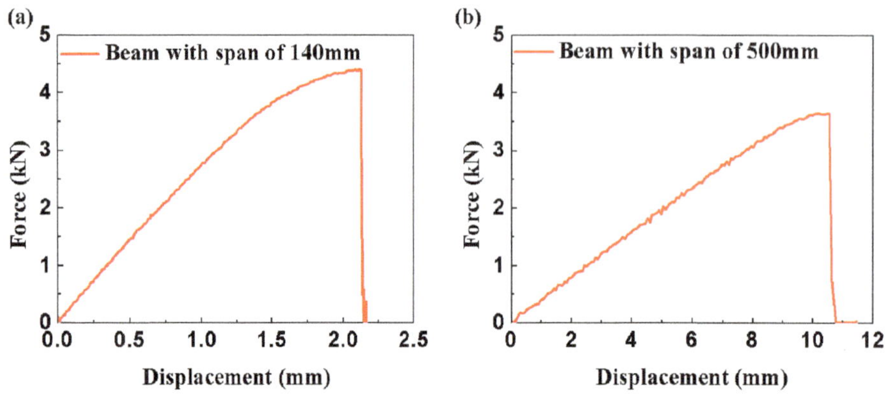

Figure 13. Force-displacement curves of the ROHACELL 110WF core sandwich beams in three point bending (a) span length of 140 mm (b) span length of 500 mm.

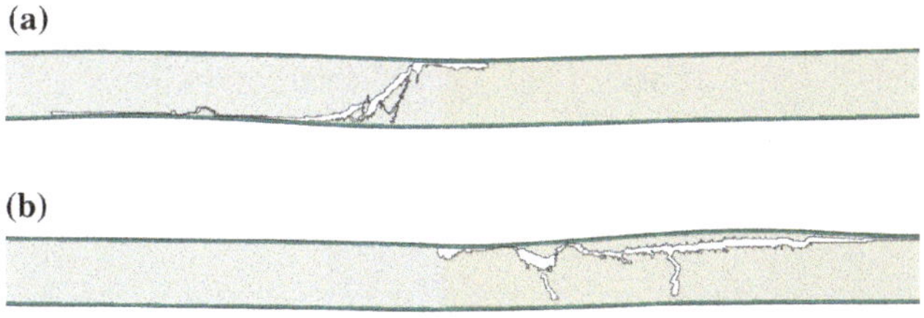

Figure 14. Numerical simulations of the failure the ROHACELL 110WF core sandwich beams in three point bending (a) span length of 140 mm (b) span length of 500 mm.

Table 5. Numerical results of the ROHACELL 110WF core sandwich beams in three point bending.

Span (mm)	Load capability (kN)	Failure mode
140	4.8	Core shearing
500	3.6	Face wrinkling

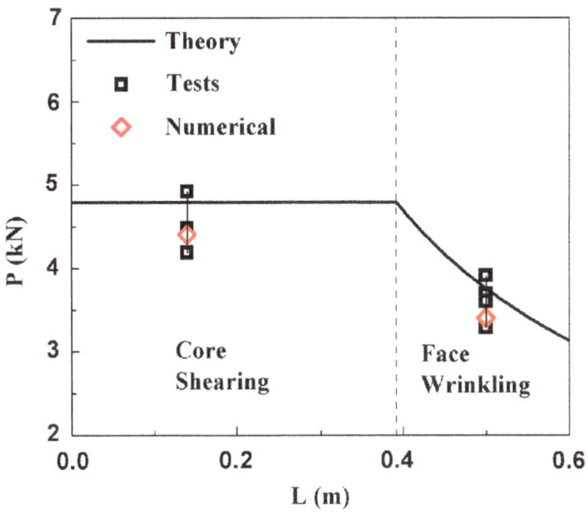

Figure 15. The failure loads of the sandwich beams with ROHACELL 110WF PMI foam core and carbon fiber skin sheets in bending.

the common nodes in the surfaces where they are bonded, as shown in Figure 11. To ensure the precision of simulation, the mesh in the failure area of the beam is refined based on the test results in Section 2. The global mesh size is 1 mm while the size of the refined mesh is 0.4 mm. The 4-node bilinear plane strain quadrilateral, reduced integration, hourglass control element (CPE4R) is used. The loading cell and supports are regard as rigid while the friction-free contact is set for the interaction between the rigid and the deformable body. All the DOFs of the supports are fixed and the loading cell is applied with a displacement load which is large enough to cause failure of the beam. The nonlinearity geometric analysis (Nlgeom) option is adopted in the whole simulating process.

The mechanical properties of the carbon fiber is described by the orthotropic material with the Young's modulus of the principal direction and thickness direction and the failure stress been given in Table 4. The mechanical property of the PMI foam is described by the Low Density Foam model. The uniaxial tension and compression test data that derived from the theoretical analysis are used as the input of model, as shown in Figure 12. Noticing that the stress-strain curve of PMI foam under uniaxial tension is treated as linear [3, 23]. The maximum principal stress criteria is used to model the failure in the PMI foam core and the elements that exceed the maximum allowable principal tensile stress, which is 3.7 MPa according to the tests [26], will be removed during the simulation process.

4.2 Simulation results

The force-displacement curves of the PMI foam core sandwich beams with span lengths of 140 mm and 500 mm are shown in Figure 13. The figure indicates that the load capabilities of the beam of 140 mm and 500 mm span length is 5.1 kN and 3.6 kN, respectively. The numerical results of PMI foam core sandwich beams tested in three point bending until failure are shown in Figure 14. It can be found that the PMI foam core sandwich beam with span length of 140 mm fails in core shearing mode, while the sandwich beam with span length of 500 mm fails in face wrinkling mode (Figure 14).

5 Results discussion

The ROHACELL 110WF foam core and carbon fiber skin sheets sandwich beams' load capabilities that derived from theoretical predictions, tests data and numerical simulations are compared in Figure 15, while the beams' failure modes based on test results and numerical simulations are shown in Table 6.

Table 6. The failure modes of sandwich beams with ROHACELL 110WF PMI foam core and carbon fiber skin sheets in three point bending.

Span length	140 mm	500 mm
Numerical		
Tests		

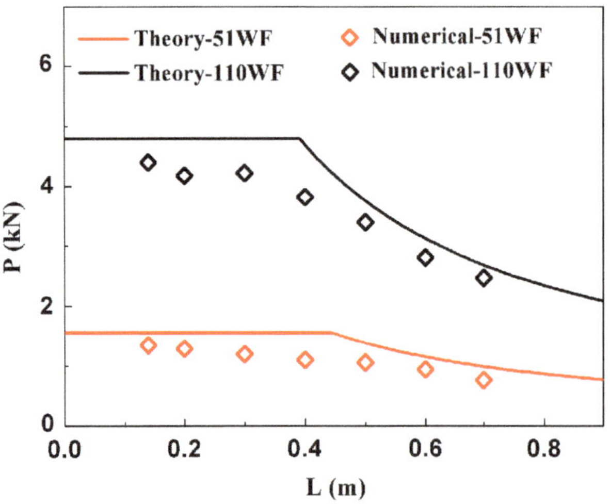

Figure 16. Numerical results of the failure loads of the sandwich beams with different PMI foam cores.

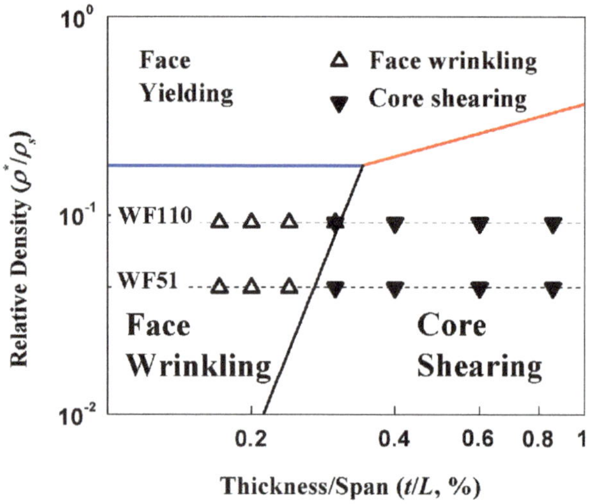

Figure 17. Failure mode map of PMI foam core/carbon-fiber epoxy resin composites sandwich beams.

From these figures it can be concluded that the results of different methods show good consistency, which validates the theoretical predictions of the failure of sandwich beams with PMI foam core in three point bending tests. It also proves the feasibility of modeling the failure of beam sandwich structure with PMI foam core based on maximum principal stress criteria in the core material.

The above conclusions give the basis to apply the virtual experiments method to the investigation of PMI foam theory. In order to verify the exactitude of the failure mode map as shown in Figure 5, considerable real tests are needed to provide convincing properties. However, by adopting the above-mentioned hybrid methods (both experimental and numerical), the large amount of tests can be reduced by conducting equivalent numerical simulations where the mechanical properties of the PMI foam are obtained from the theoretical analysis. Herein, the ROHACELL 51WF and ROHACELL 110WF PMI foam are selected as the typical core materials in the

Table 7. Numerical simulations of the failure modes of sandwich beams with different PMI foam cores.

Span (mm)	ROHACELL 51WF	ROHACELL 110WF
140		
200	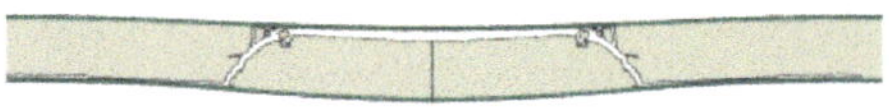	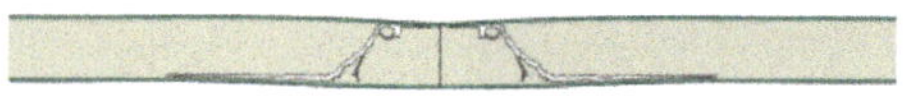
300	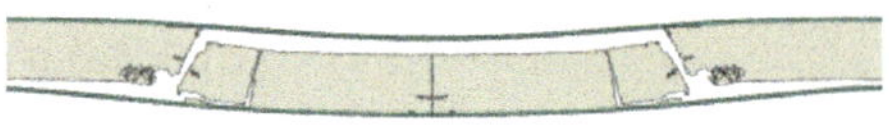	
400	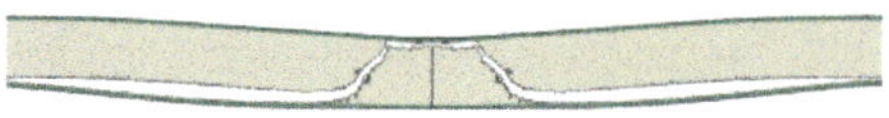	
500		
600	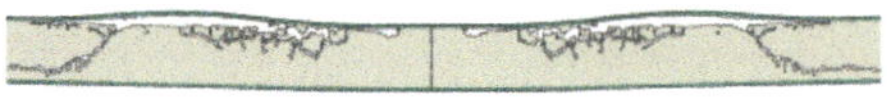	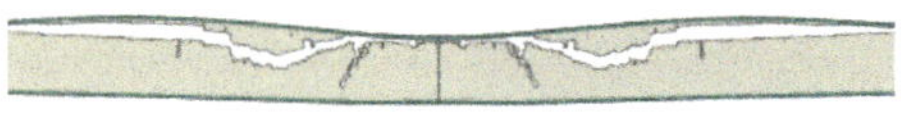
700	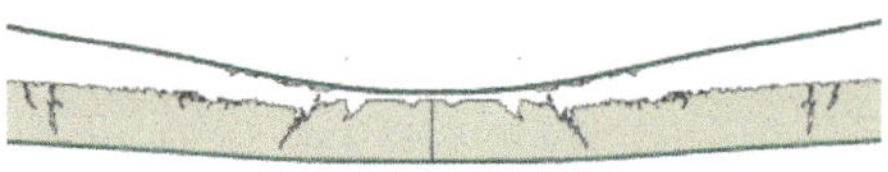	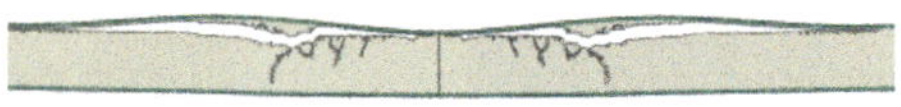

simulations and the skin sheets remain as carbon fiber whose mechanical properties are given in Table 3, a series of sandwich beams with different span lengths are modeled, in which only half of the beam is modeled while the symmetric boundary condition is applied. The results are given as below.

Figure 16 shows the numerical results of the failure limit of the sandwich beams with different type of PMI foams. The numerical results of failure limit show good consistency with the theoretical predictions. The modes and failure limit of the sandwich beams in bending tests are provided in Table 7. Noticing that for each beam its symmetric model is developed while the result is mirrored about the symmetric plane, except for the sandwich beam with ROHACELL 110WF PMI foam core and span length of 400 mm, whose span is very close to the transition span length of ROHACELL 110WF as mentioned in Section 1. The whole model of the foregoing sandwich beam is developed so that the no details will be missed in its simulation result. It can be seen from Table 7 that for the sandwich beams with ROHACELL 110WF PMI foam core, the principle failure mode is core shearing when the span length is less than 400 mm and face wrinkling when the span length is larger than 400 mm. When the span length is 400 mm, both of the core shearing and face wrinkling occur to the beam as the critical loads of the two failure modes are very close. As for the sandwich beams with ROHACELL 51WF PMI foam core, core shearing is the principle failure mode when the span length is less than 500 mm. When the span length is greater than or equal to 500 mm, the failure mode becomes face wrinkling. Put the numerical results of the sandwich beam with different PMI foam core into the corresponding positions in the failure mode map as shown in Figure 17, where the simulated failure modes of different sandwich beams agree well with the theoretical predictions.

6 Conclusions

The constitutive relation of the PMI foam under compression is studied on the basis of the cellular solids theory [23]. The obtained constitutive relation is then applied to investigate the failure mechanism of sandwich beams with PMI foam core in three point bending tests. The tests of the sandwich beams with ROHACELL 110WF foam core and carbon fiber skin sheets in three point bending tests are carried out and the corresponding numerical models are developed. Theoretical analysis shows that the PMI foam core sandwich beams fail in the mode of core shearing in small span situation while face wrinkling turns out to be the principle failure mode in large span situation, which is validated by the both experimental findings and the numerical modeling. In addition, the images captured by high speed camera show that the failure process of sandwich beams (PMI foam core) with span length of 500 mm is complex: The face adjacent to the loading cell wrinkles at first and the excessive wrinkling deformation causes the progressive failure of the nearby foam material until most of the upper face was separated from the core material. The present studies provide a theoretic tool for the design of sandwich structures with PMI foam core and give the basis to reduce the requirements of tests by numerical modeling for the validation of design of PMI foam core embedded sandwich beams.

Implications and influences

The studies provide a theoretic tool for the design of sandwich structures with PMI foam core and give the basis to reduce the requirements of tests by numerical modeling for the validation of design of PMI foam core embedded sandwich beams.

Acknowledgements. The research is supported by the 973 Program (No. 2014CB049000), National Natural Science Foundation of China (11372062), China Postdoctoral Science Foundation (2014M551070), 111 Project (B14013).

References

1. Seibert HF. 2006. Applications for PMI foams in aerospace sandwich structures. Reinforced Plastics, 50(1), 44–48.
2. Li Q, Mines R. 2002. Strain measures for rigid crushable foam in uniaxial compression. Strain, 38(4), 132–140.
3. Li Q, Mines R, Birch R. 2000. The crush behaviour of Rohacell-51WF structural foam. International Journal of Solids and Structures, 37(43), 6321–6341.
4. Chen C, Anderson W, Lakes R. 1994. Relating the properties of foam to the properties of the solid from which it is made. Cellular Polymers, 13(1), 16–32.
5. Burman M, Zenkert D. 1997. Fatigue of foam core sandwich beams – 1: undamaged specimens. International Journal of Fatigue, 19(7), 551–561.
6. Zenkert D, Burman M. 2009. Tension, compression and shear fatigue of a closed cell polymer foam. Composites Science and Technology, 69(6), 785–792.
7. Zenkert D, Burman M. 2011. Failure mode shifts during constant amplitude fatigue loading of GFRP/foam core sandwich beams. International Journal of Fatigue, 33(2), 217–222.
8. Akay M, Hanna R. 1990. A comparison of honeycomb-core and foam-core carbon-fibre/epoxy sandwich panels. Composites, 21(4), 325–331.
9. Tsang P, Dugundji J. 1992. Damage resistance of graphite/epoxy sandwich panels under low speed impacts. Journal of the American Helicopter Society, 37(1), 75–81.
10. Bernard ML, Lagace PA. 1989. Impact resistance of composite sandwich plates. Journal of Reinforced Plastics and Composites, 8(5), 432–445.
11. Wu C, Sun C. 1996. Low velocity impact damage in composite sandwich beams. Composite Structures, 34(1), 21–27.
12. Benderly D, Putter S. 2004. Characterization of the shear/compression failure envelope of Rohacell foam. Polymer Testing, 23(1), 51–57.
13. Kazemahvazi S, Tanner D, Zenkert D. 2009. Corrugated all-composite sandwich structures. Part 2: Failure mechanisms and experimental programme. Composites Science and Technology, 69(7), 920–925.
14. McGarva L, Åström B. 1999. Experimental investigation of compression moulding of glass/PA12-PMI foam core sandwich components. Composites Part A: Applied Science and Manufacturing, 30(10), 1171–1185.

15. Can W, Hao-ran C, Zhen-kun L. 2010. Experimental investigation of interfacial fracture behavior in foam core sandwich beams with visco-elastic adhesive interface. Composite Structures, 92(5), 1085–1091.
16. Mamalis A, Spentzas K, Manolakos D, Ioannidis M, Papapostolou D. 2008. Experimental investigation of the collapse modes and the main crushing characteristics of composite sandwich panels subjected to flexural loading. International Journal of Crashworthiness, 13(4), 349–362.
17. Flores-Johnson E, Li Q. 2011. Experimental study of the indentation of sandwich panels with carbon fibre-reinforced polymer face sheets and polymeric foam core. Composites Part B: Engineering, 42(5), 1212–1219.
18. Yang F, Lin Q, Jiang J. 2015. Experimental study on fatigue failure and damage of sandwich structure with PMI foam core. Fatigue & Fracture of Engineering Materials & Structures, 38(4), 456–465.
19. Rinker M, John M, Zahlen PC, Schäuble R. 2011. Face sheet debonding in CFRP/PMI sandwich structures under quasi-static and fatigue loading considering residual thermal stress. Engineering Fracture Mechanics, 78(17), 2835–2847.
20. Shipsha A, Hallström S, Zenkert D. 2003. Failure mechanisms and modelling of impact damage in sandwich beams-a 2D approach: part I-experimental investigation. Journal of Sandwich Structures and Materials, 5(1), 7–31.
21. Rizov V, Shipsha A, Zenkert D. 2005. Indentation study of foam core sandwich composite panels. Composite Structures, 69(1), 95–102.
22. Mamalis A, Manolakos D, Ioannidis M, Papapostolou D. 2005. On the crushing response of composite sandwich panels subjected to edgewise compression: experimental. Composite Structures, 71(2), 246–257.
23. Gibson LJ, Ashby MF. 1997. Cellular solids: structure and properties. Cambridge University Press.
24. Chen C, Lakes R. 1995. Analysis of the structure-property relations of foam materials. Cellular Polymers, 14(3), 186–202.
25. Simone A, Gibson L. 1998. Effects of solid distribution on the stiffness and strength of metallic foams. Acta Materialia, 46(6), 2139–2150.
26. Pei H. 2005. ROHACELL Technical Manual. Yingchuang Degussa (China) Investment Co., Ltd. Shanghai Branch: China.
27. Zenkert D. 1997. The handbook of sandwich construction. UK Engineering Materials Advisory Services Ltd/Chameleon Press Ltd: London.
28. Allen G. 1969. Analysis and Design of Structural Sandwich Panels. Pergamon Press: Oxford. p. 1969.
29. GB/T 1456-2005. 2005. Test method for flexural properties of sandwich construction. Standardization Administration of the People's Republic of China: Beijing, China.

22

Electronic band structure, optical, dynamical and thermodynamic properties of cesium chloride (CsCl) from first-principles

Suat Bingol, Bahattin Erdinc*, and Harun Akkus

Department of Physics, Faculty of Science, Yuzuncu Yil University, 65080 Van, Turkey

Abstract – The geometric structural optimization, electronic band structure, total density of states for valence electrons, density of states for phonons, optical, dynamical, and thermodynamical features of cesium chloride have been investigated by linearized augmented plane wave method using the density functional theory under the generalized gradient approximation. Ground state properties of cesium chloride are studied. The calculated ground state properties are consistent with experimental results. Calculated band structure indicates that the cesium chloride structure has an indirect band gap value of 5.46 eV and is an insulator. From the obtained phonon spectra, the cesium chloride structure is dynamically stable along the various directions in the Brillouin zone. Temperature dependent thermodynamic properties are studied using the harmonic approximation model.

Key words: Electronic band structure, Optical properties and dynamical properties, Density functional theory.

1 Introduction

To use the materials in nature, their physical and chemical properties must be known. To set information about these properties, we have to determine the electronic structure, optical and dynamical properties of materials. It requires so much effort for calculating electronic band structure and optical properties of crystals. However, in recent years, these properties could be calculated easily using computer calculation based on first-principles. Density functional theory has proved to be highly successful in describing electronic, optical, dynamical and thermodynamical properties of materials. Using density functional theory, many works has been devoted to investigate alkali halides in B2 structure in scientific and technological area because of their interesting electro-optic, electro-mechanic properties and large dielectric constants etc.

The cesium chloride structure has considerable been attracting attention due to its interesting properties. Cesium chloride crystal is classified as an insulator at normal conditions. This compound is interesting due to s-d transition at the conduction band bottom under pressure. At zero pressure, the ideal crystal structure of cesium chloride has a simple cubic lattice, where the cesium atoms are located at the corners of the cube and chloride atom at center.

The physical properties of cesium chloride are a subject of considerably theoretical and experimental interest. The dielectric constant of cesium chloride was experimentally measured depending on temperature and pressure [1]. Both experimental and theoretical measurements of the second-order Raman spectra of cesium chloride were carried out at both 300 and 80 K [2]. The ionic conductivity of cesium chloride depending on temperature and hydrostatic pressure were presented and discussed [3]. Spectra for the $Cs^{+}3d$ in cesium chloride were investigated in detail [4]. The instability of cesium chloride crystal was investigated under high pressures using a simple model calculation [5]. The electronic band structure and cohesive properties of cesium chloride compound were presented within the local density functional formalism [6]. The preparation and microstructural characterization of cesium chloride were reported in detail using combined cryogenic and room temperature ball milling [7]. The absorption spectra of cesium chloride molecule and dimer were investigated using time of-flight technique [8]. The effect of cesium chloride capping layers on semitransparent Ca/Ag cathodes was studied [9]. The thermodynamic of cesium chloride crystal between 736 K and 1077 K was investigated using Knudsen effusion mass spectrometry [10]. The structural stability of some cesium chloride structure types was studied using full potential method under the density functional theory [11]. First-principles study of the rocksalt-cesium chloride relative phase stability in alkali

*e-mail: bahattinerdinc@yyu.edu.tr

halides was investigated [12]. About 20 alkali halides crystals under hydrostatic pressure were examined the behaviors such as Cohesive and thermal properties, equations of state (EOS), thermodynamic stability ranges of the most common rocksalt (B1) and cesium chloride (B2) phases, phase transition properties and phase transition [13–24].

It is seen from this brief historical search that no a comprehensive calculation of physical properties of cesium chloride crystal from first principles have so far been done or published. In this work, the geometric structural optimization, electronic band structure, total density of valence electrons states, optical properties, dynamical properties, density of phonon states and thermodynamic properties of cesium chloride crystal structure have been examined using a pseudopotential method based on the density functional theory under the generalized gradient approximation.

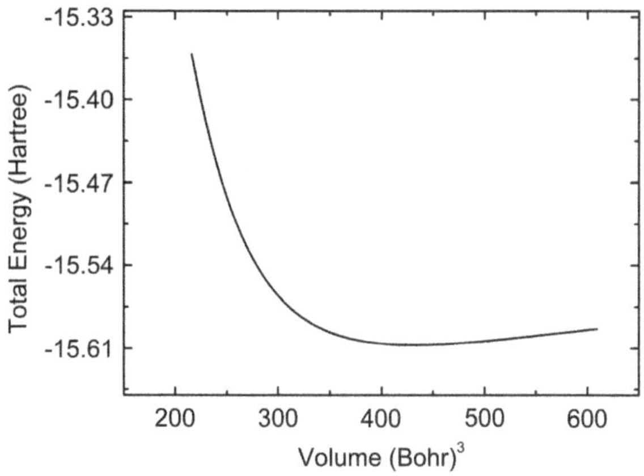

Figure 1. Total energy dependence on volume for cesium chloride.

2 Discussion and results

All the parameters of the calculations of electronic band structure, total density of the valence electron states, linear optical properties, dynamical and some thermodynamic properties for cesium chloride compound were performed using ABINIT code [25] based on density functional theory. The norm-conserving generalized gradient approximation pseudopotentials of cesium chloride structure are produced in the Perdew-Burke-Ernzerhof generalized gradient approximation functional (PBE-GGA-96) [26] and used in all calculations. For the electronic wave functions, the plane waves were chosen as the basis set. For solving Kohn-Sham equations [27], the conjugate gradient minimization method [28] was utilized. In all calculations, the exchange-correlation effects were taken into consideration with the Perdew-Burke-Ernzerhof generalized gradient approximation [26]. As the true valence, for cesium atom 6*s* electron and for chloride atom 3*s* and 3*p* electrons were taken. All calculations were performed by taking one molecule per cubic unit cell. The total energy calculations were done to a good convergence at 40 hartree of cutoff energy using $8 \times 8 \times 8$ Monkhorst-Pack mesh grid [29] for geometric structural optimization, electronic band structure, dynamical and thermodynamic properties of cesium chloride crystal structure. However, to calculate optical properties of cesium chloride crystal structure, the irreducible Brillouin zone was sampled with $10 \times 10 \times 10$ Monk-Horst-Pack mesh grid.

Using various methods [30, 31], the lattice parameters of cesium chloride crystal structure was investigated. This crystal structure has cubic structure with the space groups Pm-3m. As shown in Table 1, the calculated cubic unit cell has one molecule with two atoms in unit cell. Firstly, the total energy optimization with respect to cutoff energy has been performed. The value of cutoff energy was obtained as 40 hartree. Secondly, the total energy optimization by the number of *k* points has been performed. The number of *k* points was calculated as 20. Then, ground state energy of cubic cesium chloride crystal for different volumes around the experimental equilibrium volume is calculated. The total energy optimization as versus the volume of cesium chloride crystal structure is given in Figure 1. The calculated value of lattice constant is given in Table 1. As seen from Table 1, the theoretically calculated lattice constant value is 7.5540 bohr. In addition, the experimental result and calculated value of lattice constant were compared and given in Table 1. Table 1 indicates that the calculated value for lattice parameters for cesium chloride is 7.5540 bohr, while experimental value is 7.7669 bohr. As seen, there exists a good agreement between the calculated and experimental results. The obtained lattice constant was used for calculation of all physical properties of this structure.

The physical properties such as the electronic band structure, total density of the valence electron states, optical and dynamical for cesium chloride have been investigated using the ABINIT code. Firstly, pseudopotential method based on the density functional theory within generalized gradient approximation was used in electronic structure calculations of cesium chloride. Electronic band structure and total density of valence electron states of this compound was calculated and plotted in Figure 2. As seen from the figure that the minimum of conduction band and maximum of valence band are located at the Γ and M symmetry points on high symmetry directions of Brillouin zone, respectively. In addition, the obtained band structure of cesium chloride crystal structure has an indirect

Table 1. Crystal structure data of cesium chloride.

Space group Pm-3m (221)						
Cesium chloride	Calculated (GGA)	Experimental	Atom	*x*	*y*	*z*
a (bohr)	7.5540	7.7669 [17]	Cs	0.0	0.0	0.0
E_g (eV)	5.46	8.3 [18]	Cl	0.5	0.5	0.5

band gap value of 5.46 eV from M to Γ symmetry points of the Brillouin zone. The theoretical band gap of this compound is smaller than the experimental result [31]. It is well known that the obtained band gap by the pseudopotential method is smaller than that obtained from experiments due to imperfection of the method.

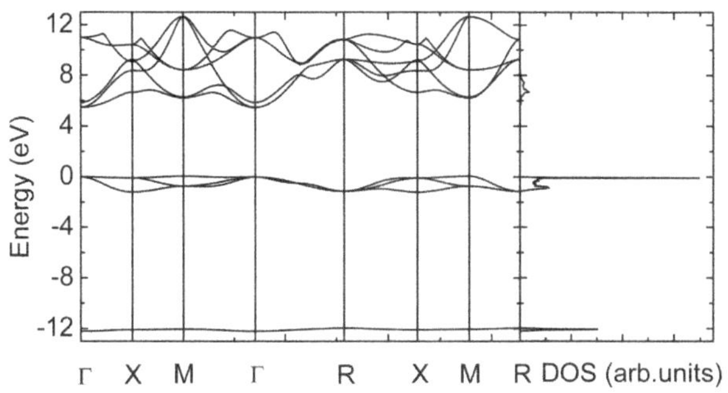

Figure 2. Electronic band structure and total density of states of cesium chloride compound from first-principles.

Secondly, using the density functional theory within generalized gradient approximation, the real (ε_1) and imaginary (ε_2) parts of complex linear dielectric function as a function of the photon energy for cesium chloride are investigated. Moreover, the optical constant such as reflectivity (R), absorption coefficient spectra (α), energy-loss functions for volume ($-\text{Im}\varepsilon^{-1}$) and for surface ($-\text{Im}(1 + \varepsilon)^{-1}$), extinction coefficient (k), refractive index (n) and effective number of electrons (N_{eff}) depending on the photon energy of this crystal structure are investigated and given in Figure 3. The calculations only pertain to electronic transitions and neglect all lattice vibrational effects. The calculations of the real and imaginary parts of complex linear dielectric function have been analyzed from 0 to 38 eV photon energy range and plotted in Figures 3a and 3b. If we analysis the Figure 3b, the imaginary part of

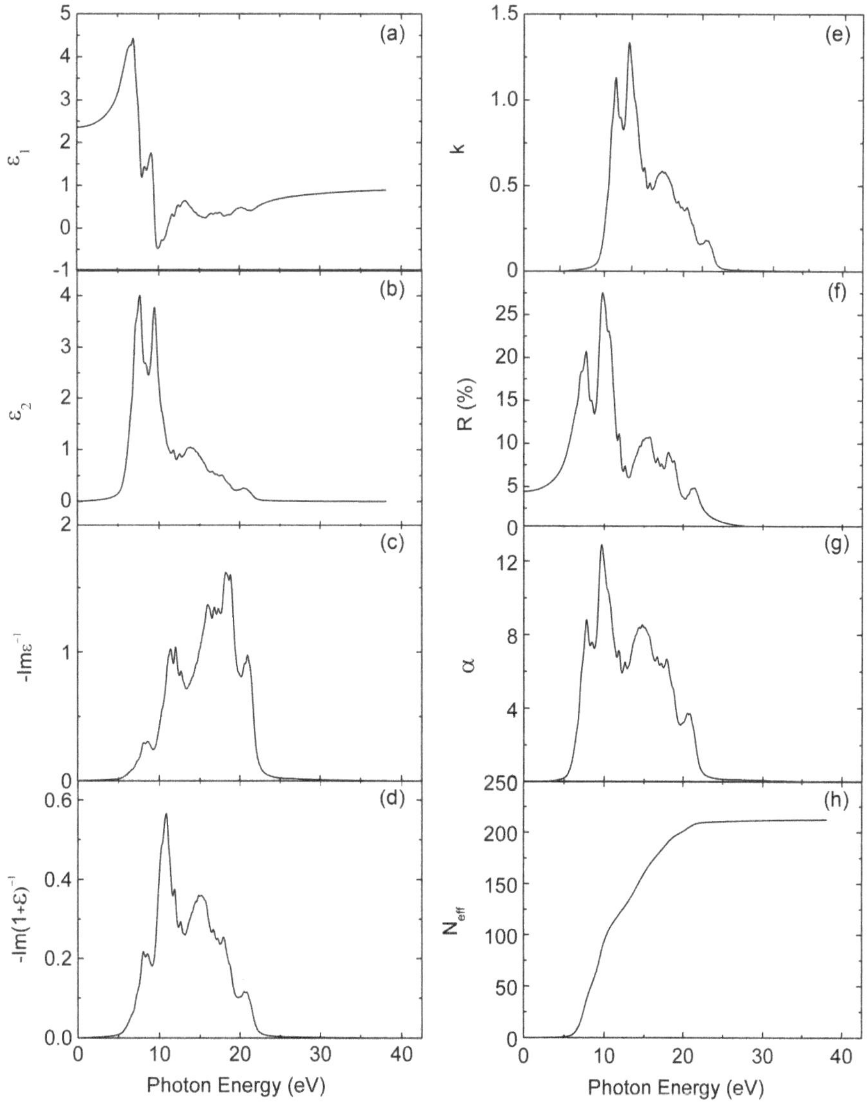

Figure 3. Optical spectra of cesium chloride crystal from first-principles.

complex linear dielectric function shows peaks which correspond to the transition from valence band to conduction band at many points. In present work, the main peak values in the calculated imaginary part of complex linear dielectric function are at 0.24, 4.00, 3.78, 1.00, 0.93, 1.05, 0.59, 0.53 and 0.28 at 5.28, 7.65, 9.43, 11.68, 12.52, 13.83, 16.56, 17.74 and 20.59 eV of photon energy. There is no absorption between 0 and 5.46 eV photon-energy range. 7.00–10.00 eV energy range is of strong absorption and appreciable reflectivity region. It means that optical absorption increases rapidly in this range. High reflectivity exists between 10.00 and 12.00 eV.

Real part of the complex linear dielectric function is presented in Figure 3a. One can see from the figure that the calculated static dielectric constant for cesium chloride crystal structure is 2.37 at 0 eV of photon energy. Figure 3a shows that the real part of complex linear dielectric function is equal to zero at 9.47 and 11.17 eV where the interband transitions consisting mostly of plasmon excitations and the scattering probabilities of volume and surface losses are directly related to the energy loss function. Furthermore, the range of 0–11 eV is classified as a normal dispersion and the range of 11–20 eV are characteristic of an anomalous dispersion.

The optical constants such as energy loss function, reflectivity spectrum, extinction coefficient, absorption coefficient and effective number of electrons were investigated using complex linear dielectric function. The energy-loss functions describing the energy loss of fast electrons making transitions in the material for volume and surface are plotted in Figures 3c and 3d, respectively. The calculated energy-loss functions for volume and surface show mainly intense maxima as presented in Figures 3c and 3d, respectively. Here, the obtained main peak values for volume loss function are at 0.3, 1.03, 1.04, 1.37, 1.62 and 0.98 at 8.30, 11.42, 12.05, 15.92, 18.20 and 20.90 eV of photon energy, while the obtained main peak values for surface loss function are at 0.22, 0.57, 0.36, 0.25 and 0.12 at 7.94, 10.76, 15.04, 17.94 and 20.67 eV of photon energy. These peak values coincide with one of zeros of linear dielectric function's real part. The sharp maxima in the energy-loss functions are related to plasma oscillations. The calculated extinction coefficient, reflectivity spectrum and absorption coefficient show mainly intense maxima as presented in Figures 3e–3g, respectively. Here, the obtained main peak values for extinction coefficient are at 1.13, 1.34, 0.60, 0.37 and 0.18 at 7.76, 9.68, 14.41, 17.84 and 20.57 eV of photon energy. The obtained main peak values for reflectivity spectrum are at 0.21, 0.28, 0.11, 0.089 and 0.049 at 7.66, 9.81, 15.42, 18.12 and 21.37 eV of photon energy. The obtained main peak values for absorption coefficient are at 8.77, 12.92, 8.56, 6.69 and 3.73 at 7.82, 9.69, 14.67, 17.87 and 20.67 eV of photon energy. The effective number of valence electron which contributes to the interband transition is investigated using the sum rule [32]. This calculation is given in Figure 3h. This figure indicates that the effective number of valence electron of cesium chloride crystal structure is zero up to 5.46 eV, then rises rapidly up to 20 eV and reaches a saturation value above 25.0 eV.

The calculated refractive index of cesium chloride over a wide range of wavelengths is shown in Figure 4. It is seen from Figure 4 that the calculated value of refractive index decreases with a transition from the intrinsic absorption region to long waves when normal dispersion takes place. A maximum value of the refractive index as a function of wavelength was obtained as a value of 2.2.

In the unit cell of cesium chloride crystal contains two atoms. The cesium atom is located at the point of (0.0, 0.0, 0.0) which chloride atom at the point of (0.5, 0.5, 0.5) in the reduced coordinates inside the unit cell. Using the generalized gradient approximation, the phonon properties of cesium chloride were investigated. The calculated phonon dispersion curves and phonon density of states of cesium chloride crystal along the high-symmetry directions are plotted in Figure 5. Six phonon branches totally appear due to two atoms per unit cell for cesium chloride crystal. While three of them are the acoustical branches, the remaining are the optical branches. This figure shows that this structure is dynamically stable because all phonon frequencies are positive. Because of differences in masses of cesium and chloride atoms, there is a gap between the acoustical and the optical phonon branches as expected. The band gap between the acoustic and optical branches is about 1.2 THz. The frequency values of longitudinal and transverse optical phonon modes in the center of Brillouin zone are calculated as 5.45 THz and 3.96 THz, respectively. Some of the acoustical and optical branches are degenerate along the symmetrical directions

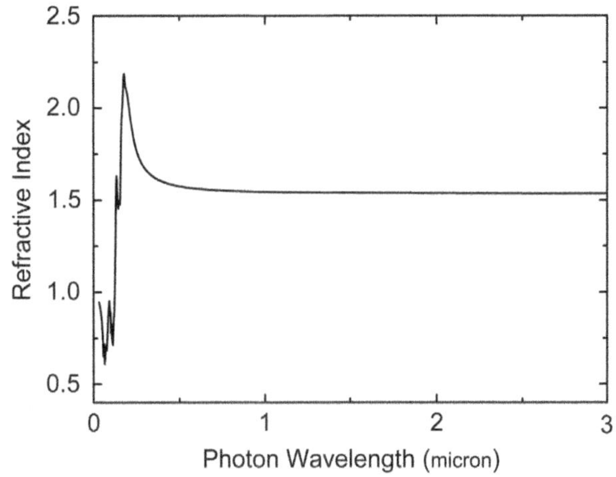

Figure 4. Refractive index dispersion of cesium chloride compound from first-principles.

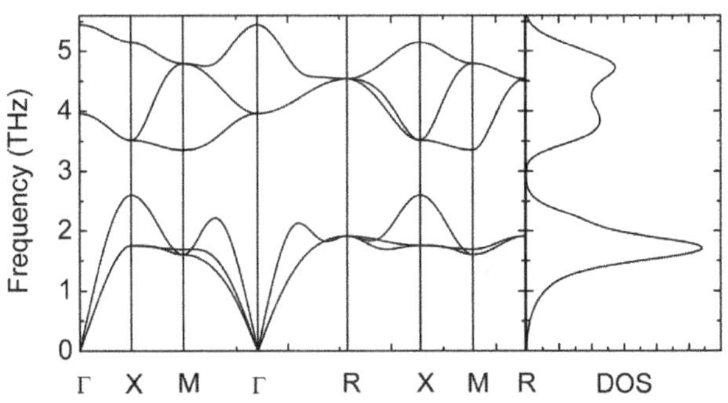

Figure 5. Dispersion curves for phonons and phonon density of states of cesium chloride crystal from first-principles.

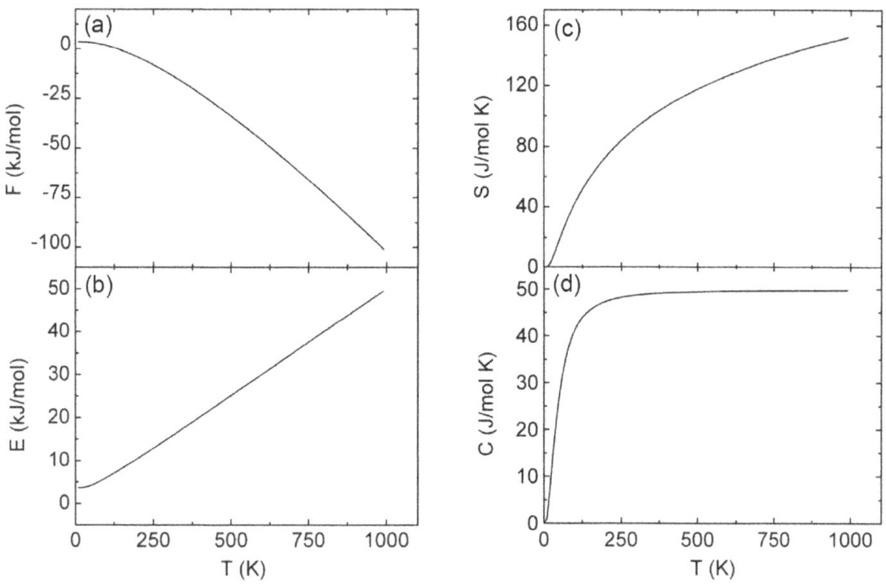

Figure 6. Thermodynamic properties of cesium chloride crystal from first-principles. (a) Contribution of phonons to free energy as a function of temperature, (b) phonon contribution to internal energy as a function of temperature, (c) the entropy as a function of temperature, and (d) specific heat as a function of temperature.

inside the Brillouin zone. As see in Figure 5, the direction from Γ to R and from Γ to X is more symmetrical than the other directions. Transverse acoustical and transverse optical phonon modes are degenerate along these symmetry directions. When the symmetry is broken, the degeneracy in the modes along the other directions disappears. In high symmetry directions of Γ to R and Γ to X, acoustical dispersion curves for phonons are the linear as a function of k for small values of k as seen in Figure 5. The phonon density of states is shown in the rightmost panel of Figure 5. While transverse acoustic phonon modes are between 0.3 THz and 1.6 THz in the phonon density of states, longitudinal acoustical phonon modes lie to between 1.6 THz and 2.6 THz in the phonon density of states. All the optical phonon modes are between 3.5 THz and 5.5 THz in the phonon density of states.

Thermodynamic properties of cesium chloride crystal have been investigated using whole phonon band structure calculations. The phonon contributions to the Helmholtz free energy (F) and internal energy (E), the entropy (S) and the specific heat (C_V) of cesium chloride crystal as a function of temperature are calculated. These properties as a function of temperature are plotted in Figure 6. The thermodynamic properties of this crystal are investigated using the harmonic approximation [33]. As seen from Figures 6a and 6b, the contribution of phonons to the internal energy and free energy do not vanish at zero temperature. At zero temperature, the contributions of phonons to the internal and free energies are same and equal to 3.8 kJ/mol. As seen from Figure 6a–6c, while the free energy of cesium chloride crystal decreases, as the temperature rises, internal energy and entropy increase. The specific heat is given in Figure 6d. The specific heat reaches a classic limit value at 830 K.

3 Conclusions

As far as we know, the structure optimization, electronic band structure, density of states, optical, dynamic and thermodynamic properties of cesium chloride have not been investigated theoretically. In the present work, by linearized augmented plane wave method using the density functional theory under the generalized gradient approximation, the geometric structural optimization, electronic band structure, total density of valence electron state, optical properties, density of phonon state, dynamical and thermodynamical properties of cesium chloride crystal structure have been investigated and discussed in detail. The calculated structure optimization of cesium chloride structure was compared with the experimental result and a good agreement between them was found. As a result of the calculations, it is seen that the fundamental gap of cesium chloride crystal structure is indirect one from M to Γ in the Brillouin zone. The found band gap is smaller than the experimental one. The real and imaginary parts of complex linear dielectric function have been investigated as a function of the photon energy. The obtained phonon spectrum of cesium chloride crystal structure indicates that this crystal is dynamically stable along the various directions in the Brillouin zone. Moreover, the temperature dependent thermodynamic properties such as Helmholtz free energy, internal energy, entropy and heat capacity are investigated using the harmonic approximation model after calculating the phonon frequencies. The detailed definition of these properties is given.

Acknowledgements. This work has been supported by The Unit of Scientific Research Projects of Yuzuncu Yil University under project No. 2011-FED-B010.

References

1. Havinga EE, Bosman AJ. 1965. Temperature dependence of dielectric constants of crystals with NaCl and CsCl structure. Phys. Rev., 140, A292–A303.
2. Agrawal BS, Kirby RD, Hardy JR. 1975. Theoretical and experimental studies of the second-order Raman spectra of CsCl, CsBr, and CsI. Phys. Rev. B, 11, 5153–5162.
3. Samara A. 1980. Pressure and temperature dependences of the ionic conductivity of CsCl. Phys. Rev. B, 22, 6476–6479.
4. Motais P, Belin E, Bonnelle C. 1982. Direct radiative recombination of core-hole highly excited states in CsCl. Phys. Rev. B, 25, 5492–5498.
5. Vohra YK, Duclos SJ, Ruoff AL. 1985. Instability of the CsC1 Structure in ionic solids at high pressures. Phys. Rev. Lett., 54, 570–573.
6. Satpathy S. 1986. Electron energy bands and cohesive properties of CsC1, CsBr, and Csl. Phys. Rev. B, 33, 8706–8715.
7. Tiwary CS, Verma A, Biswas K, Mondal AK, Chattopadhyay K. 2011. Preparation of ultrafine CsCl crystallites by combined cryogenic and room temperature ball milling. Ceram. Int., 37, 3677–3686.
8. Ruusa R, Kukk E, Huttula M, Aksela H, Aksela S. 2005. Resonant VUV absorption in CsCl monomers and dimers-experiment and theory. J. Electron Spectros. Relat. Phenomena, 144–147, 1215–1218.
9. Leea CJ, Hana JI, Choib DK, Moonc DG. 2010. Transparent organic light-emitting devices with CsCl capping layers on semitransparent Ca/Ag cathodes. J. Mater. Sci. Eng. B, 172, 76–79.
10. Lisek I, Kapala J, Miller M. 1998. Thermodynamic study of the CsCl–$NdCl_3$ system by Knudsen effusion mass spectrometry. J. Alloys Compd., 278, 113–122.
11. Novakovic N, Ivanovic N, Koteski V, Radisavljevic I, Belosevic-Cavor J, Cekic B. 2006. Structural stability of some CsCl structure HfTM (TM = Co, Rh, Ru, Fe) compounds. Intermetallics, 14, 1403–1410.
12. Florez M, Recio JM, Francisco E, Blanco MA, Martin Pendas A. 2002. First-principles study of the rocksalt–cesium chloride relative phase stability in alkali halides. Phys. Rev. B, 66, 144112–144117.
13. Born M, Huang K. 1954. Dynamical theory of crystal lattices. Oxford University Press: New York.
14. Darnell AJ, McCollum WA. 1970. Thermodynamics of the Fm3m ⇄ Pm3m transition in the potassium and rubidium halides. J. Phys. Chem. Solids, 31, 805–815.
15. Basset WA, Takahashi T, Mao HK, Weaber JS. 1968. Pressure-induced phase transformation in NaCl. J. Appl. Phys., 39, 319–325.
16. Petrunina TI, Soshnikov VI, Estrin EI. 1972. Hysteresis and the features of phase transformation in rubidium halides under pressure. Sov. Phys. Crystallogr., 17, 367–369.
17. Vaidya N, Kennedy GC. 1971. Compressibility of 27 halides to 45 kbar. J. Phys. Chem. Solids, 32, 951–964.
18. Mei WN, Boyer LL, Mehl MJ, Ossowski MM, Stokes HT. 2000. Calculation of electronic, structural, and vibrational properties in alkali halides using a density-functional method with localized densities. Phys. Rev. B, 61, 11425–11431.
19. Lacam A, Peyronneau J. 1973. Polymorphic transition of rubidium chloride under high pressure. Effect of experimental parameters on hysteresis effect. J. Phys. (Paris), 34, 1047–1053.
20. Sims CE, Barrera GD, Allan NL, Mackrodt WC. 1998. Thermodynamics and mechanism of the B1-B2 phase transition in group-I halides and group-II oxides. Phys. Rev. B, 57, 11164–11172.
21. Pistorius CWFT. 1976. Phase relations and structures of solids at high pressures. Prog. Solid State Chem., 11, 1–151.
22. Kohler U, Johannsen PG, Hozapfel WB. 1997. Equation-of-state data for CsCl-type alkali halides. J. Phys. Condens. Matter, 9, 5581–5592.
23. Yagi T. 1978. Experimental determination of thermal expansivity of several alkali halides at high pressures. J. Phys. Chem. Solids, 39, 563–571.
24. Hofmeister AM. 1997. IR spectroscopy of alkali halides at very high pressures: Calculation of equations of state and of the response of bulk moduli to the B1−B2 phase transition. Phys. Rev. B, 56, 5835–5855.
25. Gonze X, Beuken JM, Caracas R, Detraux F, Fuchs M, Rignanese GM, Sindic L, Verstrate M, Zerah G, Jollet F, Torrent M, Roy A, Mikami M, Ghosez P, Raty JY, Allan DC. 2002. First-principle computation of material properties: the ABINIT software. Comput. Mater. Sci., 25, 478–492.
26. Perdew JP, Burke K, Ernzerhof M. 1996. Generalized gradient approximation made simple. Phys. Rev. Lett., 77, 3865–3868.
27. Kohn W, Sham LJ. 1965. Self-consistent equations including exchange and correlation effects. Phys. Rev., 140, A1133.
28. Payne MC, Teter MP, Allan DC, Arias TA, Joannopoulos JD. 1992. Iterative minimization techniques for ab initio total-energy calculations: molecular dynamics and conjugate gradients. Rev. Mod. Phys., 64, 1045–1097.
29. Monkhorst HJ, Pack JD. 1976. Special points for Brillouin-zone integrations. Phys. Rev. B, 13, 5188–5192.
30. Kittel C. 1996. Introduction to solid state physics, 7th Edition. John Wiley and Sons Inc., USA.
31. Poole RT, Jenkin JG, Liesegang J, Leckey RCG. 1975. Electronic band structure of the alkali halides. I. Experimental parameters. Phys. Rev. B, 11, 5179–5185.
32. Philipp HR, Ehrenreich H. 1963. Optical properties of semiconductors. Phys. Rev., 129, 1550–1560.
33. Lee C, Gonze X. 1995. Ab initio calculation of the thermodynamic properties and atomic temperature factors of SiO_2 α-quartz and stishovite. Phys. Rev. B, 51, 8610–8163.

23

Backbone cup – a structure design competition based on topology optimization and 3D printing

Ji-Hong Zhu*, Kai-Ke Yang, and Wei-Hong Zhang

Engineering Simulation and Aerospace Computing (ESAC), Northwestern Polytechnical University, Xi'an, Shaanxi 710072, P.R. China

Abstract – This paper addresses a structure design competition based on topology optimization and 3D Printing, and proposes an experimental approach to efficiently and quickly measure the mechanical performance of the structures designed using topology optimization. Since the topology optimized structure designs are prone to be geometrically complex, it is extremely inconvenient to fabricate these designs with traditional machining. In this study, we not only fabricated the topology optimized structure designs using one kind of 3D Printing technology known as stereolithography (SLA), but also tested the mechanical performance of the produced prototype parts. The finite element method is used to analyze the structure responses, and the consistent results of the numerical simulations and structure experiments prove the validity of this new structure testing approach. This new approach will not only provide a rapid access to topology optimized structure designs verifying, but also cut the turnaround time of structure design significantly.

Key words: Structure design competition, Topology optimization, Stereolithography, Quick structure testing.

1 Introduction

Ever since the first time Bendsøe [1], Suzuki and Kikuchi [2] used homogenization based approach to solve structural topology optimization problems, many scholars had focused their attention on this appealing filed of structure optimization. The latter researches proposed the solid isotropic material with penalization (SIMP) method [3–5] which made it possible to gain practical structure design through topology optimization, the ESO method [6], and the level set method [7], etc. For recent advances of topology optimization, the readers are referred to the literature surveys of Deaton and Grandhi [8], Guo and Cheng [9]. Meanwhile, structural topology optimization has been used widely in automobiles [10], buildings [11], and aircrafts [12] to gain a structure design with the best mechanical performance. However, the structure designs obtained by topology optimization are prone to be geometrically complex such as hollow structures and porous structures, which can hardly be fabricated using traditional machining. A compromising solution is that we consider the structural topology optimization problems with manufacturing constraints [13], which naturally sacrifices the mechanical performance of the optimized designs at some level.

Over the last two decades, additive manufacturing (AM) [14] also called 3D printing thrives. Unlike the traditional machining methods, AM forms parts layer by layer through the joining of liquid, powder, or sheet materials. There are several AM technologies [15] including stereolithography, selective laser melting (SLM), three dimensional printing (3DP), and fused deposition modeling (FDM). Benefitting from its characteristics of manufacturing, AM is able to fabricate structure designs which are topologically complex, produce individualized products, and accelerate the process of manufacturing parts for no tooling needed at all. A lot of valuable researches has been done using AM technology recently. Castilho et al. [16] have explored to produce porous scaffolds for tissue engineering by means of a powder-based 3D printer. Moon et al. [17] investigate small and deployable UAVs wing designs using Objet 350 3D printer. Nevertheless the applications of stereolithography technology still are producing appearance models and durable concept models [18]. Considering the great need for massive topology optimized structure designs validation and the huge time and money cost for metal parts testing, here we propose to validate the topology optimized designs using prototype parts made by stereolithography.

In this study, we aim to develop an experimental approach to efficiently and quickly measure the mechanical performance of the structures designed using topology optimization. Taking the advantages of topology optimization and stereolithography, we can obtain an optimal structure design using topology

*e-mail: JH.Zhu_FEA@nwpu.edu.cn

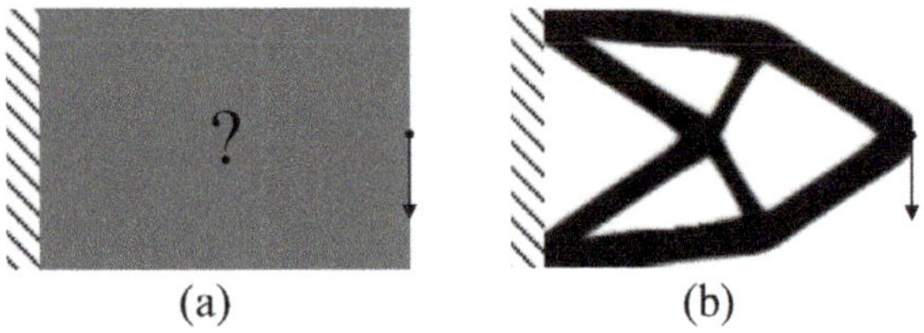

Figure 1. (a) Initial finite element model and (b) the optimal material layout.

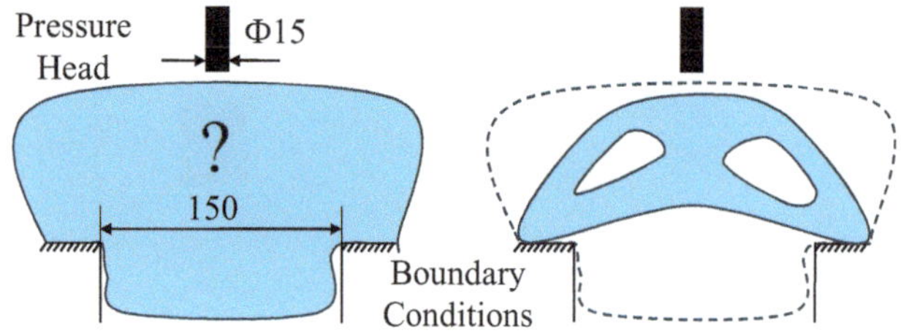

Figure 2. Illustration of the requirements of the structure competition.

optimization, fabricate it quickly and conveniently through stereolithography, and conduct a test to investigate the mechanical behavior of the design. So far we have developed a quick experimentally structure testing procedure. The finite element method is used to analyze the structure responses, and the consistent results of the numerical simulations and structure experiments prove the validity of the proposed methodology. Thus this new structure testing approach will not only provide a rapid access to topology optimized structures testifying, but also cut the turnaround time of structure design significantly.

2 Basic formulation of topology optimization

Topology optimization is used to gain the best material distribution within a design domain by seeking where to put the material (solid) and where not to (void) as shown in Figure 1. The topology of a structure can be represented by the density at different locations within the domain. The density is usually defined as a design variable and can assume 0 or 1 with 0 representing void and 1 representing the solid. The optimization method used in this work iteratively optimizes the elemental densities in the design domain to minimize the compliance (maximize the stiffness) of the final structure while satisfying the volume constraint. The mathematical form of the optimization model can be expressed as:

$$\left.\begin{aligned} \underset{\mathbf{x}}{\text{minimize}} &: C(\mathbf{x},\mathbf{u}) = \frac{1}{2}\mathbf{u}^{\mathrm{T}}\mathbf{K}\mathbf{u} \\ \text{subject to} &: \mathbf{K}\mathbf{u} = \mathbf{f} \\ &: V(\mathbf{x}) = \int_{\Omega} \mathbf{x}\, d\Omega \leq V_s \end{aligned}\right\} \quad (1)$$

Where $C(\mathbf{x})$ is the compliance, $\mathbf{x}$ is the density vector. $\mathbf{K}$ is the global stiffness matrix, Ω is the design domain. $\mathbf{f}$ and $\mathbf{u}$ are the global load vector and global displacement vector, respectfully; V_s is the volume constraint. Employing the SIMP method, the problem is relaxed for density to have any value between 0 and 1 with small lower bound of $x_{\min} = 0.001$ to avoid singularities during calculating for equilibrium. Also, a material interpolation scheme for local stiffness is defined. Thus, following is added in the problem.

$$\begin{aligned} 0 < x_{\min} \leq x_i \leq 1 \\ E(x) = x^p E_0 \end{aligned} \quad (2)$$

Where p is the penalization factor and E_0 is the Young's modulus of the material in the solid phase.

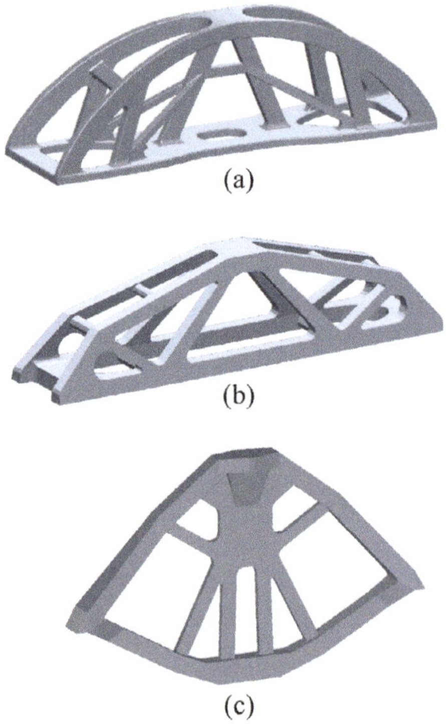

Figure 3. 3D beam design (a), (b) and (c) to the problem.

3 Structure design

3.1 A 3d beam design problem

In buildings, machineries, and bridge constructions, beams are one kind of regular parts highly used. A good beam design can not only guarantee the performance of the structure, but also satisfy mass constraint on the purpose of saving the material. So we set a 3D beam design problem as the topic of this structure design competition. Figure 2 shows the specific requirements of the problem. Between a 150 mm gap of two supporting columns, one needs to design a beam to withstand a compression force applied on the central area of the top face by a 15-mm-diameter pressure head and ensure the design satisfying the volume constraint of 30 cm^3.

3.2 Several beam designs given by contestants

Figure 3c as an example to illustrates the design process briefly. At first we simplify this structure design problem as

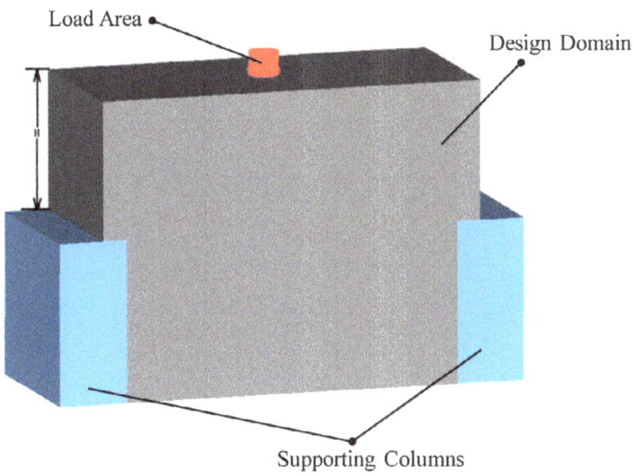

Figure 4. The minimum compliance topology optimization problem.

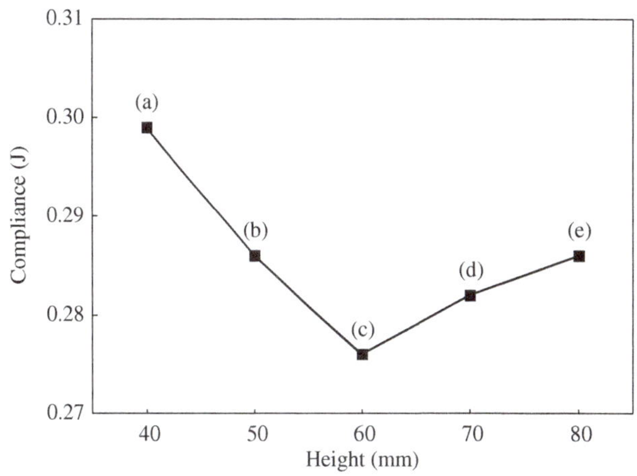

Figure 5. Compliance versus height.

Table 1. The mechanical properties of photo-polymer DSM Somos 14120.

ASTM method	Description	White 14120
D638M	Tensile strength	45.7 MPa
	Elongation at break	7.9%
	Elongation at yield	3.5%
	Modulus of elasticity	2460 MPa
	Poisson's ratio	0.23

a minimum compliance topology optimization problem and gain a minimum compliance solution, then reconstruct the design based on the optimal material layout obtained. The details are shown in the following steps.

3.2.1 Step one

Given the influence of initial domain size on the topology optimization solution, we choose height as the variation and investigate how it influences the optimal material layout as shown in Figure 4. In fact, other dimensional sizes of the design domain also have such influence, but we can eliminate that influence by giving adequate size along that orientations. The material properties are listed in Table 1, a total compression force of 1000 N is considered as the external load.

Then, topology optimization problems with different heights of design domain are solved, the relation between the compliance of the optimal material layout and the height of the design domain is shown in Figure 5.

From Figure 5 we can conclude that the optimal height for the initial design domain is 60 mm, and the corresponding compliance is 0.276 J. The optimal material layouts for the presented minimum compliance topology optimization problem are shown in Figure 6.

3.2.2 Step two

According to the optimal material layout obtained in the first step, the design is reconstructed (Figure 7), and numerically analyzed under the condition of applying 1000 N compression force using finite element method.

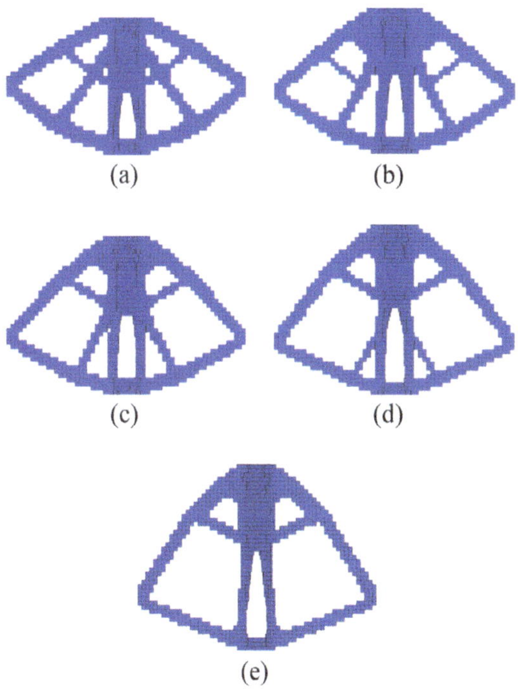

Figure 6. (a), (b), (c), (d) and (e) are the optimal material layouts for minimum compliance problems with different domain heights as in Figure 5

As shown in Figure 8, the reconstructed design well presents the load-carrying path of the topology optimized design, for the high-level similarity of the distribution of von Mises stress in two structures, and its compliance is 0.282 J which increases less than 2.2% than the optimized design. After the above two steps, a structure design with the maximum stiffness performance is obtained to this structure design problem.

4 3D printing method and material properties

4.1 3D printing method

Stereolithography is one kind of liquid based 3D printing technology which processes with ultraviolet photo-curable

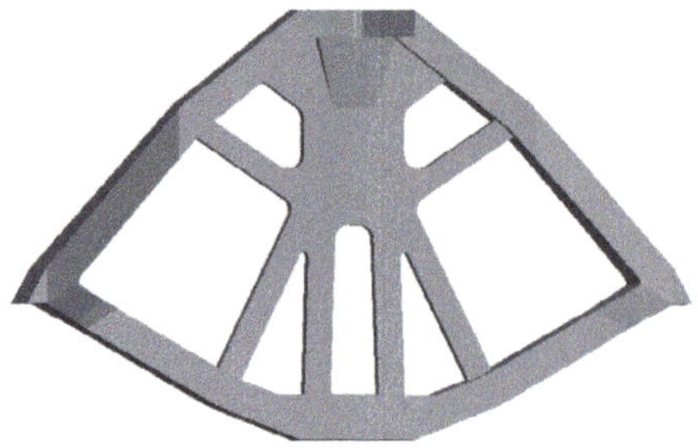

Figure 7. The reconstructed design.

(a) (b)

Figure 8. (a) von Mises stress in the optimized design and in (b) reconstructed design.

resin. Due to its capability of rapid manufacturing and the ingredient's characteristics, stereolithography has become an ideal method to produce dies of injection molding and casting. Here we choose stereolithography technology to fabricate all the structure designs for its high time efficiency and low fabrication cost. The 3D printer we use is SPS350B, which is invented by Shaanxi Hengtong Intelligent Machine Co., Ltd (SHIM) of China.

The process of stereolithography includes: modeling a solid part in computer with CAD software; converting the CAD file into Standardized Triangular Language (STL) file format to create volumetric mesh and the support structure; Slicing the volumetric-meshed model layer by layer and generate SLC file format; exporting the SLC file to computer of SLA; building the structure by curing the ultraviolet photo-curable resin layer by layer using a ultraviolet ray; removing the support structure to get the design part; post-curing the part in an ultraviolet oven [19].

The material used in this experiment is DSM Somos 14120, which is a low viscosity about 240 cps at 30 °C liquid photopolymer that produces strong, tough, water-resistant parts. The specific information about this material is given in Table 1.

4.2 Mechanical properties of cured resin material

Various process parameters affect the stereolithography process part quality characteristics, among which build parameters such as layer thickness, manufacturing orientation, hatch spacing have much influence according to the research of Raju et al. [20]. The research of Salmoria et al. [21] on somos 7110 resin reveals green samples fabricated by stereolithography process have layer ordering microstructures. Using the prototype parts produced by stereolithography to verify the mechanical performance of isotropic-material structure designs, it is vital for us to investigate the degree of anisotropy caused by the nonuniformity existed in the stereolithography produced samples.

Table 2. Parameter levels for the experiment.

Variable	Level 1	Level 2	Level 3
Manufacture orientations	0°	45°	90°

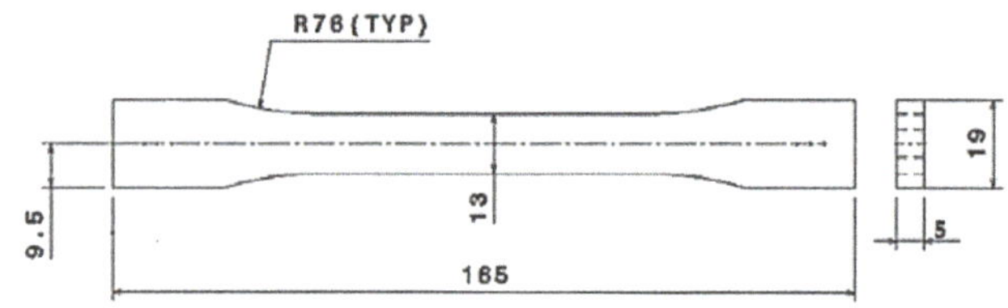

Figure 9. Tensile test specimen.

Figure 10. The setup of the standard tension test.

Fifteen tension specimens of five replications as per the ASTM standards (ASTM D638) are fabricated using DSM Somos 14120 resin in SLA SPS350B machine. The only variable between the three groups of specimens is manufacturing orientation, three levels are picked as show in Table 2 (Figures 9–11).

The elastic stiffness and yield strength of stereolithography produced samples of different manufacturing orientations are represented by the elastic modulus and yield stress shown in Figure 12.

From Figure 12, we can conclude that the anisotropy exists between the stiffness of different manufacture orientations is about 418 MPa, which is around 13% with respect to the stiffness of 90° manufacture orientation and the anisotropy exists between the yield strength of different manufacture orientations is about 21 MPa, which is around 34% with respect to the yield strength of 90° manufacture orientation. Regardless of the slight anisotropy caused by the stereolithography

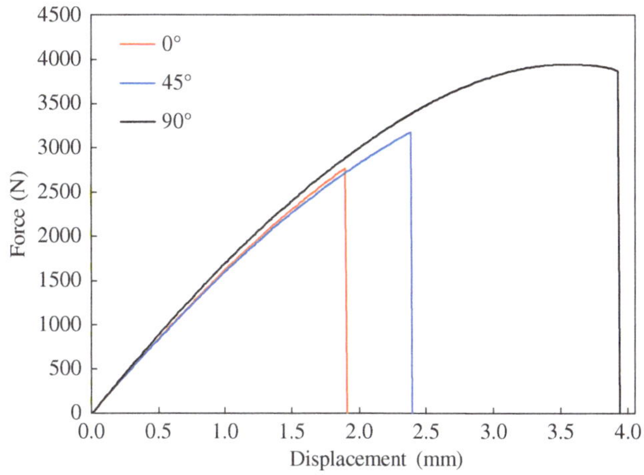

Figure 11. The load-displacement curves of specimens.

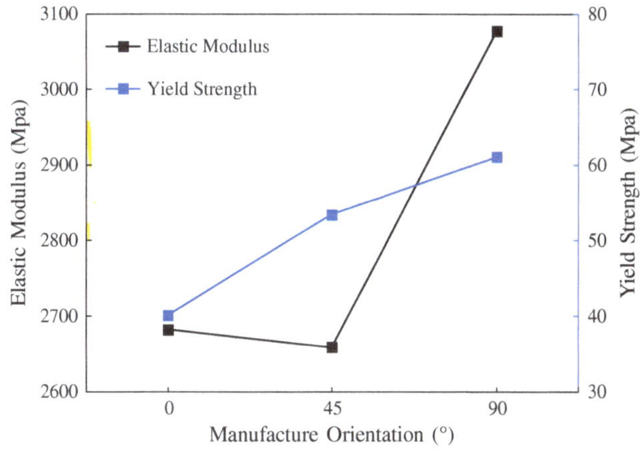

Figure 12. The elastic modulus and yield strength of different manufacture orientations.

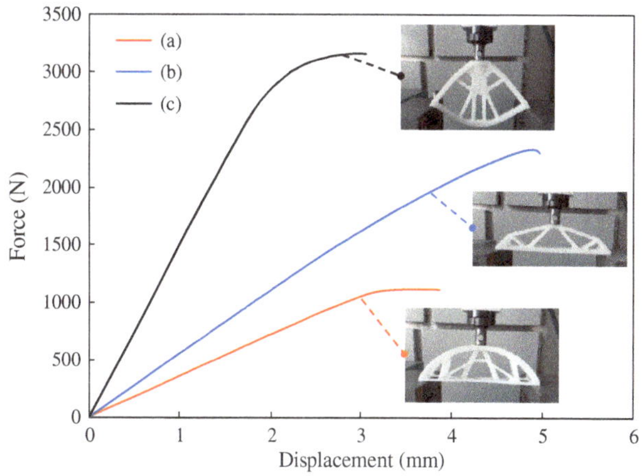

Figure 13. Load-displacement curves of the structures.

Table 3. Comparison of numerical simulations and experiment results.

No. of the structure	(a)	(b)	(c)
Compliance of reconstructed design (J)	1.22	0.85	0.28
Simulated displacement (mm)	2.64	1.74	0.59
Actual displacement (mm)	2.83	1.80	0.68

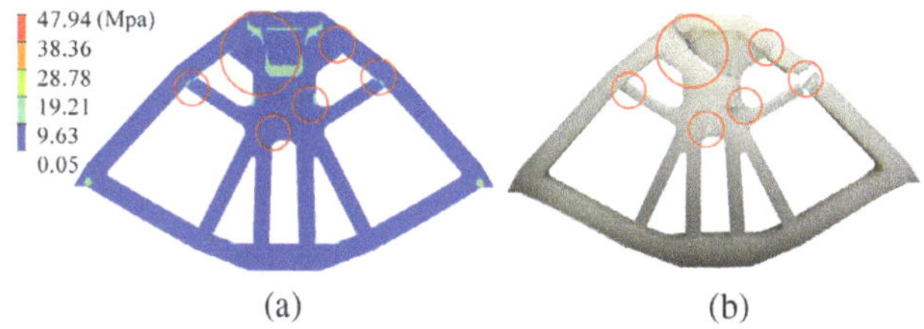

Figure 14. (a) The von Mises stress in design (c) and (b) the fragments of the tested parts.

process, we consider the material isotropic and the isotropic-material structure designs are tested using this stereolithography produced material.

5 Experiment

5.1 Structure test procedure

The static structure tests were carried out at a feed rate of 3.6 mm/min using Testresources Testing System. All the structures were tested under the same conditions and experimental parameters. The test procedures were as follows:

1. Set the parameters of the experiment.
2. Place the structure on the test platform.
3. Lower the pressure head towards upper surface of the structure such that the structure is firmly held.
4. Preset zero the load and the displacement on the measuring scale and carry out the test.
5. Record the load applied on the structure and the displacement of the pressure head simultaneously.
6. Continue the test until the structure fails.

At the end of the experiments, several load-displacement curves were generated for comparison. The results will be discussed in the next section in details.

5.2 Results and discussions

The test results of three proposed designs are shown in Figure 13. Take the load-displacement curve of structure (c) as example, It goes much steeper than two other curves, which indicates structure (c) has the better stiffness performance, and this conclusion can also be drawn from the numerical simulation results in Table 3 as the less compliance, the stiffer the design is. When compression force reaches 1000 N, the actual displacement that structure (c) generates is 0.68 mm which is 15.25% percentage bigger than 0.59 mm of the numerical simulation, and for structure (a) and (b) the relative difference between the actual displacement and the simulated displacement are 7.20% and 3.45%, respectively. The consistent displacements mean that it is advisable for us to ignore the tiny anisotropy existed in the light-cured resin material from the stereolithography process. Moreover the places where structure

(c) falls apart are very close to the places where the von Mises stress concentrates in the simulation result as presented in Figure 14. Thus from the high accordance between the numerical simulations and experiment results, we can draw the conclusion that the approach of quickly testing the mechanical performance of the structure designs using prototype parts proves valuable.

6 Conclusions

In this study, we established a quick structure testing approach for topology optimized designs using topology optimization and stereolithography technologies. Conclusions of this work are as follows:

- successful application of fabricating load-carrying structures using stereolithography;
- established a quickly structure testing approach to verify the mechanical performance of structures designed using topology optimization.

Acknowledgements. This work is supported by National Natural Science Foundation of China (90916027, 51275424, 11172236), the 111 Project (B07050), Science and Technology Research and development projects in Shaanxi Province (2014KJXX-37), the Fundamental Research Funds for the Central Universities (3102014JC02020505).

References

1. Bendsøe MP. 1988. Generating optimal topologies in structural design using a homogenization method. Comput. Methods Appl. Mech. Eng., 71(2), 197–224.
2. Suzuki K, Kikuchi N. 1991. A homogenization method for shape and topology optimization. Comput. Methods Appl. Mech. Eng., 93(3), 291–318.
3. Bendsøe MP. 1989. Optimal shape design as a material distribution problem. Struct. Optim., 1, 193–202.
4. Zhou M, Rozvany GIN. 1991. The COC algorithm, part II: topological, geometry and generalized shape optimization. Comput. Methods Appl. Mech. Eng., 89, 197–224.
5. Mlejnek HP. 1992. Some aspects of the genesis of structures. Struct Optim, 5, 64–69.
6. Xie YM, Steven GP. 1997. Basic evolutionary structural optimization. Springer: London.
7. Allaire G, Jouve F, Toader AM. 2004. Structural optimization using sensitivity analysis and a level-set method. J. Comput. Phys., 194(1), 363–393.
8. Deaton JD, Grandhi RV. 2013. A survey of structural and multidisciplinary continuum topology optimization: post 2000. Struct. Multidisc. Optim., 49(1), 1–38.
9. Guo X, Cheng GD. 2010. Recent development in structural design and optimization. Acta Mech. Sin., 26(6), 807–823.
10. Yang RJ, Chahande AI. 1995. Automotive applications of topology optimization. Struct. Optim., 9, 245–249.
11. Stromberg LL, Beghini A, Baker WF, Paulino GH. 2011. Application of layout and topology optimization using pattern gradation for the conceptual design of buildings. Struct. Multidisc. Optim., 43, 165–180.
12. Zhu JH, Zhang WH, Xia L. 2015. Topology optimization in aircraft and aerospace structures design. Arch. Comput. Methods Eng., 1–28.
13. Zhu JH, Gu XJ, Zhang WH, Beckers P. 2013. Structural design of aircraft skin stretch-forming die using topology optimization. J. Comput. Appl. Math., 246, 278–288.
14. Gibson I, Rosen DW, Stucker B. 2010. Additive manufacturing technologies. Springer: New York.
15. Pham DT, Gault RS. 1998. A comparison of rapid prototyping technologies. Int. J. Mach. Tool. Manuf., 38, 1257–1287.
16. Castilho M, Pires I, Gouveia B, Rodrigues J. 2011. Structural evaluation of scaffolds prototypes produced by three-dimensional printing. Int. J. Adv. Manuf. Technol., 56, 561–569.
17. Moon SK, Tan YE, Hwang JH, Yoon YJ. 2014. Application of 3D printing technology for designing light-weight unmanned aerial vehicle wing structures. Int. J. Precis. Eng. Manuf. Green Tech., 1(3), 223–228.
18. Melchels FPW, Feijen J, Grijpma DW. 2010. A review on stereolithography and its applications in biomedical engineering. Biomaterials, 31(24), 6121–6130.
19. Kruth JP, Leu MC, Nakagawa T. 1998. Progress in additive manufacturing and rapid prototyping. CIRP Ann. Manuf. Techn., 47, 525–540.
20. Raju BS, Chandrashekar U, Drakshayani DN. 2013. Optimization studies on improving the strength characteristic for parts made of photosensitive polymer. J. Inst. Eng. India Ser. D, 94(1), 35–41.
21. Salmoria GV, Ahrens CH, Fredel M, Soldi V, Pires ATN. 2005. Stereolithography somos 7110 resin: mechanical behavior and fractography of parts post-cured by different methods. Polymer Testing, 24(2), 157–162.

24

An adaptive dual-optimal path-planning technique for unmanned air vehicles

Clifford A. Whitfield*

The Ohio State University, 201 W. 19th Avenue, Columbus, Ohio 43210, USA

Abstract – A multi-objective technique for unmanned air vehicle path-planning generation through task allocation has been developed. The dual-optimal path-planning technique generates real-time adaptive flight paths based on available flight windows and environmental influenced objectives. The environmentally-influenced flight condition determines the aircraft optimal orientation within a downstream virtual window of possible vehicle destinations that is based on the vehicle's kinematics. The intermittent results are then pursued by a dynamic optimization technique to determine the flight path. This path-planning technique is a multi-objective optimization procedure consisting of two goals that do not require additional information to combine the conflicting objectives into a single-objective. The technique was applied to solar-regenerative high altitude long endurance flight which can benefit significantly from an adaptive real-time path-planning technique. The objectives were to determine the minimum power required flight paths while maintaining maximum solar power for continual surveillance over an area of interest (AOI). The simulated path generation technique prolonged the flight duration over a sustained turn loiter flight path by approximately 2 months for a year of flight. The potential for prolonged solar powered flight was consistent for all latitude locations, including 2 months of available flight at 60° latitude, where sustained turn flight was no longer capable.

Key words: Multi-objective optimization, Path-planning, Real-time, Unmanned air vehicle, Solar-powered flight.

1 Introduction

Unmanned Air Vehicle (UAV) classifications typically fall into one of six functional categories; however, recently unmanned systems have been developed for multi-role capabilities. The six functions include: target and decoy, reconnaissance, combat, logistics, civil and commercial, and research and development [1]. Regardless of the UAVs function, the basic essentials of unmanned vehicles include communication relay between the vehicle and ground operator for maintaining situational awareness, vehicle and payload sensors based on the UAV's function, and the use of global positioning systems to enable precise navigation. With the increase in radio frequency technology, the UAV became more sophisticated with the combined capability for remote and built-in control to perform low-level pilot duties such as flight-path stabilization. However, these vehicles were still not considered autonomous, commonly defined as the ability to make decisions without human intervention. UAV-autonomy technology falls under the following categories: sensor fusion, communications, path planning, trajectory generation (sometimes called motion planning), trajectory regulation, task allocation and scheduling, and cooperative tactics [1]. A significant amount of effort has been focused in UAV autonomy, with the recent advances accredited to the field of control science. In fact, autonomy may continue to be the driving field for UAV development and expanding the UAV market.

Path-planning and trajectory generation and regulation methods for UAV flight have been investigated extensively, which include: evolution-based, extremum-seeking, randomized, neural-network, direct collocation with nonlinear programming, B-spline, quadtree data structuring, probabilistic approaches, Chebyshev pseudospectral, and fast graph search methods. Additionally, several algorithms have been developed, for example: evolutionary, random tree, Dijkstra, Reduced-State and Hierarchical Dijkstra, and A* search. Even investigations into the problem formulation of the dynamic and objective equations have been considered, such as an nth degree Lagrange polynomial approximation. Off-line and on-line approaches have been considered implementing, including others, the methods and algorithms listed; demonstrating the diversity of approaches in the path-planning category of UAV-autonomy [2–44]. To further categorize the path planning methods and trajectory generation efforts, consider

*e-mail: whitfield.22@osu.edu

the several goals and flight conditions that were investigated. Flight in a horizontal plane 2-D was analyzed [5, 6, 17, 33, 44], along with considering the effects of constant winds [2–4]. Employing obstacle avoidance [2, 27–32] and target tracking [33–41] within the trajectory generation was addressed. Path planning for multiple UAV formation and collision free cooperative control and tactical applications have been considered [42, 43]. Expanding beyond flight vehicles, even steady turns and optimal paths for underwater gliders have been investigated [45].

The underlying commonality of the work mentioned is the single-objective analysis. Typical single-objectives to path-planning include time of flight or vehicle performance, such as endurance. Conflicting objectives when determining flight trajectories are generally unaccounted for. However, there are two main existing procedures that convert the multiple conflicting objectives into a single objective problem that have been used [2]. One method (ε-constraint method), selects one of the objectives for the analysis while the others are used as constraints, restricting the objective within a safe limit. The other, the weighted-sum approach, combines all objectives to form a single-objective. Both procedures require artificial parameters, such as limiting values in the ε-constraint method or assigning relative weights in the weighted-sum procedure. Consequently, the results are dependent on the accuracy of the chosen parameters and, in the case of UAV flight tasks, choosing the parameters are not straight forward.

Concerns for the procedures should be addressed. For conversion of multiple objectives into a single objective, the objective types need to be consistent; either all of minimization or maximization type. With a conversion procedure using a weighted-vector, each objective must be normalized in magnitude prior to determining the weighted average. This requires relying on past experience or guessing when assigning the relative weight (or ε) vector.

New frameworks have been developed accounting for multiple objectives to UAV flight management. The operational objectives, generally, include: platform safety; minimizing fuel, time, distance; and minimizing deviation from the current path. The main commonalities of the recent developments have been the multiple solution paths generated based on predefined terrain criteria called meta-paths. The meta-paths, in excess of 15 or more, provide an off-line analysis and the decision, from the available generated paths, the "best" possible route. For instance; an off-line path-planning investigation using multi-objective evolutionary algorithms was considered by Mittal and Deb [2]. The procedure introduced a three-step hybrid algorithm. The steps of the algorithm first generated several path solutions, called Pareto-optimal, each with different trade-offs to the objective functions. With the generated solutions, 8–10 were selected providing trade-offs between the objectives; and locally searched to obtain a set of solutions close to the true Pareto-optimal front, of which a single flight path was to be selected. The off-line process required prior terrain topography information and representation in determining the UAV paths. Similarly, Foo et al. and Swarzentruber et al. [46, 47] used particle swarm optimization to generate multiple solutions based on predefined criteria. A single summarized representation of the 15 total alternate paths, for a particular preference was selected to form the meta-paths in the final decision making process.

The following introduces an adaptive multi-objective technique for UAV path and trajectory autonomy generation through task allocation. The technique is an adaptive path generation for UAVs considering multiple independent conflicting objectives without the necessity of developing artificial or weighting parameters. The technique generates a single flight path based on an environmentally-influenced objectively-driven flight window for the UAV. The process is locally-optimal based and adaptive, reducing, if not eliminating the "curse of dimensionality", which is often stated when describing the significant size in decision space required for off-line path search techniques. It is not limited by individual methods or algorithms, however, certain procedures in comparison are advantageous. The goal was to develop a technique that did not require off-line analysis of meta-paths to determine the "best" possible route, but rather a single adapting flight path furthering the UAV-autonomy. Due to the nature of the optimized condition, the path generated is locally constrained based on the influenced objectives; and may not present the "best" global optimum the off-line meta-paths provide. Additionally, the objectives were to verify and validate the technique to demonstrate its capability as an adaptive multi-objective real-time path-planning procedure, not to conduct an in-depth analysis of comparing and contrasting the many different algorithms available.

2 Dual-optimal path-planning technique

The Dual-Optimal Path-Planning (DOPP) technique utilizes dynamic optimization for determining optimal flight trajectories between continually updating optimal intermediate-states for the UAV based on environmentally-influenced objectives. The environmentally-influenced optimal condition, known as the "driver" determines the next condition, within a downstream virtual window of possible vehicle destinations and orientation built from the UAV kinematics. The step results are pursued by a dynamic optimization technique to determine an optimal flight path by minimizing a cost function subject to dynamic equation constraints, control inequality constraints, interior state equality/inequality constraints, and boundary conditions with initial and final states determined by the driver-optimized results.

This sequential technique is a multi-objective optimization procedure consisting of two goals without requiring additional information to combine the conflicting objectives into a single-objective. The individual goals and objective complexity are not limited in type and are determined on an individual basis. The solving technique is also algorithm independent, allowing for versatility in the process of developing the vehicle trajectory. The technique expressed mathematically in general form is shown in Figure 1.

Notice the solution to the driver-objective determines the final time solution of $\mathbf{x}(t)$ for the path cost function. The path-objective analysis was based from classical optimal control theory and the powerful result known as the "minimum principle of Pontryagin". A visual of the basic DOPP flight

Driver:

$$Minimize\ f(\mathbf{x});\ subject\ to\ g_i(\mathbf{x}),\ (i=1,\ldots,n),$$

with solution:

$$\mathbf{x}_k^*,\ (k=1,\ldots,N)$$

Path:

$$Minimize\ J(u)=\int_0^T f_0(\mathbf{x}(t),u(t))dt,$$

where

$$\mathbf{x}(t),\ (\mathbf{x}_{k-1}^*(t),\ldots,\mathbf{x}_k^*(t)):\ [0,T]\to\Re^n$$

is the solution of the differential system with boundary conditions and with initial and final conditions:

$$\frac{d\mathbf{x}_j}{dt}=f_j(\mathbf{x},u(t)),\quad (j=1,\ldots,n);$$

$$\mathbf{x}(0)=\mathbf{x}_{k-1}^*;\quad \mathbf{x}(T)=\mathbf{x}_k^*$$

where

$$J(u^*)=\min_{u\in U}J(u)$$

with u^* *and the associated path* $\mathbf{x}$ *called 'optimal.'*

Figure 1. Dual-optimal path-planning technique.

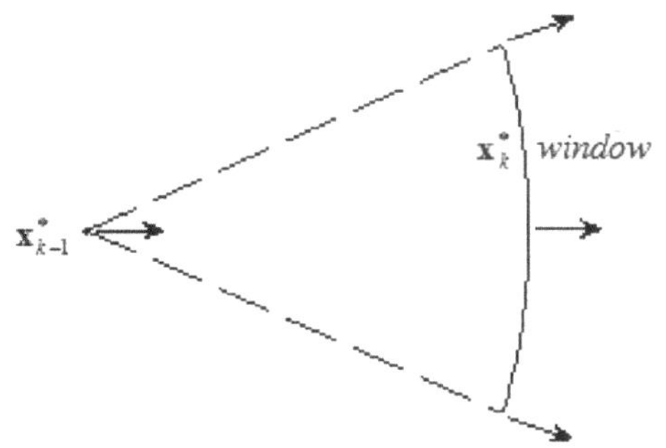

Figure 2. Dual-optimal path-planning planer flight maneuvering window.

window for the technique can be seen for planar flight in Figure 2.

The vehicles initial location $\mathbf{x}_{k-1}$ ($\mathbf{x}_{k-1}^*$ for an optimum) for planer flight consist of its location (x and y), velocity (V), bank angle (ϕ), and heading (ψ); $f(\mathbf{x}_{k-1}) = f(x_{k-1}, y_{k-1}, V_{k-1}, \phi_{k-1}, \psi_{k-1})$. In terms of three-dimensional, the vehicle's pitch angle and altitude are also required. At the initial location with the desired driver window projection distance, the available UAV flight window is projected forward of the heading direction and bounded based on the vehicle's kinematics.

For instance, looking at Figure 2, at the initial vehicle location $\mathbf{x}_{k-1}^*$, if the vehicle were to bank hard right (right wing down) would correspond to the farthest right location (with respect to initial heading) of the driver window. Similarly, a hard left bank would place the vehicle at the farthest left available position. Throughout the window, the vehicle has a $\mathbf{x}_k$ condition with corresponding location, velocity, bank and heading. The environmentally-influenced driver-objective is optimized along this available flight window determining the $\mathbf{x}_k^*$ location and vehicle orientation. With that destination, the vehicle then proceeds from the initial to intermediate-state $\mathbf{x}_k^*$ location along an optimum flight path. The vehicle's path-objective, for example could be minimizing time, maximizing endurance, or minimizing power required. The process is then continued for all intermediate-states until the overall destination or time is reached. Three dimensional flight follows a similar setup that includes altitude change and UAV pitch flight characteristics.

Even though the technique is independent of algorithm, there are methods that are more suited for on-line processing. In general for the driver-objective search algorithms are advantageous, due to generating the available flight window from the UAVs flight mechanics. Regardless, in this technique the methods are required to be adaptive within the flight window (for all N), a case in point is an adaptive-A* search algorithm.

A case study that could be benefit greatly from this technique was developed to verify and validate the technique. The case study focused on solar-regenerative extended flight (months) at high altitude for continual surveillance of an area of interest. For this example simulation, multiple objectives are predominate and developing off-line meta-paths are extremely, if not at all possible to take all variables into account through the entire flight duration. The technique was constrained using off-the-shelf technologies that are currently employed on UAVs, along with flight operations common for existing aircraft that provide high-altitude long-duration surveillance.

3 Case-study

High altitude flight above 60,000 ft (18.3 km), benefits from staying above turbulence, weather and air traffic, but comes with a price. As altitude increases the density decreases, requiring a larger lifting surface and a propulsion system that can operate in a low density atmosphere. These issues become more apparent for sustained on station persistent surveillance and reconnaissance flight. Continual surveillance over an area implies slow flight to maintain situational awareness. Slow flight in turn implies an increase in wing area and typically a decrease in wing loading, the ratio of the vehicle weight to the wings planform area. This encompasses its own challenges in vehicle design and operation. Regardless, there are current UAV platforms that reach these higher altitudes. The RQ-4 Global Hawk is one such UAV that has been used for reconnaissance, but it has a limited fuel capacity which results in a limited endurance and thus, restricted surveillance capability. Alternate methods for propulsion systems to potentially replace conventional fuel systems and increase endurance capability have been investigated. One method utilizes the photovoltaic solar panel to gather power from the Sun, and has been adapted for use on high altitude UAV platforms. An aircraft that uses solar panels and powered by the Sun during the day and stored energy through the night by solar regenerative power systems, in principle could sustain flight for weeks or months. Only mechanical wear would limit the flight duration.

The case-study application for long duration flight demonstrates the potential of the DOPP technique, and considers the change in ephemeris effects throughout the year. The case-study is limited to a constant altitude of 60,000 ft (18.3 km), and the flight window is set based on existing surveillance package capabilities to restrict the vehicle's flight paths to provide and maintain continual situational awareness over an area of interest. A summary of the requirements and limitations for a solar-regenerative high altitude long endurance UAV is discussed. A more in-depth analysis is provided in reference [48].

3.1 Requirements and limitations

There are three main factors that need to be considered to appropriately model the DOPP technique: the environment (ephemeris effects, available solar flux at altitude, and high-altitude winds at a given location), the UAV systems (photovoltaics, energy storage system, electric motor requirements, and surveillance equipment limitations at altitude), and the UAV platform (vehicle geometry and flight mechanics).

3.1.1 Flight environment

The maximum amount of available solar power depends on the vehicle's geometry and flight orientation relative to the Sun's instantaneous location. The orientation is determined by the surface's obliquity factor, $\cos(\theta)$ and is determined by the local solar coordinates (Z and A) and the tilt coordinates of the surface (Δ and Ψ). The obliquity factor is given by:

$$\cos\theta = \cos Z \cos\Delta + \sin Z \sin\Delta \cos(A - \Psi) \tag{1}$$

The first tilt coordinate (Δ) is the angle between surface normal and the vertical. The second (Ψ) is the angle between the projection of the surface normal in the horizontal plane and due south direction. The tilt coordinates fix the direction of the surface normal. This is similar in the same manner as the solar coordinates in fixing the direction of the Sun's rays.

The model used for tracing the Sun through the sky at any given location and duration of time was the described by Weider [49], and determines the local solar coordinates and the zenith angle (Z) and azimuth angle (A) for an observer through the geocentric coordinates. The geocentric coordinate system has an origin at the center of the Earth with the z-axis connecting the north-south poles. The other two axes lie on the equatorial plane with the x-axis oriented toward the local (chosen) meridian.

For UAV flight operations over a long period (months), the equation of time was implemented within the model to account for the cumulative effect of the slight variations in the solar day throughout the entire year. The variation in time can increase or decrease the total solar day by as much as 16 min, dependent on the time of year, and cannot be ignored for long duration solar powered flight.

Within Earth's atmosphere, the solar intensity felt on a surface is composed of the direct and diffuse flux components ($F^{(\text{direct})}$ and $F^{(\text{diffuse})}$, respectively). The latter term is the radiation that is scattered due to the atmosphere and the reflected radiation due to an underlying terrain. For the DOPP technique application, three assumptions will be made in order to develop a semiquantitative analysis of solar flux: (1) the diffuse flux due to the scattering of the solar intensity through the atmosphere has a significantly smaller effect than the direct flux component. This assumption is often used for approximating solar intensity at the Earth's surface. The validity of this assumption increases at high altitudes with the solar radiation traveling through less of the atmosphere, (2) the downward diffuse flux on the surface due to scattering radiation is isotropic, and (3) the upward diffuse flux on the surface will be considered negligible and disregarded in the formulation of the analysis. The upward diffuse flux is on the order of the reflective radiation component due to the underlying Earth's surface. With increase in the altitude the Earth's reflectivity radiation decreases, and for high altitude applications, the effects of the upward diffuse flux will be comparatively less to the downward diffuse flux component. With these assumptions the following equation will be used in determine the total solar flux:

$$F = F^{(\text{direct})} + \left[\frac{1+\cos\Delta}{2}\right] F^{\downarrow(\text{diffuse})} \tag{2}$$

where,

$$F^{(\text{direct})} = S_0 \mu e^{-\tau/\mu_0} \tag{3}$$

and,

$$F^{\downarrow(\text{diffuse})} = S_0 \mu_0 \left[\frac{1}{1+G}\left(Ge^{\gamma^+\tau} + e^{\gamma^-\tau}\right) - e^{-\tau/\mu_0}\right] \tag{4}$$

S_0 is the solar constant (125.6 W/ft^2 or 1352 W/m^2). The surface orientation parameters are the tilt angle $\cos(\Delta)$, obliquity of the panel to the Sun's rays $\cos(\theta)$, and the cosine of the solar zenith angle $\mu_0 = \cos(Z)$. The equations include two environmental parameters, the optical thickness (τ) and the single scattering albedo ($\tilde{\omega}_0$) within the variables G, and $\gamma^\pm$:

$$G = -\left[\frac{\gamma^- + A}{\gamma^+ + A}\right] e^{(\gamma^- - \gamma^+)\tau} \tag{5}$$

$$\gamma^\pm = 1/2(C-A) \pm 1/2\left[(C-A)^2 - 4BD\right]^{1/2} \tag{6}$$

where,

$$A = \frac{2-\tilde{\omega}_0}{2\mu_0} \qquad B = \tilde{\omega}_0$$
$$C = 2 - \tilde{\omega}_0 \qquad D = \frac{\tilde{\omega}_0}{2\mu_0} \tag{7}$$

The environmental parameters are functions of the Sun's ray wavelengths. For the parameters, Rayleigh's law is commonly used in which the theory suggests that the scattering of solar energy varies smoothly with wavelength. This law is based on the altitude conditions, mean molecular mass of air, the index of refraction of the atmosphere and the constant Avogadro's number. For a relatively clear atmosphere the averaged optical thickness and the scattering is less than 0.3 and 0.6 respectively. Considering a completely horizontal surface ($\Delta = \mu = Z = 0°$), the solar flux increases with altitude from 93 W/ft^2 (1000 W/m^2 or 1 Sun) at the Earth's surface to slightly over 121 W/ft^2 (1300 W/m^2 or 1.3 Sun) at

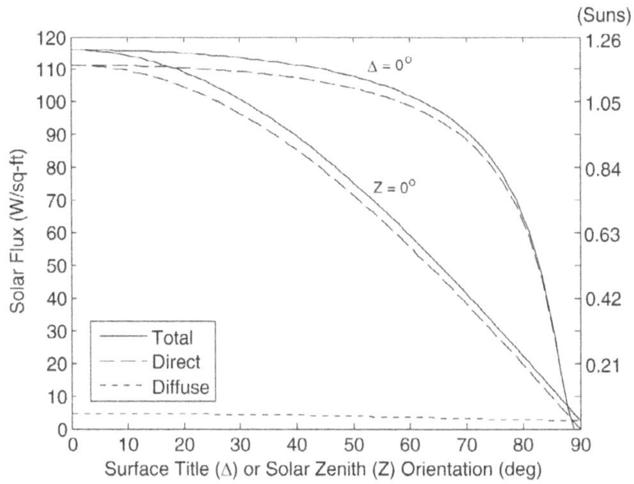

Figure 3. Solar flux with surface orientation at 60,000 ft (18.3 km) altitude.

82,000 ft (25 km) altitude. Also, note that in terms of orientation the component for the solar intensity due to the downward diffuse flux is only a function of the cosine of the solar zenith and independent of the tilt and obliquity of the surface.

For a solar powered UAV at an altitude of 60,000 ft (18.3 km) and a surface oriented horizontally, the maximum solar flux is 116.1 W/ft^2 (1250 W/m^2 or 1.25 Sun). Of which, the direct solar flux is 111.5 W/ft^2 (1200 W/m^2 or 1.2 Sun) and the downward diffuse flux is 4.6 W/ft^2 (50 W/m^2 or 0.0046 Sun), 4% of the total. The total, direct and diffuse solar flux as a function of orientation for the case-study altitude of 60,000 ft (18.3 km) is shown in Figure 3.

The two conditions in Figure 3 are variation in surface title angle Δ with Sun in a constant location, zero solar zenith angle ($Z = 0°$), and the variation in solar zenith angle from sunrise/set to solar noon with the surface normal to the Sun for all zenith angles, zero surface title angle ($\Delta = 0°$).

In addition to the available solar flux at high altitudes, and the based on the fact that high-altitude vehicles of this nature have significantly low wing-loading characteristics, the pervasive winds at those altitudes must also be considered. A study of the upper winds (ranging from 100 to 25 mb pressure altitudes) for high altitude platform design was conducted in excess of a 20 year span over four areas of interest [50]. These areas include the contiguous United States, the Norwegian area, the Mediterranean area and the Pacifica area from Alaska to Japan and the surrounding regions. Results of the investigations shown that high altitude wind speeds were at a minimum for altitude ranges of 59,000–72,000 ft (18–22 km). Table 1 summarizes the average high altitude wind speed for each of the United States, case study location.

Wind roses are provided as an appendix of reference [50] for every 10 mb pressure altitude throughout the four locations and seasons.

3.1.2 UAV systems

The UAV systems, common to high altitude surveillance aircraft, are discussed in order to set constraints for the fixed and optimized flight pattern analysis. The systems were chosen based on existing equipment that have been proven and used in both solar-regenerative and surveillance platforms; these include: photovoltaics, energy storage system, electric motors, and surveillance equipment.

Table 1. Average wind speeds.

Season	United States
Winter	30 kn
Spring	15 kn
Summer	25 kn
Fall	15 kn

A semiquantitative empirical analysis approach [49] was taken to model the photovoltaic expected available power equation (8) and photovoltaic efficiency equation (9):

$$P_A = VJ \tag{8}$$

$$\eta = \frac{P_A}{F} = \frac{VJ}{F} \tag{9}$$

where, the current (J) supplied to an external load is a combination of the photocurrent (J_p) through radiative energy or flux and the junction current (J_j) shunted by the diodes of the photovoltaic material:

$$J = J_p - J_j \tag{10}$$

Applying the modeling approach was done in conjunction to matching overall maximum power available from a representative UAV, NASA Pathfinder. This representative UAV was chosen due to its demonstrated flight tests and promising potential for high-altitude solar-powered long endurance flight.

At 60,000 ft (18.3 km) and at solar noon the maximum solar flux is 111.5 W/ft^2 (1200 W/m^2 or 1.2 Sun). With 75% of the upper wing surface covered in solar panels, the amount of power capable of being generated at solar noon was 8000 W. Physical parameters for the empirical analysis were adjusted within appropriate bounds to match recorded power available at altitude during solar noon. With the expected photovoltaic parameters modeled, Figure 4 shows the cell array solar energy characteristics at 60,000 ft (18.3 km).

Energy storage systems are key in sustaining continuous surveillance capability for solar-regenerative high altitude UAVs. This becomes particularly apparent with the decreasing daylight hours during the winter months for the northern hemisphere, and the energy per mass capacity W hr/lbs (W hr/kg) that can be stored during the daylight is critical. Lithium-sulfur batteries, from Sion Power have reported energy/mass values of 778.1 W hr/lbs (350 W hr/kg); while Lynntech's Gen5 fuel cell technology [51] has shown experimental performance of 1411 W hr/lbs (640 W hr/kg). For the case-study UAV platform, a realistic representative weight from existing aircraft for the energy storage system was selected and used throughout the example discussion for fixed and optimized flight patterns.

Lightweight brushless direct current electric motors and their performance were selected based on existing technology

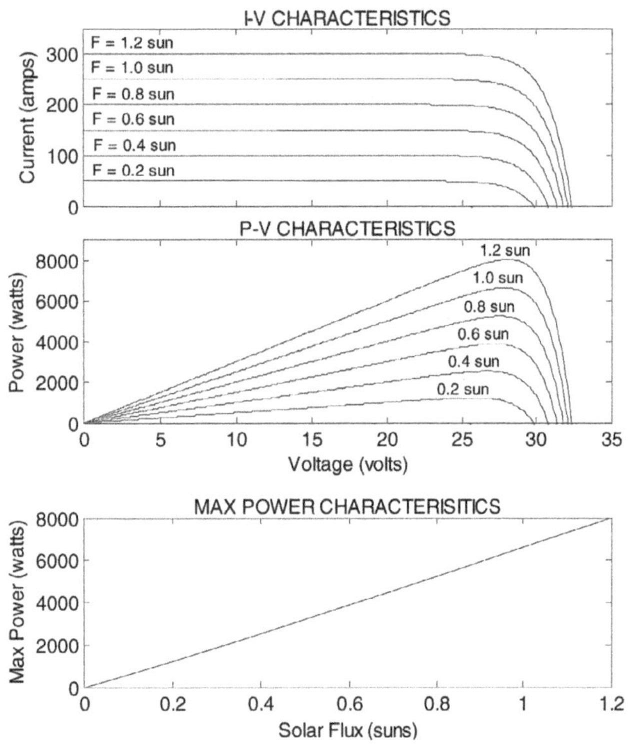

Figure 4. Solar Cell Array Characteristic at 60,000 ft (18.3 km).

that has been incorporated and tested on high-altitude UAV platforms [48]. The gearless motors are capable of producing 1.5 horsepower (1.25 kW) each.

The specifications of the surveillance equipment were used in determining the flight window constraints for the high altitude UAV operation. For instance, for a vehicle flying at an altitude of 60,000 ft (18.3 km) would have a line-of-sight that correlates to 90.260° (sunrise/set) in longitude, or a distance of up to 6220 mile (10,010.1 km) radius (at the equator). However, to process any detail on the Earth's surface from that altitude the surveillance equipment determines the maximum "range-of-sight" to maintain situational awareness of an AOI on the Earth's surface for the vehicle.

For the case-study analysis and overall program capability demonstration, the Hughes Integrated Surveillance & Reconnaissance sensor system was used [51]. This system has been carried on the Global Hawk, a UAV performing ISR missions with an on station endurance of 24–30 h, and integrates three separate systems controlled through one processor for expanding surveillance capabilities. The operational modes are; (1) wide-area that can detect movement within a radius of 62 miles (99.8 km), (2) strip-mode that provides 20 ft (6.1 m) resolution over a 23 mile (37 km) wide region, and (3) spot-mode that can provide 6 ft (1.8 m) of resolution over a 3.8 mile (6.1 km) region with sea-surveillance capability.

The solar-powered flight pattern analysis approach taken was to demonstrate optimal performance for the vehicle in range-of-sight for wide-area mode, and provided a boundary constraint for the optimal flight paths generated real-time by the DOPP technique at altitude. If a target was acquired, and more detailed surveillance required, then a pattern-transfer would be taken to a new smaller range-of-sight window constraint for the strip-mode or the spot-mode function.

Table 2. Representative UAV platform summary [48, 52].

Platform	Performance
Weight: 550 lbs	Altitude: 60,000 ft
Wing loading: 0.7 lbs/ft^2	Pressure: 151 lbs/ft^2
Power loading: 13.5 W/ft^2	Temperature: −69.7 °F
Geometry	
Span: 98.5 ft	Profile drag coefficient: 0.0170
Length: 12.0 ft	Max lift coefficient: 1.3
Chord: 8.0 ft	Efficiency factor: 0.92
Aspect ratio: 2.0	Max lift-to-drag ratio: 22.6

3.1.3 UAV platform

Similar geometric characteristics and flight mechanics of a representative UAV was selected for the case-study based on its promising flight tests [52]. The representative UAV is a flying wing with a high aspect ratio, resulting in very low wig loadings, less than one pound per square foot, limiting the on-station flight capability. Table 2 summarizes the representative UAV platform and aerodynamic performance characteristics.

With the wind tunnel and flight test data that has been published [52], this solar powered vehicle demonstrated the capability for sustained flight at high altitude.

3.2 Baseline flight pattern

The sustained turn flight, or loiter pattern, is a logical choice for maintaining continual situational awareness. This flight pattern was selected as the baseline for DOPP technique comparisons. The loitered flight has several dependent flight variables, such as the turn radius is set by the UAVs speed and bank angle. Therefore, consider the representative UAV platform required power in order to maintain a sustained turn at various bank angles shown in Figure 5. The maximum available power and the vehicle's aerodynamic stall velocity are included, showing the feasible velocity range for turning flight.

Referring to Figure 5, the minimum power required for sustained bank angles occurs at the vehicle's stall. Allowing for a 20% safety factor in vehicle speed above stall, the velocity for minimum power required for the representative UAV was set at a constant 50 kn. At this velocity the maximum aerodynamic efficiency for the vehicle was maintained at a lift coefficient of 0.85 and lift-to-drag ratio of 22 for level flight. The velocity setting also allows bank angles for the UAV up to ±20° without getting dangerously close to losing the necessary lift to maintain flight. With further increase in bank, the structural integrity along with the potential for the vehicle to stall would be of concern, especially with the vehicle's low wing loading.

At 50 kn the vehicle requires an angle-of-attack of 10°. While the bank angle is increased to its maximum limit of ±20°, the required change in lift and therefore change in angle-of-attack for constant speed is less than half of degree. This minimal change was ignored for the analysis; and the angle-of-attack was assumed a constant 10° and accounted

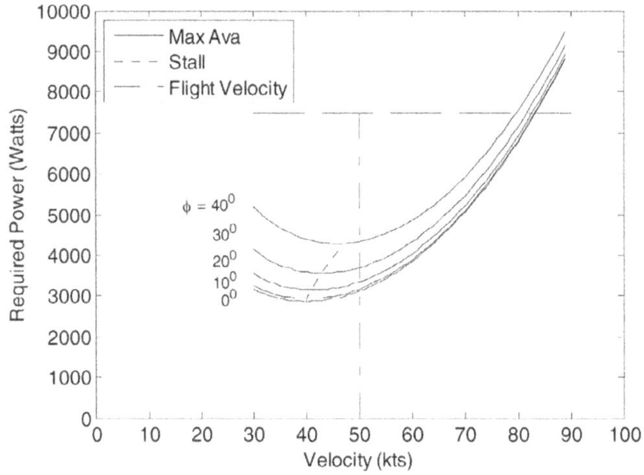

Figure 5. Sustained turn required power.

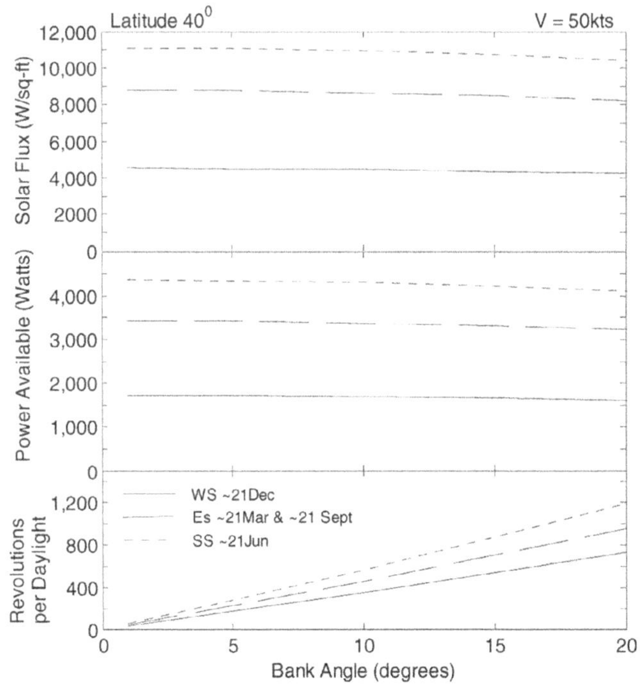

Figure 6. Total sustained turn solar flux and available power.

for in the surface obliquity tilt angles for determining the solar flux. Figure 6 shows the sustained turn analysis for the UAV for varying the bank angle at latitude of 40°.

The above figure shows the total solar flux and corresponding averaged power available for the winter and summer solstices and equinoxes, as well as the revolutions per daylight (RPDL) for constant speed flight. Comparing the averaged available power of Figure 5 to the power required for flight in Figure 6, it can be seen that the representative UAV does not gather enough solar power during the winter solstice to maintain flight, and also limited in available flight duration during the equinox times-of-year.

The other parameters of interest when considering sustained turn flight is the initial starting location at sunrise and the direction, clockwise (D_C) or counterclockwise ($D_{C\text{-}C}$). However when considering these parameter and due to the fact that the UAV has high revolutions per daylight the effects are inconsequential. This can be seen in Table 3.

The vehicle conditions were sustained ±20° bank angle at a constant 50 kn over the case-study location. The initial vehicle locations at sunrise were the four main cardinal directions; North, South, East, and West. With a turning radius of 0.10 nautical-miles (608 ft), notice the high revolutions per day light and range over the time-of-year for the UAV. This is evident by observing the solar flux variation from sunrise to sunset for the vehicle in a sustained turn. Figure 7 shows the solar flux variation for the solstices and equinoxes.

The revolutions per daylight are shown in the figure for each time-of-year. The total solar fluxes listed in Table 3 are the integrand of solar flux variation of the above figure. Changing the initial cardinal direction at sunrise the solar flux frequency and peak magnitudes are maintained, with different sunrise conditions.

Considering the changes in latitude, the total solar flux and averaged available power, along with the revolutions per daylight for the UAV can be seen in Figure 8.

The flight conditions are consistent with the previous analysis, 20° bank angle at a constant 50 kn. Recall that the power required for the vehicle, from Figure 5 is 3350 W for the bank angle. The requirement limits the possible sustained turn, loiter flight for the representative UAV in maintaining situational awareness; especially during the winter solstice, where there is only sufficient power for latitudes ranging from the equator to approximately 15°.

3.3 DOPP case study modeling

The UAV kinematic modeling considered is shown below equation (11). The vehicle is at a constant altitude, speed (V) and angle-of-attack.

$$[\mathbf{x}] = \begin{bmatrix} x \\ y \\ \psi \end{bmatrix} \qquad [\dot{\mathbf{x}}] = \begin{bmatrix} V\cos\psi + V_{Wx} \\ V\sin\psi + V_{Wy} \\ (gV)\tan\phi \end{bmatrix} \tag{11}$$

The position and orientation of the vehicle is given by the Cartesian coordinates (x and y), vehicle heading (ψ) and bank angle (ϕ).

By introducing the bang-level-bang technique for vehicle control, the dependence on inertial roll rate and its effect to surface obliquity is eliminated in the near time control for vehicle orientation to meet its "driven" determined position and orientation. This simplification assumes that the vehicle has the ability to instantaneous reach a new bank angle and is common in flight vehicle trajectory analysis. Furthermore, the nature of the UAV's slow flight does not require large instantaneous changes in vehicle bank angle, eliminating the need for large roll rates, further validating the use of the UAV's control technique.

Recall that the objective was to determine the minimum required power flight paths to the predetermined location and orientation for obtaining maximum solar flux established by the "driver". Along the available window of flight of longitude

Table 3. Total sustained turn solar flux with initial start location.

Sustained turn flight	Latitude	40°			
0.10 nm radius	V	50 kn			
	φ	20°			
	Total solar flux (W/sq-ft) [averaged Ava power (W)]				
	Initial start location at sunrise				
	N	S	E	W	RPDL
		Winter Solstice ~21 December			
D_C		4232 [1604]			728
$D_{C\text{-}C}$		4232 [1604]			728
		Equinoxes ~21 March & ~21 September			
D_C		8240 [3216]			955
$D_{C\text{-}C}$		8240 [3216]			955
		Summer Solstice ~21 June			
D_C		10,421 [4093]			1182
$D_{C\text{-}C}$		10,421 [4093]			1182

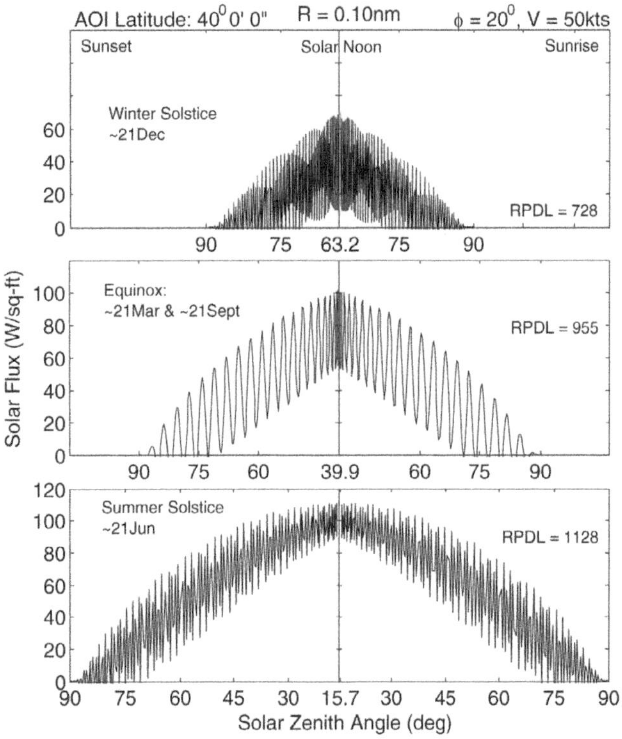

Figure 7. Sustained turn solar flux variations.

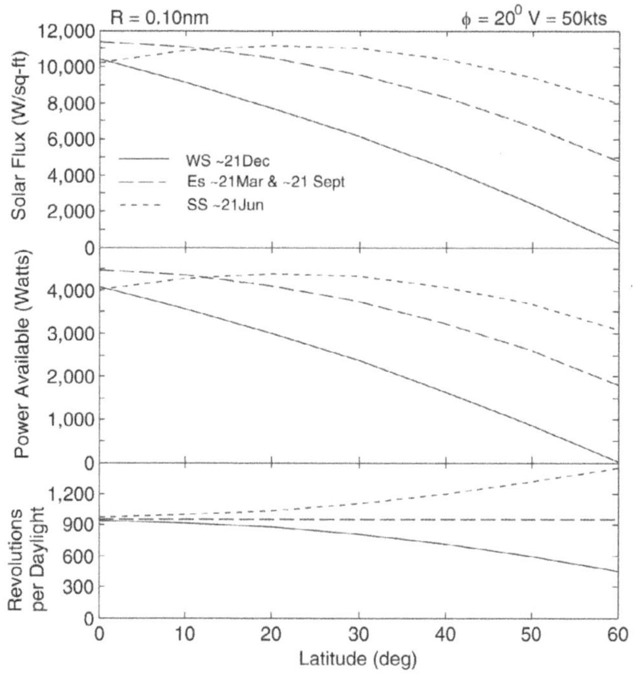

Figure 8. Total sustained turn solar flux and available power for latitude variations.

and latitude location in the inertial frame of reference, the solar flux was maximized for the corresponding vehicle bank angle and heading angle. The direct effect of solar flux was used, not accounting for the small contribution of diffusion; however, by maximizing the direct flux, the flux due to diffusion is inherently maxed, refer to Figure 3. The heading angle was transformed into the inertial frame of reference, including the effects of vehicle angle-of-attack. The flight window and objectives can be seen in Figure 9.

Referring to Figure 5, and allowing for a 20% safety factor in vehicle speed above stall, the velocity for minimum power required for the UAV was set at a constant 50 kn, consistent with sustained turn loiter flight. The constant velocity flight satisfies the minimum power required, simplifying the path-planning process and maximizing aerodynamic efficiency for the vehicle in level flight with a constant angle-of-attack of 10°. Considering the discussed modeling for the vehicle path-planning, Figure 10 shows the DOPP technique for the representative UAV for solar-regenerative high altitude long endurance flight.

The "driver" conditions are constrained by the bank angle constraint (g_1); with the objective of maximizing solar flux (solar power) throughout the vehicle's flight window when

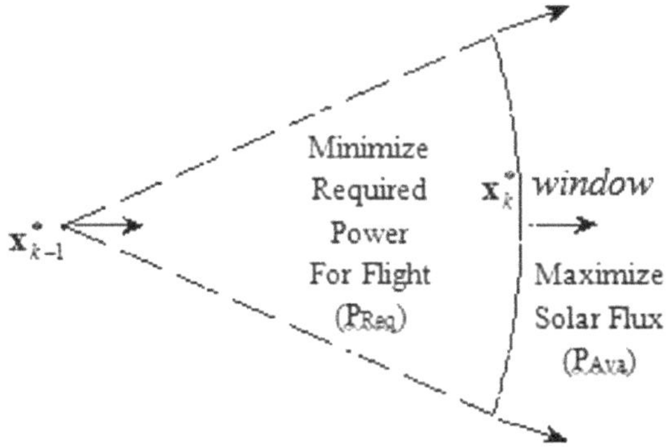

Figure 9. Flight maneuvering window and objectives.

Driver:

$$\text{Maximize } F^{(direct)} = S_0 \mu e^{-\tau/\mu_0};$$

$$S_0 \equiv \text{Solar Constant}$$

$$\tau \equiv \text{Optical Thickness}$$

$$\mu_o = \cos Z \equiv \text{Solar Zenith}$$

$$\mu = \cos\theta \equiv \text{Surface Obliquity}$$

$$\cos\theta = \cos Z \cos\Delta + \sin Z \sin\Delta \cos(A-\psi)$$

$$\text{Subject to: } \quad g_1 = |\phi_{mx} - \phi| \geq 0;$$

with b.c.

$$g_2 = (\text{Longitude} - \text{Longitude}_{AOI})^2 - (\text{Latitude} - \text{Latitude}_{AOI})^2 \leq R^2_{R-o-S}$$

with solution:

$$\mathbf{x}^*_k = (\phi^*_k, \psi^*_k, \Psi^*_k, \text{Longitude}^*_k, \text{Latitude}^*_k), \quad (k = 1, \ldots, N)$$

Path:

$$\text{Minimize } P_{req}(u) = J(u) = \int_0^T f_0(\mathbf{x}(t), u(t))\,dt,$$

where

$$\mathbf{x}(t), (\mathbf{x}^*_{k-1}(t), \ldots, \mathbf{x}^*_k(t)): \; [0, T] \rightarrow \Re^n$$

is the solution of the differential system with boundary conditions and with initial and final conditions

$$\frac{d\mathbf{x}_j}{dt} = f_j(\mathbf{x}, u(t)), \quad (j = 1, \ldots, n);$$

$$\mathbf{x}(0) = \mathbf{x}^*_{k-1}; \quad \mathbf{x}(T) = \mathbf{x}^*_k$$

where

$$J(u^*) = \min_{u \in U} J(u)$$

with u^* and the associated path $\mathbf{x}$ called 'optimal.'

Figure 10. DOPP technique for solar-regenerative high altitude long endurance flight.

considering surface tilt (μ) at that specific time of day (μ_0). The second constraint (g_2), boundary condition, assures that the vehicle stays within the defined range-of-sight window over the area of interest. The maximum solar power is determined along the "driver" window at each time step and sets the destination for the vehicle's path. The "path", in the case of the representative UAV is simplified by the constant velocity condition for minimizing power required for flight. The initial ($\mathbf{x}^*_{k-1}$) and final (step, $\mathbf{x}^*_k$) destination for the vehicle is flown by a constant radius of curvature trajectory set by the ascertained bank angle (ϕ^*_k) for the vehicle's "driver" optimum destination.

The procedure for the on-line analysis requires a total of six updating adaptive processes from sunrise to sunset. The flow of processes can be seen in Figure 11.

The first requires the definition of area of interest (AOI), with equation of time, the solar and geocentric coordinates, and hours of daylight including sunrise/sunset location for the time-of-year of interest. The second, independent process is the definition of the available flight window based on the representative UAV kinematics and the user defined radius of flight window for the "driver" window objective analysis. The first "driver" objective constraint (g_1) limits the maximum bank angle of the vehicle. Next, the flight window, area of interest, and time-of-year information is combined and transferred into the Earth inertial frame of reference. The second boundary condition constraint (g_2) for the "driver" objective insures that the UAV stays within the available range-of-sight for continual surveillance over the area. Parameters for solar flux measurements are determined along "driver" window. Following, the available solar flux along with the vehicle flight window is determined. The UAV maximum solar flux "driver" objective is then determined using an adaptive-A* search procedure in determining the necessary parameters $\mathbf{x}^*_k$ conditions to be met. That information is then used to determine flight path for minimum power required in order to meet the forward step destination and vehicle orientation. This process is iterated from initial location at sunrise through sunset.

This process is "locally" optimal for increasing the adaptability and on-line capability for multi-objective approach for solar powered flight. The DOPP technique only projects forward one step at a time until the overall goal is met. As an example of the local optimality and the pros and cons consider the UAV directly over the area of interest and initially heading north at sunrise. Figure 12 shows path of the vehicle for the first 50 min of flight.

The vehicle has a northerly initial heading for the example figure, during sunrise. In order to obtain a maximum amount of solar flux, the UAV banks a positive 20°, leaning the upper solar panel wing surface into the Sun's intensity. Even though the technique is maximizing solar flux at each step, because it does not project meta-paths, like an off-line analysis, it is limited to a single flight trajectory. The result with starting on a single path may require compromise to reach the end goal. For the flight duration shown above the vehicle starts a sustained turn flight banking the solar panel surface into the Sun, but as it projects further along the flight path there is a decrease in maximum solar flux. However, the DOPP technique adapts and quickly brings a change to the flight trajectory to further increase the maximum solar flux. This can be seen in Figure 13, for the maximum solar flux variation from the example flight of Figure 12.

The small amount of max obtained solar flux is due to the nature of the given example during sunrise when the Sun is close to the horizon and around a 90° zenith angle.

Predicting further steps ahead to determine, by selecting an alternate (non-optimal) destination, there may be meta-paths in which overall produces more solar power. However, this would require a time intensive and limited off-line analysis, and would void the purpose of introducing real-time path-planning adaptability in furthering UAV-autonomy.

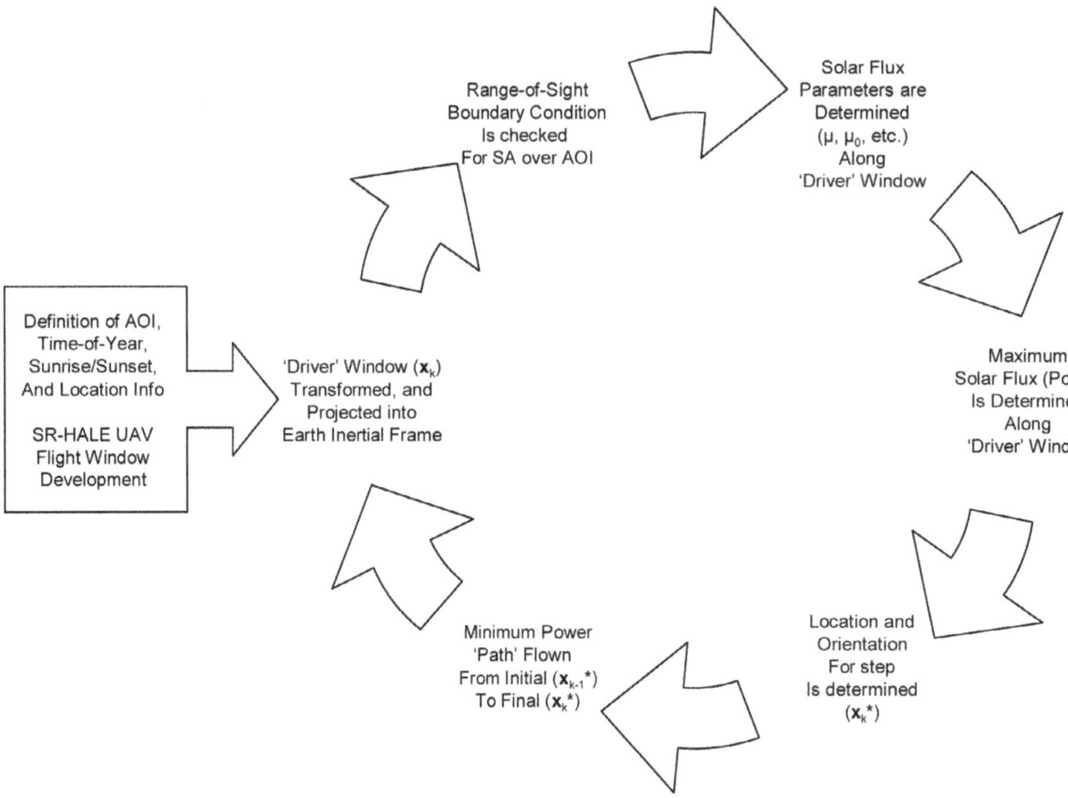

Figure 11. DOPP flow chart for solar-regenerative high altitude long endurance flight.

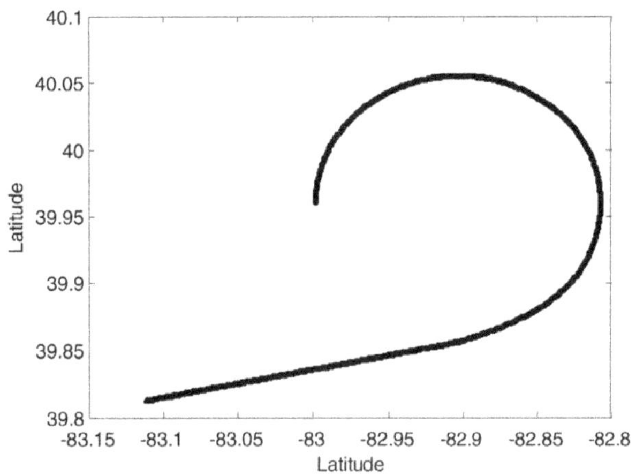

Figure 12. DOPP example flight at sunrise.

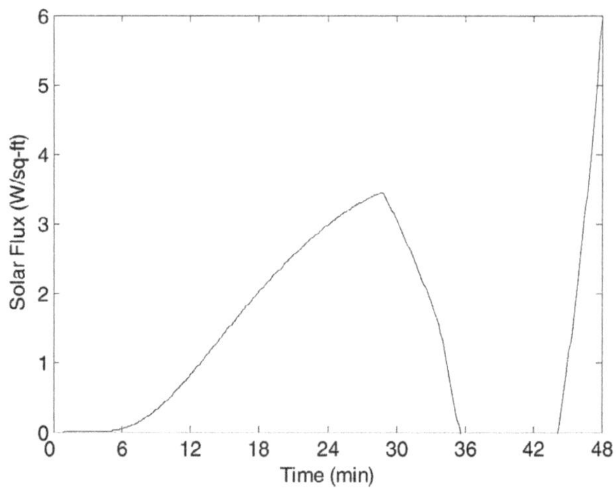

Figure 13. DOPP example of local optimality process: sunrise.

4 Case-study results

Consider the full day of flight on the summer solstice (~21 June) over the area of interest at 40° latitude, and with the initial location of the vehicle was set directly over the area and heading north. Figure 14 shows the flight pattern, with and without the range-of-sight boundary correction of the UAV for the entire day. The path outside of the range-of-sight circle is in violation of the second boundary condition constraint. Implementing the constraint requires the vehicle to bank hard right and maintains continual situational awareness over the area of interest.

As discussed previously, the figure shows that during sunrise ($Z = 90°$) the vehicle is in a sustained bank. As the solar zenith angle decreases the vehicle path requires less sustained banking flight to orient the solar panel surface into the Sun's rays for solar power. The total solar flux gather for the day was 12,838 W/ft^2 (1193 W/m^2), an averaged available power of 5100 W. When compared to the baseline loiter, sustained turn flight at a 20° bank angle, this resulted in a 19.7% increase in averaged available power; and a 2.0% decrease in power required for flight.

To obtain a better appreciation for the technique's capability, consider the solar flux variation from sunrise to sunset for

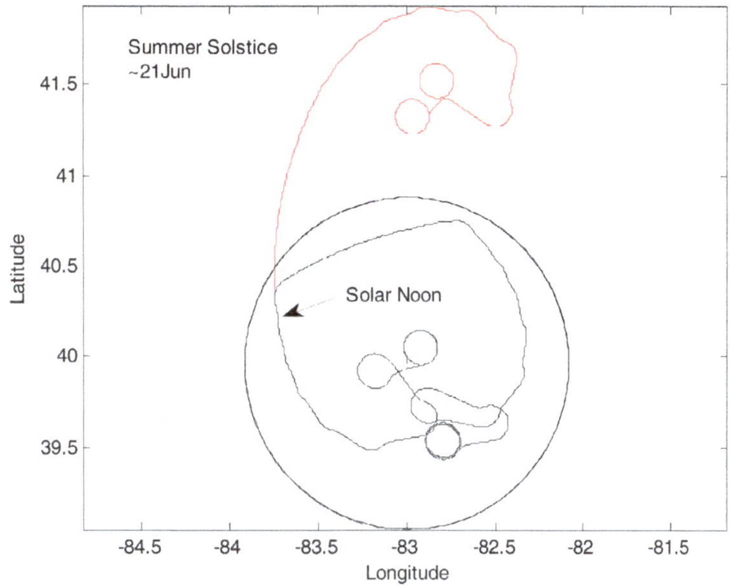

Figure 14. DOPP flight pattern on summer solstice.

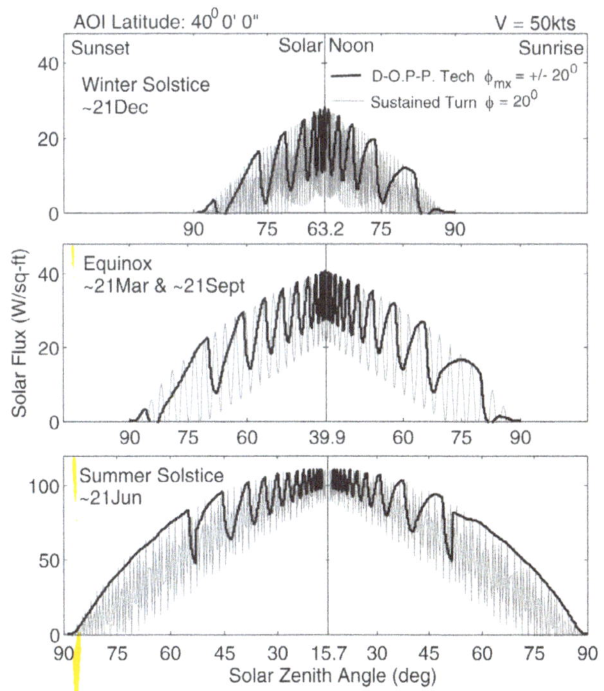

Figure 15. DOPP technique solar flux variation comparisons with the baseline.

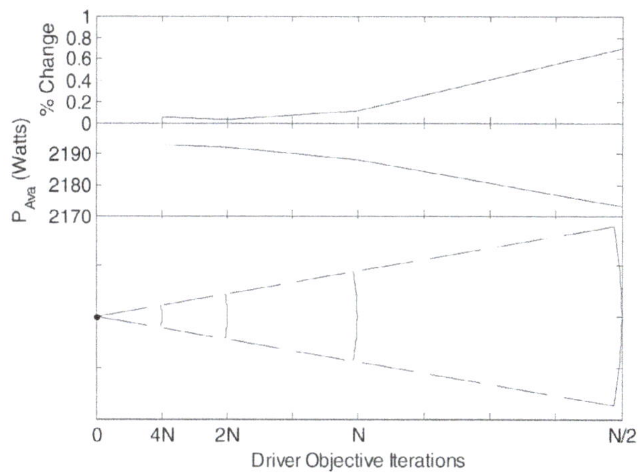

Figure 16. DOPP driver window projection effects on available power.

the DOPP flight pattern compared to the baseline sustained turn (Figure 15). The winter and summer solstice and equinox(s) times-of-year were considered.

Figure 15 shows that as the Sun approaches solar noon, the vehicle's bank angle becomes more sensitive since the variation of solar flux across the "driver" window becomes less.

Before continuing the analysis, consider two conditions that have an effect on the optimal flight technique. First, consider the effects of distance in which the "driver" window is projected forward of the UAV. The results shown above were for a projection window distance of 0.01 nautical-miles with N iterations. The effects on the maximum total available power and percent differences for varying the number of iterations is shown in Figure 16. The conditions shown in the figure are during the winter solstice, the worst case condition, showing the largest differences and the potential for most improvement.

The figure shows that with further increase in number of iterations, decreasing projection of "driver" window, that the improvements determined are within less than 0.1% difference. The adverse effects of adding computation time required for decreasing the project window outweighs the minimal gain in performance. Other considerations must be accounted for when increasing the projection distance, decreasing the number of iterations, other than the significant loss in improvement shown above. As the "driver" window is projected farther, the "path" for the vehicle does not become apparent; meaning, that the actual path for the vehicle to meet its destination may not be simply a sustained turn to meet its objective. For the

Table 4. Effects of varying UAV bank angle limit.

		Case study location		
Longitude:	82° 59′ 56″	West	(−82.99889 W)	$V = 50$ kn
Latitude:	39° 57′ 40″	North	(39.96111 N)	
D-O.P-P. Technique		Summer solstice ~21 June		% Increase
φ_{mx} limit (°)	F_{total} (W/sq-ft)	Ave P_{Ava} (W)	Ave P_{Req} (W)	Ava/Req
±10	11,000	4325	3157	
±20	12,838	5065	3259	14%/3.1%
±30	14,349	5672	3530	10%/8%

Table 5. Flight maneuvering comparisons.

		Case study location		
Longitude	82° 59′ 56″ West	(−82.99889 W)	Sustained turn $\varphi = 20°$	50 kn
Latitude	39° 57′ 40″ North	(39.96111 N)	D-O.P-P. $\varphi = 0° - \pm 20°$	50 kn
Maneuver	Averaged available power		Averaged power required	
		Winter Solstice ~21 December		
Sustained turn	1604 W		3346 W	
D-O.P-P.	2214 W	27.6%	3268 W	−2.4%
		Equinoxes ~21 March & ~21 September		
Sustained turn	3221 W		3346 W	
D-O.P-P.	3838 W	16.1%	3259 W	−2.7%
		Summer Solstice ~21 June		
Sustained turn	4093 W		3346 W	
D-O.P-P.	5100 W	19.7%	3280 W	−2.0%

remainder of the analysis, the projection window was maintained at its initial location.

Next, consider the effects of varying the bank angle limit for the UAV. Table 4 shows the effects on total available solar flux, and both averaged power available and required for the flight using the DOPP technique.

By increasing the bound from ±10° to ±20° there is a significant improvement (14%) in averaged available solar power with a relatively small increase in required power (3.1%). With further increase in bank angle limit, the tradeoff becomes less significant. If the vehicle structural integrity allowed for a bank angle limit of ±30°, there would be a 10% improvement in available power, but subsequently an 8% increase in required power as well.

Continuing the analysis and comparisons, Table 5 are the summary results comparing the maximum obtainable solar power and the power required for sustained turn, and the average power required for the vehicle flying with the adaptive DOPP process. The summary results are during the equinoxes and summer and winter solstices over the case-study area of interest.

The DOPP technique increased the available solar power by 27% during the winter solstice, the time-of-year with minimum amount of daylight, while decreasing the power required for flight by 2% throughout the day. During the equinoxes and summer solstice similar improvements resulted, demonstrating the benefits of the path-planning technique and its application to solar regenerative high altitude long endurance surveillance flight.

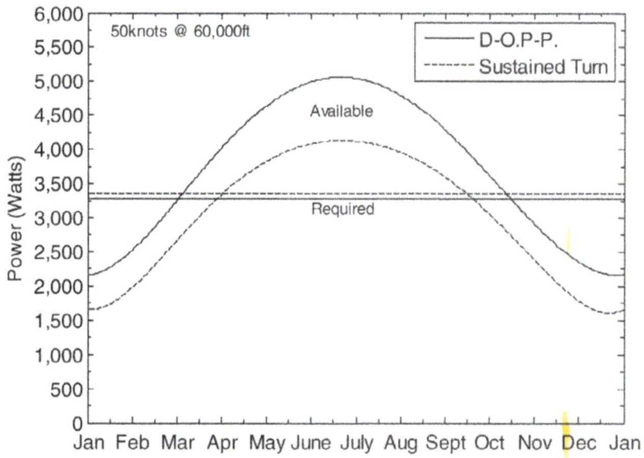

Figure 17. Averaged power comparison for a year at 40° latitude.

The optimum flight patterns and the daily averaged solar flux and solar power gathered was analyzed and compared to the baseline sustained turn flight for an entire year over the area of interest (Figure 17).

For the UAV flying a sustained turn loiter over the area of interest, the available flight time throughout the year for the vehicle to obtain enough solar power to maintain flight would be approximately between 1 April and 13 September, 165 days total. Using the DOPP technique for generating the flight path, the available flight window over the area increased to total of 222 days, 4 March–12 October. This is an additional 57 days

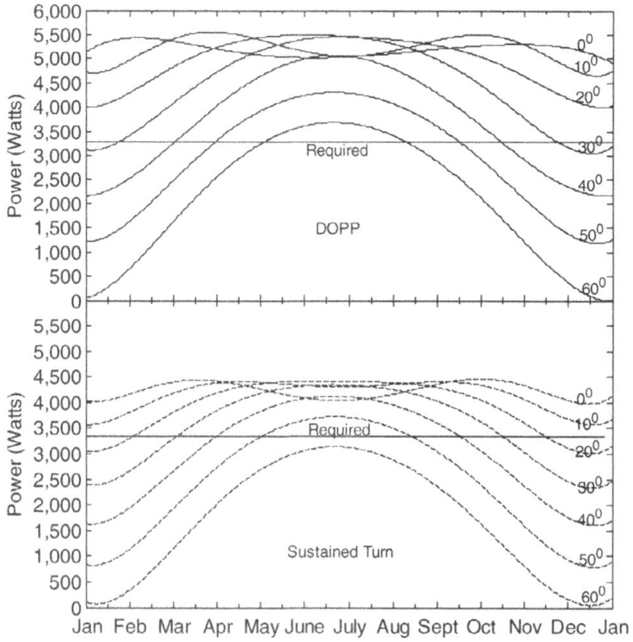

Figure 18. Latitude location effects on the averaged power variation.

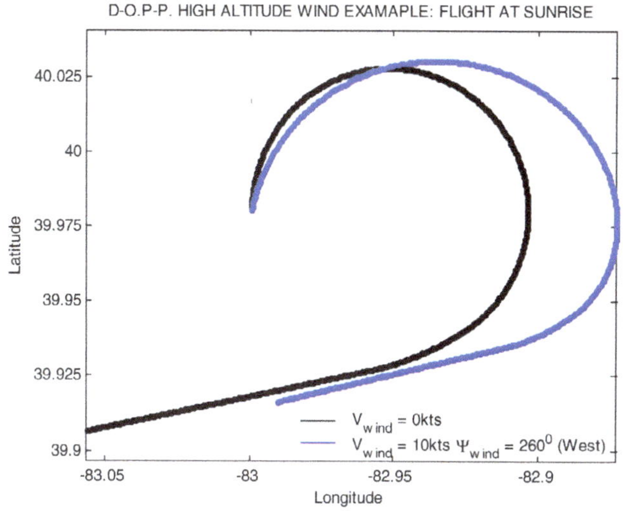

Figure 19. High altitude wind example flight at sunrise.

of available flight for the year, a 26% improvement over the sustained turn loiter flight.

Similarly, the daily averaged available power for the optimum flight patterns was compared with the baseline sustained turn for variations in latitude from the equator to 60° (Figure 18).

The analysis was limited to latitudes less than 60°. For latitudes greater than 60°, there is not enough sufficient averaged available power to meet the required power for the representative UAV to maintain flight.

From Figure 18, flying a sustained turn loiter pattern was limited to 50° latitude or less. However when flying the DOPP pattern, the vehicle is capable of increasing the flight time at each latitude by approximately 2 months, including an approximate 2 months flight period at 60° latitude.

4.1 High altitude wind effects

Accounting for high altitude winds (V_w) in the UAV's kinematic modeling equation (9), Figure 19 is the corresponding example of the first hour of flight. This was compared with the no wind example (Figure 12) during sunrise.

The effects of a wind velocity of 10 kn at a W-S-W heading (Ψ_w = 260° Inertial Earth frame) are shown, typical conditions that would be seen during the spring and summer months. The vehicle has a northerly initial heading and the high altitude wind affected "driver" window is shown to project the flight window, "pushing" the vehicle further east. When considering the entire day, for this example, the total solar flux available decreased (compared to Table 4) by 1.8%, and increased the averaged power required by less than 0.1%.

A few important comments on wind effects: first, the high altitude wind effects and the resultant available solar flux and averaged power required for UAV flight were determined to be a case-by-case study. There were no observable trends for either the increase or decrease in available solar flux with variations in time-of-year or high altitude wind conditions, supporting the need for on-line adaptive capability for a high altitude solar powered flight path generation. Second, as the wind speed increased, an increase in vehicle speed was needed to compensate. For this condition, the increase in vehicle speed allowed for flexibility in vehicle movement for orientation increasing available solar power. However, this is not without the cost of increasing power required. Finally, and most significant, there was always a noticeable significant increase in available solar flux for flight with DOPP process over the baseline sustained turn flight, regardless of wind conditions.

5 Conclusions

The following introduced a multi-objective dual-optimal path-planning technique for unmanned air vehicle path and trajectory autonomy generation, through task allocation and sensor fusion. The dynamic optimization technique generates online adaptive flight paths for vehicle based on available flight windows and environmental influenced objectives. The real-time path generation for aircraft considered multiple independent conflicting objectives without the necessity of developing artificial parameters. The on-line process was locally-optimal based and adaptive, reducing the significant size in decision space required for off-line path search techniques. The technique was not limited to individual methods or path planning algorithms, however, certain procedures, like A* search proved advantageous in real-time analysis.

Solar powered high altitude long endurance flight was investigated as a case-study of the dual-optimal path-planning technique. The flight mechanics of a flight test proven representative aircraft was used, and a case-study example location in which to support continual situational awareness was chosen. The vehicle was limited to a constant altitude of 60,000 ft (18.3 km) and high altitude winds for the area of interest were considered. At that altitude the range-of-sight, i.e. the limited geographical flight window, was determined by existing surveillance equipment common onboard high altitude aircraft. The surveillance system and the determined flight constraint window assured continual surveillance capability over the area of interest. Solar panels, covering 80% of the upper surface of the UAV wing, electric motors, and energy storage system were selected based on existing flight test demonstrated equipment.

A common sustained turn loiter pattern was considered a baseline fixed flight maneuver for comparisons. Sustained turn analysis consisted of varying location and time-of-year as well as the vehicle's bank angle, flight velocity, turn radius, direction of flight, and initial starting location. The variation of each flight condition was individually considered to determine the trends, if any, and their effects for solar powered flight. It was shown, that the latitude location and time-of-year varied the available solar flux significantly. The other, vehicle dependent flight conditions were determined over the case-study area of interest at 40° latitude. It was determined that variations in flight velocity and the turning radius trends were consistent, a result that was utilized not only in determining baseline conditions but in dual-optimal path-planning modeling and analysis as well.

The objectives of the optimization technique were to determine the minimum required power flight paths to the predetermined location and orientation for obtaining maximum solar flux established by the "driver". This was done for still wind and constant wind conditions at altitude over the area of interest and for variations in latitude. The aircraft kinematics was used to develop the flight maneuvering window and was constrained by a maximum bank angle of ±20°. It was determined that minimum power required occurred at the vehicles stall speed. To account for this the flight velocity was set to a constant 50 kn, 20% safety margin, while maintaining minimum power required. The projection of the flight window was set at one nautical-mile, assuring sufficient response time for the vehicle.

The path-planning technique increased the total available solar flux by as much as 20%, and decreased the average power required for flight compared to the baseline sustained turn loiter flight for the case-study area of interest over the time span of one year. This increased the total flight time over the area throughout the year by an additional 57 days, a 26% improvement over the baseline loiter flight. Considering latitude variations from the equator through 60°, the technique increased the available flight time consistently by an approximate 2 months, including flight at 60° latitude where the baseline flight pattern was unattainable. Finally, when considering high altitude wind effects, the results showed that the physical flight path were altered; however, regardless if high altitude wind effects were considered the DOPP technique significantly improved the flight capability over the baseline conditions.

Implication and influences

The published work has an impact in providing continual development of real-time autonomous optimal path-planning and trajectory generation capability for unmanned vehicle systems. The technique developed and demonstrated is an adaptive path generation for vehicles considering multiple independent conflicting objectives without the necessity of developing artificial or weighting parameters. The technique can be applied to unmanned air, ground, and sea based vehicles. The technique is capable of adapting existing optimization algorithms into the dual-optimal path-planning model. Additional implications of the technique to real-time optimal path-planning for unmanned air systems that could have a significant impact include, but not limited to: cooperative tactics between unmanned air vehicles, dynamic target pursuit, and evading potential threats.

References

1. Trefz JL Jr. 2003. From persistent ISR to precision strikes: the expanding role of UAVS, LCDR, US Navy, Navy War College: Newport, RI.
2. Mittal S, Deb K. 2007. Three-dimensional offline path planning for UAVs using multiobjective evolutionary algorithms. Indian Institute of Technology: Kanpur, India.
3. McGee TG, Spry S, Hedrick K. 2005. Optimal path planning in a constant wind with a bounded turning rate. Center for Collaborative Control for Unmanned Vehicles, University of California: Berkeley, CA.
4. McGee TG, Hedrick K. 2007. Optimal path planning with a kinematic airplane model. Journal of Guidance, Control, and Navigation, 30(2), 629–633.
5. Bestaoui Y, Dahmani H, Belharet K. 2009. Geometry of translational trajectories for an autonomous aerospace vehicle with wind effect, in 47th AIAA Aerospace Sciences Meeting Including the New Horizons Forum and Aerospace Exposition, Orlando, FA.
6. Shapira I, Ben-Asher JZ. 1997. Near-optimal horizontal trajectories for autonomous air vehicles. Journal of Guidance, Control, and Navigation, 20(4), 735–741.
7. Erzberger H, Lee H. 1971. Optimum horizontal guidance techniques for aircraft. Journal of Aircraft, 8(2), 95–101.
8. Tsitsiklis JN. 1995. Efficient algorithms for globally optimal trajectories. IEEE Transactions on Automatic Control, 40(9), 1528–1538.
9. McManus IA, Walker RA. 2006. Multidisciplinary approach to intelligent unmanned-airborne-vehicle mission planning. Journal of Aircraft, 43(2), 318–335.
10. Fahroo F, Ross IM. 2002. Direct trajectory optimization by a Chebyshev pseudospectral method. Journal of Guidance, Control, and Dynamics, 25(1), 3860–3864.
11. Dogan A. 2003. Probabilistic path planning for UAVs, in 2nd AIAA "Unmanned Unlimited" Systems, Technologies, and Operations, San Diego, CA.
12. Lennon JA, Atkins EM. 2004. Optimal path-planning with behavior based cost definition, in AIAA 1st Intelligent Systems Technical Conference, Chicago, IL.
13. Amin JN, Boskovic JD, Mehra RK. 2006. A fast and efficient approach to path planning for unmanned vehicles, in AIAA Guidance, Navigation, and Control Conference and Exhibit, Keystone, CO.
14. Geiger BR, Horn JF, DeLullo AM, Long LN. 2006. Optimal path planning of UAVs using direct collocation with nonlinear programming, in AIAA Guidance, Navigation, and Control Conference and Exhibit, Keystone, CO.

15. Geiger BR, Horn JF. 2009. Neural network based trajectory optimization for unmanned aerial vehicles, in 47th AIAA Aerospace Sciences Meeting Including the New Horizons Forum and Aerospace Exposition, Orlando, FA.
16. Narayan P, Campbell D, Walker R. 2008. Multi-objective UAS flight management in time constrained low altitude local environments, in 46th AIAA Aerospace Sciences Meeting and Exhibit, Reno, NV.
17. Yokoyama N, Ochi Y. 2008. Optimal path planning for skid-to-turn unmanned aerial vehicle, in AIAA Guidance, Navigation, and Control Conference and Exhibit, Honolulu, Hawaii.
18. Rippel E, Bar-Gill A, Shimkin N. 2005. Fast graph-search algorithms for general aviation flight trajectory generation. Technion - Israel Institute of Technology: Israel.
19. Chang W, Hsiao F, Sheu D. 2006. Two-point flight path planning using a fast graph-search algorithm. Journal of Aerospace Computing, Information, and Communication, 3, 453–488.
20. Schwartzentruber L, Foo JL, Winer EH. 2008. Three-dimensional multi-objective UAV path planner using meta-paths for decision making and visualization, in 12th AAIA/ISSMO Multidisciplinary Analysis and Optimization Conference, Victoria, British Columbia, Canada.
21. Miele A, Lee YL, Wu GD. Optimal trajectories for an aerospace plane, part 1: formulation, results, and analysis, Aero-Astronautics Report No. 247, Rice University, 1990.
22. Chieng W-Y. 2007. The study of flight path planning for multiple target visitations, Dissertation. Aeronautics & Astronautics, etd-0613107-115826.
23. Frazzoli E, Dahleh MA, Feron E. 2002. Real-time motion planning for agile autonomous vehicles. Journal of Guidance, Control, and Navigation, 25(1), 1–48.
24. Pongpunwattana A, Rysdyk R. 2004. Real-time planning for multiple autonomous vehicles in dynamic uncertain environments. University of Washington: Seattle, WA.
25. Richards N, Sharma N, Ward D. 2002. A hybrid A*/automaton approach to on-line path planning with obstacle avoidance, in AIAA 1st Intelligent Systems Technical Conference, Chicago IL.
26. Howlett JK, McLain TW, Goodrich MA. 2006. Learning real-time A* path planner for unmanned air vehicle target sensing. Journal of Aerospace Computing, Information, and Communication, 3, 1–22.
27. Rathbun D, Capozzi B. 2003. Evolutionary approaches to path planning through uncertain environments, in AIAA's 1st Technical Conference and Workshop on Unmanned Aerospace Vehicles, Portsmouth, VA.
28. Pettit RL, Homer ML. 2004. An autonomous threat evasion response algorithm for unmanned air vehicles during low altitude flight, in AIAA 1st Intelligent Systems Technical Conference, Chicago, IL.
29. Geyer MS, Johnson EN. 2006. 3D obstacle avoidance in adversarial environments for unmanned aerial vehicles, in AIAA Guidance, Navigation, and Control Conference and Exhibit, Keystone, CO.
30. Kabamba PT, Meerkov SM, Zeitz FH III. 2006. Optimal path planning for unmanned combat aerial vehicles to defeat radar tracking. Journal of Guidance, Control, and Navigation, 29(2), 279–288.
31. Park JB, Vorsmann P. 2007. Strategies for the implementation of a sense and avoid system for unmanned air vehicles, in 3rd US-European Competition and Workshop on Micro Air Vehicle Systems & European Micro Air Vehicles, Conference and Flight Competition, Toulouse, France.
32. Tooren J, Heni M, Knoll A, Beck J. 2007. Development of an autonomous avoidance algorithm for UAVs in general airspace, EADS Defense & Security, Military Air Systems: Munich, Germany.
33. Jorris TR, Cobb RG. 2008. Multiple method 2-D trajectory optimization satisfying waypoints and No-Fly zone constraints. Journal of Guidance, Control, and Dynamics, 31(3), 543–553.
34. Zengin U, Dogan A. 2004. Dynamic target pursuit by UAVs in probabilistic threat exposure map, in AIAA 3rd Unmanned Unlimited Technical Conference, Workshop and Exhibit, Chicago IL.
35. Caveney DK, Hedrick JK. 2005. Path planning for targets in close proximity with a bounded turn-rate aircraft, in AIAA Guidance, Navigation, and Control Conference and Exhibit, San Francisco CA.
36. Maj. Burns BS, Blue PA, Capt. Zollars MD. 2007. Simulation of real-time trajectory generator for automated aerial refueling with a required time of arrival, in AIAA Modeling and Simulation Technologies Conference and Exhibit, Hilton Head SC.
37. Sattigeri RJ, Johnson E, Calise AJ, Ha J. 2007. Vision-based target tracking with adaptive target state estimator, in AIAA Guidance, Navigation and Control Conference and Exhibit, Hilton Head SC.
38. Cloutier JR, Lin CF, Yang C. 1993. Maneuvering target tracking via smoothing and filtering through measurement concatenation. Journal of Guidance, Control, and Dynamics, 16(2), 377–384.
39. Han JD, Campbell M. 2003. Artificial potential guided evolutionary path plan for target pursuit and obstacle avoidance, in AIAA Guidance, Navigation, and Control Conference and Exhibit, Austin TX.
40. Larson RA, Mears MJ, Maj. Blue PA. 2005. Path planning for unmanned air vehicles to goal states in operational environments, in AIAA Infotech@Aerospace, Arlington VA.
41. Ousingsawat J, Campbell ME. 2004. On-line estimation and path planning for multiple vehicles in an uncertain environment. International Journal of Robust and Nonlinear Control, 14(8), 741–756.
42. Harl N. 2008. Coordinated rendezvous of unmanned air vehicles to a formation: a sliding mode approach, in AIAA Guidance, Navigation and Control Conference and Exhibit, Honolulu, Hawaii.
43. Bollino KP, Lewis LR. 2008. Collision-free multi-UAV optimal path planning and cooperative control for tactical applications, in AIAA Guidance, Navigation and Control Conference and Exhibit, Honolulu, Hawaii.
44. Boissonnat JD, Cerezo A, Leblond J. 1994. Shortest paths of bounded curvature in the plane. Journal of Intelligent and Robotic Systems, 11, 5–20.
45. Mahmoudian N, Woolsey CA, Geisbert J. 2007. Steady turns and optimal paths for underwater gliders, in AIAA Guidance, Navigation and Control Conference and Exhibit, Hilton Head, SC.
46. Foo JL, Knutzon JS, Oliver JH, Winer EH. 2007. Three-dimensional multi-objective path planning of unmanned aerial vehicles using particle swarm optimization, in AIAA/ASME/ASCE/ASC Structures, Structural Dynamics, and Materials Conference, Honolulu, Hawaii.
47. Swartzentruber L, Foo JL, Winer EH. 2009. Three-dimensional multi-objective UAV path planner using terrain information, in

AIAA/ASME/ASCE/ASC Structures, Structural Dynamics, and Materials Conference, Palm Springs, CA.
48. Whitfield CA. An adaptive dual-optimal path-planning technique for unmanned air vehicles with applications to solar-regenerative high altitude long endurance flight, Dissertation, The Ohio State University, Columbus, OH, 2009.
49. Weider S. 1982. An introduction to solar energy for scientists and engineers. John Wiley & Sons, New York.
50. Strganac TW. Wind study for high altitude platform design, NASA-RP-1044, N80-12661, 1979.
51. Pearson J, Gregorek GM, Whitfield CA. High altitude morphing aircraft (HAMAC). AFRL-VA-WP-TR-2007-XXX– Final Report to USAF for Phase I Small Business Initiative, 2007.
52. Alemayehu D, Eaton E, Faruque I. 2004. HALE UAV: aerovironment pathfinder: aerodynamic and stability analysis, case-study: planform optimization. http://www.aoe.vt.edu/~mason/Mason_f/pathfinder.pdf

Permissions

All chapters in this book were first published in IJSMDO, by EDP Sciences; hereby published with permission under the Creative Commons Attribution License or equivalent. Every chapter published in this book has been scrutinized by our experts. Their significance has been extensively debated. The topics covered herein carry significant findings which will fuel the growth of the discipline. They may even be implemented as practical applications or may be referred to as a beginning point for another development.

The contributors of this book come from diverse backgrounds, making this book a truly international effort. This book will bring forth new frontiers with its revolutionizing research information and detailed analysis of the nascent developments around the world.

We would like to thank all the contributing authors for lending their expertise to make the book truly unique. They have played a crucial role in the development of this book. Without their invaluable contributions this book wouldn't have been possible. They have made vital efforts to compile up to date information on the varied aspects of this subject to make this book a valuable addition to the collection of many professionals and students.

This book was conceptualized with the vision of imparting up-to-date information and advanced data in this field. To ensure the same, a matchless editorial board was set up. Every individual on the board went through rigorous rounds of assessment to prove their worth. After which they invested a large part of their time researching and compiling the most relevant data for our readers.

The editorial board has been involved in producing this book since its inception. They have spent rigorous hours researching and exploring the diverse topics which have resulted in the successful publishing of this book. They have passed on their knowledge of decades through this book. To expedite this challenging task, the publisher supported the team at every step. A small team of assistant editors was also appointed to further simplify the editing procedure and attain best results for the readers.

Apart from the editorial board, the designing team has also invested a significant amount of their time in understanding the subject and creating the most relevant covers. They scrutinized every image to scout for the most suitable representation of the subject and create an appropriate cover for the book.

The publishing team has been an ardent support to the editorial, designing and production team. Their endless efforts to recruit the best for this project, has resulted in the accomplishment of this book. They are a veteran in the field of academics and their pool of knowledge is as vast as their experience in printing. Their expertise and guidance has proved useful at every step. Their uncompromising quality standards have made this book an exceptional effort. Their encouragement from time to time has been an inspiration for everyone.

The publisher and the editorial board hope that this book will prove to be a valuable piece of knowledge for researchers, students, practitioners and scholars across the globe.

List of Contributors

Husam Hamad
Electronics and Communications Engineering Department, College of Engineering at Al-Lith, Umm Al-Qura University, Makkah Al-Mukarramah, KSA

Awad Al-Zaben
Biomedical Engineering Department, Hijjawi College of Engineering Technology, Yarmouk University, Irbid, Jordan

Rami Owies
Biomedical Engineering Department, College of Engineering, Jordan University of Science and Technology, Irbid, Jordan
Abdelkader Rami, Habib Hamdaoui, Houari Sayah, and Abdelkader Zeblah*
University of Djillali Liabes, Faculty of Technology, Sidi Bel Abbes, Algeria

Hanen Jamai and Habib Sammouda
LabEM (LR11ES34), University of Sousse, ESSTH Sousse, Rue Lamine Abbassi, 4011 Hammam Sousse, Tunisia

Bernard Pateyron
CNRS-SPCTS/CEC, 12 rue Atlantis, 87068 Limoges, France

M. El Ganaoui
Université de Lorraine, Lab. Energe´tique de Longwy – LERMAB/FJV, Longwy, France

Abdelkader Rami, Habib Hamdaoui, Houari Sayah and Abdelkader Zeblah
University of Djillali Liabes, Faculty of Technology, Sidi Bel Abbes, Algeria

Souad Messai and Slimane Gabsi
Unité de Recherche Environnement, Catalyse et Analyse des Procédés URECAP, Ecole Nationale d'Ingénieurs de Gabés, Rue Omar Ibn-Elkhattab, 6029 Gabés, Tunisie

Mohamed El Ganaoui and Laurent Chrusciel
Laboratoire d'Etudes et de Recherche sur le Matériau Bois LERMAB, Université de Lorraine-Faculté des Sciences et Technologies, Boulevard des Aiguillettes, 54506 Vandoeuvre-lés-Nancy Cedex, France

Jalila Sghaier
Laboratoire d'Energé tique et des Transferts Thermiques et Massiques, Faculté des Sciences de Tunis, Campus Universitaire, 2092 El Manar, Tunisie

Narasimha R. Nagaiah and Christopher D. Geiger
Department of Industrial Engineering and Management Systems, University of Central Florida, Orlando, Florida 32816, USA

Riadh Chaari, Moez Abennadher, Jamel Louati and Mohamed Haddar
Mechanics Modelling and Production Research Unit (U2MP), University of Sfax, National Engineering School of Sfax (E.N.I.S.), B.P. 3038 Sfax, Tunisia

Scott A. Burton
American Optimization LLC, Springboro, Ohio 45066, USA

Edward J. Alyanak and Raymond M. Kolonay
Aerospace Systems Directorate Air Force Research Laboratory, Wright-Patterson Air Force Base, Ohio 45433, USA

Nacer Lamrous, Said Makhlouf and Nora Belkaid
Mouloud Mammeri University, L.M.S.E. Laboratory, PO Box 17, RP 15000, Tizi-Ouzou, Algeria

Rudy Chocat
IFMA, EA3867, Laboratoires de Mécanique et Ingénieries, Clermont Université, CP 104488, 63000 Clermont-Ferrand, France

Loïc Brevault
Onera – The French Aerospace Lab, BP 80100, 91123 Palaiseau Cedex, France
CNES – Launchers Directorate, 52 rue Jacques Hillairet, 75612 Paris, France

Mathieu Balesdent and Sébastien Defoort
Onera – The French Aerospace Lab, BP 80100, 91123 Palaiseau Cedex, France

Ji-Hong Zhu and Wei-Hong Zhang
Engineering Simulation and Aerospace Computing, Northwestern Polytechnical University, 710072, Xi'an, Shaanxi, P.R. China

Magesh Chandramouli
School of Technology, Purdue University, Calumet 46323, USA

Gary R. Bertoline
School of Technology, Purdue University, West Lafayette 47907, USA

Stanislaw Raczynski
Universidad Panamericana, Facultad de Ingeniería, Augusto Rodin 498, 03920 Mexico DF, Mexico

Ji-Hong Zhu, Qing Li and Wei-Hong Zhang
Engineering Simulation and Aerospace Computing (ESAC), The Key Laboratory of Contemporary Design and Integrated Manufacturing Technology, Northwestern Polytechnical University, Xi'an, Shaanxi 710072, P.R. China

T. Radha Ramanan
Assistant Professor, Department of Mechanical Engineering, National Institute of Technology Calicut, Kozhikode 673601, Kerala, India

Muhammed Iqbal
Senior Lecturer, Department of Mechanical Engineering, MEA Engineering College, Perinthalmanna 679325, Kerala, India

K. Umarali
Research Scholar, Department of Mechanical Engineering, National Institute of Technology Calicut, Kozhikode 673601, Kerala, India

Karim Lahmer and Rachid Bessaïh
Département de Génie Mécanique, (LEAP), Université Constantine 1, Route d'Ain El Bey, Constantine 25000, Algeria

Angel Scipioni and Mohammed El Ganaoui
Universitéde Lorraine, Institut Universitaire de Technologie Henri Poincaré de Longwy (LERMAB & GREEN), 186 Route de Lorraine, 54400 Cosnes-et-Romain, France

Vimal Savsani
Department of Mechanical Engineering, Pandit Deendayal Petroleum University, Raysan, Gandhinagar 382 007, Gujarat, India

Sinem Erden Gulebaglan
Department of Electric Program, Vacational School of Van, Yuzuncu Yil University, 65080 Van, Turkey

Emel Kilit Dogan, Mehmet Nurullah Secuk, Murat Aycibin, Bahattin Erdinc and Harun Akkus
Department of Physics, Faculty of Science, Yuzuncu Yil University, 65080 Van, Turkey

Qi Xia and Tielin Shi
The State Key Laboratory of Digital Manufacturing Equipment and Technology, Huazhong University of Science and Technology, Wuhan, P.R. China

Moirangthem Deben Singh, Ram Krishna Mehta and Arvind Kumar Singh
Department of Electrical Engineering, North Eastern Regional Institute of Science & Technology, 791 109 Nirjuli, Arunachal Pradesh, India

Sofiane Guessasma
INRA, UR1268 Biopolymères Interactions Assemblages, 44300 Nantes, France

Weihong Zhang and Jihong Zhu
Laboratory of Engineering Simulation & Aerospace Computing-ESAC, Northwestern Polytechnical University, 710072 Xian, Shaanxi, P.R. China

Sofiane Belhabib
LUNAM University of Nantes Angers Le Mans, CNRS, GEPEA, UMR 6144, IUT de Nantes, 2 avenue du Professeur Jean Rouxel, 44475 Carquefou Cedex, France

Hedi Nouri
Department of Polymers and Composites Technology & Mechanical Engineering, Mines Douai, 941 rue Charles Bourseul, CS 10838, 59508 Douai, France

Bo Wang, Yunfeng Shi and Caihua Zhou
Faculty of Vehicle Engineering and Mechanics, Dalian University of Technology, Dalian 116024, China

Tong Li
School of Chemistry, Physics and Mechanical Engineering, Queensland University of Technology, 2 George St., Brisbane, QLD 4001, Australia

Suat Bingol, Bahattin Erdinc and Harun Akkus
Department of Physics, Faculty of Science, Yuzuncu Yil University, 65080 Van, Turkey

Ji-Hong Zhu, Kai-Ke Yang and Wei-Hong Zhang
Engineering Simulation and Aerospace Computing (ESAC), Northwestern Polytechnical University, Xi'an, Shaanxi 710072, P.R. China

Clifford A. Whitfield
The Ohio State University, 201 W. 19th Avenue, Columbus, Ohio 43210, USA

Index